PICK-UPS/LAND CRUISER/4RUNNER
1989-96 REPAIR MANUAL

CHILTON'S

Covers all U.S. and Canadian models of Toyota Pick-up, Tacoma, T100, Land Cruiser and 4 Runner; 2 and 4 wheel drive

by Dawn M. Hoch, S.A.E.

CHILTON *Automotive Books*

PUBLISHED BY **HAYNES NORTH AMERICA**, Inc.

Printed in Malaysia
© 1996 Haynes North America, Inc.
ISBN 0-8019-8682-6
Library of Congress Catalog Card No. 96-84181
0123456789 9876543210

Haynes Publishing Group
Sparkford Nr Yeovil
Somerset BA22 7JJ England

Haynes North America, Inc
859 Lawrence Drive
Newbury Park
California 91320 USA

ABCDE
FGHIJ
KLMNO
PQRS
3

6P3

Contents

Contents

DRIVE TRAIN   7

SUSPENSION AND STEERING   8

BRAKES   9

BODY AND TRIM   10

GLOSSARY

MASTER INDEX

SAFETY NOTICE

Proper service and repair procedures are vital to the safe, reliable operation of all motor vehicles, as well as the personal safety of those performing repairs. This manual outlines procedures for servicing and repairing vehicles using safe, effective methods. The procedures contain many NOTES, CAUTIONS and WARNINGS which should be followed, along with standard procedures to eliminate the possibility of personal injury or improper service which could damage the vehicle or compromise its safety.

It is important to note that repair procedures and techniques, tools and parts for servicing motor vehicles, as well as the skill and experience of the individual performing the work vary widely. It is not possible to anticipate all of the conceivable ways or conditions under which vehicles may be serviced, or to provide cautions as to all possible hazards that may result. Standard and accepted safety precautions and equipment should be used when handling toxic or flammable fluids, and safety goggles or other protection should be used during cutting, grinding, chiseling, prying, or any other process that can cause material removal or projectiles.

Some procedures require the use of tools specially designed for a specific purpose. Before substituting another tool or procedure, you must be completely satisfied that neither your personal safety, nor the performance of the vehicle will be endangered.

Although information in this manual is based on industry sources and is complete as possible at the time of publication, the possibility exists that some car manufacturers made later changes which could not be included here. While striving for total accuracy, the authors or publishers cannot assume responsibility for any errors, changes or omissions that may occur in the compilation of this data.

PART NUMBERS

Part numbers listed in this reference are not recommendations by Haynes North America, Inc. for any product brand name. They are references that can be used with interchange manuals and aftermarket supplier catalogs to locate each brand supplier's discrete part number.

SPECIAL TOOLS

Special tools are recommended by the vehicle manufacturer to perform their specific job. Use has been kept to a minimum, but where absolutely necessary, they are referred to in the text by the part number of the tool manufacturer. These tools can be purchased, under the appropriate part number, from your local dealer or regional distributor, or an equivalent tool can be purchased locally from a tool supplier or parts outlet. Before substituting any tool for the one recommended, read the SAFETY NOTICE at the top of this page.

ACKNOWLEDGMENTS

The publisher expresses appreciation to Toyota Motor Corporation for their generous assistance.

1

GENERAL INFORMATION AND MAINTENANCE

HOW TO USE THIS BOOK

This Chilton's Total Car Care manual is intended to help you learn more about the inner workings of your vehicle while saving you money on its upkeep and operation.

The beginning of the book will likely be referred to the most, since that is where you will find information for maintenance and tune-up. The other sections deal with the more complex systems of your vehicle. Systems (from engine through brakes) are covered to the extent that the average do-it-yourselfer can attempt. This book will not explain such things as rebuilding a differential because the expertise required and the special tools necessary make this uneconomical. It will, however, give you detailed instructions to help you change your own brake pads and shoes, replace spark plugs, and perform many more jobs that can save you money and help avoid expensive problems.

A secondary purpose of this book is a reference for owners who want to understand their vehicle and/or their mechanics better.

Where to Begin

Before removing any bolts, read through the entire procedure. This will give you the overall view of what tools and supplies will be required. So read ahead and plan ahead. Each operation should be approached logically and all procedures thoroughly understood before attempting any work.

If repair of a component is not considered practical, we tell you how to remove the part and then how to install the new or rebuilt replacement. In this way, you at least save labor costs.

Avoiding Trouble

Many procedures in this book require you to "label and disconnect . . . " a group of lines, hoses or wires. Don't be think you can remember where everything goes—you won't. If you hook up vacuum or fuel lines incorrectly, the vehicle may run poorly, if at all. If you hook up electrical wiring incorrectly, you may instantly learn a very expensive lesson.

You don't need to know the proper name for each hose or line. A piece of masking tape on the hose and a piece on its fitting will allow you to assign your own label. As long as you remember your own code, the lines can be reconnected by matching your tags. Remember that tape will dissolve in gasoline or solvents; if a part is to be washed or cleaned, use another method of identification. A permanent felt-tipped marker or a metal scribe can be very handy for marking metal parts. Remove any tape or paper labels after assembly.

Maintenance or Repair?

Maintenance includes routine inspections, adjustments, and replacement of parts which show signs of normal wear. Maintenance compensates for wear or deterioration. Repair implies that something has broken or is not working. A need for a repair is often caused by lack of maintenance. for example: draining and refilling automatic transmission fluid is maintenance recommended at specific intervals. Failure to do this can shorten the life of the transmission/transaxle, requiring very expensive repairs. While no maintenance program can prevent items from eventually breaking or wearing out, a general rule is true: MAINTENANCE IS CHEAPER THAN REPAIR.

TOOLS AND EQUIPMENT

See Figures 1 thru 15

Without the proper tools and equipment it is impossible to properly service your vehicle. It would be virtually impossible to catalog every tool that you would need to perform all of the operations in this book. It would be unwise for the amateur to rush out and buy an expensive set of tools on the theory that he/she may need one or more of them at some time.

The best approach is to proceed slowly, gathering a good quality set of those tools that are used most frequently. Don't be misled by the low cost of bargain tools. It is far better to spend a little more for better quality. Forged wrenches, 6 or 12-point sockets and fine tooth ratchets are by far preferable to their less expensive counterparts. As any good mechanic can tell you, there are few worse experiences than trying to work on a vehicle with bad tools. Your monetary savings will be far outweighed by frustration and mangled knuckles.

Two basic mechanic's rules should be mentioned here. First, whenever the left side of the vehicle or engine is referred to, it means the driver's side. Conversely, the right side of the vehicle means the passenger's side. Second, screws and bolts are removed by turning counterclockwise, and tightened by turning clockwise unless specifically noted.

Safety is always the most important rule. Constantly be aware of the dangers involved in working on an automobile and take the proper precautions. Please refer to the information in this section regarding SERVICING YOUR VEHICLE SAFELY and the SAFETY NOTICE on the acknowledgment page.

Avoiding the Most Common Mistakes

Pay attention to the instructions provided. There are 3 common mistakes in mechanical work:

1. Incorrect order of assembly, disassembly or adjustment. When taking something apart or putting it together, performing steps in the wrong order usually just costs you extra time; however, it CAN break something. Read the entire procedure before beginning. Perform everything in the order in which the instructions say you should, even if you can't see a reason for it. When you're taking apart something that is very intricate, you might want to draw a picture of how it looks when assembled in order to make sure you get everything back in its proper position. When making adjustments, perform them in the proper order. One adjustment possibly will affect another.

2. Overtorquing (or undertorquing). While it is more common for overtorquing to cause damage, undertorquing may allow a fastener to vibrate loose causing serious damage. Especially when dealing with aluminum parts, pay attention to torque specifications and utilize a torque wrench in assembly. If a torque figure is not available, remember that if you are using the right tool to perform the job, you will probably not have to strain yourself to get a fastener tight enough. The pitch of most threads is so slight that the tension you put on the wrench will be multiplied many times in actual force on what you are tightening.

There are many commercial products available for ensuring that fasteners won't come loose, even if they are not torqued just right (a very common brand is Loctite®). If you're worried about getting something together tight enough to hold, but loose enough to avoid mechanical damage during assembly, one of these products might offer substantial insurance. Before choosing a threadlocking compound, read the label on the package and make sure the product is compatible with the materials, fluids, etc. involved.

3. Crossthreading. This occurs when a part such as a bolt is screwed into a nut or casting at the wrong angle and forced. Crossthreading is more likely to occur if access is difficult. It helps to clean and lubricate fasteners, then to start threading the bolt, spark plug, etc. with your fingers. If you encounter resistance, unscrew the part and start over again at a different angle until it can be inserted and turned several times without much effort. Keep in mind that many parts have tapered threads, so that gentle turning will automatically bring the part you're threading to the proper angle. Don't put a wrench on the part until it's been tightened a couple of turns by hand. If you suddenly encounter resistance, and the part has not seated fully, don't force it. Pull it back out to make sure it's clean and threading properly.

Be sure to take your time and be patient, and always plan ahead. Allow yourself ample time to perform repairs and maintenance.

Begin accumulating those tools that are used most frequently: those associated with routine maintenance and tune-up. In addition to the normal assortment of screwdrivers and pliers, you should have the following tools:

• Wrenches/sockets and combination open end/box end wrenches in sizes 1/8–3/4 in. and/or 3mm–19mm 13/16 in. or 5/8 in. spark plug socket (depending on plug type).

➡**If possible, buy various length socket drive extensions. Universal-joint and wobble extensions can be extremely useful, but be careful when using them, as they can change the amount of torque applied to the socket.**

• Jackstands for support.
• Oil filter wrench.
• Spout or funnel for pouring fluids.

Fig. 1 All but the most basic procedures will require an assortment of ratchets and sockets

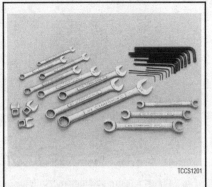

Fig. 2 In addition to ratchets, a good set of wrenches and hex keys will be necessary

Fig. 3 A hydraulic floor jack and a set of jackstands are essential for lifting and supporting the vehicle

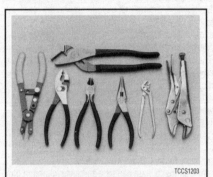

Fig. 4 An assortment of pliers, grippers and cutters will be handy for old rusted parts and stripped bolt heads

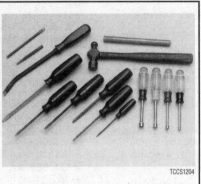

Fig. 5 Various drivers, chisels and prybars are great tools to have in your toolbox

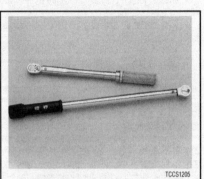

Fig. 6 Many repairs will require the use of a torque wrench to assure the components are properly fastened

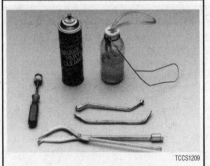

Fig. 7 Although not always necessary, using specialized brake tools will save time

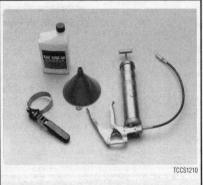

Fig. 8 A few inexpensive lubrication tools will make maintenance easier

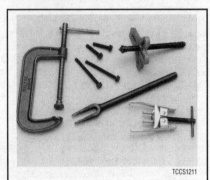

Fig. 9 Various pullers, clamps and separator tools are needed for many larger, more complicated repairs

Fig. 10 A variety of tools and gauges should be used for spark plug gapping and installation

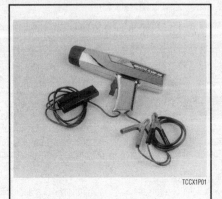

Fig. 11 Inductive type timing light

Fig. 12 A screw-in type compression gauge is recommended for compression testing

TCCX1P03

Fig. 13 A vacuum/pressure tester is necessary for many testing procedures

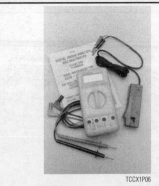

TCCX1P06

Fig. 14 Most modern automotive multimeters incorporate many helpful features

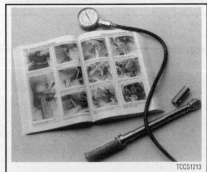

TCCS1213

Fig. 15 Proper information is vital, so always have a Chilton Total Car Care manual handy

• Grease gun for chassis lubrication (unless your vehicle is not equipped with any grease fittings)

• Hydrometer for checking the battery (unless equipped with a sealed, maintenance-free battery).

• A container for draining oil and other fluids.

• Rags for wiping up the inevitable mess.

In addition to the above items there are several others that are not absolutely necessary, but handy to have around. These include an equivalent oil absorbent gravel, like cat litter, and the usual supply of lubricants, antifreeze and fluids. This is a basic list for routine maintenance, but only your personal needs and desire can accurately determine your list of tools.

After performing a few projects on the vehicle, you'll be amazed at the other tools and non-tools on your workbench. Some useful household items are: a large turkey baster or siphon, empty coffee cans and ice trays (to store parts), a ball of twine, electrical tape for wiring, small rolls of colored tape for tagging lines or hoses, markers and pens, a note pad, golf tees (for plugging vacuum lines), metal coat hangers or a roll of mechanic's wire (to hold things out of the way), dental pick or similar long, pointed probe, a strong magnet, and a small mirror (to see into recesses and under manifolds).

A more advanced set of tools, suitable for tune-up work, can be drawn up easily. While the tools are slightly more sophisticated, they need not be outrageously expensive. There are several inexpensive tach/dwell meters on the market that are every bit as good for the average mechanic as a professional model. Just be sure that it goes to a least 1200–1500 rpm on the tach scale and that it works on 4, 6 and 8-cylinder engines. The key to these purchases is to make them with an eye towards adaptability and wide range. A basic list of tune-up tools could include:

• Tach/dwell meter.

• Spark plug wrench and gapping tool.

• Feeler gauges for valve adjustment.

• Timing light.

The choice of a timing light should be made carefully. A light which works on the DC current supplied by the vehicle's battery is the best choice; it should have a xenon tube for brightness. On any vehicle with an electronic ignition system, a timing light with an inductive pickup that clamps around the No. 1 spark plug cable is preferred.

In addition to these basic tools, there are several other tools and gauges you may find useful. These include:

• Compression gauge. The screw-in type is slower to use, but eliminates the possibility of a faulty reading due to escaping pressure.

• Manifold vacuum gauge.

• 12V test light.

• A combination volt/ohmmeter

• Induction Ammeter. This is used for determining whether or not there is current in a wire. These are handy for use if a wire is broken somewhere in a wiring harness.

As a final note, you will probably find a torque wrench necessary for all but the most basic work. The beam type models are perfectly adequate, although the newer click types (breakaway) are easier to use. The click type torque wrenches tend to be more expensive. Also keep in mind that all types of torque wrenches should be periodically checked and/or recalibrated. You will have to decide for yourself which better fits your pocketbook, and purpose.

Special Tools

Normally, the use of special factory tools is avoided for repair procedures, since these are not readily available for the do-it-yourself mechanic. When it is possible to perform the job with more commonly available tools, it will be pointed out, but occasionally, a special tool was designed to perform a specific function and should be used. Before substituting another tool, you should be convinced that neither your safety nor the performance of the vehicle will be compromised.

Special tools can usually be purchased from an automotive parts store or from your dealer. In some cases special tools may be available directly from the tool manufacturer.

SERVICING YOUR VEHICLE SAFELY

▶ **See Figures 16, 17 and 18**

It is virtually impossible to anticipate all of the hazards involved with automotive maintenance and service, but care and common sense will prevent most accidents.

The rules of safety for mechanics range from "don't smoke around gasoline," to "use the proper tool(s) for the job." The trick to avoiding injuries is to develop safe work habits and to take every possible precaution.

Do's

• Do keep a fire extinguisher and first aid kit handy.

• Do wear safety glasses or goggles when cutting, drilling, grinding or prying, even if you have 20–20 vision. If you wear glasses for the sake of vision, wear safety goggles over your regular glasses.

• Do shield your eyes whenever you work around the battery. Batteries contain sulfuric acid. In case of contact with, flush the area with water or a mixture of water and baking soda, then seek immediate medical attention.

• Do use safety stands (jackstands) for any undervehicle service. Jacks are for raising vehicles; jackstands are for making sure the vehicle stays raised until you want it to come down.

• Do use adequate ventilation when working with any chemicals or hazardous materials. Like carbon monoxide, the asbestos dust resulting from some brake lining wear can be hazardous in sufficient quantities.

• Do disconnect the negative battery cable when working on the electrical system. The secondary ignition system contains EXTREMELY HIGH VOLTAGE. In some cases it can even exceed 50,000 volts.

• Do follow manufacturer's directions whenever working with potentially hazardous materials. Most chemicals and fluids are poisonous.

• Do properly maintain your tools. Loose hammerheads, mushroomed punches and chisels, frayed or poorly grounded electrical cords, excessively worn screwdrivers, spread wrenches (open end), cracked sockets, slipping ratchets, or faulty droplight sockets can cause accidents.

• Likewise, keep your tools clean; a greasy wrench can slip off a bolt head, ruining the bolt and often harming your knuckles in the process.

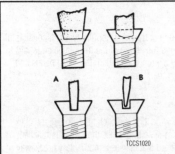

Fig. 16 Screwdrivers should be kept in good condition to prevent injury or damage which could result if the blade slips from the screw

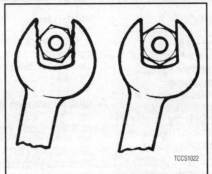

Fig. 17 Using the correct size wrench will help prevent the possibility of rounding off a nut

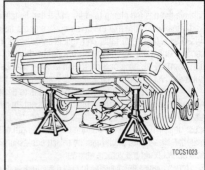

Fig. 18 NEVER work under a vehicle unless it is supported using safety stands (jackstands)

• Do use the proper size and type of tool for the job at hand. Do select a wrench or socket that fits the nut or bolt. The wrench or socket should sit straight, not cocked.

• Do, when possible, pull on a wrench handle rather than push on it, and adjust your stance to prevent a fall.

• Do be sure that adjustable wrenches are tightly closed on the nut or bolt and pulled so that the force is on the side of the fixed jaw.

• Do strike squarely with a hammer; avoid glancing blows.

• Do set the parking brake and block the drive wheels if the work requires a running engine.

Don'ts

• Don't run the engine in a garage or anywhere else without proper ventilation—EVER! Carbon monoxide is poisonous; it takes a long time to leave the human body and you can build up a deadly supply of it in your system by simply breathing in a little at a time. You may not realize you are slowly poisoning yourself. Always use power vents, windows, fans and/or open the garage door.

• Don't work around moving parts while wearing loose clothing. Short sleeves are much safer than long, loose sleeves. Hard-toed shoes with neoprene soles protect your toes and give a better grip on slippery surfaces. Watches and jewelry is not safe working around a vehicle. Long hair should be tied back under a hat or cap.

• Don't use pockets for toolboxes. A fall or bump can drive a screwdriver deep into your body. Even a rag hanging from your back pocket can wrap around a spinning shaft or fan.

• Don't smoke when working around gasoline, cleaning solvent or other flammable material.

• Don't smoke when working around the battery. When the battery is being charged, it gives off explosive hydrogen gas.

• Don't use gasoline to wash your hands; there are excellent soaps available. Gasoline contains dangerous additives which can enter the body through a cut or through your pores. Gasoline also removes all the natural oils from the skin so that bone dry hands will suck up oil and grease.

• Don't service the air conditioning system unless you are equipped with the necessary tools and training. When liquid or compressed gas refrigerant is released to atmospheric pressure it will absorb heat from whatever it contacts. This will chill or freeze anything it touches.

• Don't use screwdrivers for anything other than driving screws! A screwdriver used as an prying tool can snap when you least expect it, causing injuries. At the very least, you'll ruin a good screwdriver.

• Don't use an emergency jack (that little ratchet, scissors, or pantograph jack supplied with the vehicle) for anything other than changing a flat! These jacks are only intended for emergency use out on the road; they are NOT designed as a maintenance tool. If you are serious about maintaining your vehicle yourself, invest in a hydraulic floor jack of at least a 1½ ton capacity, and at least two sturdy jackstands.

FASTENERS, MEASUREMENTS AND CONVERSIONS

Bolts, Nuts and Other Threaded Retainers

▶ **See Figures 19 and 20**

Although there are a great variety of fasteners found in the modern car or truck, the most commonly used retainer is the threaded fastener (nuts, bolts, screws, studs, etc.). Most threaded retainers may be reused, provided that they are not damaged in use or during the repair. Some retainers (such as stretch bolts or torque prevailing nuts) are designed to deform when tightened or in use and should not be reinstalled.

Whenever possible, we will note any special retainers which should be replaced during a procedure. But you should always inspect the condition of a retainer when it is removed and replace any that show signs of damage. Check all threads for rust or corrosion which can increase the torque necessary to

Fig. 19 There are many different types of threaded retainers found on vehicles

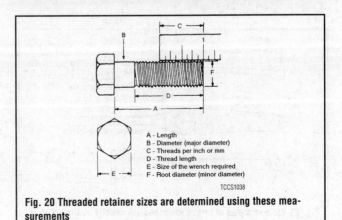

A - Length
B - Diameter (major diameter)
C - Threads per inch or mm
D - Thread length
E - Size of the wrench required
F - Root diameter (minor diameter)

Fig. 20 Threaded retainer sizes are determined using these measurements

achieve the desired clamp load for which that fastener was originally selected. Additionally, be sure that the driver surface of the fastener has not been compromised by rounding or other damage. In some cases a driver surface may become only partially rounded, allowing the driver to catch in only one direction. In many of these occurrences, a fastener may be installed and tightened, but the driver would not be able to grip and loosen the fastener again.

If you must replace a fastener, whether due to design or damage, you must ALWAYS be sure to use the proper replacement. In all cases, a retainer of the same design, material and strength should be used. Markings on the heads of most bolts will help determine the proper strength of the fastener. The same material, thread and pitch must be selected to assure proper installation and safe operation of the vehicle afterwards.

Thread gauges are available to help measure a bolt or stud's thread. Most automotive and hardware stores keep gauges available to help you select the proper size. In a pinch, you can use another nut or bolt for a thread gauge. If the bolt you are replacing is not too badly damaged, you can select a match by finding another bolt which will thread in its place. If you find a nut which threads properly onto the damaged bolt, then use that nut to help select the replacement bolt.

✳✳ WARNING

Be aware that when you find a bolt with damaged threads, you may also find the nut or drilled hole it was threaded into has also been damaged. If this is the case, you may have to drill and tap the hole, replace the nut or otherwise repair the threads. NEVER try to force a replacement bolt to fit into the damaged threads.

Torque

Torque is defined as the measurement of resistance to turning or rotating. It tends to twist a body about an axis of rotation. A common example of this would be tightening a threaded retainer such as a nut, bolt or screw. Measuring torque is one of the most common ways to help assure that a threaded retainer has been properly fastened.

When tightening a threaded fastener, torque is applied in three distinct areas, the head, the bearing surface and the clamp load. About 50 percent of the measured torque is used in overcoming bearing friction. This is the friction between the bearing surface of the bolt head, screw head or nut face and the base material or washer (the surface on which the fastener is rotating). Approximately 40 percent of the applied torque is used in overcoming thread friction. This leaves only about 10 percent of the applied torque to develop a useful clamp load (the force which holds a joint together). This means that friction can account for as much as 90 percent of the applied torque on a fastener.

TORQUE WRENCHES

▶ See Figure 21

In most applications, a torque wrench can be used to assure proper installation of a fastener. Torque wrenches come in various designs and most automotive supply stores will carry a variety to suit your needs. A torque wrench should be used any time we supply a specific torque value for a fastener. Again, the general rule of "if you are using the right tool for the job, you should not have to strain to tighten a fastener" applies here.

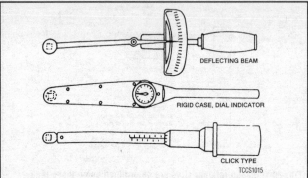

DEFLECTING BEAM

RIGID CASE, DIAL INDICATOR

CLICK TYPE

TCCS1015

Fig. 21 Various styles of torque wrenches are usually available at your local automotive supply store

Beam Type

The beam type torque wrench is one of the most popular types. It consists of a pointer attached to the head that runs the length of the flexible beam (shaft) to a scale located near the handle. As the wrench is pulled, the beam bends and the pointer indicates the torque using the scale.

Click (Breakaway) Type

Another popular design of torque wrench is the click type. To use the click type wrench you pre-adjust it to a torque setting. Once the torque is reached, the wrench has a reflex signaling feature that causes a momentary breakaway of the torque wrench body, sending an impulse to the operator's hand.

Pivot Head Type

▶ See Figure 22

Some torque wrenches (usually of the click type) may be equipped with a pivot head which can allow it to be used in areas of limited access. BUT, it must be used properly. To hold a pivot head wrench, grasp the handle lightly, and as you pull on the handle, it should be floated on the pivot point. If the handle comes in contact with the yoke extension during the process of pulling, there is a very good chance the torque readings will be inaccurate because this could alter the wrench loading point. The design of the handle is usually such as to make it inconvenient to deliberately misuse the wrench.

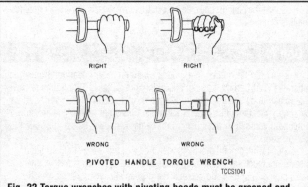

RIGHT   RIGHT

WRONG   WRONG

PIVOTED HANDLE TORQUE WRENCH

TCCS1041

Fig. 22 Torque wrenches with pivoting heads must be grasped and used properly to prevent an incorrect reading

➡ **It should be mentioned that the use of any U-joint, wobble or extension will have an effect on the torque readings, no matter what type of wrench you are using. For the most accurate readings, install the socket directly on the wrench driver. If necessary, straight extensions (which hold a socket directly under the wrench driver) will have the least effect on the torque reading. Avoid any extension that alters the length of the wrench from the handle to the head/driving point (such as a crow's foot). U-joint or wobble extensions can greatly affect the readings; avoid their use at all times.**

Rigid Case (Direct Reading)

A rigid case or direct reading torque wrench is equipped with a dial indicator to show torque values. One advantage of these wrenches is that they can be held at any position on the wrench without affecting accuracy. These wrenches are often preferred because they tend to be compact, easy to read and have a great degree of accuracy.

TORQUE ANGLE METERS

Because the frictional characteristics of each fastener or threaded hole will vary, clamp loads which are based strictly on torque will vary as well. In most applications, this variance is not significant enough to cause worry. But, in certain applications, a manufacturer's engineers may determine that more precise clamp loads are necessary (such is the case with many aluminum cylinder heads). In these cases, a torque angle method of installation would be specified. When installing fasteners which are torque angle tightened, a predetermined seating torque and standard torque wrench are usually used first to remove any compliance from the joint. The fastener is then tightened the specified additional portion of a turn measured in degrees. A torque angle gauge (mechanical protractor) is used for these applications.

Standard and Metric Measurements

▶ See Figure 23

Throughout this manual, specifications are given to help you determine the condition of various components on your vehicle, or to assist you in their installation. Some of the most common measurements include length (in. or cm/mm), torque (ft. lbs., inch lbs. or Nm) and pressure (psi, in. Hg, kPa or mm Hg). In most cases, we strive to provide the proper measurement as determined by the manufacturer's engineers.

Though, in some cases, that value may not be conveniently measured with what is available in your toolbox. Luckily, many of the measuring devices which are available today will have two scales so the Standard or Metric measurements may easily be taken. If any of the various measuring tools which are available to you do not contain the same scale as listed in the specifications, use the accompanying conversion factors to determine the proper value.

The conversion factor chart is used by taking the given specification and multiplying it by the necessary conversion factor. For instance, looking at the first line, if you have a measurement in inches such as "free-play should be 2 in." but your ruler reads only in millimeters, multiply 2 in. by the conversion factor of 25.4 to get the metric equivalent of 50.8mm. Likewise, if the specification was given only in a Metric measurement, for example in Newton Meters (Nm), then look at the center column first. If the measurement is 100 Nm, multiply it by the conversion factor of 0.738 to get 73.8 ft. lbs.

CONVERSION FACTORS

LENGTH–DISTANCE

Inches (in.)	x 25.4	= Millimeters (mm)	x .0394	= Inches
Feet (ft.)	x .305	= Meters (m)	x 3.281	= Feet
Miles	x 1.609	= Kilometers (km)	x .0621	= Miles

VOLUME

Cubic Inches (in3)	x 16.387	= Cubic Centimeters	x .061	= in3
IMP Pints (IMP pt.)	x .568	= Liters (L)	x 1.76	= IMP pt.
IMP Quarts (IMP qt.)	x 1.137	= Liters (L)	x .88	= IMP qt.
IMP Gallons (IMP gal.)	x 4.546	= Liters (L)	x .22	= IMP gal.
IMP Quarts (IMP qt.)	x 1.201	= US Quarts (US qt.)	x .833	= IMP qt.
IMP Gallons (IMP gal.)	x 1.201	= US Gallons (US gal.)	x .833	= IMP gal.
Fl. Ounces	x 29.573	= Milliliters	x .034	= Ounces
US Pints (US pt.)	x .473	= Liters (L)	x 2.113	= Pints
US Quarts (US qt.)	x .946	= Liters (L)	x 1.057	= Quarts
US Gallons (US gal.)	x 3.785	= Liters (L)	x .264	= Gallons

MASS–WEIGHT

Ounces (oz.)	x 28.35	= Grams (g)	x .035	= Ounces
Pounds (lb.)	x .454	= Kilograms (kg)	x 2.205	= Pounds

PRESSURE

Pounds Per Sq. In. (psi)	x 6.895	= Kilopascals (kPa)	x .145	= psi
Inches of Mercury (Hg)	x .4912	= psi	x 2.036	= Hg
Inches of Mercury (Hg)	x 3.377	= Kilopascals (kPa)	x .2961	= Hg
Inches of Water (H_2O)	x .07355	= Inches of Mercury	x 13.783	= H_2O
Inches of Water (H_2O)	x .03613	= psi	x 27.684	= H_2O
Inches of Water (H_2O)	x .248	= Kilopascals (kPa)	x 4.026	= H_2O

TORQUE

Pounds–Force Inches (in–lb)	x .113	= Newton Meters (N·m)	x 8.85	= in–lb
Pounds–Force Feet (ft–lb)	x 1.356	= Newton Meters (N·m)	x .738	= ft–lb

VELOCITY

Miles Per Hour (MPH)	x 1.609	= Kilometers Per Hour (KPH)	x .621	= MPH

POWER

Horsepower (Hp)	x .745	= Kilowatts	x 1.34	= Horsepower

FUEL CONSUMPTION*

Miles Per Gallon IMP (MPG)	x .354	= Kilometers Per Liter (Km/L)
Kilometers Per Liter (Km/L)	x 2.352	= IMP MPG
Miles Per Gallon US (MPG)	x .425	= Kilometers Per Liter (Km/L)
Kilometers Per Liter (Km/L)	x 2.352	= US MPG

*It is common to covert from miles per gallon (mpg) to liters/100 kilometers (1/100 km), where mpg (IMP) x 1/100 km = 282 and mpg (US) x 1/100 km = 235.

TEMPERATURE

Degree Fahrenheit (°F)	= (°C x 1.8) + 32
Degree Celsius (°C)	= (°F – 32) x .56

TCCS1044

Fig. 23 Standard and metric conversion factors chart

MODEL IDENTIFICATION

▶ See Figure 24

Toyota's have a model identification plate which is located under the hood on the firewall. The models are identified by a code. These codes explain exactly what your truck is and has. For example; model number VZN170L–CRMDKAB is a 1996 Tacoma 4WD, 5 speed manual transmission, Xtracab Deluxe V–6. When you walk into a dealer or your local parts store to purchase a part for your truck, and are not sure exactly the type of vehicle you drive, the model number and vehicle identification numbers are the key.

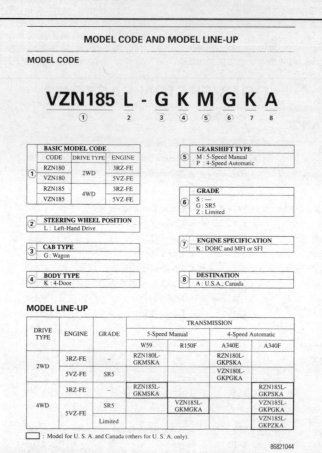

MODEL CODE AND MODEL LINE-UP

MODEL CODE

VZN185 L - G K M G K A
① 2 ③ ④ ⑤ ⑥ 7 8

① BASIC MODEL CODE

CODE	DRIVE TYPE	ENGINE
RZN180	2WD	3RZ-FE
VZN180		5VZ-FE
RZN185	4WD	3RZ-FE
VZN185		5VZ-FE

② STEERING WHEEL POSITION
L : Left-Hand Drive

③ CAB TYPE
G : Wagon

④ BODY TYPE
K : 4-Door

⑤ GEARSHIFT TYPE
M : 5-Speed Manual
P : 4-Speed Automatic

⑥ GRADE
S : —
G : SR5
Z : Limited

⑦ ENGINE SPECIFICATION
K : DOHC and MFI or SFI

⑧ DESTINATION
A : U.S.A., Canada

MODEL LINE-UP

DRIVE TYPE	ENGINE	GRADE	TRANSMISSION			
			5-Speed Manual		4-Speed Automatic	
			W59	R150F	A340E	A340F
2WD	3RZ-FE	–	RZN180L-GKMSKA		RZN180L-GKPSKA	
	5VZ-FE	SR5			VZN180L-GKPGKA	
4WD	3RZ-FE	–	RZN185L-GKMSKA			RZN185L-GKPSKA
	5VZ-FE	SR5		VZN185L-GKMGKA		VZN185L-GKPGKA
		Limited				VZN185L-GKPZKA

☐ : Model for U. S. A. and Canada (others for U. S. A. only).

86821044

Fig. 24 The model code is helpful for obtaining the correct part

SERIAL NUMBER IDENTIFICATION

Vehicle

▶ See Figures 25 and 26

All models have the vehicle identification number stamped on a plate attached to the left side of the instrument panel. The plate is visible by looking through the windshield from the outside.

All trucks and 4Runner's also have the VIN stamped into the metal of the outer face of the right front frame rail. The number also appears on the Certification Label attached to the left door pillar.

In addition to the instrument panel number, all Land Cruisers display the VIN on the Certification Label on the left front door post. The 1991–96 Land Cruisers also have the number stamped into the outer face of the right front frame rail.

The serial number consists of a series of 17 digits including the six digit serial or production number. The first three digits are the World Manufacturer Identification number. The next five digits are the Vehicle Description Section. The remaining nine digits are the production numbers including various codes on body style, trim level (base, luxury, etc.) and safety equipment or other information.

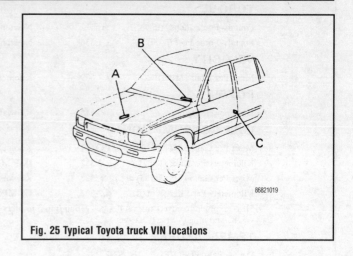

86821019

Fig. 25 Typical Toyota truck VIN locations

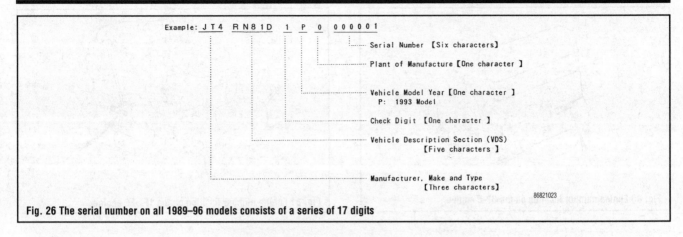

```
Example: J T 4  R N 8 1 D  1  P  0  000001
                                          └······· Serial Number [Six characters]

                                     └··········· Plant of Manufacture [One character]

                                └················ Vehicle Model Year [One character]
                                                 P:  1993 Model

                            └···················· Check Digit [One character]

                      └·········(··············· Vehicle Description Section (VDS)
                                                 [Five characters]

              └··························(········ Manufacturer, Make and Type
                                                 [Three characters]
                                                              86821023
```

Fig. 26 The serial number on all 1989–96 models consists of a series of 17 digits

VEHICLE IDENTIFICATION CHART

Engine Code						Model Year	
Code	**Liters**	**Cu. cc.**	**Cyl.**	**Fuel Sys.**	**Eng. Mfg.**	**Code**	**Year**
5 th digit						**10th digit**	
R (22R)	2.4	2366	4	2 BBL	TMC	K	1989
R (22R-E)	2.4	2366	4	EFI	TMC	L	1990
U (2RZ-FE)	2.4	2438	4	EFI	TMC	M	1991
U (3RZ-FE)	2.7	2693	4	EFI	TMC	N	1992
V (3VZ-E)	3.0	2959	6	EFI	TMC	P	1993
V (3VZ-FE)	3.0	2959	6	EFI	TMC	R	1994
F (3F-E)	4.0	241	6	EFI	TMC	S	1995
U (5VZ-FE)	3.4	3378	6	EFI	TMC	T	1996
F (1FZ-FE)	4.5	4477	6	EFI	TMC		
6 th digit							
L (2RZ-FE)	2.4	2366	4	EFI	TMC		
N (3RZ-FE)	2.7	2693	4	EFI	TMC		
M (5VZ-FE)	3.4	3378	6	EFI	TMC		
J (1FZ-FE)	4.5	4477	6	EFI	TMC		

BBL - Carbureted
EFI - Electronic Fuel Injected
TMC - Toyota Motor Corporation

86821035

Engine

▶ See Figures 27, 28, 29, 30 and 31

Each engine is referred to by both its family designation, such as 22R-E, and its production or serial number. The serial number can be important when ordering parts. Certain changes may have been made during production of the engine; different parts will be required if the engine was assembled before or after the change date. Generally, parts stores and dealers list this data in their catalogs, so have the engine number handy when you go. Refer to the illustrations to determine the engine number location.

It's a good idea to record the engine number while the vehicle is new. Jotting it inside the cover of the owner's manual or similar easy-to-find location will prevent having to scrape many years of grime off the engine when the number is finally needed.

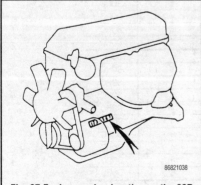

Fig. 27 Engine number location on the 22R and 22R-E engines

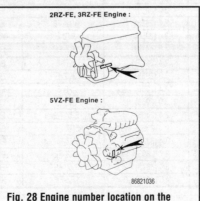

Fig. 28 Engine number location on the 2RZ-FE, 3RZ-FE and 5VZ-FE engines

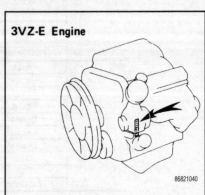

Fig. 29 Engine number location on the 3VZ-E engine

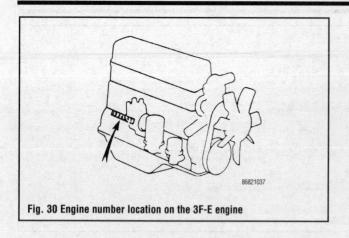

Fig. 30 Engine number location on the 3F-E engine

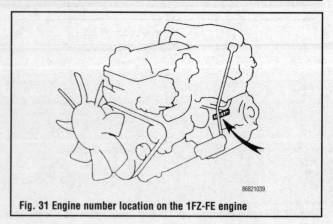

Fig. 31 Engine number location on the 1FZ-FE engine

ENGINE IDENTIFICATION

Year	Model	Engine Displacement Liters (cc)	Engine Series (ID/VIN)	Fuel System	No. of Cylinders	Engine Type
1989	4Runner	2.4 (2366)	22R	2 BBL	4	SOHC
	4Runner	2.4 (2366)	22R-E	EFI	4	SOHC
	4Runner	3.0 (2959)	3VZ-E	EFI	6	SOHC
	Pick-up	2.4 (2366)	22R	2 BBL	4	SOHC
	Pick-up	2.4 (2366)	22R-E	EFI	4	SOHC
	Pick-up	3.0 (2959)	3VZ-FE	EFI	6	SOHC
	Land Cruiser	4.0 (3956)	3F-E	EFI	6	OHV
1990	4Runner	2.4 (2366)	22R-E	EFI	4	SOHC
	4Runner	3.0 (2959)	3VZ-E	EFI	6	SOHC
	Pick-up	2.4 (2366)	22R	2 BBL	4	SOHC
	Pick-up	2.4 (2366)	22R-E	EFI	4	SOHC
	Pick-up	3.0 (2959)	3VZ-FE	EFI	6	SOHC
	Land Cruiser	4.0 (3956)	3F-E	EFI	6	OHV
1991	4Runner	2.4 (2366)	22R-E	EFI	4	SOHC
	4Runner	3.0 (2959)	3VZ-E	EFI	6	SOHC
	Pick-up	2.4 (2366)	22R-E	EFI	4	SOHC
	Pick-up	3.0 (2959)	3VZ-E	EFI	6	SOHC
	Land Cruiser	4.0 (3956)	3F-E	EFI	6	OHV
1992	4Runner	2.4 (2366)	22R-E	EFI	4	SOHC
	4Runner	3.0 (2959)	3VZ-FE	EFI	6	SOHC
	Pick-up	2.4 (2366)	22R-E	EFI	4	SOHC
	Pick-up	3.0 (2959)	3VZ-FE	EFI	6	SOHC
	Land Cruiser	4.0 (3956)	3F-E	EFI	6	OHV
1993	4Runner	2.4 (2366)	22R-E	EFI	4	SOHC
	4Runner	3.0 (2959)	3VZ-E	EFI	6	SOHC
	Pick-up	2.4 (2366)	22R-E	EFI	4	SOHC
	Pick-up	3.0 (2959)	3VZ-E	EFI	6	SOHC
	T100	3.0 (2959)	3VZ-E	EFI	6	SOHC
	Land Cruiser	4.5 (4477)	1FZ-FE	EFI	6	DOHC
1994	4Runner	2.4 (2366)	22R-E	EFI	4	SOHC
	4Runner	3.0 (2959)	3VZ-E	EFI	6	SOHC
	Pick-up	2.4 (2366)	22R-E	EFI	4	SOHC
	Pick-up	3.0 (2959)	3VZ-E	EFI	6	SOHC
	T100	2.7 (2693)	3RZ-FE	EFI	4	DOHC
	T100	3.0 (2959)	3VZ-E	EFI	6	SOHC
	Land Cruiser	4.5 (4477)	1FZ-FE	EFI	6	DOHC
1995	4Runner	2.4 (2366)	22R-E	EFI	4	SOHC
	4Runner	3.0 (2959)	3VZ-E	EFI	6	SOHC
	Pick-up	2.4 (2366)	22R-E	EFI	4	SOHC
	Pick-up	3.0 (2959)	3VZ-E	EFI	6	SOHC
	Tacoma	2.4 (2438)	2RZ-FE	EFI	4	DOHC
	Tacoma	2.7 (2693)	3RZ-FE	EFI	4	DOHC
	Tacoma	3.4 (3378)	5VZ-FE	EFI	6	DOHC
	T100	2.7 (2693)	3RZ-FE	EFI	4	DOHC
	T100	3.4 (3378)	5VZ-FE	EFI	6	DOHC
	Land Cruiser	4.5 (4477)	1FZ-FE	EFI	6	DOHC

ENGINE IDENTIFICATION

Year	Model	Engine Displacement Liters (cc)	Engine Series (ID/VIN)	Fuel System	No. of Cylinders	Engine Type
1996	4Runner	2.7 (2693)	3RZ-FE	EFI	4	DOHC
	4Runner	3.4 (3378)	5VZ-FE	EFI	6	DOHC
	Tacoma	2.4 (2438)	2RZ-FE	EFI	4	DOHC
	Tacoma	2.7 (2693)	3RZ-FE	EFI	4	DOHC
	Tacoma	3.4 (3378)	5VZ-FE	EFI	6	DOHC
	T100	2.7 (2693)	3RZ-FE	EFI	4	DOHC
	T100	3.4 (3378)	5VZ-FE	EFI	6	DOHC
	Land Cruiser	4.5 (4477)	1FZ-FE	EFI	6	DOHC

BBL - Barrel carburetor

EFI - Electronic fuel injection

SOHC - Single overhead camshaft

DOHC - Double overhead camshaft

OHV - Overhead valve

86821043

Transmissions

Automatic and manual transmissions are used in all of the trucks excluding the Land Cruiser. The Land Cruiser only uses an automatic. The identity of the transmission is on the driver's door tag. You can also obtain the type of transmission from the model number.

MANUAL TRANSMISSION CHART

Transmission Type	Years	Models
G40	1989-90	Truck
G57	1989-92	Truck
G58	1989-96	Truck, 4Runner, Tacoma
W46	1989-90	Truck
W55	1989-95	Truck
W56	1989-96	Truck, 4Runner, T100
W59	1996	Tacomoa, T100
R150	1989-96	Truck, 4Runner, Tacoma
R150F	1989-96	Truck, 4Runner, Tacoma, T100

86821045

AUTOMATIC TRANSMISSION CHART

Transmission Type	Years	Models
A43D	1989-96	Truck, Tacoma
A340E	1989-96	Truck, 4Runner, T100, Tacoma
A340F	1989-96	Truck, 4Runner
A340H	1989-95	Truck, 4Runner
A440F	1989-92	Land Cruiser
A442F	1993-95	Land Cruiser
A343F	1996	Land Cruiser

86821046

ROUTINE MAINTENANCE

General Information

◗ **See Figures 32 thru 38**

Proper maintenance is the key to long and trouble-free vehicle life. As a conscientious owner and driver, set aside a Saturday morning, say once a month, to check or replace items which could cause major problems later.

Keep your own personal log to jot down which services you performed, how much the parts cost you, the date, and the exact odometer reading at the time. Keep all receipts for such items as engine oil and filters, so that they may be referred to in case of related problems or to determine operating expenses. As a do-it-yourselfer, these receipts are the only proof you have that the required maintenance was performed. In the event of a warranty problem, these receipts will be invaluable.

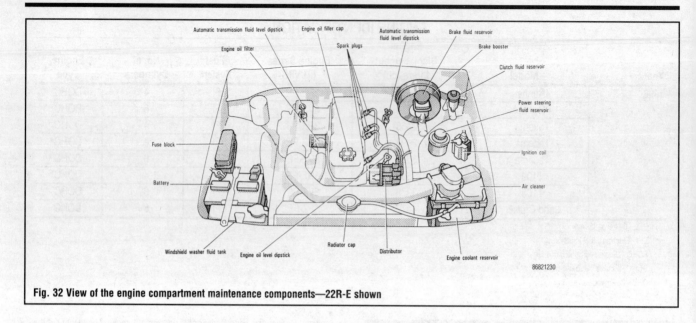

Fig. 32 View of the engine compartment maintenance components—22R-E shown

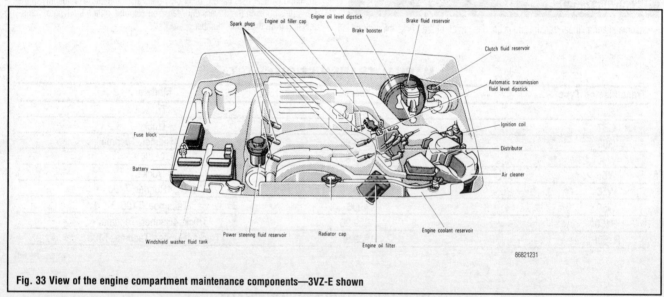

Fig. 33 View of the engine compartment maintenance components—3VZ-E shown

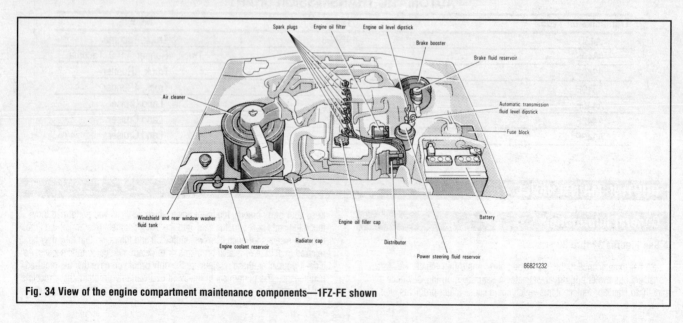

Fig. 34 View of the engine compartment maintenance components—1FZ-FE shown

The literature provided with your vehicle when it was originally delivered includes the factory recommended maintenance schedule. If you no longer have this literature, replacement copies are usually available from the dealer. A maintenance schedule is provided later in this section, in case you do not have the factory literature.

These checks and inspections can be done either by yourself a reputable shop, or the Toyota dealer.

Here are a few of the scheduled maintenance items that need to be checked frequently:

OUTSIDE THE VEHICLE
- Tire pressure—use a gauge to check the pressure

- Tire surfaces and lug nuts—check the tread depth and ensure all the lug nuts are in place
- Tire rotation—rotate every 6200 miles (1000 km)
- Fluid leaks—check the underneath for leaks of any kind
- Doors and the engine hood—check the latches ensuring they are securing properly

INSIDE THE VEHICLE
- Lights—make sure all the lights are in working order
- Reminder indicators—ensue all the warning lights and buzzers function properly

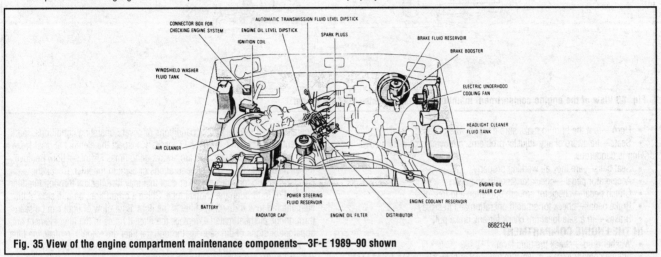

Fig. 35 View of the engine compartment maintenance components—3F-E 1989–90 shown

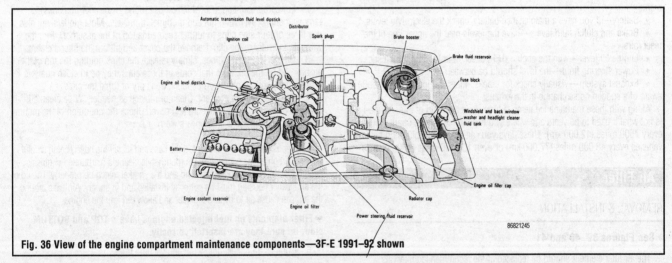

Fig. 36 View of the engine compartment maintenance components—3F-E 1991–92 shown

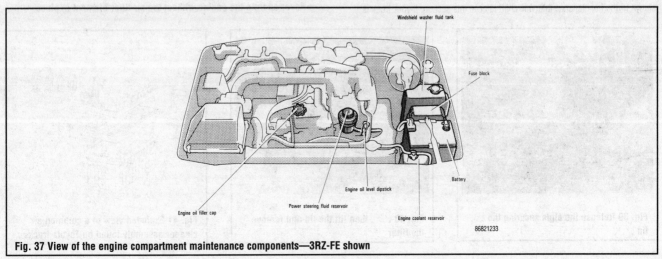

Fig. 37 View of the engine compartment maintenance components—3RZ-FE shown

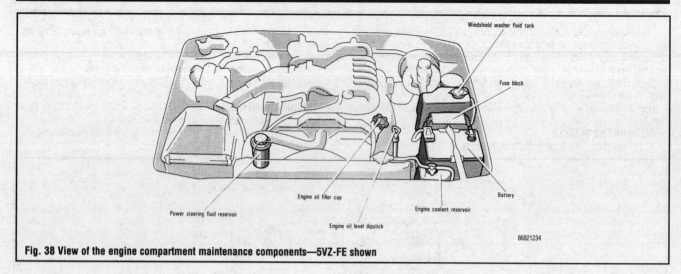

Fig. 38 View of the engine compartment maintenance components—5VZ-FE shown

- Horn—toot the horn to make sure it works when needed
- Seats—be aware of any adjuster problems, a moving seat while you're driving is dangerous
- Seat belts—are they all working properly
- Accelerator pedal—check for smooth operation
- Clutch pedal—check for smooth operation and free-play
- Brake pedal—check for smooth operation and free-play
- Brakes—in a safe location, check for any brake pull

IN THE ENGINE COMPARTMENT
- Washer fluid—check the fluid level
- Engine coolant level—make sure the level is between the FULL and LOW marks
- Battery—if you have a maintenance battery, check the electrolyte levels
- Brake and clutch fluid levels—have the levels near the upper line of the reservoirs
- Engine oil level—with the engine **OFF**, check fluid level on the dipstick
- Power steering fluid—the level should be between HOT and COLD
- Exhaust system—visually check for cracks, holes and loose supports. Be aware of a sudden noise change in the exhaust

Along with these maintenance items, a tune-up is also part of this. A tune-up is not what it used to be years ago where you need to replace the spark plugs every 7500 miles (12,000 km). These days you can replace the plugs on some vehicles every 48,000 miles (77,000 km) or even 100,000 miles (160,900 km).

Air Filter

REMOVAL & INSTALLATION

▶ **See Figures 39, 40 and 41**

The air filter element should be replaced at the recommended intervals shown in the Maintenance Intervals chart later in this section. If your truck is operated under severely dusty conditions or severe operating conditions, more frequent changes will certainly be necessary. Inspect the element at least twice a year. Early spring and early fall are always good times for inspection. Remove the element and check for any perforations or tears in the filter. Check the cleaner housing for signs of dirt or dust that may have leaked through the filter element or in through the snorkel tube. Shine a bright light on one side of the element and look through the filter at the light. If no glow of light can be seen through the element material, replace the filter. If holes in the filter element are apparent or signs of dirt seepage through the filter are evident, replace the filter.

1. Disconnect all hoses, ducts and vacuum tubes which would block removal of the top of the air cleaner assembly.

2. For carbureted engines (which all use round air cleaner housings), remove the top cover wing nut and grommet, if present. Most models will also use three or four side clips to further secure the top of the assembly. Pull the wire tab and release the clip. Remove the cover and lift out the filter element.

3. On fuel injected engines, simply release the clips holding the top of the air box and lift the lid. Note that some of these clips may be in close quarters against bodywork or other components; don't pry or force the clips.

4. Remove the filter element. Clean or replace as needed. Wipe clean all surfaces of the air cleaner housing and cover. Check the condition of the mounting gasket and replace it if it appears worn or broken.

To install:

5. Reposition the filter element in the case and install the cover, being careful not to overtighten the wingnut(s). On round-style cleaners (carbureted engines), be certain that the arrows on the cover lid and the snorkel match up properly. The lid of the air cleaner housing must be correctly installed and fit snugly. Air leaks around the top can cause air to bypass the filter and allow dirt into the engine.

➡**Filter elements on fuel injected engines have a TOP and BOTTOM side, be sure they are inserted correctly.**

6. Connect all hoses, duct work and vacuum lines.

➡**Never operate the engine without the air filter element in place.**

Fig. 39 Release the clips securing the lid . . .

Fig. 40 . . . then lift the lid and remove the filter

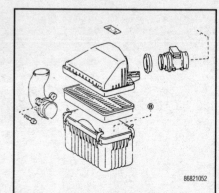

Fig. 41 Exploded view of a common air cleaner assembly found on Toyota trucks

Fuel Filter

REMOVAL & INSTALLATION

✳✳ CAUTION

Never smoke when working around or near gasoline. Make sure that there are no active ignition sources (heaters, electric motors or fans, welders, anything with sparks or open flame.) in the area. Have a fire extinguisher within arm's reach at all times.

Carbureted Engines

▶ **See Figure 42**

The fuel filter on carbureted engines is a plastic cylinder with two ports for the fuel. It is usually located on right rear frame rail, near the fuel tank.

1. Using a pair of pliers, expand the hose clamp on one side of the filter, and slide the clamp further down the hose, past the point to which the filter pipe extends. Remove the other clamp in the same manner.
2. Grasp the hoses near the ends and twist them gently to pull them free from the filter pipe.
3. Pull the filter from the clip and discard.

To install:

4. Install the new filter into the clip. The arrow must point towards the hose that runs to the carburetor.

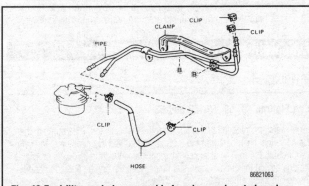

Fig. 42 Fuel filter and pipe assembly found on carbureted engines

5. Push the hoses onto the filter pipes, then slide the clamps back into position.
6. Start the engine and check for leaks.

Fuel Injected Engines

▶ **See Figures 43, 44 and 45**

The fuel filters on injected engines are a metal cylinder which is located in the rear frame rail or in the engine compartment. You may find it under the injection manifold on some models. Each model varies.

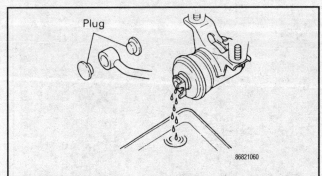

Fig. 43 Place a pan under the delivery pipe to catch the dripping fuel

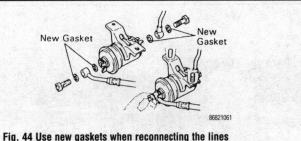

Fig. 44 Use new gaskets when reconnecting the lines

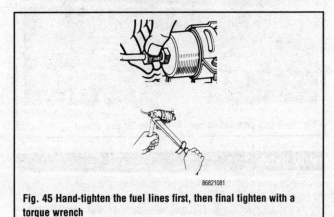

Fig. 45 Hand-tighten the fuel lines first, then final tighten with a torque wrench

1. Unbolt the retaining screws and remove the protective shield from the fuel filter.
2. Place a pan under the delivery pipe to catch the dripping fuel and SLOWLY loosen the union bolt to bleed off the fuel pressure. The fuel system is under pressure. Release pressure slowly and contain spillage. Observe "no smoking/no open flame precautions".
3. Remove the union bolt and drain the remaining fuel.
4. Disconnect and plug the inlet line.
5. Unbolt and remove the fuel filter.

To install:

➡ **When tightening the fuel line bolts to the fuel filter, you must use a torque wrench. The tightening torque is very important, as under or over tightening may cause fuel leakage. Insure that there is no fuel line interference and that there is sufficient space between the fuel lines and other components.**

6. Coat the flare unit, union nut and all bolt threads with engine oil.
7. Hand-tighten the inlet line to the fuel filter.
8. Install the fuel filter and then tighten the inlet line nut to 22 ft. lbs. (29 Nm).
9. Reconnect the delivery pipe using new gaskets and then tighten the union bolt to 22 ft. lbs. (29 Nm).
10. Run the engine for a short period and check for any fuel leaks.
11. Install the protective shield.

Fuel Cap Gasket

REMOVAL & INSTALLATION

▶ **See Figure 46**

All vehicles require the replacement of the fuel filler cap gasket at 60,000 miles (96,558 km). The gasket is important in sealing the filler neck and keeping the vapors from the tank routed through the vapor emission system. On 1989 Land Cruisers the gasket is held in with 4 small screws and a retaining plate which must be removed. The later models eliminate the screws and retainer; gently pry the gasket off with your fingers or a tool. Install the new gasket and make certain it is not twisted or crimped.

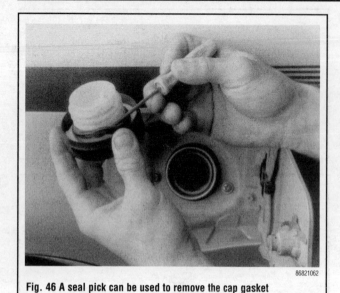

Fig. 46 A seal pick can be used to remove the cap gasket

PCV Valve

The PCV valve regulates crankcase ventilation during various engine operating conditions. At high vacuum (idle speed and partial load range) it will open slightly and at low vacuum (full throttle) it will open fully. This causes vapor to be removed from the crankcase by the engine vacuum and then sucked into the combustion chamber where it is burned along with the fuel.

➡The PCV system will not function properly unless the oil filler cap is tightly sealed. Check the gasket on the cap and be certain it is not leaking. Replace the cap or gasket or both if necessary to ensure proper sealing.

REMOVAL & INSTALLATION

▸ **See Figures 47 and 48**

1. Check the ventilation hoses and lines for leaks or clogging. Clean or replace as necessary.
2. Locate the PCV valve in the cylinder head cover and remove it by pulling it upward.
3. Test the valve by attaching a clean tube to the crankcase end of the valve, then blow through it. There should be free passage of air through the valve.
4. Move the tube to the other end of the valve. Blow into the tube. There should be little or no passage of air through the valve.

Fig. 47 Locate the PCV valve in the cylinder head cover and remove it by pulling it upward

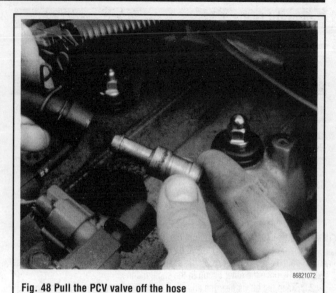

Fig. 48 Pull the PCV valve off the hose

5. If the PCV valve failed either of the preceding two checks, it will require replacement.
6. Pull the PCV valve off the hose.
To install:
7. Slip the hose back onto the proper end of the PCV valve.
8. Press the valve into the retaining grommet in the cylinder head cover.

Evaporative Canister (Charcoal Canister)

SERVICING

▸ **See Figures 49, 50, 51, 52 and 53**

The canister cycles the fuel vapor from the fuel tank and, if carbureted, the carburetor float chambers into the intake manifold and eventually into the cylinders for combustion. The activated charcoal element within the canister acts as a storage device for the fuel vapors at times when the engine operating conditions do not allow efficient burning of the vapors.

The only required service for the canister is inspection at the intervals specified in the Maintenance Chart at the end of this section. If the charcoal element is saturated (possibly from carburetor flooding), the entire canister will require replacement. Label and disconnect the canister purge hoses, loosen the retaining bracket bolt(s) and lift out the canister. Installation is simply the reverse of the removal process. To check the canister:

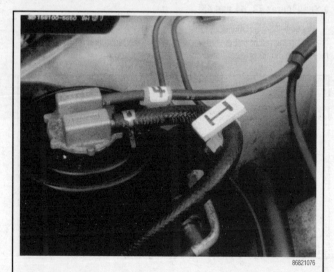

Fig. 49 Label the vacuum lines leading to the canister

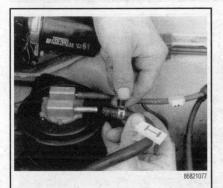

Fig. 50 Remove the vacuum lines attached to the canister

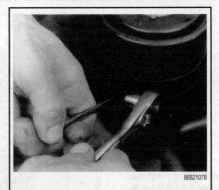

Fig. 51 Unfasten the retaining bolts from the canister . . .

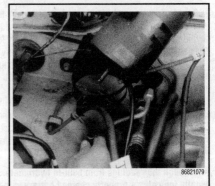

Fig. 52 . . . then lift the canister up and remove the lower hose attached to the unit

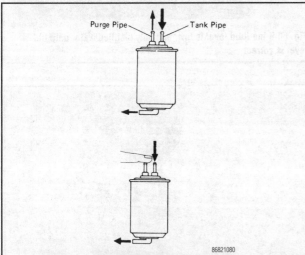

Fig. 53 When applying the air, hold a finger over the purge pipe to force all the air out the bottom port

1. Label the vacuum lines leading to the canister.
2. Remove the vacuum lines attached to the canister.
3. Unfasten the retaining bolts from the canister.
4. Lift the canister up and remove the lower hose attached to the unit.
5. Inspect the case for any cracking or damage.
6. Using low pressure compressed air, blow into the tank pipe (flanged end) and check that air flows freely from the other ports.
7. Blow into the purge pipe (next to tank pipe) and check that air does not flow from the other ports. If air does flow, the check valve has failed and the canister must be replaced.
8. Never attempt to flush the canister with fluid or solvent. Low pressure air 43 psi (294 kPa) maximum may be used to evaporate any vapors within the canister. When applying the air, hold a finger over the purge pipe to force all the air out the bottom port.
9. No carbon should come out of the filter at any time. Loose charcoal is a sign of internal failure in the canister.

Battery

PRECAUTIONS

Always use caution when working on or near the battery. Never allow a tool to bridge the gap between the negative and positive battery terminals. Also, be careful not to allow a tool to provide a ground between the positive cable/terminal and any metal component on the vehicle. Either of these conditions will cause a short circuit, leading to sparks and possible personal injury.

Do not smoke or all open flames/sparks near a battery; the gases contained in the battery are very explosive and, if ignited, could cause severe injury or death.

All batteries, regardless of type, should be carefully secured by a battery hold-down device. If not, the terminals or casing may crack from stress during vehicle operation. A battery which is not secured may allow acid to leak, making it discharge faster. The acid can also eat away at components under the hood.

Always inspect the battery case for cracks, leakage and corrosion. A white corrosive substance on the battery case or on nearby components would indicate a leaking or cracked battery. If the battery is cracked, it should be replaced immediately.

GENERAL MAINTENANCE

Always keep the battery cables and terminals free of corrosion. Check and clean these components about once a year.

Keep the top of the battery clean, as a film of dirt can help discharge a battery that is not used for long periods. A solution of baking soda and water may be used for cleaning, but be careful to flush this off with clear water. DO NOT let any of the solution into the filler holes. Baking soda neutralizes battery acid and will de-activate a battery cell.

Batteries in vehicles which are not operated on a regular basis can fall victim to parasitic loads (small current drains which are constantly drawing current from the battery). Normal parasitic loads may drain a battery on a vehicle that is in storage and not used for 6–8 weeks. Vehicles that have additional accessories such as a phone or an alarm system may discharge a battery sooner. If the vehicle is to be stored for longer periods in a secure area and the alarm system is not necessary, the negative battery cable should be disconnected to protect the battery.

Remember that constantly deep cycling a battery (completely discharging and recharging it) will shorten battery life.

BATTERY FLUID

▶ See Figure 54

Check the battery electrolyte level at least once a month, or more often in hot weather or during periods of extended vehicle operation. On non-sealed batteries, the level can be checked either through the case (if translucent) or by removing the cell caps. The electrolyte level in each cell should be kept filled to the split ring inside each cell, or the line marked on the outside of the case.

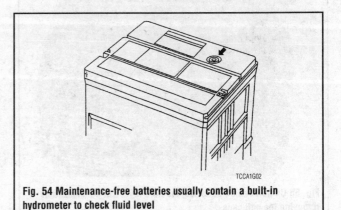

Fig. 54 Maintenance-free batteries usually contain a built-in hydrometer to check fluid level

If the level is low, add only distilled water through the opening until the level is correct. Each cell must be checked and filled individually. Distilled water should be used, because the chemicals and minerals found in most drinking water are harmful to the battery and could significantly shorten its life.

If water is added in freezing weather, the vehicle should be driven several miles to allow the water to mix with the electrolyte. Otherwise, the battery could freeze.

Although some maintenance-free batteries have removable cell caps, the electrolyte condition and level on all sealed maintenance-free batteries must be checked using the built-in hydrometer "eye." The exact type of eye will vary. But, most battery manufacturers, apply a sticker to the battery itself explaining the readings.

➡**Although the readings from built-in hydrometers will vary, a green eye usually indicates a properly charged battery with sufficient fluid level. A dark eye is normally an indicator of a battery with sufficient fluid, but which is low in charge. A light or yellow eye usually indicates that electrolyte has dropped below the necessary level. In this last case, sealed batteries with an insufficient electrolyte must usually be discarded.**

Checking the Specific Gravity

▶ **See Figures 55, 56 and 57**

A hydrometer is required to check the specific gravity on all batteries that are not maintenance-free. On batteries that are maintenance-free, the specific gravity is checked by observing the built-in hydrometer "eye" on the top of the battery case.

⁑⁑ CAUTION

Battery electrolyte contains sulfuric acid. If you should splash any on your skin or in your eyes, flush the affected area with plenty of clear water. If it lands in your eyes, get medical help immediately.

The fluid (sulfuric acid solution) contained in the battery cells will tell you many things about the condition of the battery. Because the cell plates must be kept submerged below the fluid level in order to operate, the fluid level is extremely important. And, because the specific gravity of the acid is an indication of electrical charge, testing the fluid can be an aid in determining if the battery must be replaced. A battery in a vehicle with a properly operating charging system should require little maintenance, but careful, periodic inspection should reveal problems before they leave you stranded.

At least once a year, check the specific gravity of the battery. It should be between 1.20 and 1.26 on the gravity scale. Most auto stores carry a variety of inexpensive battery hydrometers. These can be used on any non-sealed battery to test the specific gravity in each cell.

The battery testing hydrometer has a squeeze bulb at one end and a nozzle at the other. Battery electrolyte is sucked into the hydrometer until the float is lifted

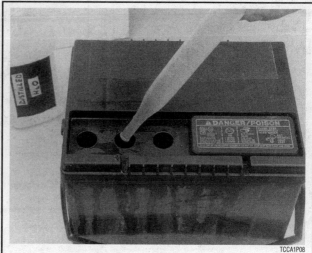

Fig. 56 If the fluid level is low, add only distilled water until the level is correct

Fig. 57 Check the specific gravity of the battery's electrolyte with a hydrometer

from its seat. The specific gravity is then read by noting the position of the float. If gravity is low in one or more cells, the battery should be slowly charged and checked again to see if the gravity has come up. Generally, if after charging, the specific gravity between any two cells varies more than 50 points (0.50), the battery should be replaced, as it can no longer produce sufficient voltage to guarantee proper operation.

CABLES

▶ **See Figures 58, 59, 60 and 61**

Once a year (or as necessary), the battery terminals and the cable clamps should be cleaned. Loosen the clamps and remove the cables, negative cable first. On top post batteries, the use of a puller specially made for this purpose is recommended. These are inexpensive and available in most parts stores. Side terminal battery cables are secured with a small bolt.

Clean the cable clamps and the battery terminal with a wire brush, until all corrosion, grease, etc., is removed and the metal is shiny. It is especially important to clean the inside of the clamp thoroughly (an old knife is useful here), since a small deposit of oxidation there will prevent a sound connection and inhibit starting or charging. Special tools are available for cleaning these parts, one type for conventional top post batteries and another type for side terminal batteries. It is also a good idea to apply some dielectric grease to the terminal, as this will aid in the prevention of corrosion.

Fig. 55 On non-sealed batteries, the fluid level can be checked by removing the cell caps

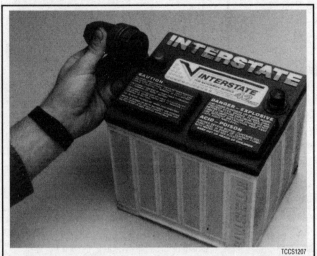

Fig. 58 The underside of this special battery tool has a wire brush to clean post terminals

Fig. 59 Place the tool over the battery posts and twist to clean until the metal is shiny

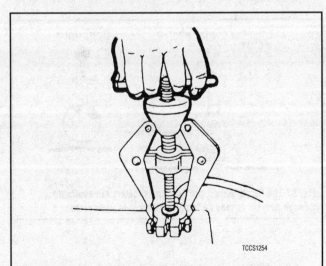

Fig. 60 A special tool is available to pull the clamp from the post

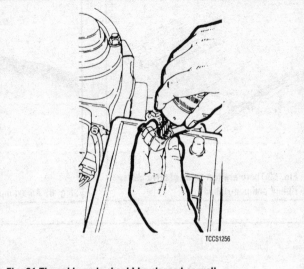

Fig. 61 The cable ends should be cleaned as well

After the clamps and terminals are clean, reinstall the cables, negative cable last; DO NOT hammer the clamps onto battery posts. Tighten the clamps securely, but do not distort them. Give the clamps and terminals a thin external coating of grease after installation, to retard corrosion.

Check the cables at the same time that the terminals are cleaned. If the cable insulation is cracked or broken, or if the ends are frayed, the cable should be replaced with a new cable of the same length and gauge.

CHARGING

❋❋ CAUTION

The chemical reaction which takes place in all batteries generates explosive hydrogen gas. A spark can cause the battery to explode and splash acid. To avoid personal injury, be sure there is proper ventilation and take appropriate fire safety precautions when working with or near a battery.

A battery should be charged at a slow rate to keep the plates inside from getting too hot. However, if some maintenance-free batteries are allowed to discharge until they are almost "dead," they may have to be charged at a high rate to bring them back to "life." Always follow the charger manufacturer's instructions on charging the battery.

REPLACEMENT

When it becomes necessary to replace the battery, select one with an amperage rating equal to or greater than the battery originally installed. Deterioration and just plain aging of the battery cables, starter motor, and associated wires makes the battery's job harder in successive years. This makes it prudent to install a new battery with a greater capacity than the old.

Belts

INSPECTION

▶ **See Figures 62, 63, 64, 65 and 66**

Check the condition of the drive belts and check the belt tension at least every 15,000 miles (24,000 km). Inspect the belts for signs of glazing or cracking. A glazed belt will be perfectly smooth from slippage, while a good belt will have a slight texture of fabric visible. Cracks will generally start at the inner edge of the belt and run outward. Replace the belt at the first sign of cracking or if the glazing is severe.

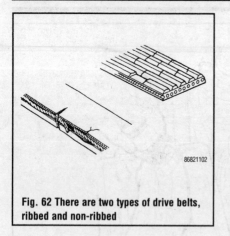

Fig. 62 There are two types of drive belts, ribbed and non-ribbed

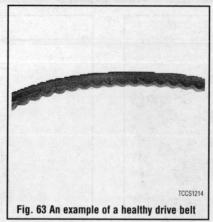

Fig. 63 An example of a healthy drive belt

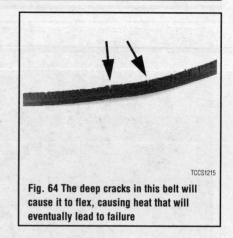

Fig. 64 The deep cracks in this belt will cause it to flex, causing heat that will eventually lead to failure

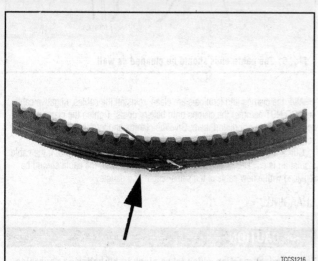

Fig. 65 The cover of this belt is worn exposing the critical reinforcing cords to excessive wear

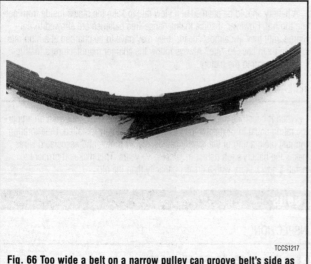

Fig. 66 Too wide a belt on a narrow pulley can groove belt's side as well as damage the cover and seat

Belt tension does not refer to play or droop. By placing your thumb midway between the two pulleys, it should be possible to depress the belt ¼–½ in. (6–13mm). If any of the belts can be depressed more than this, or cannot be depressed this much, adjust the tension. While this is an inaccurate test, it provides a quick reference. Inadequate tension will always result in slippage or wear, while excessive tension will damage pulley bearings and cause belts to fray and crack. A belt should be just tight enough to perform without slipping or squealing.

Its not a bad idea to replace all drive belts at 60,000 miles (96,000 km) regardless of their condition.

ADJUSTMENT

▶ **See Figure 67**

Toyota measures belt tension in pounds of force as determined by a belt tension tester. The Nippondenso and Burroughs testers are available through dealers or may be found at retail auto parts stores. The tester slips over a short section of the drive belt, and, when tightened, reads the deflection pressure on a dial. This is one of the most exact ways of setting tension and purchase of this tool or its equivalent is recommended.

Specifications for new belts are slightly higher than for used belts. A new belt is one which has not been run under tension for more than 5 minutes. Anything else is a used belt.

22R and 22R-E Engines

- Alternator with air pump: New—125 ft. lbs. (169 Nm) Used—80 ft. lbs. (108 Nm)
- Power Steering: New—125 ft. lbs. (169 Nm) Used—80 ft. lbs. (108 Nm)
- Air Conditioning: New—125 ft. lbs. (169 Nm) Used—80 ft. lbs. (169 Nm)

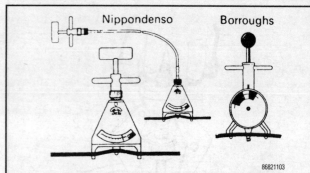

Fig. 67 The Nippondenso and Burroughs testers are available through dealers or may be found at retail auto parts stores

3VZ-E Engine

- Alternator: New—160 ft. lbs. (217 Nm) Used—100 ft. lbs. (136 Nm)
- Power Steering: New—125 ft. lbs. (169 Nm) Used—80 ft. lbs. (108 Nm)
- Air Conditioning: New—125 ft. lbs. (169 Nm) Used—80 ft. lbs. (108 Nm)

5VZ-E Engine

- Alternator: New—160 ft. lbs. (217 Nm) Used—100 ft. lbs. (136 Nm)
- Power Steering: New—158 ft. lbs. (214 Nm) Used—103 ft. lbs. (140 Nm)
- Air Conditioning: New—160 ft. lbs. (217 Nm) Used—100 ft. lbs. (136 Nm)

2RZ-FE and 3RZ-FE Engines

- Alternator: New—165 ft. lbs. (224 Nm) Used—115 ft. lbs. (156 Nm)
- Power Steering: New—158 ft. lbs. (214 Nm) Used—103 ft. lbs. (140 Nm)
- Air Conditioning: New—160 ft. lbs. (217 Nm) Used—100 ft. lbs. (136 Nm)

3F-E Engine

- Alternator: New—145 ft. lbs. (196 Nm) Used—100 ft. lbs. (136 Nm)
- Power Steering and air pump: New—145 ft. lbs. (196 Nm) Used—100 ft. lbs. (136 Nm)
- Air conditioning: New—125 ft. lbs. (169 Nm) Used—80 ft. lbs. (108 Nm)

1FZ-FE Engine

- Alternator: New—110 ft. lbs. (149 Nm) Used—67 ft. lbs. (91 Nm)
- Air Conditioning: New—125 ft. lbs. (169 Nm) Used—80 ft. lbs. (108 Nm)

REMOVAL & INSTALLATION

▶ **See Figures 68 thru 73**

When buying replacement belts, remember that the fit is critical according to the length of the belt ("diameter"), the width of the belt, the depth of the belt and the angle or profile of the V shape. The belt shape should exactly match the shape of the pulley; belts that are not an exact match can cause noise, slippage and premature failure.

If a belt must be replaced, the driven unit must be loosened and moved to its extreme loosest position, generally by moving it toward the center of the motor. After removing the old belt, check the pulleys for dirt or built-up material which could affect belt contact. Carefully install the new belt, remembering that it is new and unused-it may appear to be just a little too small to fit over the pulley flanges.

Fit the belt over the largest pulley (usually the crankshaft pulley at the bottom center of the motor) first, then work on the smaller one(s). Gentle pressure in the direction of rotation is helpful. Some belts run around a third or idler pulley, which acts as an additional pivot in the belt's path. It may be possible to loosen the idler pulley as well as the main component, making your job much easier. Depending on which belt(s) you are changing, it may be necessary to loosen or remove other interfering belts to get at the one(s) you want.

After the new belt is installed, draw tension on it by moving the driven unit away from the motor and tighten its mounting bolts. This is sometimes a three or four-handed job; you may find an assistant helpful. Make sure that all the bolts you loosened get re-tightened and that any other loosened belts also have the correct tension. A new belt can be expected to stretch a bit after installation so be prepared to re-adjust your new belt, if needed, within the first hundred miles/kilometers of use.

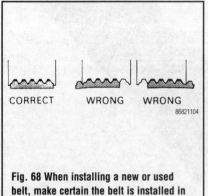

Fig. 68 When installing a new or used belt, make certain the belt is installed in the grooves correctly

VP : Vane Pump
AL : Alternator
CK : Crankshaft
CC : Cooler Compressor
IP : Idle Pulley
WP : Water Pump
AP : Air Pump

Fig. 69 Listing of pulley abbreviations

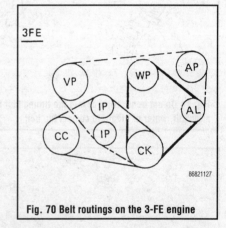

Fig. 70 Belt routings on the 3-FE engine

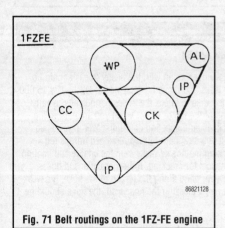

Fig. 71 Belt routings on the 1FZ-FE engine

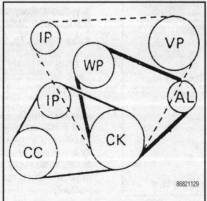

Fig. 72 Belt routings on the 3RZ-FE engine

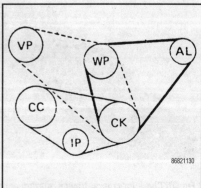

Fig. 73 Belt routings on the 3VZ-FE and 5VZ-FE engines

Timing Belts

INSPECTION

▶ **See Figures 74, 75, 76 and 77**

Toyota recommends that the timing belt be replaced on vehicles that are used in extensive idling or low speed driving for long distances. Police, taxi and door-to-door deliveries are commonly used in this manner. The timing belt should be replaced every 60,000 miles (96,000 km) in these cases. Toyota trucks do not have interference engines; where engine damage will occur if the belt snaps. If your vehicle has high milage, you may want to consider replacing the belt to prevent the possibility of having it snap. Or if your engine is being overhauled, inspect the belt for wear and replace if needed. In the event the belt does snap while you are driving, turn the engine **OFF** immediately. Section 3 has removal and installation procedures available.

Fig. 76 Damage on only one side of the belt could indicate a faulty guide

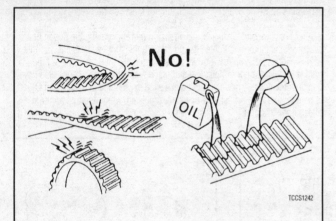

Fig. 74 Do not bend, twist or turn the timing belt inside out. Never allow oil, water or steam to contact the belt

Fig. 77 ALWAYS replace the timing belt at the interval specified by the manufacturer

Hoses

▶ **See Figures 78, 79, 80 and 81**

Upper and lower radiator hoses along with the heater hoses should be checked for deterioration, leaks and loose hose clamps at least every 15,000 miles (24,000 km). It is also wise to check the hoses periodically in early spring and at the beginning of the fall or winter when you are performing other maintenance. A quick visual inspection could discover a weakened hose which might have left you stranded if it had remained unrepaired.

Whenever you are checking the hoses, make sure the engine and cooling system is cold. Visually inspect for cracking, rotting or collapsed hoses, replace as necessary. Run your hand along the length of the hose. If a weak or swollen spot is noted when squeezing the hose wall, the hose should be replaced.

Fig. 75 Check for cracks, fraying, glazing or damage of any kind

Fig. 78 The cracks developing along this hose are the result of age and hardening

Fig. 79 A hose clamp that is too tight can cause older hoses to separate and tear

Fig. 80 Check for soft, spongy hoses like this is (swollen at the clamp)—this hose will eventually burst

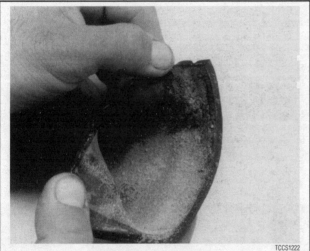

Fig. 81 Debris or contaminants in the cooling system can cause a hose to weaken from the inside out

REMOVAL & INSTALLATION

♦ See Figures 82 thru 87

✳✳ CAUTION

Never remove the pressure cap while the engine is running or personal injury from scalding hot coolant or steam may result. If possi-

ble, wait until the engine has cooled to remove the pressure cap. If this is not possible, wrap a thick cloth around the pressure cap and turn it slowly to the stop. Step back while the pressure is released from the cooling system. When you are sure all the pressure has been released, still using the cloth, turn and remove the cap.

1. Remove the radiator pressure cap.
2. Position a clean container under the radiator and/or engine petcock or plug, then open the drain and allow the cooling system to drain to an appropriate level. For some upper hoses only a little coolant must be drained. To remove hoses positioned lower on the engine, such as a lower radiator hose, the entire cooling system must be emptied.

✳✳ CAUTION

When draining coolant, keep in mind that cats and dogs are attracted by ethylene glycol antifreeze, and are quite likely to drink any that is left in an uncovered container or in puddles on the ground. This will prove fatal in sufficient quantity. Always drain coolant into a sealable container. Coolant may be reused unless it is contaminated or several years old.

3. Loosen the hose clamps at each end of the hose requiring replacement. Clamps are usually either of the spring tension type (which require pliers to squeeze the tabs and loosen) or of the screw tension type (which require screw or hex drivers to loosen). Pull the clamps back on the hose away from the connection.
4. Twist, pull and slide the hose off the fitting taking care not to damage the neck of the component from which the hose is being removed.

➡If the hose is stuck at the connection, do not try to insert a screwdriver or other sharp tool under the hose end in an effort to free it, as

Fig. 82 When the engine is cool, remove the radiator cap

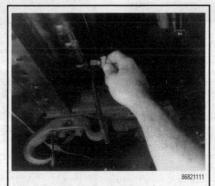

Fig. 83 Attach a short length of hose to the petcock before loosening the nut

Fig. 84 Using a pair of pliers, squeeze the tabs to loosen the upper hose

Fig. 85 As you grasp the hose clamp slide it back

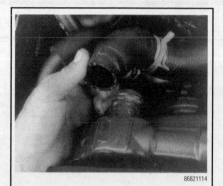

Fig. 86 Carefully twist, then pull the hose off the neck

Fig. 87 When sliding the lower hose off, remember there will be a small amount of fluid still in the hose

the connection and/or hose may become damaged. Heater connections especially may be easily damaged by such a procedure. If the hose is to be replaced, use a single-edged razor blade to make a slice along the portion of the hose which is stuck on the connection, perpendicular to the end of the hose. Do not cut deep so as to prevent damaging the connection. The hose can then be peeled from the connection and discarded.

5. Clean both hose mounting connections. Inspect the condition of the hose clamps and replace them, if necessary.

To install:

6. Dip the ends of the new hose into clean engine coolant to ease installation.

7. If a clamp shows signs of any damage (bent, too loose, hard to tighten, etc.) now is the time to replace it. A good rule of thumb is that a new hose is always worth new clamps. Slide the clamps over the replacement hose, then slide the hose ends over the connections into position.

8. Position and secure the clamps at least ¼ in. (6.35mm) from the ends of the hose. Make sure they are located inside the raised bead of the connector.

9. Reinstall the petcock and properly refill the cooling system with the clean drained engine coolant or a suitable mixture of ethylene glycol coolant and water.

10. If available, install a pressure tester and check for leaks. If a pressure tester is not available, run the engine until normal operating temperature is reached (allowing the system to naturally pressurize), then check for leaks.

✳✳ CAUTION

If you are checking for leaks with the system at normal operating temperature, BE EXTREMELY CAREFUL not to touch any moving or hot engine parts. Once temperature has been reached, shut the engine OFF, and check for leaks around the hose fittings and connections which were removed earlier.

CV-Boots

INSPECTION

▶ **See Figures 88 and 89**

The CV (Constant Velocity) boots should should be checked for damage each time the oil is changed and any other time the vehicle is raised for service. These boots keep water, grime, dirt and other damaging matter from entering the CV-joints. Any of these could cause early CV-joint failure which can be expensive to repair. Heavy grease thrown around the inside of the front wheel(s) and on the brake caliper can be an indication of a torn boot. Thoroughly check the boots for missing clamps and tears. If the boot is damaged, have it replaced immediately.

Fig. 88 CV-Boots must be inspected periodically for damage

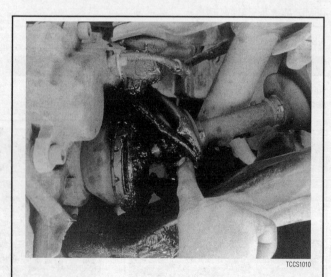

Fig. 89 A torn boot should be replaced immediately

Spark Plugs

▶ See Figure 90

A typical spark plug consists of a metal shell surrounding a ceramic insulator. A metal electrode extends downward through the center of the insulator and protrudes a small distance. Located at the end of the plug and attached to the side of the outer metal shell is the side electrode. The side electrode bends in at a 90° angle so that its tip is just past and parallel to the tip of the center electrode. The distance between these two electrodes (measured in thousandths of an inch or hundredths of a millimeter) is called the spark plug gap.

The spark plug does not produce a spark but instead provides a gap across which the current can arc. The coil produces anywhere from 20,000 to 50,000 volts (depending on the type and application) which travels through the wires to the spark plugs. The current passes along the center electrode and jumps the gap to the side electrode, and in doing so, ignites the air/fuel mixture in the combustion chamber.

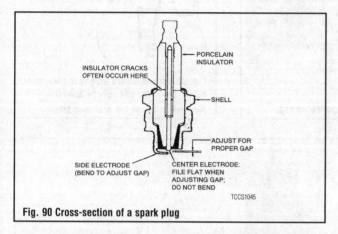

Fig. 90 Cross-section of a spark plug

SPARK PLUG HEAT RANGE

▶ See Figure 91

Spark plug heat range is the ability of the plug to dissipate heat. The longer the insulator (or the farther it extends into the engine), the hotter the plug will operate; the shorter the insulator (the closer the electrode is to the block's cooling passages) the cooler it will operate. A plug that absorbs little heat and remains too cool will quickly accumulate deposits of oil and carbon since it is not hot enough to burn them off. This leads to plug fouling and consequently to misfiring. A plug that absorbs too much heat will have no deposits but, due to

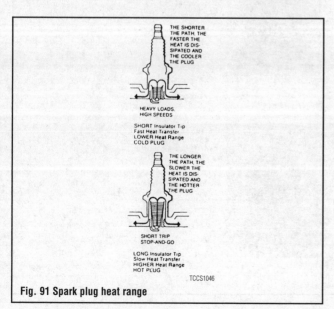

Fig. 91 Spark plug heat range

the excessive heat, the electrodes will burn away quickly and might possibly lead to preignition or other ignition problems. Preignition takes place when plug tips get so hot that they glow sufficiently to ignite the air/fuel mixture before the actual spark occurs. This early ignition will usually cause a pinging during low speeds and heavy loads.

The general rule of thumb for choosing the correct heat range when picking a spark plug is: if most of your driving is long distance, high speed travel, use a colder plug; if most of your driving is stop and go, use a hotter plug. Original equipment plugs are generally a good compromise between the 2 styles and most people never have the need to change their plugs from the factory-recommended heat range.

REMOVAL & INSTALLATION

➡Remove the spark plugs and wires one at a time to avoid confusion and miswiring during installation.

1. Disconnect the negative battery cable, and if the vehicle has been run recently, allow the engine to thoroughly cool.
2. Carefully twist the spark plug wire boot to loosen it, then pull upward and remove the boot from the plug. Be sure to pull on the boot and not on the wire, otherwise the connector located inside the boot may become separated.
3. Using compressed air, blow any water or debris from the spark plug well to assure that no harmful contaminants are allowed to enter the combustion chamber when the spark plug is removed. If compressed air is not available, use a rag or a brush to clean the area.

➡Remove the spark plugs when the engine is cold, if possible, to prevent damage to the threads. If removal of the plugs is difficult, apply a few drops of penetrating oil or silicone spray to the area around the base of the plug, and allow it a few minutes to work.

4. Using a spark plug socket that is equipped with a rubber insert to properly hold the plug, turn the spark plug counterclockwise to loosen and remove the spark plug from the bore.

✳✳ WARNING

Be sure not to use a flexible extension on the socket. Use of a flexible extension may allow a shear force to be applied to the plug. A shear force could break the plug off in the cylinder head, leading to costly and frustrating repairs.

To install:
5. Inspect the spark plug boot for tears or damage. If a damaged boot is found, the spark plug wire must be replaced.
6. Using a wire feeler gauge, check and adjust the spark plug gap. When using a gauge, the proper size should pass between the electrodes with a slight drag. The next larger size should not be able to pass while the next smaller size should pass freely.
7. Carefully thread the plug into the bore by hand. If resistance is felt before the plug is almost completely threaded, back the plug out and begin threading again. In small to reach areas, an old spark plug wire and boot could be used as a threading tool. The boot will hold the plug while you twist the end of the wire and the wire is supple enough to twist before it would allow the plug to crossthread.

✳✳ WARNING

Do not use the spark plug socket to thread the plugs. Always carefully thread the plug by hand or using an old plug wire to prevent the possibility of cross-threading and damaging the cylinder head bore.

8. Carefully tighten the spark plug. If the plug you are installing is equipped with a crush washer, seat the plug, then tighten about ¼ turn to crush the washer. If you are installing a tapered seat plug, tighten the plug to 11–15 ft. lbs. (15–20 Nm).
9. Apply a small amount of silicone dielectric compound to the end of the spark plug lead or inside the spark plug boot to prevent sticking, then install the boot to the spark plug and push until it clicks into place. The click may be felt or heard, then gently pull back on the boot to assure proper contact.

INSPECTION & GAPPING

▶ **See Figures 92, 93, 94 and 95**

Check the plugs for deposits and wear. If they are not going to be replaced, clean the plugs thoroughly. Remember that any kind of deposit will decrease the efficiency of the plug. Plugs can be cleaned on a spark plug cleaning machine, which can sometimes be found in service stations, or you can do an acceptable job of cleaning with a stiff brush. If the plugs are cleaned, the electrodes must

be filed flat. Use an ignition points file, not an emery board or the like, which will leave deposits. The electrodes must be filed perfectly flat with sharp edges; rounded edges reduce the spark plug voltage by as much as 50%.

Check spark plug gap before installation. The ground electrode (the L-shaped one connected to the body of the plug) must be parallel to the center electrode and the specified size wire gauge (please refer to the Tune-Up Specifications chart for details) must pass between the electrodes with a slight drag.

➡ **NEVER adjust the gap on a used platinum type spark plug.**

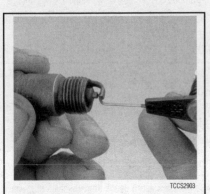

Fig. 92 Checking the spark plug gap with a feeler gauge

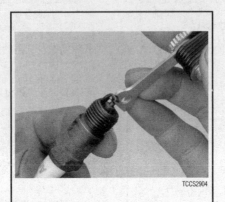

Fig. 93 Adjusting the spark plug gap

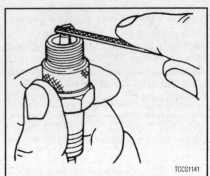

Fig. 94 If the standard plug is in good condition, the electrode may be filed flat— WARNING: do not file platinum plugs

A **normally worn** spark plug should have light tan or gray deposits on the firing tip.

A **carbon fouled** plug, identified by soft, sooty, black deposits, may indicate an improperly tuned vehicle. Check the air cleaner, ignition components and engine control system.

This spark plug has been **left in the engine too long,** as evidenced by the extreme gap- Plugs with such an extreme gap can cause misfiring and stumbling accompanied by a noticeable lack of power.

An **oil fouled** spark plug indicates an engine with worn poston rings and/or bad valve seals allowing excessive oil to enter the chamber.

A **physically damaged** spark plug may be evidence of severe detonation in that cylinder. Watch that cylinder carefully between services, as a continued detonation will not only damage the plug, but could also damage the engine.

A **bridged or almost bridged** spark plug, identified by a build-up between the electrodes caused by excessive carbon or oil build-up on the plug.

Fig. 95 Inspect the spark plug to determine engine running conditions

Always check the gap on new plugs as they are not always set correctly at the factory. Do not use a flat feeler gauge when measuring the gap on a used plug, because the reading may be inaccurate. A round-wire type gapping tool is the best way to check the gap. The correct gauge should pass through the electrode gap with a slight drag. If you're in doubt, try one size smaller and one larger. The smaller gauge should go through easily, while the larger one shouldn't go through at all. Wire gapping tools usually have a bending tool attached. Use that to adjust the side electrode until the proper distance is obtained. Absolutely never attempt to bend the center electrode. Also, be careful not to bend the side electrode too far or too often as it may weaken and break off within the engine, requiring removal of the cylinder head to retrieve it.

Spark Plug Wires

TESTING

▶ **See Figures 96 and 97**

Every 36,000 miles (58,000 km) or so, the resistance of the wires should be checked with an ohmmeter. Wires with excessive resistance will cause misfiring, and may make the engine difficult to start in damp weather. Generally, the useful life of the cables is 36,000-50,000 miles (58,000–80,000 km).

To check resistance, remove the distributor cap, leaving the wires attached to the cap but removing them from the spark plugs. Look at each contact inside the cap for any sign of cracking or burning. A small amount of discoloration is normal but there should be no heavy burn marks or contact marks. Connect one lead of an ohmmeter to an electrode within the cap; connect the other lead to the corresponding spark plug terminal. Replace any wire which shows a resistance over 25,000 ohms.

Test the high tension lead from the coil by connecting the ohmmeter between the center contact in the distributor cap and either of the primary terminals of the coil. If resistance is more than 25,000 ohms, remove the cable from the coil and check the resistance of the cable alone. Anything over 15,000 ohms is cause for replacement. It should be remembered that resistance is also a function of length; the longer the cable, the greater the resistance. Thus, if the cables on your truck are longer than the factory originals, resistance will be higher, quite possibly outside these limits. Toyota recommends the 25,000 ohm limit be observed in all cases.

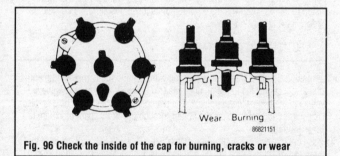

Fig. 96 Check the inside of the cap for burning, cracks or wear

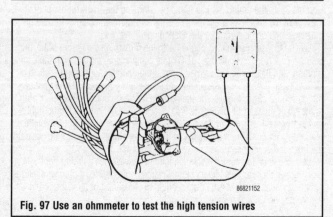

Fig. 97 Use an ohmmeter to test the high tension wires

REMOVAL & INSTALLATION

▶ **See Figure 98**

At every tune-up, visually inspect the spark plug cables for burns, cuts, or breaks in the insulation. Check the boots and the nipples on the distributor cap and coil. Replace any damaged wiring. Always replace spark plug wiring in sets, with a coil wire as well. Length is important; get the correct set for your vehicle.

When installing new cables, replace them one at a time to avoid mix-ups. Start by replacing the longest one first. Install the boot firmly over the spark plug. Route the wire over the same path as the original. Insert the nipple firmly into the tower on the cap or the coil.

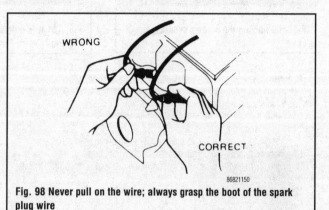

Fig. 98 Never pull on the wire; always grasp the boot of the spark plug wire

Distributor Cap and Rotor

REMOVAL & INSTALLATION

▶ **See Figures 99, 100, 101 and 102**

1. Disconnect the negative battery cable.
2. Remove the distributor cap rubber boot if equipped.
3. Loosen the screws securing the cap on the distributor.
4. Usually it is helpful to tag the wires leading to the cap for easy identification on installation.
5. Lift the cap off the distributor. Pulling from the wire boot, remove the plug wires from the cap.
6. Once the cap and wires are off, lift the rotor off the shaft.

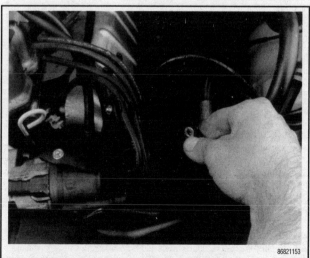

Fig. 99 Unsnap and remove the distributor boot

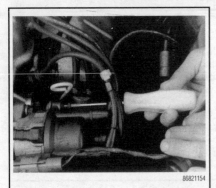

Fig. 100 Loosen the screws securing the cap

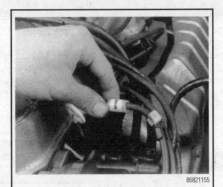

Fig. 101 It is helpful to tag the wires on the cap

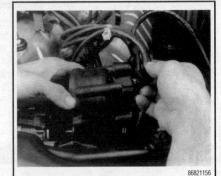

Fig. 102 Pull the wires off the cap from the boot, not the wire itself

To install:

7. Install the rotor onto the distributor shaft. The rotor only goes on one way so there should be no mix-up in replacement.

8. Apply a small amount of dielectric grease on the tip of the rotor and the inside of the cap carbon ends.

9. Attach the tagged wires to their proper locations on the cap.

10. Fit the cap onto the distributor, then tighten the bolts.

11. Make sure all the wires are secure on the cap and spark plugs. Install the negative battery cable.

12. Start the vehicle and check for any misses.

INSPECTION

When inspecting a cap and rotor, look for signs of cracks, burns and wear. The inside of the cap may be burnt or have wear on the carbon ends. On the rotor look at the tip for burning and excessive wear.

Ignition Timing

GENERAL INFORMATION

Ignition timing is the measurement in degrees of crankshaft rotation at the instant the spark plug fires while the piston is on its compression stroke.

Ignition timing is adjusted by loosening the distributor locking device and turning the distributor in its mount.

Ideally, the air/fuel mixture in the cylinder will be ignited by the spark plug and just beginning its rapid expansion as the piston passes Top Dead Center (TDC) of the compression stroke. If this happens, the piston will be beginning the power stroke just as the compressed air/fuel mixture starts to burn and expand. The expansion (explosion) of the air/fuel mixture will then force the piston down on the power stroke and turn the crankshaft.

It takes a fraction of a second for the spark from the plug to completely ignite the mixture in the cylinder. Because of this, the spark plug must fire before the piston reaches TDC, if the mixture is to be completely ignited as the piston passes TDC. This measurement is given in degrees of crankshaft rotation before the piston reaches top dead center (BTDC). If the ignition timing setting for your engine is seven (7°) BTDC, this means that the spark plug must fire at the time when the piston for that cylinder is 7° before reaching the top of its compression stroke. However, this only holds true while your engine is at idle speed.

As the engine accelerates from idle, the speed of the engine (rpm, or revolutions per minute) increases. The increase in rpm means that the pistons are now traveling up and down much faster. Because of this, the spark plugs will have to fire even sooner if the mixture is to be completely ignited as the piston passes TDC. To accomplish this, the distributor incorporates means to advance the timing of the spark as engine speed increases.

The distributor in your Toyota has at least two means of advancing the ignition timing. One is called centrifugal advance and is actuated by weights in the distributor. The other is called vacuum advance and is controlled by that larger circular housing on the side of the distributor. In many later models, the vacuum advance is replaced by full computerized control, thanks to the Engine Control Unit or ECU.

This function is known as Electronic Spark Advance (ESA) and is accomplished by Toyota's programming of the ECU. By monitoring the rpm, intake air volume, engine temperature, throttle position and other variables, the microprocessor decides the correct moment to trigger the spark for any engine operating condition from cold idle through wide open throttle. The system is simple, non-adjustable and reliable.

In addition, some early distributors have a vacuum retard mechanism which is contained in the same housing on the side of the distributor as the vacuum advance. The function of this mechanism is to retard the timing of the ignition spark under certain engine conditions. This causes more complete burning of the air/fuel mixture in the cylinder and consequently lowers exhaust emissions. Both the vacuum advance and the vacuum retard have vacuum hoses connected to them. Don't get them mixed up during removal.

Because these mechanisms change ignition timing, it is necessary to disconnect and plug the one or two vacuum lines from the distributor when setting the basic ignition timing.

If ignition timing is set too far advanced (BTDC), the ignition and expansion of the air/fuel mixture in the cylinder will try to force the piston down in the cylinder while it is still traveling upward. This causes engine "ping", a sound which resembles marbles being dropped into an empty tin can. If the ignition timing is too far retarded (after, or ATDC), the piston will have already started down on the power stroke when the air/fuel mixture ignites. This will cause the piston to be forced down only a portion of its travel, resulting in poor engine performance and lack of power.

INSPECTION AND ADJUSTMENT

Ignition timing adjustment is checked with a timing light. This instrument is inductively connected to the number one (No. 1) spark plug of the engine as well as to the two battery terminals. The timing light flashes every time an electrical current is sent from the distributor through the No. 1 spark plug wire to the spark plug.

Except for the 3F-E engine, the crankshaft pulley and the front cover of the engine are marked with a timing pointer and a timing scale. On Land Cruisers with the 3F-E, the timing mark is found be removing the small rubber plug at the rear of the engine where it meets the transmission. The plug is located on the same side of the engine as the spark plugs.

When the timing pointer is aligned with the **0** mark on the timing scale, the piston in the No. 1 cylinder is at TDC of its compression stroke. With the engine running, and the timing light aimed at the timing pointer and timing scale, the stroboscopic flashes from the timing light will allow you to check the ignition timing setting of the engine. The timing light flashes every time the spark plug in the No. 1 cylinder of the engine fires. Since the flash from the timing light makes the crankshaft pulley seem stationary for a moment you will be able to read the exact position of the piston in the No. 1 cylinder on the timing scale on the front of the engine.

Because your Toyota has electronic ignition, always use a timing light with an inductive pickup. This pickup simply clamps onto the No. 1 plug wire, eliminating any adapters. It is not as susceptible to crossfiring or false triggering, which may occur with a conventional light.

22-R Engine

1. Warm-up the engine, then turn the ignition **OFF**. Connect a tachometer to both battery terminals and to the service connector coming from the distributor.

2. Clean off the timing marks. The marks are on the crankshaft pulley and timing cover. The timing notches in the crankshaft pulley are normally marked at the factory with red or white paint. You may want to retouch them if they are dark, using chalk or paint. Fluorescent (dayglow) paint is excellent for this purpose. You might have to bump the engine around with the starter (just touch the key to the **START** position very briefly) to find the pulley marks.

3. Connect a timing light according to the manufacturer's instructions.

4. Disconnect the vacuum line(s) from the distributor vacuum unit. Clamp or plug the line(s); golf tees are excellent for this job. Any vacuum leak will make the engine run poorly.

➡On 22R engines with High Altitude Compensation (HAC) systems, there are two vacuum hoses which connect to the distributor. Both must be disconnected and plugged.

5. Be sure that the all wires are clear of the fan and moving belts, pulleys etc. Start the engine.

✖✖ CAUTION

Keep fingers, clothes, tools, hair, and wiring leads clear of the all moving parts. Be sure that you are running the engine in a well ventilated area.

6. Allow the engine to run at idle speed with the gear shift in Neutral (manual) and Park (P) with automatic transmission. Use the tachometer to check idle speed and adjust the idle if necessary.

✖✖ CAUTION

Be sure that the parking brake is set and that the front wheels are blocked to prevent the vehicle from rolling.

7. Point the timing light at the marks indicated in the chart and illustrations. With the engine at idle, timing should be at the specification given on the "Tune-Up Specifications" chart.

8. If the timing is not at the specification, loosen the pinch bolt (hold-down bolt) at the base of the distributor just enough so that the distributor can be turned. Turn the distributor to advance or retard the timing as required. Once the proper marks are seen to align, timing is correct. Tighten the hold-down bolt.

➡Remember that you are using metal tools on a running engine. Watch out for moving parts and don't touch any of the spark plug wires with the wrench.

9. Recheck the timing. Stop the engine; disconnect the tachometer and timing light. Except for engines equipped with HAC, connect the vacuum line(s) to the distributor vacuum unit.

10. On engines with HAC (identified in the NOTE earlier), after setting the initial timing, reconnect the vacuum hose at the distributor. Recheck the timing. It should now be about 13° BTDC.

11. If the advance is still about 8°, pinch the hose between the HAC valve and the three way connector. It should now be about 13°. If not, the HAC valve should be checked for proper operation.

22R-E, 3VZ-E and 3F-E Engines

▶ See Figures 103, 104, 105 and 106

1. Warm up the engine, then turn it **OFF**. Connect a timing light to the engine following the manufacturer"s instructions.

2. Connect a tachometer to both battery terminals. Connect the tachometer probe to the IG−terminal in the check connector box. The check connector box is located next to the underhood fuse and relay box on trucks and 4Runner's; for Land Cruisers, the check connector box is on the firewall. Use the correct adapter to insure a tight fit on the terminal. Not all tachometers, particularly older ones are electrically compatible with this system. Read instructions for your unit before using.

3. Start the engine and run it at idle. Check the idle speed on the tachometer; adjust it to specification if necessary.

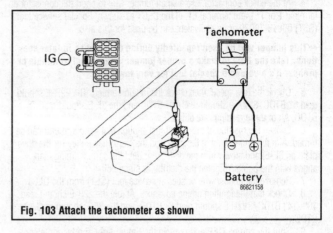

Fig. 103 Attach the tachometer as shown

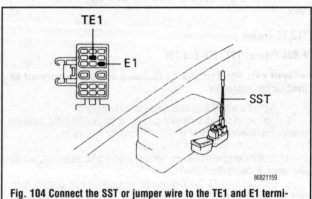

Fig. 104 Connect the SST or jumper wire to the TE1 and E1 terminals of the DLC1

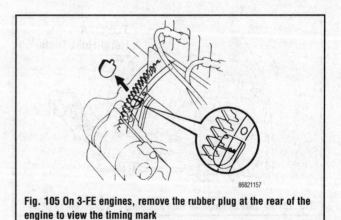

Fig. 105 On 3-FE engines, remove the rubber plug at the rear of the engine to view the timing mark

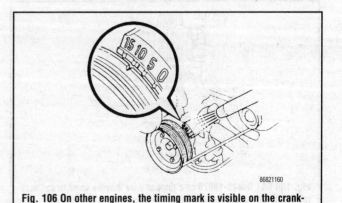

Fig. 106 On other engines, the timing mark is visible on the crankshaft pulley

4. At the check connector, use a small jumper wire to short the connector at terminal T or TE1 and terminal E1, of the DLC1 as shown. Special service tool (SST) 09843-18020 or its equivalent can be used for this also.

➡This jumper will be used repeatedly during diagnostics in later sections. Take the time to make a proper jumper with correct terminals or probes. It's a valuable special tool for very low cost.

5. Check the idle speed, then check the ignition timing. The 22R-E should read 5° BTDC, 3VZ-FE should read 10° BTDC and the 3F-E should read 7° BTDC. All of these readings are at idle.

6. Loosen the distributor pinch bolt just enough that the distributor can be turned. Aim the timing light at the marks on the crankshaft pulley (or the drive plate on 3F-E) and slowly turn the distributor until the correct timing mark aligns with the pointer. Tighten the distributor pinch bolt.

7. Remove the jumper wire or special service tool (SST) from the DLC1.

8. Check that the ignition timing advances. At idle the 22R-E should read 10°–14° BTDC, 3VZ-FE should read 8° BTDC, and the 3F-E should read 12° BTDC.

9. Shut the engine **OFF** and remove the timing light and tachometer leads.

1FZ-FE Engine

▶ See Figures 107, 108 and 109

➡Toyota's hand-held tester or an equivalent OBD-II scan tool must be used for this procedure.

1. Warm the engine to normal operating temperature.
2. Connect an OBD-II compliant scan tool to the DLC3 located under the dash on the driver's side. Refer to Section 4 for more information.
3. Connect the timing light to the engine.
4. Race the engine to 2500 rpm for approximately 90 seconds. Check the idle speed. It should read 600–700 rpm.
5. Using SST 09843-18020 or its equivalent jumper wire, connect terminals TE1 and E1 of the DLC1 under the hood.
6. With the transmission in neutral and the A/C off, check the timing. It should read 3° BTDC at idle.

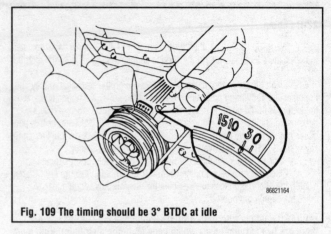

Fig. 109 The timing should be 3° BTDC at idle

7. Loosen the hold-down bolt, and adjust it by turning the distributor. Tighten the hold-down bolt to 13 ft. lbs. (18 Nm), then recheck the ignition timing again.

8. Remove the jumper wire from the DLC1.

9. Check the timing, the reading should be 2°–13° BTDC at idle. The timing mark will move in this range.

10. Disconnect the timing light from the engine.
11. Disconnect the scan tool.

2RZ-FE and 3RZ-FE Engines

▶ See Figures 110 and 111

➡Toyota's hand-held tester or an equivalent OBD-II scan tool must be used for this procedure.

1. Warm the engine to normal operating temperature.
2. Connect an OBD-II compliant scan tool to the DLC3 located under the dash on the driver's side. Refer to Section 4 for more information.
3. Connect the timing light to the engine.

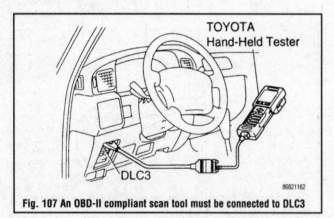

Fig. 107 An OBD-II compliant scan tool must be connected to DLC3

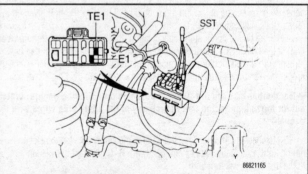

Fig. 110 Using the SST 09843-18020 or a jumper wire, connect terminals TE1 and E1 of the DLC1

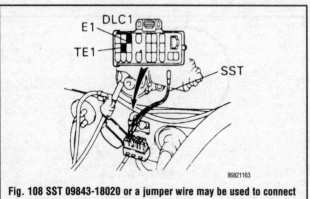

Fig. 108 SST 09843-18020 or a jumper wire may be used to connect terminals TE1 and E1 of the DLC1

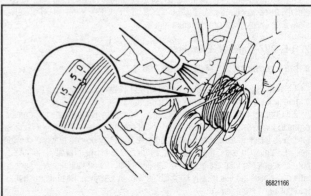

Fig. 111 The timing should be between 3° and 7° BTDC at idle

4. Using SST 09843-18020 or its equivalent jumper wire, connect terminals TE1 and E1 of the DLC1 under the hood.

5. After the engine speed is kept at about 1000 rpm for 5 seconds, check that it returns to idle speed.

6. Check the ignition timing, the reading should be 3°–7° BTDC at idle.

7. Remove the jumper wire from the DLC1.

8. Recheck the timing, the mark ranges from 7°–18° BTDC at idle.

9. Disconnect the scan tool.

10. Disconnect the timing light.

Valve Lash

▸ See Figures 112 and 113

As part of every major tune-up or service interval, the valve clearance should be checked and adjusted if necessary. For all trucks and 4Runner's covered by this book, the specification is every 30,000 miles (48,000 km) or 36 months. For all Land Cruisers, the requirement is 15,000 miles (24,000 km) or 18 months.

If the valve clearance is too large, part of the lift of the camshaft will be used up in removing the excessive clearance, thus the valves will not be opened far enough. This condition makes the valve train noisy as they take up the excessive clearance, and the engine will perform poorly, since a smaller amount of air/fuel mixture will be admitted to the cylinders. The exhaust valves will not open far enough to vent the cylinder completely; retained pressure (back pressure) will restrict the entry of the next air/fuel charge. If the valve clearance is too small, the intake and exhaust valves will not fully seat on the cylinder head when they close. This causes internal cylinder leakage and prevents the hot valve from transferring some heat to the head and cooling off. Therefore, the engine will run poorly (due to gases escaping from the combustion chamber), and the valves will overheat and warp (since they cannot transfer heat unless they are firmly touching the seat in the cylinder head).

➡ While all valve adjustments must be as accurate as possible, it is better to have the valve adjustment slightly loose than slightly tight, as burnt valves may result from overly tight adjustments.

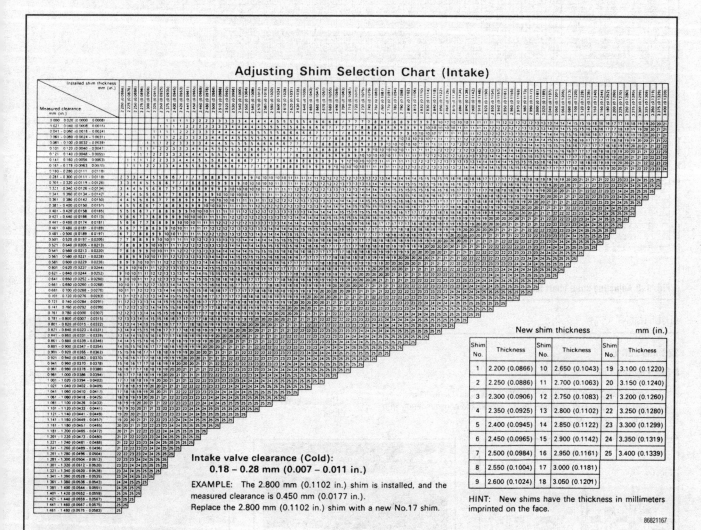

Intake valve clearance (Cold):
0.18 – 0.28 mm (0.007 – 0.011 in.)

EXAMPLE: The 2.800 mm (0.1102 in.) shim is installed, and the measured clearance is 0.450 mm (0.0177 in.).
Replace the 2.800 mm (0.1102 in.) shim with a new No.17 shim.

New shim thickness   mm (in.)

Shim No.	Thickness	Shim No.	Thickness	Shim No.	Thickness
1	2.200 (0.0866)	10	2.650 (0.1043)	19	3.100 (0.1220)
2	2.250 (0.0886)	11	2.700 (0.1063)	20	3.150 (0.1240)
3	2.300 (0.0906)	12	2.750 (0.1083)	21	3.200 (0.1260)
4	2.350 (0.0925)	13	2.800 (0.1102)	22	3.250 (0.1280)
5	2.400 (0.0945)	14	2.850 (0.1122)	23	3.300 (0.1299)
6	2.450 (0.0965)	15	2.900 (0.1142)	24	3.350 (0.1319)
7	2.500 (0.0984)	16	2.950 (0.1161)	25	3.400 (0.1339)
8	2.550 (0.1004)	17	3.000 (0.1181)		
9	2.600 (0.1024)	18	3.050 (0.1201)		

HINT: New shims have the thickness in millimeters imprinted on the face.

86821167

Fig. 112 Adjusting shim chart for the intake valves

Adjusting Shim Selection Chart (Exhaust)

Measured clearance mm (in.) \ Installed shim thickness mm (in.)	2.500 (0.0984)	2.520 (0.0992)	2.540 (0.1000)	2.550 (0.1004)	2.560 (0.1008)	2.580 (0.1016)	2.600 (0.1024)	2.620 (0.1031)	2.640 (0.1039)	2.650 (0.1043)	2.660 (0.1047)	2.680 (0.1055)	2.700 (0.1063)	2.710 (0.1067)	2.720 (0.1071)	2.730 (0.1075)	2.740 (0.1079)	2.750 (0.1083)	2.760 (0.1087)	2.770 (0.1091)	2.780 (0.1094)	2.790 (0.1098)	2.800 (0.1102)	2.810 (0.1106)	2.820 (0.1110)	2.830 (0.1114)	2.840 (0.1118)	2.850 (0.1122)	2.860 (0.1126)	2.870 (0.1130)	2.880 (0.1134)
0.000 – 0.030 (0.0000 – 0.0012)																		1	1	1	1	1	1	1	2	2	2	2	2	3	3
0.031 – 0.050 (0.0012 – 0.0020)															1	1	1	1	1	1	1	2	2	2	2	2	3	3	3	3	3
0.051 – 0.070 (0.0020 – 0.0028)													1	1	1	1	1	1	1	2	2	2	2	2	3	3	3	3	3	4	4
0.071 – 0.090 (0.0028 – 0.0035)												1	1	1	1	1	1	2	2	2	2	2	3	3	3	3	3	4	4	4	4
0.091 – 0.110 (0.0036 – 0.0043)											1	1	1	1	1	2	2	2	2	2	3	3	3	3	3	4	4	4	4	4	5
0.111 – 0.130 (0.0044 – 0.0051)									1	1	1	1	1	2	2	2	2	2	3	3	3	3	3	4	4	4	4	4	5	5	5
0.131 – 0.150 (0.0052 – 0.0059)								1	1	1	1	1	2	2	2	2	3	3	3	3	3	4	4	4	4	4	5	5	5	5	5
0.151 – 0.170 (0.0059 – 0.0067)							1	1	1	1	1	2	2	2	3	3	3	3	3	4	4	4	4	4	5	5	5	5	5	6	6
0.171 – 0.190 (0.0067 – 0.0075)						1	1	1	1	2	2	2	3	3	3	3	3	4	4	4	4	4	5	5	5	5	5	6	6	6	6
0.191 – 0.210 (0.0075 – 0.0083)					1	1	1	1	2	2	2	3	3	3	3	4	4	4	4	4	5	5	5	5	5	6	6	6	6	6	7
0.211 – 0.230 (0.0083 – 0.0091)			1	1	1	1	1	2	2	2	3	3	4	4	4	4	4	4	5	5	5	5	5	6	6	6	6	6	7	7	7
0.231 – 0.249 (0.0091 – 0.0098)		1	1	1	1	1	2	2	3	3	3	3	4	4	4	4	5	5	5	5	5	6	6	6	6	6	7	7	7	7	7
0.250 – 0.350 (0.0098 – 0.0138)																															
0.351 – 0.370 (0.0138 – 0.0146)	2	3	3	3	3	4	4	5	5	5	5	6	6	6	7	7	7	7	7	8	8	8	8	8	9	9	9	9	9	10	10
0.371 – 0.390 (0.0146 – 0.0154)	3	3	3	4	4	4	5	5	5	6	6	6	7	7	7	7	7	8	8	8	8	8	9	9	9	9	9	10	10	10	10
0.391 – 0.410 (0.0154 – 0.0161)	3	3	4	4	4	5	5	5	6	6	6	7	7	7	7	8	8	8	8	8	9	9	9	9	9	10	10	10	10	10	11
0.411 – 0.430 (0.0162 – 0.0169)	3	4	4	4	5	5	5	6	6	6	7	7	7	8	8	8	8	8	9	9	9	9	9	10	10	10	10	10	11	11	11
0.431 – 0.450 (0.0170 – 0.0177)	4	4	5	5	5	5	6	6	7	7	7	7	8	8	8	8	9	9	9	9	9	10	10	10	10	10	11	11	11	11	11
0.451 – 0.470 (0.0178 – 0.0185)	4	5	5	5	5	6	6	7	7	7	7	8	8	8	9	9	9	9	9	10	10	10	10	10	11	11	11	11	11	12	12
0.471 – 0.490 (0.0185 – 0.0193)	5	5	5	6	6	6	7	7	7	8	8	8	9	9	9	9	9	10	10	10	10	10	11	11	11	11	11	12	12	12	12
0.491 – 0.510 (0.0193 – 0.0201)	5	5	6	6	6	7	7	7	8	8	8	9	9	9	9	10	10	10	10	10	11	11	11	11	11	12	12	12	12	12	13
0.511 – 0.530 (0.0201 – 0.0209)	5	6	6	6	7	7	7	8	8	8	9	9	9	10	10	10	10	10	11	11	11	11	11	12	12	12	12	12	13	13	13
0.531 – 0.550 (0.0209 – 0.0217)	6	6	7	7	7	7	8	8	9	9	9	9	10	10	10	10	11	11	11	11	11	12	12	12	12	12	13	13	13	13	13
0.551 – 0.570 (0.0217 – 0.0224)	6	7	7	7	7	8	8	9	9	9	9	10	10	10	11	11	11	11	11	12	12	12	12	12	13	13	13	13	13	14	14
0.571 – 0.590 (0.0225 – 0.0232)	7	7	7	8	8	8	9	9	9	10	10	10	11	11	11	11	11	12	12	12	12	12	13	13	13	13	13	14	14	14	14
0.591 – 0.610 (0.0233 – 0.0240)	7	7	8	8	8	9	9	9	10	10	10	11	11	11	11	12	12	12	12	12	13	13	13	13	13	14	14	14	14	14	15
0.611 – 0.630 (0.0241 – 0.0248)	7	8	8	8	9	9	9	10	10	10	11	11	11	12	12	12	12	12	13	13	13	13	13	14	14	14	14	14	15	15	15
0.631 – 0.650 (0.0248 – 0.0256)	8	8	9	9	9	9	10	10	11	11	11	11	12	12	12	12	13	13	13	13	13	14	14	14	14	14	15	15	15	15	15
0.651 – 0.670 (0.0256 – 0.0264)	8	9	9	9	9	10	10	11	11	11	11	12	12	12	13	13	13	13	13	14	14	14	14	14	15	15	15	15	15	16	16
0.671 – 0.690 (0.0264 – 0.0272)	9	9	9	10	10	10	11	11	11	12	12	12	13	13	13	13	13	14	14	14	14	14	15	15	15	15	15	16	16	16	16
0.691 – 0.710 (0.0272 – 0.0280)	9	9	10	10	10	11	11	11	12	12	12	13	13	13	13	14	14	14	14	14	15	15	15	15	15	16	16	16	16	16	17
0.711 – 0.730 (0.0280 – 0.0287)	9	10	10	10	11	11	11	12	12	12	13	13	13	14	14	14	14	14	15	15	15	15	15	16	16	16	16	16	17	17	17
0.731 – 0.750 (0.0288 – 0.0295)	10	10	11	11	11	11	12	12	13	13	13	13	14	14	14	14	15	15	15	15	15	16	16	16	16	16	17	17	17	17	17
0.751 – 0.770 (0.0296 – 0.0303)	10	11	11	11	11	12	12	13	13	13	13	14	14	14	15	15	15	15	15	16	16	16	16	16	17	17	17	17	17	17	17
0.771 – 0.790 (0.0304 – 0.0311)	11	11	11	12	12	12	13	13	13	14	14	14	14	15	15	15	15	16	16	16	16	16	17	17	17	17	17	17	17		
0.791 – 0.810 (0.0311 – 0.0319)	11	11	12	12	12	13	13	13	14	14	14	15	15	15	15	16	16	16	16	16	17	17	17	17	17	17	17				
0.811 – 0.830 (0.0319 – 0.0327)	11	12	12	12	13	13	13	14	14	14	15	15	15	16	16	16	16	16	17	17	17	17	17	17	17						
0.831 – 0.850 (0.0327 – 0.0335)	12	12	13	13	13	13	14	14	15	15	15	15	16	16	16	16	17	17	17	17	17	17	17								
0.851 – 0.870 (0.0335 – 0.0343)	12	13	13	13	13	14	14	15	15	15	15	16	16	16	17	17	17	17	17	17	17										
0.871 – 0.890 (0.0343 – 0.0350)	13	13	13	14	14	14	15	15	15	16	16	16	16	17	17	17	17	17	17												
0.891 – 0.910 (0.0351 – 0.0358)	13	13	14	14	14	15	15	15	16	16	16	17	17	17	17	17	17														
0.911 – 0.930 (0.0359 – 0.0366)	13	14	14	14	15	15	15	16	16	16	17	17	17	17	17																
0.931 – 0.950 (0.0367 – 0.0374)	14	14	15	15	15	15	16	16	17	17	17	17	17																		
0.951 – 0.970 (0.0374 – 0.0382)	14	15	15	15	15	16	16	17	17	17	17	17																			
0.971 – 0.990 (0.0382 – 0.0390)	15	15	15	16	16	16	17	17	17	17	17																				
0.991 – 1.010 (0.0390 – 0.0398)	15	15	16	16	16	17	17	17	17																						
1.011 – 1.030 (0.0398 – 0.0406)	15	16	16	16	17	17	17	17																							
1.031 – 1.050 (0.0406 – 0.0413)	16	16	17	17	17	17	17																								
1.051 – 1.070 (0.0414 – 0.0421)	16	17	17	17	17	17																									
1.071 – 1.090 (0.0422 – 0.0429)	17	17	17	17	17																										
1.091 – 1.110 (0.0430 – 0.0437)	17	17	17																												
1.111 – 1.130 (0.0437 – 0.0445)	17	17																													
1.131 – 1.150 (0.0445 – 0.0413)	17																														

Measured clearance mm (in.) \ Installed shim thickness mm (in.)	2.890 (0.1138)	2.900 (0.1142)	2.910 (0.1146)	2.920 (0.1150)	2.930 (0.1154)	2.940 (0.1157)	2.950 (0.1161)	2.960 (0.1166)	2.970 (0.1169)	2.980 (0.1173)	2.990 (0.1177)	3.000 (0.1181)	3.010 (0.1185)	3.020 (0.1189)	3.030 (0.1193)	3.040 (0.1197)	3.050 (0.1201)	3.060 (0.1205)	3.070 (0.1209)	3.080 (0.1213)	3.090 (0.1217)	3.100 (0.1220)	3.120 (0.1228)	3.140 (0.1236)	3.150 (0.1240)	3.160 (0.1244)	3.180 (0.1252)	3.200 (0.1260)	3.220 (0.1268)	3.240 (0.1276)	3.250 (0.1280)	3.260 (0.1283)	3.280 (0.1291)	3.300 (0.1299)
0.000 – 0.030 (0.0000 – 0.0012)	3	3	3	4	4	4	4	4	5	5	5	5	5	6	6	6	6	6	7	7	7	7	8	8	8	8	9	9	10	10	10	10	11	11
0.031 – 0.050 (0.0012 – 0.0020)	4	4	4	4	4	5	5	5	5	5	6	6	6	6	6	7	7	7	7	7	8	8	8	9	9	9	9	10	10	11	11	11	11	12
0.051 – 0.070 (0.0020 – 0.0028)	4	4	4	5	5	5	5	5	6	6	6	6	6	7	7	7	7	7	8	8	8	8	9	9	9	9	10	10	11	11	11	11	12	12
0.071 – 0.090 (0.0028 – 0.0035)	4	5	5	5	5	5	6	6	6	6	6	7	7	7	7	7	8	8	8	8	8	9	9	9	10	10	10	11	11	11	12	12	12	13
0.091 – 0.110 (0.0036 – 0.0043)	5	5	5	5	6	6	6	6	6	7	7	7	7	7	8	8	8	8	8	9	9	9	9	10	10	10	11	11	11	12	12	12	13	13
0.111 – 0.130 (0.0044 – 0.0051)	5	5	6	6	6	6	6	7	7	7	7	7	8	8	8	8	8	9	9	9	9	10	10	10	10	11	11	11	12	12	12	13	13	13
0.131 – 0.150 (0.0052 – 0.0059)	6	6	6	6	6	7	7	7	7	7	8	8	8	8	8	9	9	9	9	9	10	10	10	11	11	11	11	12	12	13	13	13	13	14
0.151 – 0.170 (0.0059 – 0.0067)	6	6	6	7	7	7	7	7	8	8	8	8	8	9	9	9	9	9	10	10	10	10	11	11	11	11	12	12	13	13	13	13	14	14
0.171 – 0.190 (0.0067 – 0.0075)	6	7	7	7	7	7	8	8	8	8	8	9	9	9	9	9	10	10	10	10	10	11	11	11	12	12	12	13	13	13	14	14	14	15
0.191 – 0.210 (0.0075 – 0.0083)	7	7	7	7	8	8	8	8	8	9	9	9	9	9	10	10	10	10	11	11	11	11	11	12	12	12	13	13	13	14	14	14	15	15
0.211 – 0.230 (0.0083 – 0.0091)	7	7	8	8	8	8	8	9	9	9	9	9	10	10	10	10	10	11	11	11	11	12	12	12	12	13	13	13	14	14	14	15	15	15
0.231 – 0.249 (0.0091 – 0.0098)	8	8	8	8	8	9	9	9	9	9	10	10	10	10	10	11	11	11	11	11	12	12	12	13	13	13	13	14	14	15	15	15	15	16
0.250 – 0.350 (0.0098 – 0.0138)																																		
0.351 – 0.370 (0.0138 – 0.0146)	10	10	10	11	11	11	11	11	12	12	12	12	12	13	13	13	13	13	14	14	14	14	14	15	15	15	16	16	17	17	17	17	17	
0.371 – 0.390 (0.0146 – 0.0154)	10	11	11	11	11	11	12	12	12	12	12	13	13	13	13	13	14	14	14	14	14	15	15	15	16	16	16	17	17	17	17	17		
0.391 – 0.410 (0.0154 – 0.0161)	11	11	11	11	12	12	12	12	12	13	13	13	13	13	14	14	14	14	14	15	15	15	15	16	16	16	17	17	17	17				
0.411 – 0.430 (0.0162 – 0.0169)	11	11	12	12	12	12	12	13	13	13	13	13	14	14	14	14	14	15	15	15	15	15	16	16	16	17	17	17	17					
0.431 – 0.450 (0.0170 – 0.0177)	12	12	12	12	12	13	13	13	13	13	14	14	14	14	14	15	15	15	15	15	16	16	16	17	17	17	17	17						
0.451 – 0.470 (0.0178 – 0.0185)	12	12	12	13	13	13	13	13	14	14	14	14	14	15	15	15	15	15	16	16	16	16	17	17	17	17	17							
0.471 – 0.490 (0.0185 – 0.0193)	12	13	13	13	13	13	14	14	14	14	14	15	15	15	15	15	16	16	16	16	16	17	17	17	17	17								
0.491 – 0.510 (0.0193 – 0.0201)	13	13	13	13	14	14	14	14	14	15	15	15	15	15	16	16	16	16	16	17	17	17	17	17										
0.511 – 0.530 (0.0201 – 0.0209)	13	13	14	14	14	14	14	15	15	15	15	15	16	16	16	16	16	17	17	17	17	17	17											
0.531 – 0.550 (0.0209 – 0.0217)	14	14	14	14	14	15	15	15	15	15	16	16	16	16	16	17	17	17	17	17	17	17												
0.551 – 0.570 (0.0217 – 0.0224)	14	14	14	15	15	15	15	15	16	16	16	16	16	17	17	17	17	17	17	17														
0.571 – 0.590 (0.0225 – 0.0232)	14	15	15	15	15	15	16	16	16	16	16	17	17	17	17	17	17	17																
0.591 – 0.610 (0.0233 – 0.0240)	15	15	15	15	16	16	16	16	16	17	17	17	17	17	17	17																		
0.611 – 0.630 (0.0241 – 0.0248)	15	15	16	16	16	16	16	17	17	17	17	17	17	17																				
0.631 – 0.650 (0.0248 – 0.0256)	16	16	16	16	16	17	17	17	17	17	17	17																						
0.651 – 0.670 (0.0256 – 0.0264)	16	16	16	17	17	17	17	17	17	17																								
0.671 – 0.690 (0.0264 – 0.0272)	16	17	17	17	17	17	17	17																										
0.691 – 0.710 (0.0272 – 0.0280)	17	17	17	17	17	17																												
0.711 – 0.730 (0.0280 – 0.0287)	17	17	17	17																														
0.731 – 0.750 (0.0288 – 0.0295)	17	17																																

Exhaust valve clearance (Cold):
0.25 – 0.35 mm (0.010 – 0.014 in.)

EXAMPLE: The 2.800 mm (0.1102 in.) shim is installed, and the measured clearance is 0.440 mm (0.0173 in.). Replace the 2.800 mm (0.1102 in.) shim with a No. 10 shim.

New shim thickness mm (in.)

Shim No.	Thickness	Shim No.	Thickness
1	2.500 (0.0984)	10	2.950 (0.1161)
2	2.550 (0.1004)	11	3.000 (0.1181)
3	2.600 (0.1024)	12	3.050 (0.1201)
4	2.650 (0.1043)	13	3.100 (0.1220)
5	2.700 (0.1063)	14	3.150 (0.1240)
6	2.750 (0.1083)	15	3.200 (0.1260)
7	2.800 (0.1102)	16	3.250 (0.1280)
8	2.850 (0.1122)	17	3.300 (0.1299)
9	2.900 (0.1142)		

HINT: New shims have the thickness in millimeters imprinted on the face.

86821168

Fig. 113 Adjusting shim chart for the exhaust valves

ADJUSTMENT

22R and 22R-E

➧ **See Figures 114, 115, 116 and 117**

1. Start the engine and warm it up to normal operating temperature.
2. Turn the engine **OFF**. Remove the air cleaner and housing, along with the hot air and cold air intake ducts.

⁂ CAUTION

Components will be hot. The engine head, block and radiator will be very hot.

3. Remove any other hoses, cables, or wires attached to the valve cover. The valve cover (or cylinder head cover) is the domed steel item with the oil filler in it.

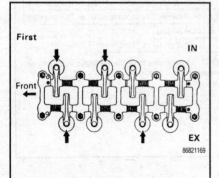

Fig. 114 Adjust the clearance of half the valves, do the arrowed ones first

86821170

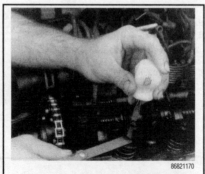

Fig. 115 Use a gauge to measure the distance between the stem and the rocker arm

86821171

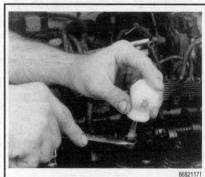

Fig. 116 Hold the adjusting screw in position, then tighten the locknut

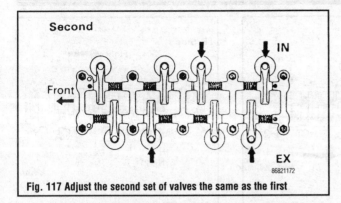

Fig. 117 Adjust the second set of valves the same as the first

4. Remove the small nuts holding the valve cover, then lift the cover off. Retrieve the rubber gasket and put it aside; it can be reused if not damaged or crushed out of shape. Beware of hot oil dripping from the inside of the cover.

5. Use a large wrench on the crankshaft pulley bolt to turn the engine clockwise until the timing mark on the pulley to **0** on the scale. Turning the engine will be easier if the spark plugs are removed, but this is not required.

❊❊ CAUTION

Do not attempt to align the engine by using the ignition switch to turn the engine. Doing will splash hot oil onto everything in the area, including you.

6. Check that the rockers on No.1 cylinder are loose and the rockers on No.4 are under tension. (No.1 is closest to the radiator; No. 4 is closest to the firewall.) If this is true, the engine is aligned with No.1 piston at top dead center. If it is not true, turn the engine one full revolution clockwise and realign the timing mark at zero; recheck the rockers.

7. Adjust the clearance 0.008 in. (0.20mm) intake and 0.012 in. (0.30mm) exhaust. Insert a feeler gauge and check for proper clearance between the top of the valve stem and the bottom of the rocker arm on the No. 1 intake valve. To adjust, loosen the locknut on the end of the rocker arm and turn the adjusting screw until the clearance is correct. Tighten the locknut and recheck the clearance; there should be a slight drag felt when the feeler gauge is pulled through the gap. Repeat the procedure for No 1 exhaust, No 2 intake and No. 3 exhaust.

8. Turn the crankshaft pulley one full rotation clockwise until the marks align at **0** and for the remaining valves.

9. Clean the valve cover thoroughly with a lint-free rag. Wipe any oil off the cylinder head edges in the area of the valve cover gasket.

10. Fit the gasket into the valve cover, making sure it is not crimped or twisted. If the half-moon rubber plugs came out of the valve cover, clean them and apply sealant to the part of the plug contacting the valve cover; install the half-moon plug.

➡The use of sealants on the valve cover gasket is not recommended.

11. Install the valve cover onto the head. Make certain is squarely seated and not pinching any adjacent wires or cables.

12. Install the valve cover retaining nuts. Tighten them to 43–60 inch lbs. (5–7 Nm) This is little more than finger-tight; overtightening will deform the cover and cause leaks.

13. Connect the lines, hoses and cables which were removed for access. Make certain electrical and ignition wires are firmly held by their clips or brackets.

14. Install the air cleaner with the hoses and duct work.

15. If still in place, remove the wrench and socket from the crankshaft pulley.

3VZ-E Engines

◆ See Figures 118 thru 127

The valves should be checked with the engine cold. Generally, this means the vehicle has not been driven in at least 3 hours. Use a longer interval for hot climates. Overnight cold is best.

➡The use of valve shim removing tools and a micrometer is required for this procedure. Toyota lists both shim removing tools as a set, SST 09248-55010; equivalent tools may be purchased through retail outlets. A micrometer is a precision measuring device. Do not attempt the procedure if the tools are not available.

1. Disconnect the spark plug wires, then remove the plugs.
2. Remove the air intake chamber by tagging and disconnecting:
   a. The throttle position sensor.
   b. The canister vacuum hose from the throttle body.
   c. The vacuum and fuel hoses from the pressure regulator.
   d. The PCV hose at the union.
   e. The No. 4 water by-pass hose from the union of the intake manifold.
   f. The No. 5 water by-pass hose from the water by-pass pipe.
   g. The cold start injector wire.
   h. The vacuum hose at the fuel filter.
   i. Remove the union bolt and two gaskets and then remove the cold start injector tube.
   j. The EGR gas temperature sensor wire.
   k. The EGR vacuum hoses at the air pipe and vacuum modulator.
   l. On California vehicles, remove the EGR gas sensor connector.
   m. Remove the nut, bolt and intake chamber stay.
   n. Remove the 5 nuts, EGR valve with the pipes still connected and the 2 gaskets.
   o. Disconnect the No. 1 air hose at the reed valve.
   p. Remove the six bolts and two nuts and lift off the air intake chamber and gasket.
3. Disconnect the following wiring:
   a. Knock sensor
   b. No. 2 water temperature switch
   c. Start injector time switch
   d. Water temperature switch
   e. Right side ground strap
   f. Injector connectors.
4. After the connectors are removed, unfasten the two bracket bolts and move the engine wire harness to an out of the way location.
5. Remove the left and right valve covers (cylinder head covers).
6. Use a wrench and turn the crankshaft clockwise until the notch in the pulley aligns with the timing mark **0** of the lower (No. 1) timing belt cover.

➡Check that the valve lifters on the No. 1 cylinder are loose and those on No. 4 cylinder are tight. If not, turn the crankshaft one complete revolution clockwise and realign the marks.

7. Using a flat feeler gauge, measure the clearance between the bottom of the camshaft lobe and the top surface of the valve lifter (It is the top surface of the shim). This measurement should correspond to the one given in the Tune-Up Specifications chart. Check only the No. 6 intake valve and the No. 2 exhaust valve.

➡If the measurement is within specifications, go on to the next step. If not, record the measurement taken for each individual valve. They will be used later to determine the required replacement shim sizes.

8. Turn the crankshaft ⅓ of a revolution (120°). One-third revolution is the equivalent of from 12 to 4 on a clock face. The No. 1 intake and the No. 3 exhaust valves should not be under tension from the cam lobes. Measure the clearances for these 2 valves only. Record any measurement that is out of specification.

9. Turn the crankshaft ⅓ of a revolution (120°) and measure the clearance at the No. 2 intake and the No. 4 exhaust valves. Record any measurement that is out of specification.

10. Turn the crankshaft ⅓ of a revolution (120°) and measure the clearance at the No. 3 intake and the No. 5 exhaust valves. Record any measurement that is out of specification.

11. Turn the crankshaft ⅓ of a revolution (120°) and measure the clearance at the No. 4 intake and the No. 6 exhaust valves. Record any measurement that is out of specification.

12. Turn the crankshaft ⅓ of a revolution (120°) and measure the clearance at the No. 5 intake and the No. 1 exhaust valves. Record any measurement that is out of specification.

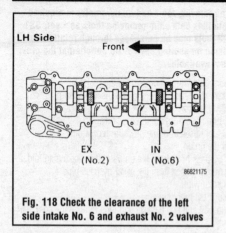

Fig. 118 Check the clearance of the left side intake No. 6 and exhaust No. 2 valves

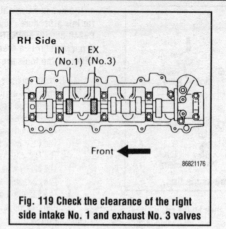

Fig. 119 Check the clearance of the right side intake No. 1 and exhaust No. 3 valves

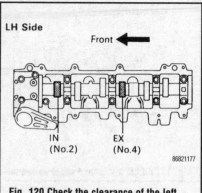

Fig. 120 Check the clearance of the left side intake No. 2 and exhaust No. 4 valves

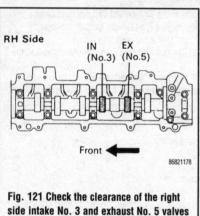

Fig. 121 Check the clearance of the right side intake No. 3 and exhaust No. 5 valves

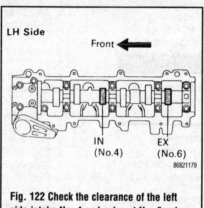

Fig. 122 Check the clearance of the left side intake No. 4 and exhaust No. 6 valves

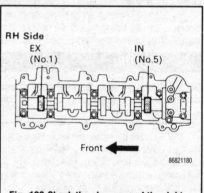

Fig. 123 Check the clearance of the right side intake No. 5 and exhaust No. 1 valves

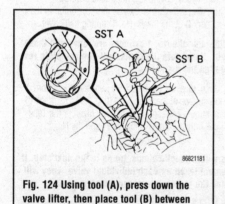

Fig. 124 Using tool (A), press down the valve lifter, then place tool (B) between the camshaft and the valve lifter

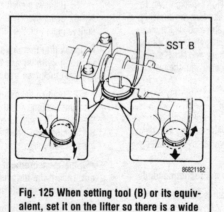

Fig. 125 When setting tool (B) or its equivalent, set it on the lifter so there is a wide space in the removal direction

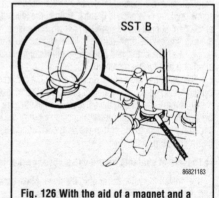

Fig. 126 With the aid of a magnet and a small flatbladded tool, remove the shim

➡ If the measurements for the six sets of valves are within specifications, you need go no further; the procedure is finished. If any clearance is not within specification, continue with the next step.

13. Turn the crankshaft clockwise until the camshaft lobe of the valve to be adjusted faces upward, taking tension off the valve.

14. Using a small awl, turn the valve lifter so that the notch is easily accessible.

15. Install the SST 09248-05410 (A) or its equivalent scissors-type tool on the lifter, then squeeze the handle so that the tool presses down the valve lifter evenly. Hold the valve lifter down with the other SST 09248-05420 (B) or its equivalent single-bladed tool. Remove tool (A) or equivalent.

➡ For easy removal of the shim, set the SST (B) or equivalent on the lifter so there is adequate space in the removal direction.

16. Using a small flatbladded tool and a magnet, remove the valve shim.

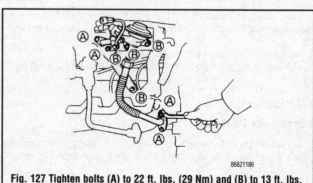

Fig. 127 Tighten bolts (A) to 22 ft. lbs. (29 Nm) and (B) to 13 ft. lbs. (18 Nm)

17. Carefully measure the thickness of the old shim with a micrometer. Locate that particular measurement in the "Installed Shim Thickness" column of the accompanying charts, then locate the recorded clearance measurement (from Step 6–11) for that valve in the "Measured Clearance " column of the charts. Index the two columns to arrive at the proper replacement shim thickness.

➡ **Replacement shims are available in 25 sizes, in increments of 0.002 in. (0.05mm), from 0.0866 in. (2.200mm) to 0.133 in. (3.400mm).**

18. Install the new shim, remove the special tool and then recheck the valve clearance.

19. Consult the clearances recorded earlier and perform Steps 12–17 for any other valve needing adjustment. Always turn the engine clockwise as you face the front of the motor.

To install:

20. The balance of the procedure is installation of the components removed in order to access the valves.

21. Start installing the spark plugs. Now is a good time to replace them if they have seen some miles.

22. Install the valve covers. Remember to place the gasket on the cover and install small amounts of sealant. Attach the covers, then tighten the 11 bolts only to 4–5 ft. lbs. (5–7 Nm). Do not overtighten.

23. When installing the air intake chamber, use a new gasket on the intake manifold and tighten the fasteners evenly to 13 ft. lbs. (18 Nm).

24. Be careful during installation of the air intake chamber stay. Refer to the illustration for bolt designation. Tighten bolts (A) to 22 ft. lbs. (29 Nm). Tighten bolts (B) to 13 ft. lbs. (18 Nm).

➡ **Use a new gasket when installing the cold start injector and tighten the union (banjo) bolt to 13 ft. lbs. (18 Nm).**

2RZ-FE and 3RZ-FE Engines

▸ **See Figures 128, 129, 130 and 131**

➡ **Adjust the valve clearance when the engine is cold.**

1. Disconnect the negative battery cable.
2. Remove the cylinder head covers.
3. Turn the crankshaft pulley and align its groove with the timing mark **0** of the No. 1 timing cover.
4. Check that the timing marks on the camshaft sprockets are in alignment with the marks on the No. 4 timing cover. If not, turn the crankshaft 1 complete revolution (360°).
5. Measure the clearance between the valve lifter and the camshaft. Record the measurements on the intake valves No. 1 and 2. Measure the exhaust valves at the 1 and 3.
   a. The intake valve clearance cold is 0.006–0.010 in. (0.15–0.25mm).
   b. The exhaust valve clearance cold is 0.010–0.014 in. (0.25–0.35mm).
6. Turn the crankshaft pulley 1 revolution (360°) and align the groove with the timing mark **0** of the No. 1 timing belt cover.
7. Measure the clearance between the valve lifter and the camshaft. Record the measurements on the intake valves No. 3 and 4. Measure the exhaust valves at the 2 and 4.
   a. The intake valve clearance cold is 0.006–0.010 in. (0.15–0.25mm).
   b. The exhaust valve clearance cold is 0.010–0.014 in. (0.25–0.35mm).
8. To adjust the intake valve clearance:
   a. Remove the intake camshaft.
   b. Using a small flatbladed tool and a magnet, remove the adjusting shim.
   c. Determine the replacement adjusting shim size by either using the chart or the following formula:
- Intake—$N=T+(A-0.008$ in./0.20mm$)$
- Exhaust—$N=T+(A-0.013$ in./0.32mm$)$
- T=Thickness of removed shim
- A=Measured valve clearance
- N=Thickness of new shim
   d. Install a new shim.
   e. Install the intake camshaft.
   f. Recheck the valve clearance.
9. To adjust the exhaust valve clearance:
   a. Turn the crankshaft to position on the cam lobe of the camshaft on the valve to be adjusted, upward.

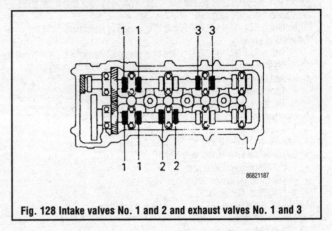

Fig. 128 Intake valves No. 1 and 2 and exhaust valves No. 1 and 3

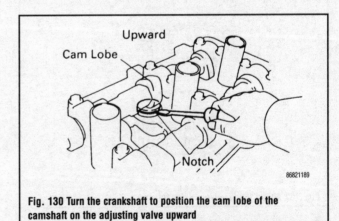

Fig. 129 Intake valves No. 3 and 4 and exhaust valves No. 2 and 4

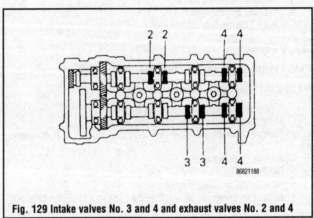

Fig. 130 Turn the crankshaft to position the cam lobe of the camshaft on the adjusting valve upward

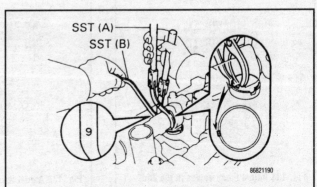

Fig. 131 Apply the SST (B) at a slight angle on the side marked with a "9", at the position shown in the illustration

b. Turn the valve lifter so that the notch is perpendicular to the camshaft and facing the spark plug side.

c. Using SST 09248-55040 (valve lifter press) or equivalent, hold the camshaft in place.

d. Using SST 09248-55040 (valve lifter press) or equivalent, press down the valve lifter and place the SST 09248-05420 (valve lifter stopper) or equivalent between the camshaft and the valve lifter.

e. Remove the tool 09248-44040 or equivalent.

f. Using a small bladed tool and a magnet, remove the adjusting shim.

10. Determine the replacement adjusting shim size by either using the chart or the following formula:
- Intake—N=T+(A-0.008 in./0.20mm)
- Exhaust—N=T+(A-0.013 in./0.32mm)
- T=Thickness of removed shim
- A=Measured valve clearance
- N=Thickness of new shim

11. Install a new shim.
12. Install the intake camshaft.
13. Recheck the valve clearance.
14. Install the cylinder head covers.
15. Connect the negative battery cable.

5VZ-FE Engine

♦ See Figures 124, 125, 126, 132, 133, 134 and 135

➡Adjust the valves when the engine is cold.

1. Disconnect the negative battery cable.
2. Disconnect the throttle cable from the linkage.
3. Remove the air cleaner cover, air flow meter, and the air duct assemblies.
4. Remove the V-bank cover.
5. Remove the emission control valve set.
6. Remove the air intake chamber.
7. Disconnect the engine harness from the injectors and the ignition coils.

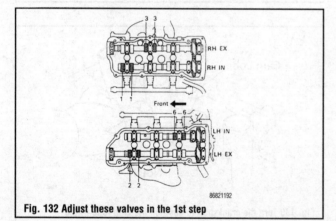

Fig. 132 Adjust these valves in the 1st step

8. Detach the ignition coils and keep them in order for reassembly.
9. Remove the spark plugs.
10. Remove the cylinder head covers.
11. Turn the crankshaft pulley and align its groove with the timing mark **0** of the No. 1 timing cover.
12. Check that the valve lifters on the No. 1 intake are loose and the No. 1 exhaust are tight. If not, turn the crankshaft one complete revolution (360°).

➡**All measurements should be written down. These recorded measurements will need to be used in conjunction with a mathematical formula to determine the thickness of the replacement shims.**

13. Measure the clearance between the valve lifters and the camshaft. Record the measurements on valves No. 1 and 6 intake; 2 and 3 exhaust.

a. The intake valve clearance cold is 0.006–0.009 in. (0.15–0.23mm).

b. The exhaust valve clearance cold is 0.011–0.014 in. (0.25–0.37mm).

14. Turn the crankshaft ⅔ of a revolution (240°). Record the measurements on valves No. 2 and 3 intake; 4 and 5 exhaust.

15. Turn the crankshaft another ⅔ of a revolution (240°). Record the measurements on valves No. 4 and 5 intake; 1 and 6 exhaust.

16. Remove the adjusting shim by turning the crankshaft to position the cam lobe of the camshaft in the up position on the valve to be adjusted. Using a small thin flatbladed tool, turn the valve lifters so that the notches are perpendicular to the camshaft. Press down the valve lifter with SST 09248-55010 (A) or equivalent. Place SST 09248-55010 (B) or equivalent, between the camshaft and the valve lifter. Remove part (A).

17. Remove the adjusting shim with a magnet and a small flatbladed tool.

18. Determine the replacement adjusting shim size by either using charts or the following formula:
- Intake—N=T+(A-0.008 in./0.20mm)
- Exhaust—N=T+(A-0.013 in./0.32mm)
- T=Thickness of removed shim
- A=Measured valve clearance
- N=Thickness of new shim

19. Select a new shim with the thickness as close as possible to the calculated value. Install the new replacement shim.

➡**Shims are available in 17 sizes in increments of 0.0020 in. (0.050mm) from 0.06984 in. (2.500mm) to 0.1299 in. (3.300mm).**

20. Recheck the valve clearance.
21. Install the cylinder head covers.
22. Install the spark plugs and the ignition coils.
23. Connect the engine wiring harness to the injectors and the coils.
24. Install the intake chamber.
25. Install the emission control valve set.
26. Install the V-bank cover.
27. Install the air flow meter, air duct, and the air cleaner cover.
28. Connect the accelerator cable to the throttle linkage.
29. Connect the negative battery cable.

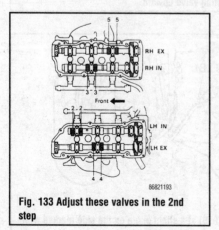

Fig. 133 Adjust these valves in the 2nd step

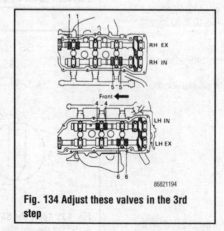

Fig. 134 Adjust these valves in the 3rd step

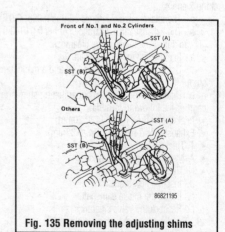

Fig. 135 Removing the adjusting shims

3F-E Engine

▶ See Figures 136 and 137

1. Start the engine and allow it to reach normal operating temperature.
2. Stop the engine. Remove any cables, hoses, wires, etc., which are attached to the valve cover and remove the valve cover.

✳✳ CAUTION

Components will be hot. The cylinder head, block and radiator will be very hot.

3. Use a wrench to turn the crankshaft pulley bolt clockwise until the timing marks are aligned at **0**. Since the timing marks are at the back of the engine, an assistant is useful.
4. Once the timing marks are aligned, check the two rocker arms (No. 1) closest to the front of the motor; they should be loose. The No. 6 rocker arms should be tight. If not, rotate the engine one full rotation (360°) and realign the timing marks at **0**.
5. Check the clearance between each of the rocker arms and valve stems for the first set of valves, as shown, using a feeler gauge. Hot the intake should be 0.008 in. (0.20mm) and exhaust 0.014 in. (0.35mm).
6. If the clearance is incorrect, loosen the locknut and turn the adjusting screw as required to gain correct clearance. The feeler gauge should pass with just the slightest drag. Tighten the locknut and recheck the clearance.
7. When all valves in the first set are adjusted, turn the crankshaft one full rotation clockwise. The two rockers closest to the firewall or back of the motor should be loose. Repeat the measuring and adjusting procedure for each valve in the second set, as shown.
8. After adjusting all of the valves, install a new valve cover gasket on the cover. Install the valve cover, tightening the bolts to 6 ft. lbs. (8 Nm). Install any other components which were removed during Step 2. Make certain all wires and cables are properly held by the clamps or brackets.
9. Recheck the engine idle speed and adjust if necessary.

1FZ-FE Engine

▶ See Figures 138, 139, 140, 141 and 142

➡ Adjust the valves when the engine is cold.

1. Drain the engine coolant from the vehicle.
2. Remove the throttle body:
   a. Disconnect the PCV hose.
   b. Detach the air cleaner hose.
   c. Disconnect the control cables from the throttle body.
   d. Remove the throttle position sensor connection.
   e. Disconnect the idle air control valve wiring.
3. Disconnect the engine wiring harness and heater valve from the cowl panel.
4. Tag and disconnect the spark plug wires.
5. Remove the valve cover.
6. Set the No. 1 cylinder to TDC/compression stroke by turning the crankshaft pulley, and aligning its groove with the timing mark **0** of the timing chain cover.
7. Check that the timing marks (one and two dots) of the camshaft drive and driven gears are in a straight line on the cylinder head surface. If they are not, turn the crankshaft 1 revolution (360°) and align the marks.
8. Inspect the valve clearance by using a thickness gauge. Measure the clearance between the valve lifter and the camshaft.
9. Record the out of specification valve clearance measurements. These will be used later to determine the replacement adjusting shims. Intake is 0.006–0.010 in. (0.15–0.25mm) and exhaust is 0.010–0.014 in. (0.25–0.35mm).
10. Turn the crankshaft pulley 1 revolution (360°) and align its groove with the timing mark **0**.
11. Check only the valves shown in the first illustration.
12. Adjust the clearance except for rear valves of the No. 6 cylinder. Remove the adjusting shim by turning the crankshaft to position the cam lobe of the camshaft on thew adjusting valve upward.

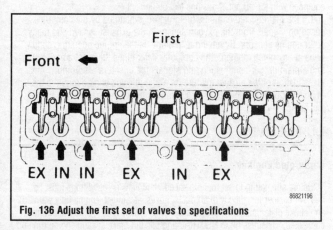

Fig. 136 Adjust the first set of valves to specifications

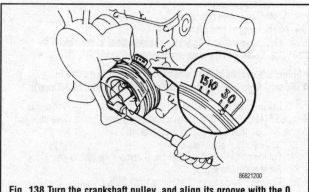

Fig. 138 Turn the crankshaft pulley, and align its groove with the 0 mark of the timing chain cover

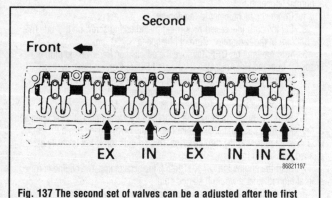

Fig. 137 The second set of valves can be a adjusted after the first set is checked

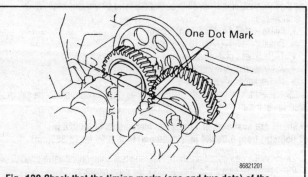

Fig. 139 Check that the timing marks (one and two dots) of the camshaft drive and driven gears are in a straight line on the cylinder head surface

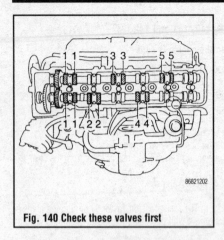

Fig. 140 Check these valves first

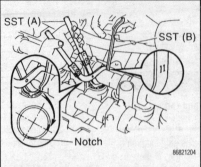

Fig. 141 Note the area of the mark "11", this is where you will apply the SST (B) or its equivalent

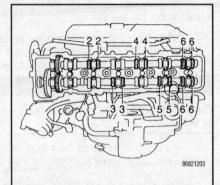

Fig. 142 Adjust the valve clearance for the second set

13. Position the notch of the valve lifter toward the spark plug side. Using SST (A) or equivalent, press down the valve lifter and place the SST (B) or equivalent between the camshaft and the valve lifter flange. Remove the SST (A) or its equivalent.

14. Apply SST (B) or equivalent at a slight angle on the side marked with "11", at the position shown in thew illustration.

15. When the SST (B) or equivalent are inserted too deeply, it will get pinched by the shim. To prevent this from occurring, insert it shallowly from the outside of the cylinder head, at a slight angle.

16. Remove the adjusting shim with a small flatbladded tool and a magnet.

17. Determine the replacement shim size by following the formula or charts. Using a micrometer, measure the thickness of the removed shim. Calculate the thickness of the new shim so that the valve clearance comes within specifications.

- Intake—$N=T+(A-0.008 \text{ in.}/0.20\text{mm})$
- Exhaust—$N=T+(A-0.013 \text{ in.}/0.32\text{mm})$
- T=Thickness of removed shim
- A=Measured valve clearance
- N=Thickness of new shim

18. Select a new shim with the thickness as close as possible to the calculated value. Install the new replacement shim.

➡**Shims are available in 17 sizes in increments of 0.0020 in. (0.050mm) from 0.06984 in. (2.500mm) to 0.1299 in. (3.300mm).**

19. Install a new adjusting shim. Place the new shim on the valve lifter. Using SST (A) 09248-06020 or equivalent, press down on the valve lifter and remove the SST (B) 09248-55050 or equivalent.

20. Recheck the valve clearance.

21. For the rear valves of cylinder No. 6, remove the distributor.

22. Remove the camshafts.

23. Remove the adjusting shim with a small flatbladded tip.

24. Determine the replacement adjusting shim size by following the formula or charts.

25. Using a micrometer, measure the thickness of the removed shim. Calculate the thickness of the new shim so that the valve clearance comes within specifications.

- Intake—$N=T+(A-0.008 \text{ in.}/0.20\text{mm})$
- Exhaust—$N=T+(A-0.013 \text{ in.}/0.32\text{mm})$
- T=Thickness of removed shim
- A=Measured valve clearance
- N=Thickness of new shim

26. Select a new shim with the thickness as close as possible to the calculated value. Install the new replacement shim.

➡**Shims are available in 17 sizes in increments of 0.0020 in. (0.050mm) from 0.06984 in. (2.500mm) to 0.1299 in. (3.300mm).**

27. Install a new adjusting shim. Place the new shim on the valve lifter.

28. Install the camshafts, then recheck the valve clearance.

29. Install the remaining components which were removed for access to the valve cover.

30. Refill the radiator with engine coolant and water mixture.

31. Check and adjust the timing. Check for coolant leaks.

Idle Speed And Mixture Adjustments

MIXTURE ADJUSTMENT

To meet emissions regulations, the mixture adjusting screw on the carbureted 22R engine is preadjusted and plugged by Toyota. Adjusting this screw is not part of routine maintenance. Access to the screw requires removal of the carburetor. Full details of re-setting the air/fuel mixture after a carburetor overhaul are covered in Section 5. In virtually all cases, the air/fuel mixture screw is NOT the cause of a problem. Even if the engine is tested with an emissions analyzer, incorrect mixture is almost always caused by air leaks in the system or faulty internal carburetor parts (clogged jets, mis-adjusted floats, etc.). We recommend all mixture adjustments be handled by a professional mechanic equipped with the proper emissions test equipment.

On the fuel injected engines, fuel mixture is handled by the computer, based on signals from the oxygen sensor in the exhaust system and many other engine sensors. By having a "picture" of the engine operating conditions at any given moment, the computer determines the exact amount of fuel to maintain proper combustion and signals the injectors accordingly. Any investigation of the air/fuel mixture on a fuel injected engine requires the use of an HC/CO emissions tester, which is well out of the reach of the home mechanic.

IDLE ADJUSTMENT

Carbureted Engines

Before attempting to set the idle speed, the following conditions must be met: Air cleaner installed with all ducts, hoses, etc. intact, engine fully warmed and choke plate open, all electric accessories switched OFF, all vacuum lines connected, transmission in NEUTRAL and parking brake set and engine running at normal temperature.

1. Connect a tachometer to the engine.

2. Set the curb idle speed by turning the adjusting screw on the side (not the bottom) of the carburetor. Correct idle speed is 700 rpm.

3. Turn the ignition **OFF**. Remove the air cleaner. Remove and plug the hoses for both the Hot Air Intake (HAI) and Mixture Control (MC) systems.

4. Disconnect the hose from the choke opener diaphragm and plug the hose end. This shuts off the choke opener system.

5. Disconnect the hose from the EGR valve.

6. Hold the throttle plate (in the bottom of the carburetor) slightly open and at the same time push the choke plate (top of the carb) closed. Hold the choke closed while releasing the throttle plate. This sounds tricky but is easily accomplished with a thin tool.

7. Start the engine but DO NOT touch the accelerator. The engine is running at high idle, as it would if started cold.

8. Set the fast idle speed by adjusting the lower screw on the carburetor. Correct idle speeds are: 49 State and Canada, 3000 rpm and California, 2600 rpm.

9. Turn the engine **OFF**. Remove the tachometer. Connect the vacuum hoses to the correct locations.

10. Install the air cleaner assembly.

Fuel Injected Engines

The idle speed on the fuel injected engines are electronically controlled by the Engine Control Module (ECM) computer. The Idle Speed Control (ISC) system electrically oversees the proper idle speed for all operating conditions. This is a good news/bad news arrangement. The bad news is that you can't adjust it; the good news is that it rarely, if ever, needs adjusting. Any roughness or uneven idle is almost always due to other causes such as a clogged injector, air leak or loose, corroded terminals in the wiring harness.

Air Conditioning System

SYSTEM SERVICE & REPAIR

➡ **It is recommended that the A/C system be serviced by an EPA Section 609 certified automotive technician utilizing a refrigerant recovery/recycling machine.**

The do-it-yourselfer should not service his/her own vehicle's A/C system for many reasons, including legal concerns, personal injury, environmental damage and cost.

According to the U.S. Clean Air Act, it is a federal crime to service or repair (involving the refrigerant) a Motor Vehicle Air Conditioning (MVAC) system for money without being EPA certified. It is also illegal to vent R-12 and R-134a refrigerants into the atmosphere. State and/or local laws may be more strict than the federal regulations, so be sure to check with your state and/or local authorities for further information.

➡ **Federal law dictates that a fine of up to $25,000 may be levied on people convicted of venting refrigerant into the atmosphere.**

When servicing an A/C system you run the risk of handling or coming in contact with refrigerant, which may result in skin or eye irritation or frostbite. Although low in toxicity (due to chemical stability), inhalation of concentrated refrigerant fumes is dangerous and can result in death; cases of fatal cardiac arrhythmia have been reported in people accidentally subjected to high levels of refrigerant. Some early symptoms include loss of concentration and drowsiness.

➡ **Generally, the limit for exposure is lower for R-134a than it is for R-12. Exceptional care must be practiced when handling R-134a.**

Also, some refrigerants can decompose at high temperatures (near gas heaters or open flame), which may result in hydrofluoric acid, hydrochloric acid and phosgene (a fatal nerve gas).

It is usually more economically feasible to have a certified MVAC automotive technician perform A/C system service on your vehicle.

R-12 Refrigerant Conversion

If your vehicle still uses R-12 refrigerant, one way to save A/C system costs down the road is to investigate the possibility of having your system converted to R-134a. The older R-12 systems can be easily converted to R-134a refrigerant by a certified automotive technician by installing a few new components and changing the system oil.

The cost of R-12 is steadily rising and will continue to increase, because it is no longer imported or manufactured in the United States. Therefore, it is often possible to have an R-12 system converted to R-134a and recharged for less than it would cost to just charge the system with R-12.

If you are interested in having your system converted, contact local automotive service stations for more details and information.

PREVENTIVE MAINTENANCE

Although the A/C system should not be serviced by the do-it-yourselfer, preventive maintenance should be practiced to help maintain the efficiency of the vehicle's A/C system. Be sure to perform the following:

• The easiest and most important preventive maintenance for your A/C system is to be sure that it is used on a regular basis. Running the system for five minutes each month (no matter what the season) will help ensure that the seals and all internal components remain lubricated.

➡ **Some vehicles automatically operate the A/C system compressor whenever the windshield defroster is activated. Therefore, the A/C system would not need to be operated each month if the defroster was used.**

• In order to prevent heater core freeze-up during A/C operation, it is necessary to maintain proper antifreeze protection. Be sure to properly maintain the engine cooling system.

• Any obstruction of or damage to the condenser configuration will restrict air flow which is essential to its efficient operation. Keep this unit clean and in proper physical shape.

➡ **Bug screens which are mounted in front of the condenser (unless they are original equipment) are regarded as obstructions.**

• The condensation drain tube expels any water which accumulates on the bottom of the evaporator housing into the engine compartment. If this tube is obstructed, the air conditioning performance can be restricted and condensation buildup can spill over onto the vehicle's floor.

SYSTEM INSPECTION

Although the A/C system should not be serviced by the do-it-yourselfer, system inspections should be performed to help maintain the efficiency of the vehicle's A/C system. Be sure to perform the following:

The easiest and often most important check for the air conditioning system consists of a visual inspection of the system components. Visually inspect the system for refrigerant leaks, damaged compressor clutch, abnormal compressor drive belt tension and/or condition, plugged evaporator drain tube, blocked condenser fins, disconnected or broken wires, blown fuses, corroded connections and poor insulation.

A refrigerant leak will usually appear as an oily residue at the leakage point in the system. The oily residue soon picks up dust or dirt particles from the surrounding air and appears greasy. Through time, this will build up and appear to be a heavy dirt impregnated grease.

For a thorough visual and operational inspection, check the following:

• Check the surface of the radiator and condenser for dirt, leaves or other material which might block air flow.

• Check for kinks in hoses and lines. Check the system for leaks.

• Make sure the drive belt is properly tensioned. During operation, make sure the belt is free of noise or slippage.

• Make sure the blower motor operates at all appropriate positions, then check for distribution of the air from all outlets.

➡ **Remember that in high humidity, air discharged from the vents may not feel as cold as expected, even if the system is working properly. This is because moisture in humid air retains heat more effectively than dry air, thereby making humid air more difficult to cool.**

Windshield Wipers

ELEMENT (REFILL) CARE & REPLACEMENT

♦ **See Figures 143, 144 and 145**

For maximum effectiveness and longest element life, the windshield and wiper blades should be kept clean. Dirt, tree sap, road tar and so on will cause streaking, smearing and blade deterioration if left on the glass. It is advisable to wash the windshield carefully with a commercial glass cleaner at least once a month. Wipe off the rubber blades with the wet rag afterwards. Do not attempt to move wipers across the windshield by hand; damage to the motor and drive mechanism will result.

To inspect and/or replace the wiper blade elements, place the wiper switch in the **LOW** speed position and the ignition switch in the **ACC** position. When the wiper blades are approximately vertical on the windshield, turn the ignition switch to **OFF**.

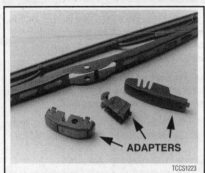

Fig. 143 Most aftermarket blades are available with multiple adapters to fit different vehicles

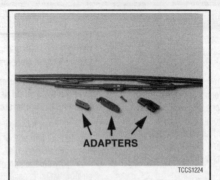

Fig. 144 Choose a blade which will fit your vehicle, and that will be readily available next time you need blades

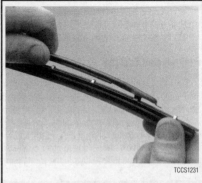

Fig. 145 When installed, be certain the blade is fully inserted into the backing

Examine the wiper blade elements. If they are found to be cracked, broken or torn, they should be replaced immediately. Replacement intervals will vary with usage, although ozone deterioration usually limits element life to about one year. If the wiper pattern is smeared or streaked, or if the blade chatters across the glass, the elements should be replaced. It is easiest and most sensible to replace the elements in pairs.

If your vehicle is equipped with aftermarket blades, there are several different types of refills and your vehicle might have any kind. Aftermarket blades and arms rarely use the exact same type blade or refill as the original equipment.

Regardless of the type of refill used, be sure to follow the part manufacturer's instructions closely. Make sure that all of the frame jaws are engaged as the refill is pushed into place and locked. If the metal blade holder and frame are allowed to touch the glass during wiper operation, the glass will be scratched.

Tires and Wheels

Common sense and good driving habits will afford maximum tire life. Make sure that you don't overload the vehicle or run with incorrect pressure in the tires. Either of these will increase tread wear. Fast starts, sudden stops and sharp cornering are hard on tires and will shorten their useful life span.

➡**For optimum tire life, keep the tires properly inflated, rotate them often and have the wheel alignment checked periodically.**

Inspect your tires frequently. Be especially careful to watch for bubbles in the tread or sidewall, deep cuts or underinflation. Replace any tires with bubbles in the sidewall. If cuts are so deep that they penetrate to the cords, discard the tire. Any cut in the sidewall of a radial tire renders it unsafe. Also look for uneven tread wear patterns that may indicate the front end is out of alignment or that the tires are out of balance.

TIRE ROTATION

◆ See Figure 146

Tires must be rotated periodically to equalize wear patterns that vary with a tire's position on the vehicle. Tires will also wear in an uneven way as the front steering/suspension system wears to the point where the alignment should be reset.

Rotating the tires will ensure maximum life for the tires as a set, so you will not have to discard a tire early due to wear on only part of the tread. Regular rotation is required to equalize wear.

When rotating "unidirectional tires," make sure that they always roll in the same direction. This means that a tire used on the left side of the vehicle must not be switched to the right side and vice-versa. Such tires should only be rotated front-to-rear or rear-to-front, while always remaining on the same side of the vehicle. These tires are marked on the sidewall as to the direction of rotation; observe the marks when reinstalling the tire(s).

Some styled or "mag" wheels may have different offsets front to rear. In these cases, the rear wheels must not be used up front and vice-versa. Furthermore, if these wheels are equipped with unidirectional tires, they cannot be rotated unless the tire is remounted for the proper direction of rotation.

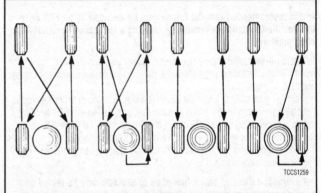

Fig. 146 Common tire rotation patterns for 4 and 5-wheel rotations

➡**The compact or space-saver spare is strictly for emergency use. It must not be included in the tire rotation or placed on the vehicle for everyday use.**

TIRE DESIGN

◆ See Figure 147

For maximum satisfaction, tires should be used in sets of four. Mixing of different brands or types (radial, bias-belted, fiberglass belted) should be avoided. In most cases, the vehicle manufacturer has designated a type of tire on which the vehicle will perform best. Your first choice when replacing tires should be to use the same type of tire that the manufacturer recommends.

When radial tires are used, tire sizes and wheel diameters should be selected to maintain ground clearance and tire load capacity equivalent to the original specified tire. Radial tires should always be used in sets of four.

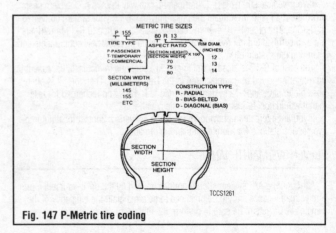

Fig. 147 P-Metric tire coding

✳✳ CAUTION

Radial tires should never be used on only the front axle.

When selecting tires, pay attention to the original size as marked on the tire. Most tires are described using an industry size code sometimes referred to as P-Metric. This allows the exact identification of the tire specifications, regardless of the manufacturer. If selecting a different tire size or brand, remember to check the installed tire for any sign of interference with the body or suspension while the vehicle is stopping, turning sharply or heavily loaded.

Snow Tires

Good radial tires can produce a big advantage in slippery weather, but in snow, a street radial tire does not have sufficient tread to provide traction and control. The small grooves of a street tire quickly pack with snow and the tire behaves like a billiard ball on a marble floor. The more open, chunky tread of a snow tire will self-clean as the tire turns, providing much better grip on snowy surfaces.

To satisfy municipalities requiring snow tires during weather emergencies, most snow tires carry either an M + S designation after the tire size stamped on the sidewall, or the designation "all-season." In general, no change in tire size is necessary when buying snow tires.

Most manufacturers strongly recommend the use of 4 snow tires on their vehicles for reasons of stability. If snow tires are fitted only to the drive wheels, the opposite end of the vehicle may become very unstable when braking or turning on slippery surfaces. This instability can lead to unpleasant endings if the driver can't counteract the slide in time.

Note that snow tires, whether 2 or 4, will affect vehicle handling in all non-snow situations. The stiffer, heavier snow tires will noticeably change the turning and braking characteristics of the vehicle. Once the snow tires are installed, you must re-learn the behavior of the vehicle and drive accordingly.

➡**Consider buying extra wheels on which to mount the snow tires. Once done, the "snow wheels" can be installed and removed as needed. This eliminates the potential damage to tires or wheels from seasonal removal and installation. Even if your vehicle has styled wheels, see if inexpensive steel wheels are available. Although the look of the vehicle will change, the expensive wheels will be protected from salt, curb hits and pothole damage.**

TIRE STORAGE

If they are mounted on wheels, store the tires at proper inflation pressure. All tires should be kept in a cool, dry place. If they are stored in the garage or basement, do not let them stand on a concrete floor; set them on strips of wood, a mat or a large stack of newspaper. Keeping them away from direct moisture is of paramount importance. Tires should not be stored upright, but in a flat position.

INFLATION & INSPECTION

▶ **See Figures 148 thru 153**

The importance of proper tire inflation cannot be overemphasized. A tire employs air as part of its structure. It is designed around the supporting strength of the air at a specified pressure. For this reason, improper inflation drastically reduces the tire's ability to perform as intended. A tire will lose some air in day-to-day use; having to add a few pounds of air periodically is not necessarily a sign of a leaking tire.

Two items should be a permanent fixture in every glove compartment: an accurate tire pressure gauge and a tread depth gauge. Check the tire pressure (including the spare) regularly with a pocket type gauge. Too often, the gauge on the end of the air hose at your corner garage is not accurate because it suffers too much abuse. Always check tire pressure when the tires are cold, as pressure increases with temperature. If you must move the vehicle to check the tire inflation, do not drive more than a mile before checking. A cold tire is generally one that has not been driven for more than three hours.

A plate or sticker is normally provided somewhere in the vehicle (door post, hood, tailgate or trunk lid) which shows the proper pressure for the tires. Never counteract excessive pressure build-up by bleeding off air pressure (letting some air out). This will cause the tire to run hotter and wear quicker.

✳✳ CAUTION

Never exceed the maximum tire pressure embossed on the tire! This is the pressure to be used when the tire is at maximum loading, but it is rarely the correct pressure for everyday driving. Consult the owner's manual or the tire pressure sticker for the correct tire pressure.

Once you've maintained the correct tire pressures for several weeks, you'll be familiar with the vehicle's braking and handling personality. Slight adjustments in tire pressures can fine-tune these characteristics, but never change the cold pressure specification by more than 2 psi. A slightly softer tire pressure will give a softer ride but also yield lower fuel mileage. A slightly harder tire will give crisper dry road handling but can cause skidding on wet surfaces. Unless you're fully attuned to the vehicle, stick to the recommended inflation pressures.

All automotive tires have built-in tread wear indicator bars that show up as ½ in. (13mm) wide smooth bands across the tire when ¹⁄₁₆ in. (1.5mm) of tread remains. The appearance of tread wear indicators means that the tires should be replaced. In fact, many states have laws prohibiting the use of tires with less than this amount of tread.

You can check your own tread depth with an inexpensive gauge or by using a Lincoln head penny. Slip the Lincoln penny (with Lincoln's head upside-down) into several tread grooves. If you can see the top of Lincoln's head in 2 adjacent grooves, the tire has less than ¹⁄₁₆ in. (1.5mm) tread left and should be replaced. You can measure snow tires in the same manner by using the "tails" side of the Lincoln penny. If you can see the top of the Lincoln memorial, it's time to replace the snow tire(s).

Fig. 148 Tires with deep cuts, or cuts which bulge, should be replaced immediately

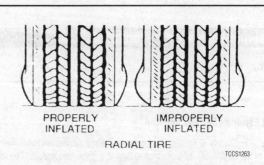

Fig. 149 Radial tires have a characteristic sidewall bulge; don't try to measure pressure by looking at the tire. Use a quality air pressure gauge

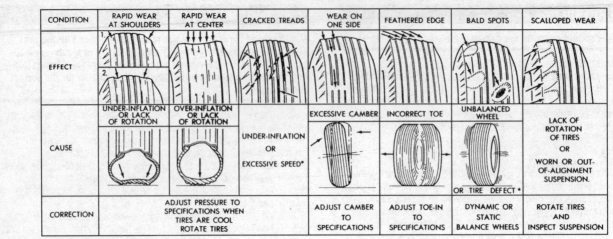

CONDITION	RAPID WEAR AT SHOULDERS	RAPID WEAR AT CENTER	CRACKED TREADS	WEAR ON ONE SIDE	FEATHERED EDGE	BALD SPOTS	SCALLOPED WEAR
EFFECT							
CAUSE	UNDER-INFLATION OR LACK OF ROTATION	OVER-INFLATION OR LACK OF ROTATION	UNDER-INFLATION OR EXCESSIVE SPEED*	EXCESSIVE CAMBER	INCORRECT TOE	UNBALANCED WHEEL OR TIRE DEFECT *	LACK OF ROTATION OF TIRES OR WORN OR OUT-OF-ALIGNMENT SUSPENSION.
CORRECTION	ADJUST PRESSURE TO SPECIFICATIONS WHEN TIRES ARE COOL ROTATE TIRES			ADJUST CAMBER TO SPECIFICATIONS	ADJUST TOE-IN TO SPECIFICATIONS	DYNAMIC OR STATIC BALANCE WHEELS	ROTATE TIRES AND INSPECT SUSPENSION

*HAVE TIRE INSPECTED FOR FURTHER USE.

TCCS1267

Fig. 150 Common tire wear patterns and causes

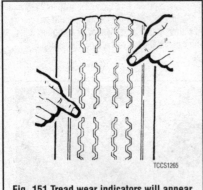

Fig. 151 Tread wear indicators will appear when the tire is worn

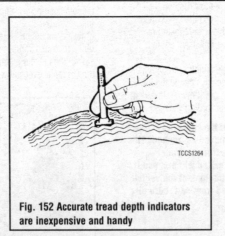

Fig. 152 Accurate tread depth indicators are inexpensive and handy

Fig. 153 A penny works well for a quick check of tread depth

FLUIDS AND LUBRICANTS

Fluid Disposal

Used fluids such as engine oil, transmission fluid, antifreeze and brake fluid are hazardous wastes and must be disposed of properly. Before draining any fluids, consult with your local authorities; in many areas, waste oil, etc. is being accepted as a part of recycling programs. A number of service stations and auto parts stores are also accepting waste fluids for recycling.

Be sure of the recycling center's policies before draining any fluids, as many will not accept different fluids that have been mixed together, such as oil and antifreeze.

Fuel and Engine Oil Recommendations

GENERAL INFORMATION

Oils

♦ See Figures 154, 155 and 156

The SAE (Society of Automotive Engineers) grade number indicates the viscosity of the engine oil; its resistance to flow at a given temperature. The lower

Fig. 154 Look for the API oil identification label when choosing your engine oil

Fig. 155 Before installing a new oil filter, coat the rubber gasket with clean oil

the SAE grade number, the lighter the oil. For example, the mono-grade oils begin with SAE 5 weight, which is a thin, light oil, and continue in viscosity up to SAE 80 or 90 weight, which are heavy gear lubricants. These oils are also known as "straight weight", meaning they are of a single viscosity, and do not vary with engine temperature.

Multi-viscosity oils offer the important advantage of being adaptable to temperature extremes. These oils have designations such as 10W–40, 20W–50, etc. The "10W–40" means that in winter (the "W" in the designation) the oil acts like a thin 10 weight oil, allowing the engine to spin easily when cold and offering rapid lubrication. Once the engine has warmed up, however, the oil acts like a straight 40 weight, maintaining good lubrication and protection for the engine's internal components. A 20W–50 oil would therefore be slightly heavier than and not as ideal in cold weather as the 10W–40, but would offer better protection at higher rpm and temperatures because when warm it acts like a 50 weight oil. Whichever oil viscosity you choose when changing the oil, make sure you are anticipating the temperatures your engine will be operating in until the oil is changed again. Refer to the oil viscosity chart for oil recommendations according to temperature.

The API (American Petroleum Institute) designation indicates the classification of engine oil used under certain given operating conditions. Only oils designated for use "Service SG" or greater should be used. Oils of the SG type perform a variety of functions inside the engine in addition to the basic function as a lubricant. Through a balanced system of metallic detergents and polymeric dispersants, the oil prevents the formation of high and low temperature deposits and also keeps sludge and particles of dirt in suspension. Acids, particularly sulfuric acid, as well as other by-products of combustion, are neutralized. Both the SAE grade number and the API designation bottle be found on the oil can. For recommended oil viscosities, refer to the chart.

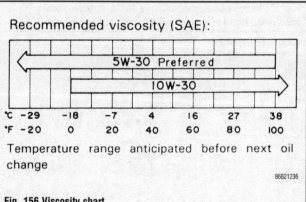

Recommended viscosity (SAE):

| 5W-30 Preferred |
| 10W-30 |

| °C | -29 | -18 | -7 | 4 | 16 | 27 | 38 |
| °F | -20 | 0 | 20 | 40 | 60 | 80 | 100 |

Temperature range anticipated before next oil change

86821236

Fig. 156 Viscosity chart

SYNTHETIC OIL

There are many excellent synthetic oils currently available that can provide better gas mileage, longer service life, and in some cases better engine protection. These benefits do not come without a few hitches, however; the main one being the price of synthetic oils, which is three or four times the price per quart of conventional oil.

Synthetic oil is not for every truck and every type of driving, so you should consider your engine's condition and your type of driving. Also, check your truck's warranty conditions regarding the use of synthetic oils.

Both brand new engines and older, high mileage engines are often the wrong candidates for synthetic oil. A synthetic oil can be so slippery that they can prevent the proper break-in of new engines; most manufacturers recommend that you wait until the engine is properly broken in 3000 miles (4830 km) before using synthetic oil. Older engines with wear have a different problem with synthetics: they leak more oil as they age. Slippery synthetic oils get past worn parts easily. If your truck is leaking oil past old seals you'll most probably have a much greater leak problem with synthetics.

Consider your type of driving. If most of your accumulated mileage is high speed, highway type driving, the more expensive synthetic oils may be a benefit. Extended highway driving gives the engine a chance to warm up, accumulating less acids in the oil and putting less stress on the engine over the long run. Trucks with synthetic oils may show increased fuel economy in highway driving, due to less internal friction.

If synthetic oil is used, it should still be replaced at regular intervals as stated in the maintenance schedule. While the oil itself will last much longer than regular oil, pollutants such as soot, water and unburned fuel still accumulate within the oil. These are the damaging elements within a motor and must be drained regularly to prevent damage.

Trucks used under harder circumstances, such as stop-and-go, city type 1driving, short trips, or extended idling, should be serviced more frequently. For the engines in these trucks, the much greater cost of synthetic or fuel-efficient oils may not be worth the investment. Internal wear increases much quicker on these trucks, causing greater oil consumption and leakage.

Fuel

It is important to use fuel of the proper octane rating in your truck. Octane rating is based on the quantity of anti-knock compounds added to the fuel and it determines the speed at which the gas will burn. The lower the octane rating, the faster it burns. The higher the octane, the slower the fuel will burn and a greater percentage of compounds in the fuel prevent spark ping (knock), detonation and pre-ignition (dieseling).

As the temperature of the engine increases, the air/fuel mixture exhibits a tendency to ignite before the spark plug is fired. If fuel of an octane rating too low for the engine is used, this will allow combustion to occur before the piston has completed its compression stroke, thereby creating a very high pressure very rapidly.

Fuel of the proper octane rating, for the compression ratio and ignition timing of your truck, will slow the combustion process sufficiently to allow the spark plug enough time to ignite the mixture completely and smoothly. The use of some super-premium fuel is no substitute for a properly tuned and maintained engine. Chances are that if your engine exhibits any signs of spark ping, detonation or pre-ignition when using regular fuel, the ignition timing should be checked against specifications or the cylinder head should be removed for decarbonizing.

Vehicles equipped with catalytic converters must use UNLEADED GASOLINE only. Use of leaded fuel shortens the life of spark plugs, exhaust systems and EGR valves and can damage the catalytic converter. Most converter equipped models are designed to operate using unleaded gasoline with a minimum rating of 91 octane. Use of unleaded gas with octane ratings lower than 91 can cause persistent spark knock which could lead to engine damage.

Light spark knock may be noticed when accelerating or driving up hills, particularly with a carbureted engine. The slight knocking may be considered normal (with 91 octane) because the maximum fuel economy is obtained under condition of occasional light spark knock. Gasoline with an octane rating higher than 91 may be used, but it is not necessary (in most cases) for proper operation.

➡ **Your engine's fuel requirement can change with time, mainly due to carbon buildup, which changes the compression ratio. If your engine pings, knocks or runs on, switch to a higher grade of fuel. Sometimes just changing brands will cure the problem. If it becomes necessary to** retard the timing from specifications, don't change it more than a few degrees. Retarded timing will reduce power output and fuel mileage and will increase the engine temperature.

Engine

OIL LEVEL CHECK

▶ See Figure 157

✖✖ CAUTION

Prolonged and repeated skin contact with used engine oil, with no effort to remove the oil, may cause skin cancer. Always follow these simple precautions when handling used motor oil:

- Avoid prolonged skin contact with used motor oil.
- Remove oil from skin by washing thoroughly with soap and water or waterless hand cleaner. Do not use gasoline, thinners or other solvents.
- Avoid prolonged skin contact with oil-soaked clothing.

Every time you stop for fuel, check the engine oil as follows:

1. Park the truck on level ground.
2. When checking the oil level it is best for the engine to be at operating temperature, although checking the oil immediately after a stopping will lead to a false reading. Wait a few minutes after turning off the engine to allow the oil to drain back into the oil pan (crankcase).
3. Open the hood and locate the dipstick. Pull the dipstick from its tube, wipe it clean and reinsert it.
4. Pull the dipstick out again and, holding it horizontally, read the oil level. The oil should be between the **F** and **L** marks on the dipstick. If the oil is below the **L** mark, add oil of the proper viscosity through the capped opening on the top of the cylinder head cover.
5. Reinsert the dipstick and check the oil level again after adding any oil. Be careful not to overfill the crankcase. Approximately one quart of oil will raise the level from the **L** to the **F**. Excess oil will generally be consumed at an accelerated rate as well as hampering engine operation.

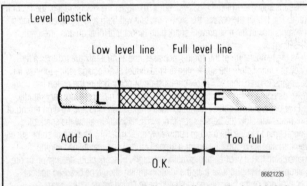

Fig. 157 Read the dipstick carefully, add oil when the level is below the L mark; do not overfill

OIL AND FILTER CHANGE

▶ See Figures 158, 159, 160, 161 and 162

The oil and filter should be changed every 7500 miles (12,000 km).

✖✖ CAUTION

Prolonged and repeated skin contact with used engine oil, with no effort to remove the oil, may cause skin cancer. Always follow these simple precautions when handling used motor oil:

- Avoid prolonged skin contact with used motor oil.
- Remove oil from skin by washing thoroughly with soap and water or waterless hand cleaner. Do not use gasoline, thinners or other solvents.
- Avoid prolonged skin contact with oil-soaked clothing.

The oil drain plug is located on the bottom, rear of the oil pan (bottom of the engine, underneath the truck). The oil filter is located on the side of the engine.

The mileage figures given are the Toyota recommended intervals assuming normal driving and conditions. Normal driving requires that the vehicle be driven far enough to warm up the oil; usually this is about 10 miles (16 km) or so. If your everyday use is shorter than this (one way), your use qualifies as severe duty.

Severe duty includes dusty, polluted or off-road conditions, as well as stop-and-go short haul uses. Regularly towing a trailer also puts the truck in this category, as does constant operation with a near capacity load. Change the oil and filter at ½ the normal interval. Half of 7500 equals 3250 miles (5229 km); round it down to the easily remembered 3000 mile (5000 km) interval. For some owners, that may be once a month; for others, it may be six months.

Always drain the oil after the engine has been running long enough to bring it to normal operating temperature. Hot oil will flow easier and more contaminants will be removed along with the oil than if it were drained cold. To change the oil and filter:

1. Run the engine until it reaches normal operating temperature.
2. Jack up the front of the truck and support it on safety stands.
3. Slide a drain pan of at least 6 quarts capacity under the oil pan.
4. Loosen the drain plug with a wrench. Turn the plug out by hand. By keeping an inward pressure on the plug as you unscrew it, oil won't escape past the threads and you can remove it without being burned by hot oil.
5. Allow the oil to drain completely and then install the drain plug. Don't overtighten the plug, or you'll be buying a new pan or a replacement plug for stripped threads.
6. Using a filter wrench, remove the oil filter. Keep in mind that it's holding about one quart of dirty, hot oil. Make certain the old gasket comes off with the filter and is not stuck to the block.
7. Empty the old filter into the drain pan and dispose of the filter.
8. Using a clean rag, wipe off the filter adapter on the engine block. Be sure that the rag doesn't leave any lint which could clog an oil passage.
9. Coat the rubber gasket on the filter with fresh oil. Spin it onto the engine by hand; when the gasket touches the adapter surface give it another ½–¾ turn. No more, or you'll squash the gasket and it will leak.

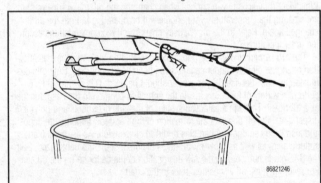

Fig. 158 Keeping an upward pressure on the drain plug while unscrewing it will prevent oil leakage until you're ready for it

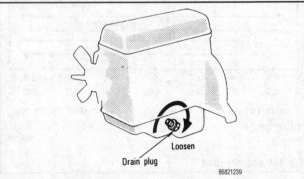

Fig. 159 Oil drain plugs are located near the lowest part of the pan, threaded either at an angle . . .

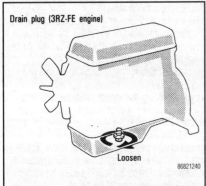

Fig. 160 . . . or threaded straight up into the bottom of the pan

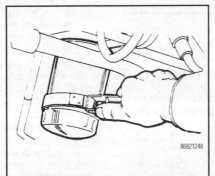

Fig. 161 An oil filter wrench is very handy. Try not to bang any components; damaging one can be costly

Fig. 162 Apply a light film of oil to the filter O-ring prior to installation

10. Refill the engine with the correct amount of fresh oil. See the "Capacities" chart.

11. Check the oil level on the dipstick. It is normal for the level to be a bit above the full mark. Start the engine and allow it to idle for a few minutes.

✳✳ WARNING

Do not run the engine above idle speed until it has built up oil pressure, indicated when the oil light goes out.

12. Shut **OFF** the engine, allow the oil to drain for a minute, and check the oil level. Check around the filter and drain plug for any leaks, and correct as necessary.

Manual Transmission

FLUID RECOMMENDATIONS

All vehicles use a multipurpose gear oil. The recommended types are: API GL-3, GL-4 or GL-5 with the viscosity SAE 75W–90 or 80W–90.

LEVEL CHECK

▶ See Figures 163, 164, 165 and 166

The oil in the manual transmission should be checked at least every 7500 miles (12,000 km) and replaced every 25,000–30,000 miles (40,000–48,000 km). Even more frequently if driven in deep water.

1. With the truck parked on a level surface, remove the filler plug (17mm) from the side of the transmission housing. A gasket will also have to be removed and discarded.

2. If the lubricant begins to trickle out of the hole, there is enough. Otherwise, carefully insert your finger (watch out for sharp threads) and check to see if the oil is up to the edge of the hole.

3. If not, add oil through the hole until the level is at the edge of the hole. Most gear lubricants come in a plastic squeeze bottle with a nozzle; making additions simple. You can also use a common everyday kitchen faster.

4. Install the filler plug with a new gasket. Run the engine and check for leaks.

Fig. 163 Use a wrench to remove the 17mm plug on the side of the case

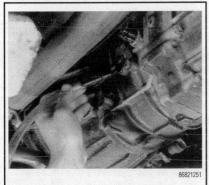

Fig. 164 A gear pump with a nozzle is also a type used to fill the case

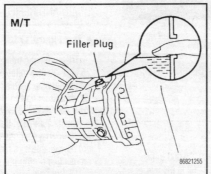

Fig. 165 The fill plug and drain plug locations on the manual transmission used with the 22R-E and the 3VZ-E engines

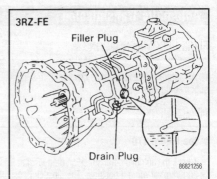

Fig. 166 The fill plug and drain plug locations on the transmission for the 3RZ-FE engine

DRAIN AND REFILL

▶ **See Figures 167, 168 and 169**

The oil in the manual transmission should be changed at least every 30,000 miles (48,000 km).

1. The transmission oil should be hot before it is drained. If the engine is at normal operating temperature, the transmission oil should be hot enough.

2. Raise the truck and support it properly on jackstands so that you can safely work underneath. You will probably not have enough room to work if the truck is not raised.

3. The drain plug is located on the bottom of the transmission. It is usually on the passenger side on four speeds, and on the bottom center of five speeds. Place a pan under the drain plug, then remove it. Keep a slight upward pressure on the plug while unscrewing it, this will keep the oil from pouring out until the plug is removed.

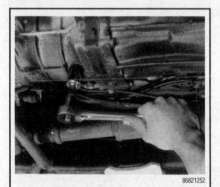
Fig. 167 Use a short extension to aid in removing the drain plug

Fig. 168 Quickly remove the plug to avoid getting oil on your hands

Fig. 169 Allow the fluid to drain completely into a pan or container, before reinstalling the plug

❋❋ CAUTION

The oil will be HOT. Be careful when you remove the plug so that you don't take a bath in hot gear oil.

4. Allow the oil to drain completely.

5. Clean off the plug then install it, tightening it until just snug.

6. Remove the filler plug from the side of the transmission case. It is usually on the driver's side of four speeds, and on the passenger side on five speeds. There will be a gasket underneath this plug. Replace it every time the bolt is removed.

7. Fill the transmission with gear oil through the filler plug hole.

8. The oil level should come right up to the edge of the hole. You can stick your finger in to verify this. Watch out for sharp threads.

9. Install the filler plug and gasket, lower the truck, then check for leaks. Dispose of the old oil in the proper manner.

Automatic Transmission

FLUID RECOMMENDATIONS

All automatic transmissions covered in this manual use DEXRON® II or its superceding fluid type.

LEVEL CHECK

▶ **See Figure 170**

Check the automatic transmission fluid level at least every 15,000 miles (24,000 km) or more if possible. The dipstick is usually located in the rear of the engine compartment. The fluid level should be checked only when the transmission is hot (normal operating temperature). The transmission is considered hot after about 20 miles (32 km) of highway driving.

1. Park the truck on a level surface with the engine idling. Shift the transmission into Neutral or Park and set the parking brake.

2. Remove the dipstick, wipe it clean and reinsert if firmly. Be sure that it has been pushed all the way in. Remove the dipstick and check the fluid level while holding it horizontally. With the engine running, the fluid level should be between the second and third notches on the dipstick.

3. If the fluid level is below the second notch, add fluid with the aid of a funnel.

4. Check the level often as you are filling the transmission. Be extremely careful not to overfill it. Overfilling will cause slippage, seal damage and overheating. Approximately one pint of ATF will raise the level from one notch to the other.

The fluid on the dipstick should always be a bright red color. If it is discolored (brown or black), or smells burnt, serious transmission troubles, probably due to overheating, should be suspected. The transmission should be inspected by a qualified service technician to locate the cause of the burnt fluid.

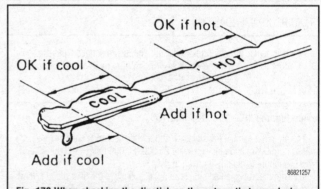
Fig. 170 When checking the dipstick on the automatic transmission, read the stick carefully

DRAIN AND REFILL

▶ **See Figure 171**

The automatic transmission fluid should be changed at least every 25,000–30,000 miles (40,000–48,000 km). If the truck is normally used in severe service, such as stop-and-go driving, trailer towing or the like, the interval should be halved. The fluid should be hot before it is drained; a 20 minute drive will accomplish this.

Toyota automatic transmissions are equipped with a drain plug in the pan, so that fluid may be drained without removing the pan. The filter within the pan is not replaceable as a maintenance item; removing the pan is not required or recommended during normal transmission maintenance.

1. With the truck safely supported on stands, position a large catch pan below the drain plug.

2. Remove the plug and gasket slowly, and be prepared for a rush of **HOT** fluid. Allow the unit to drain for some minutes; when the flow of oil is down to individual drops.

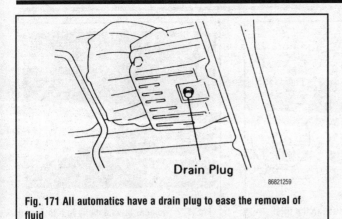

Fig. 171 All automatics have a drain plug to ease the removal of fluid

3. Install the drain plug with a new gasket if used, and tighten just snug. Add transmission fluid through the dipstick tube under the hood.

4. Reinsert the dipstick after filling, then start the engine and allow it to idle. DO NOT race the engine.

5. After the engine has idled for a few minutes, shift the transmission slowly through the gears, then return it to Park. With the engine still idling, check the fluid level on the dipstick. If necessary, add more fluid to raise the level.

PAN AND FILTER SERVICE

♦ See Figures 172, 173 and 174

1. Jack up the front end of the vehicle and support it on jackstands.

2. Place a container under the transmission drain plug and drain the transmission fluid.

3. Remove the pan securing bolts and remove the pan and gasket. On all but the A43/A44 family, this will require the use of a thin, flat bladed tool to cut the sealant bead. Do NOT use a screwdriver (you'll deform the pan lip) and do

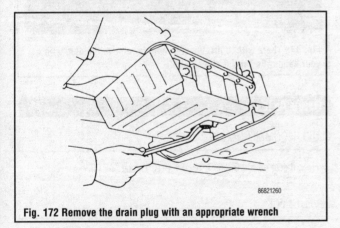

Fig. 172 Remove the drain plug with an appropriate wrench

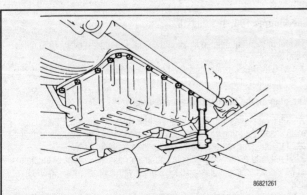

Fig. 173 Most of the pan bolts should be accessible, you may need to remove another part to obtain access for some

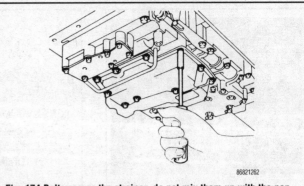

Fig. 174 Bolts secure the strainer, do not mix them up with the pan bolts

NOT drive the tool straight into the seam; if it goes too far, internal damage will result. Work on an angle to the pan and tap the tool just deeply enough to cut the sealer.

4. The pan may be washed in solvent for cleaning but must be absolutely dry when it is reinstalled. Do not wipe it out with a rag; the lint from the rag can damage the transmission. Additionally, the pans for the A43D, A44D and A440F units contain magnets; clean any shavings from the magnets and make certain the magnet(s) are correctly positioned before reinstallation.

5. Remove all traces of the old gasket or sealer from the pan and from the transmission. On A43/A44 series, install a new gasket on the pan but do not use any sealer. On all other units, apply a thin, even bead of sealant around the pan, staying inboard of the screw holes and roughly centered on the pan flange.

6. Inspect, clean or replace the transmission filter or strainer at this time, if necessary. If the filter is removed, be ready for an additional rush of fluid. For all transmissions except the A440F, tighten the bolts to 48 inch lbs. (6 Nm). On the A440F, tighten the 8mm bolts to 48 inch lbs. (6 Nm) and the 10mm bolts to 7 ft. lbs. (10 Nm).

7. Install the pan, tightening the securing bolts to the proper torque in a crisscrossing pattern. Correct pan bolt tightness:
- A43D and A44D—48 inch lbs. (6 Nm)
- A340E, F and H—65 inch lbs. (7 Nm)
- A440F—61 inch lbs. (7 Nm)

✴✴ WARNING

The pan bolts must be tightened evenly; the bolts break easily if overtightened.

8. Fill the transmission to the correct level with the specified fluid.

Transfer Case

FLUID RECOMMENDATIONS

- Trucks and 4Runner's w/MT—multipurpose gear oil API GL-3, GL-4 or GL-5; SAE 75W-90 or 80W-90
- Trucks and 4Runner's w/AT—ATF DEXRON®II or its superceding fluid type
- Land Cruiser—multipurpose gear oil API GL-4 or GL-5; SAE 90W.

LEVEL CHECK

♦ See Figures 175, 176 and 177

The oil in the transfer case should be checked at least every 7500 miles (12,000 km) and replaced every 25,000–30,000 miles (40,000–48,000 km), even more frequently if driven in deep water.

1. With the truck parked on a level surface, remove the filler plug from the side of the transfer housing.

2. If the lubricant begins to trickle out of the hole, there is enough. Otherwise, carefully insert your finger (watch out for sharp threads) and check to see if the oil is up to the edge of the hole.

Fig. 175 When removing the filler plug, remember there is a gasket, usually metal

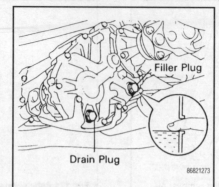

Fig. 176 Watch the threads in the filler plug, they can be sharp

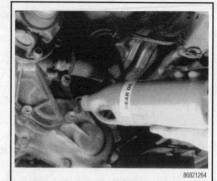

Fig. 177 The gear oil is usually in a long container with a built-in spout

3. If not, add oil through the hole until the level is at the edge of the hole. Most gear lubricants come in a plastic squeeze bottle with a nozzle; making additions simple. You can also use a common everyday kitchen faster.

4. Install the filler plug and gasket, then run the engine and check for leaks.

DRAIN AND REFILL

▶ **See Figures 178 and 179**

1. The transfer case oil should be hot before it is drained. If the engine is at normal operating temperature, the oil should be hot enough.

2. Raise the truck and support it properly on jackstands so that you can safely work underneath.

3. The drain plug is usually located on the bottom of the transfer case. Place a pan under the drain plug and remove it.

✳✳ CAUTION

The oil will be HOT. Be careful when you remove the plug so that you don't take a bath in hot gear oil.

4. Allow the oil to drain completely. Clean off the plug and install it, tightening it until it is just snug.

5. Remove the filler plug from the side of the case.

6. Fill the transfer case with the correct oil through the filler plug hole as detailed previously. Refer to the "Capacities" chart for the amount of oil needed to refill your transfer case.

7. The oil level should come right up to the edge of the hole. You can stick your finger in to verify this.

8. Install the filler plug with a new gasket, lower the truck, and check for leaks. Dispose of the old oil in the proper manner.

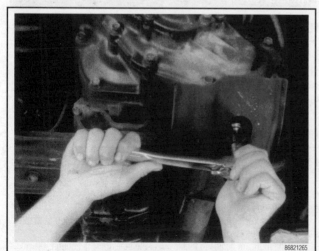

Fig. 178 This truck has a bracket/shield where you access the plug through. You may need an extension

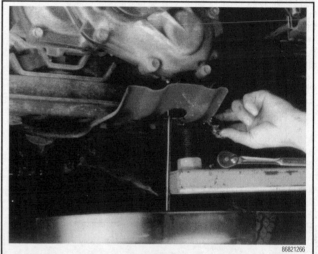

Fig. 179 There will be just enough room to remove the plug with your hands. Be careful of the hot oil

Drive Axles

FLUID RECOMMENDATIONS

Hypoid gear oil API GL–5. Below 0°F (–18°C) use SAE 90W, above 0°F (–18°C) use SAE 80W or 80W–90.

LEVEL CHECK

Front

▶ **See Figures 180, 181 and 182**

The oil in the front differential should be checked at least every 7500 miles (12,000 km).

1. With the truck parked on a level surface, remove the filler plug from the back of the differential.

➡ **The plug on the bottom is the drain plug.**

2. If the oil begins to trickle out of the hole, there is enough. Otherwise, carefully insert your finger (watch out for sharp threads) into the hole and check to see if the oil is up to the bottom edge of the filler hole.

3. If not, add oil through the hole until the level is at the edge of the hole. Most gear oils come in a plastic squeeze bottle with a nozzle, making additions simple.

4. Install the filler plug and drive the truck for a short distance. Stop the truck and check for leaks.

Fig. 180 Locate and remove the filler plug . . .

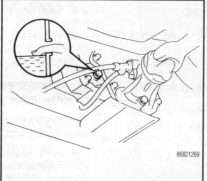

Fig. 181 . . . then check the level with your finger

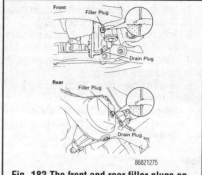

Fig. 182 The front and rear filler plugs on most of your trucks are located in these areas shown

Rear

▶ See Figure 182

The oil in the rear differential should be checked at least every 7500 miles (12,000 km).

1. With the truck parked on a level surface, remove the filler plug from the back of the differential.

➡**The plug on the bottom is the drain plug.**

2. If the oil begins to trickle out of the hole, there is enough. Otherwise, carefully insert your finger (watch out for sharp threads) into the hole and check to see if the oil is up to the bottom edge of the filler hole.

3. If not, add oil through the hole until the level is at the edge of the hole. Most gear oils come in a plastic squeeze bottle with a nozzle, making additions simple.

4. Install the filler plug and drive the truck for a short distance. Stop the truck and check for leaks.

DRAIN AND REFILL

Front

▶ See Figures 183, 184 and 185

The gear oil in the front axle should be changed at least every 25,000–30,000 miles (40,000–48,000 km) or immediately if driven in deep water.

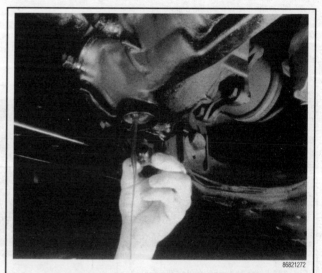

Fig. 184 Remove the drain (lower) plug and gasket, if so equipped

Fig. 183 Have a pan ready for the fluid about to flow from the axle

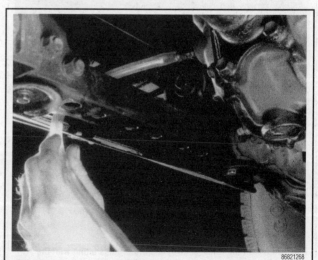

Fig. 185 Slide a the pump's tube (shown) into the filler hole, or use bottled fluid with a built-in funnel

1. Park the vehicle on a level surface. Set the parking brake.
2. Clean the area around the drain plug.
3. Remove the filler (upper) plug. Place a container which is large enough to catch all of the differential oil under the drain plug.
4. Remove the drain (lower) plug and gasket, if so equipped. Allow all of the oil to drain into the container.
5. Install the drain plug. Tighten it so that it will not leak, but do not over-tighten.
6. Refill with the proper grade and viscosity of axle lubricant. Be sure that the level reaches the bottom of the filler plug. DO NOT overfill.
7. Install the filler plug and check for leakage.

Rear

▶ **See Figures 186, 187, 188, 189 and 190**

The gear oil in the rear axle should be changed at least every 25,000–30,000 miles (40,000–48,000 km) or immediately if driven in deep water.
1. Park the vehicle on a level surface. Set the parking brake.
2. Clean the area around the drain plug.
3. Remove the filler (upper) plug. Place a container which is large enough to catch all of the differential oil under the drain plug.
4. Remove the drain (lower) plug and gasket, if so equipped. Allow all of the oil to drain into the container.

5. Install the drain plug. Tighten it so that it will not leak, but do not over-tighten.
6. Refill with the proper grade and viscosity of axle lubricant. Be sure that the level reaches the bottom of the filler plug. DO NOT overfill.
7. Install the filler plug and check for leakage.

Cooling System

FLUID RECOMMENDATIONS

When additional coolant is required to maintain the proper level, always add a 50/50 mixture of ethylene glycol antifreeze/coolant and water. The use of supplementary inhibitors or additives is neither required nor recommended.

➡**Do not use alcohol type antifreeze or plain water alone.**

LEVEL CHECK

▶ **See Figures 191, 192 and 193**

Dealing with the cooling system can be a tricky matter unless the proper precautions are observed. All vehicles use a coolant reservoir tank (expansion tank)

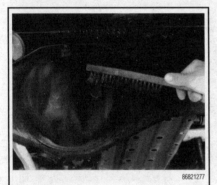

Fig. 186 Clean the area around the drain plug—rear axle shown

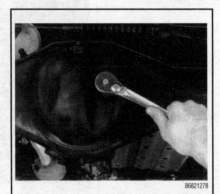

Fig. 187 Unscrew the filler plug

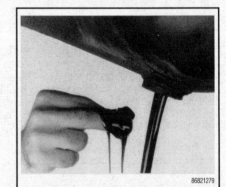

Fig. 188 If you move quick enough, this won't happen to you

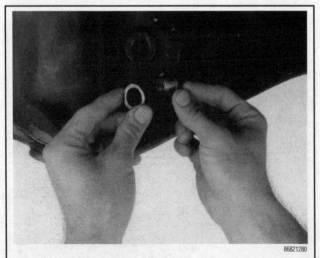

Fig. 189 Install the filler (shown) and drain plugs along with the gaskets, securely

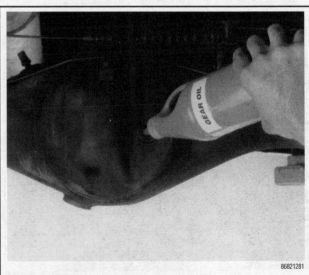

Fig. 190 Refill the axle with the proper gear oil

Fig. 191 The engine coolant reservoir can be used for checking fluid level and adding coolant

Fig. 192 If the coolant is low, add an equal amount of ethylene glycol based antifreeze

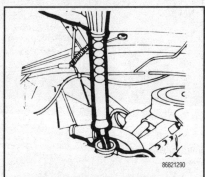

Fig. 193 A coolant hydrometer will show you the boiling and freezing points of the coolant present

connected to the radiator by a small hose. Look through the plastic tank; the fluid level should be between FULL and LOW lines. Note that the FULL line is below the very top of the tank, allowing room for expansion. If the level in the tank is low, remove the cap of the tank and add coolant to the FULL line.

✳✳ CAUTION

Never remove the radiator cap to check fluid level; there is a high risk of scalding from escaping hot fluid.

If the coolant level is low, add an equal amount of ethylene glycol based antifreeze. Avoid using water that is known to have a high alkaline content or is very hard, except in emergency situations. Drain and flush the cooling system as soon as possible after using such water.

The radiator hoses and clamps and the radiator cap should be checked at the same time as the coolant level. Hoses which are brittle, cracked, or swollen should be replaced. Clamps should be checked for tightness (screwdriver tight only.) Do not allow the clamp to cut into the hose or crush the fitting. The radiator cap gasket should be checked for any obvious tears, cracks or swelling, or any signs of incorrect seating in the radiator neck.

A 50/50 mix of coolant concentrate and water will usually provide protection to −35°F (−37°C). Freeze protection may be checked by using a cooling system hydrometer. Inexpensive hydrometers may be obtained from a local auto supply store. Follow the directions packaged with the coolant hydrometer when checking protection.

DRAIN AND REFILL

▶ See Figures 194 thru 199

✳✳ CAUTION

When draining coolant, keep in mind that cats and dogs are attracted by ethylene glycol antifreeze, and are quite likely to drink any that is left in an uncovered container or in puddles on the ground. This will prove fatal in sufficient quantity. Always drain coolant into a sealable container. Coolant may be reused unless it is contaminated or several years old.

Completely draining and refilling the cooling system every two years will remove accumulated rust, scale and other deposits.

1. Make certain the engine is COLD. Generally, this means at least 3 hours from last operation, longer in hot weather. Overnight cold is best.
2. Remove the radiator cap. Drain the existing antifreeze and coolant. Open the radiator and engine drain petcocks, or disconnect the bottom radiator hose at the radiator outlet. Set the heater temperature controls to the full HOT position.

➡**Before opening the radiator petcock, spray it with some penetrating lubricant.**

3. Close the petcocks or reconnect the lower hose and fill the system with the correct mixture of antifreeze and water in the correct amount. It will take a bit of time for the level to stabilize within the radiator.
4. Double check the petcocks for closure.
5. Install the radiator cap.
6. Flush the reservoir with water, empty it, then install fresh coolant mixture to the FULL line.
7. Start the engine, allowing it to idle. As it warms up, keep a close watch on the petcocks and/or hose fittings used to drain the system. If any leakage is seen, shut the engine OFF and fix the leak.
8. After the engine has warmed up, shut the engine OFF. Check the level in the reservoir tank and adjust the fluid level as needed.

➡**Fresh antifreeze has a detergent quality. Fluid in the reservoir may appear muddy or discolored due to the cleaning action of the new coolant. If the system has been neglected for a period of time, it may be necessary to redrain the system to eliminate this sludge.**

Fig. 194 When the engine is cool, remove the radiator cap

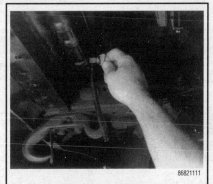

Fig. 195 The petcock or plug can be removed by turning it clockwise

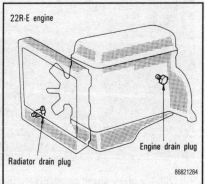

Fig. 196 Radiator and engine coolant plugs—22R-E

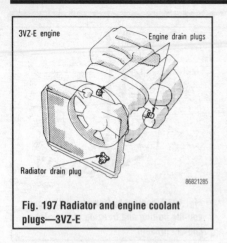

Fig. 197 Radiator and engine coolant plugs—3VZ-E

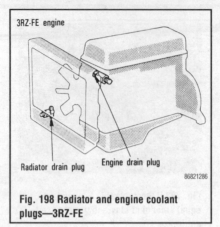

Fig. 198 Radiator and engine coolant plugs—3RZ-FE

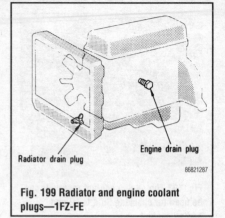

Fig. 199 Radiator and engine coolant plugs—1FZ-FE

FLUSHING AND CLEANING THE SYSTEM

Several aftermarket radiator flushing and cleaning kits can be purchased at your local auto parts store. It is recommended that the radiator be cleaned and flushed of sludge and any rust build-up once a year. Manufacturers directions for proper use, and safety precautions, come in each kit.

CLEAN RADIATOR OF DEBRIS

Periodically clean any debris such as leaves, paper, insects, etc., from the radiator fins. Pick the large pieces off by hand. The smaller pieces can be washed away with water pressure from a hose.

Carefully straighten any bent radiator fins with a pair of needle nose pliers. Be careful, the fins are very soft. Don't wiggle the fins back and forth too much. Straighten them once and try not to move them again.

Master Cylinder (Brake and Clutch)

♦ **See Figures 200 and 201**

All models have the brake master cylinder reservoir on the left side of the firewall under the hood. Trucks and 4Runner's with manual transmissions use a hydraulic (fluid actuated) clutch. The clutch fluid reservoir is located in the same area as brake reservoir. The clutch master cylinder uses the same fluid as the brakes, and should be checked at the same time as the brake master cylinder.

Fig. 200 The brake master cylinder is attached to the booster

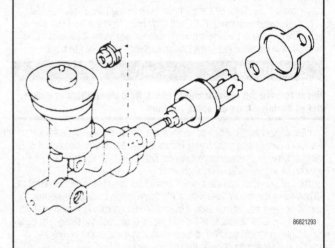

Fig. 201 The clutch master cylinder is located next to the brake master

FLUID RECOMMENDATIONS

Use only Heavy Duty Brake fluid meeting DOT 3 or SAE J1703 specifications.

LEVEL CHECK

♦ **See Figures 202, 203, 204 and 205**

The fluid in the brake and/or clutch master cylinders should be checked every 6 months or 6000 miles (9600 km). Check the fluid level at the side of the reservoir. If fluid is required, thoroughly wipe the top of the reservoir and the sides of the plastic housing with a rag before opening. The fluid must be kept free from dirt and grit. Fill the reservoir to the MAX line in the reservoir. Never overfill the reservoir. Install the reservoir cap, making sure the gasket is properly seated in the cap.

✸✸ WARNING

Brake fluid damages painted surfaces. If any fluid is spilled, immediately clean it off with plenty of clean water. It also absorbs moisture from the air; never leave a container of fluid or the master cylinder or the clutch cylinder uncovered any longer than necessary.

It is normal for the fluid level to fall as the disc brake pads wear. However, if the master cylinder requires filling frequently, you should check the system for leaks in the hoses, master cylinder, or wheel cylinders.

FLUID RECOMMENDATIONS

Use only DEXRON®II or its superceding type automatic transmission fluid in the power steering system.

Fig. 202 Check the fluid level on the side of the reservoir. If necessary, lift the cap . . .

Fig. 203 . . . then fill to the appropriate line with DOT 3 fluid

Fig. 204 As you can see on this brake master cylinder reservoir the level is slightly low

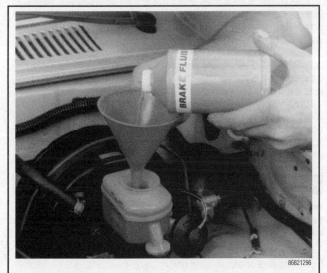

Fig. 205 With the aid of a funnel, there will be little or no mess

Fig. 206 Remove the filler cap on the power steering fluid reservoir and wipe the dipstick clean

FLUID LEVEL CHECK

▶ **See Figure 206**

Check the power steering fluid level every 6 months or 6000 miles (9600 km).

1. Park the vehicle on a level surface. Run the engine until normal operating temperature is reached.
2. Turn the steering all the way to the left and then all the way to the right several times. Center the steering wheel and shut off the engine.
3. Open the hood.
4. Remove the filler cap on the power steering fluid reservoir and wipe the dipstick clean.

5. Reinsert the dipstick and tighten the cap. Remove the dipstick and note the fluid level indicated on the dipstick.
6. The level should be at any point below the Full mark, but not below the Add mark (in the HOT or COLD ranges).
7. Add fluid as necessary. Do not overfill.

Steering Gear

FLUID RECOMMENDATIONS

Use standard hypoid-type gear oil GL–4, SAE 90W when refilling the steering gear.

FLUID LEVEL CHECK

▶ **See Figure 207**

Each year or 15,000 miles (24,000 km) you should check the steering gear housing lubricating oil. The filler plug is on top of the housing and requires a

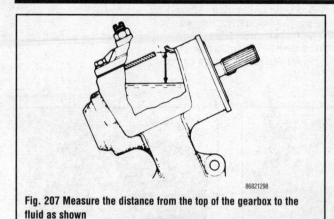

Fig. 207 Measure the distance from the top of the gearbox to the fluid as shown

14mm wrench for removal. The level should be ¾–1 inch (18–28mm) from the top on 2WD models and ⁹⁄₁₆–1¹⁄₁₆ inch (14–17mm) from the top on 4WD models.

Chassis Greasing

▶ See Figures 208, 209 and 210

Complete chassis greasing should include an inspection of all rubber suspension bushings, as well as proper greasing of the front suspension upper and lower ball joints and control arm bushings. To provide correct operation, the chassis should be greased with every oil and filter change or 7500 miles (12,070 km).

If you wish to perform this operation yourself, purchase a cartridge type grease gun and several cartridges of multipurpose lithium grease. Push the nozzle of the grease gun down firmly onto the fitting and while applying pressure, force the new grease into the fitting. Force sufficient grease into the fitting to cause some old grease to be expelled. Use an old rag to wipe off expelled grease; keep the fittings and joints free of globs of grease.

Four wheel drive vehicles have several lubrication points within the driveshaft and driveline. Some of these fittings are tucked into hard to see places; make certain each one is accounted for. Additionally, 1989–90 Land Cruisers have lube fittings at the ends of the relay rods in the steering system. The 1991–96 Land Cruisers use permanently sealed joints; no lubrication is necessary.

MANUAL TRANSMISSION AND CLUTCH LINKAGE

On models so equipped, apply a small amount of chassis grease to the pivot points of the transmission and clutch linkage.

AUTOMATIC TRANSMISSION LINKAGE

On models so equipped, apply a small amount of 10W engine oil to the kickdown and shift linkage at the pivot points.

PARKING BRAKE LINKAGE

At yearly intervals or whenever binding is noticeable in the parking brake linkage, lubricate the cable guides, levers and linkage with a suitable chassis grease.

Body Lubrication and Maintenance

Door handles, hinges, and locks should be lubricated at least once a year. Use a light lithium grease for hinges and handles. Use spray graphite for door locks. Do not inject oil into door locks; it attracts and holds dirt and grit, causing binding and stiff operation, particularly in cold weather.

Additionally, a small amount of grease should be applied to the hood and trunk latches and hinges periodically. Other moving components such as seat back pivots or cargo compartments should be attended to on as-needed basis.

➡**When performing lubrication of any item in the passenger area (doors, locks, etc.) apply lubricant sparingly and thoroughly wipe up the excess to prevent lubricant getting onto upholstery and/or clothing.**

EXTERIOR CARE

Keep the exterior of the vehicle clean. Dirt can cause small scratches in the paint and the chemicals in road dirt and air pollution can cause paint damage. Frequent washing is recommended; this is particularly important for vehicles used in snowy areas, near salt water or any vehicle used off the road.

Wash the vehicle in the shade when the body is not hot to the touch. Wash with mild car-wash soap and rinse thoroughly before the soap dries on the paint. It may be necessary to wash and rinse the vehicle in sections to prevent the soap from streaking the paint. Never use gasoline or strong solvents to clean painted surfaces.

Use a chamois or soft, lint-free rag to dry the exterior. Wring the drying rag out frequently and keep it away from areas in which it may pick up dirt or grit. If the rag or chamois is dropped on the ground, rinse it thoroughly before continuing; it may carry grit which will scratch the finish.

After the paint is completely dry, the bodywork may be waxed with any of the commercial car polishes or waxes. Read the car carefully before purchase; use good judgment in deciding the condition of the finish. A wax for older finishes will have more abrasive to remove oxidized paint. A "new car" wax will have much less abrasive.

Wheel Bearings

➡**Front wheel bearings on 4WD vehicles are covered in Section 7. Do not attempt to disassemble the bearings and hubs on these vehicles without referring to the appropriate procedures.**

Only the front wheel bearings require periodic service. The lubricant to use is high temperature disc brake wheel bearing grease meeting NLGI No. 2 specifications. This service is recommended at 30,000 miles (48,270 km) on trucks, 4Runner's and 1991–96 Land Cruisers. On 1989–90 Land Cruisers, the interval is 26,500 miles (42,638 km). In either case, if the mileage is not achieved, the bearings should be packed every 36 months or whenever the truck has been driven in water up to the hub.

Before handling the bearings there are a few things that you should remember:

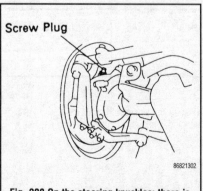

Fig. 208 On the steering knuckles; there is a screw plug to be removed and lubricated

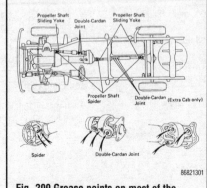

Fig. 209 Grease points on most of the truck and 4Runner's

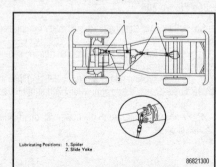

Fig. 210 Land Cruiser grease points—note that 1989–90 models also required periodic ball joint greasing on the front suspension

- Remove all outside dirt from the housing before exposing the bearing.
- Treat a used bearing as gently as you would a new one.
- Work with clean tools in clean surroundings.
- Use clean, dry canvas gloves, or at least clean, dry hands.
- Clean solvents and flushing fluids are a must.
- Use clean paper when laying out the bearings to dry.
- Protect disassembled bearings from rust and dirt. Cover them up.
- Use clean rags to wipe bearings.
- Keep the bearings in oil-proof paper when they are to be stored or are not in use.
- Clean the inside of the housing before replacing the bearings.

Do NOT do the following:
- Don't work in dirty surroundings.
- Don't use dirty, chipped, or damaged tools.
- Try not to work on wooden work benches or use wooden mallets.
- Don't handle bearings with dirty or moist hands.
- Do not use gasoline for cleaning; use a safe solvent.
- Do not spin dry bearings with compressed air. They will be damaged.
- Do not spin unclean bearings.
- Avoid using cotton waste or dirty cloths to wipe bearings.
- Do not to scratch or nick bearing surfaces.
- Do not allow the bearing to come in contact with dirt or rust at any time.

REMOVAL, PACKING, AND INSTALLATION

2 Wheel Drive

▶ See Figures 211, 212, 213, 214 and 215

1. Loosen the lugnuts.
2. Elevate and safely support the front of the truck on jackstands.
3. Remove the wheels.
4. Remove the brake caliper from its mount. Use stiff wire to suspend it out of the way; do not loosen the brake hose at the caliper. Remove the brake pads.

✳✳ CAUTION

Brake pads contain asbestos, which has been determined to be a cancer causing agent. Never clean the brake surfaces with compressed air. Avoid inhaling any dust from brake surfaces. When cleaning brakes, use commercially available brake cleaning fluids.

5. Remove the grease cap. Remove the cotter pin and locknut. Use a socket of the correct size to loosen and remove the axle nut.
6. Remove the hub and brake disc together with the outer bearing and thrust washer. Use your thumbs to keep the pieces inside the hub as the unit is removed. Be careful not to drop the outer bearing.
7. Use a small prying tool to remove the inner bearing seal. Once the seal is out, remove the bearing from the hub.
8. Place all both bearings and all the nuts, caps, etc., into a wide container of cleaning solvent. Cleanliness is essential to wheel bearing maintenance. Use a soft bristle brush to clean every bit of grease from every component. Place each cleaned component on a clean, lint-free cloth and allow them to air dry.
9. Inspect the bearings for pitting, flat spots, rust, and rough areas. Check the races (inner surfaces of the hub) for the same conditions. If any damage is seen, the components must be replaced. As a general rule, if either a bearing or race is dam-

aged, the matching part that contacts it should also be replaced. Replacement bearings, seals and other required parts can be bought at an auto parts store. The old parts that are to be replaced should be taken along to be compared with the replacement part to insure a perfect match.

To pack:

10. Pack the wheel bearings with grease. There are special devices made for the specific purpose of greasing bearings, but if one is not available, pack the wheel bearings by hand. Put a large dab of grease in the palm of your hand and push the bearing through it with a sliding motion. The grease must be forced through the side of the bearing and in between each roller. Continue until the grease begins to ooze out the other side and through the gaps between the rollers; the bearing must be completely packed with grease.

11. Coat the inside of the hub and cap with grease, but do not pack it solid. Remember that the spindle (axle) has to pass through the center.

To install:

12. Clean the spindle thoroughly and inspect it for any sign of damage. Coat it with a very light layer of bearing grease.

13. Install the inner bearing into the race. Use a seal driver of the correct diameter to install a NEW grease seal over the bearing. Reusing the old seal may cost more money than it saves; if the grease leaks out and the bearing fails, the wheel may seize while in motion.

14. Place the hub and disc onto the spindle. Install the outer bearing and the flat thrust washer.

15. Install the locknut and use a torque wrench to adjust the nut to 25 ft. lbs. (34 Nm).

16. Turn the hub in each direction several times to seat and snug the bearing. Loosen the locknut until it can be turned by hand without the wrench. Don't loosen it any more than necessary to be finger-loose. (Loosening the nut takes the pre-tension off the bearing.)

17. Use a spring tension scale to measure the force needed to turn the hub. Record the number.

18. Tighten the nut until the force required to turn the hub is:

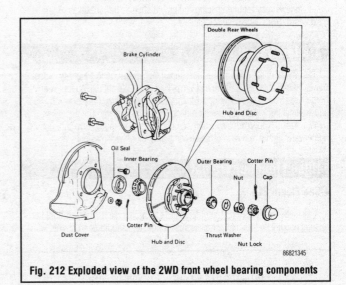

Fig. 212 Exploded view of the 2WD front wheel bearing components

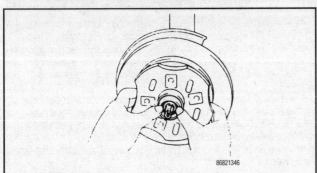

Fig. 211 Use your thumbs to keep the parts inside the hub as the unit is removed

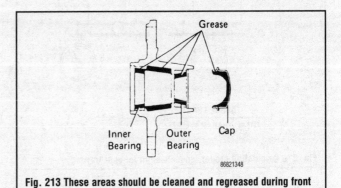

Fig. 213 These areas should be cleaned and regreased during front wheel bearing maintenance

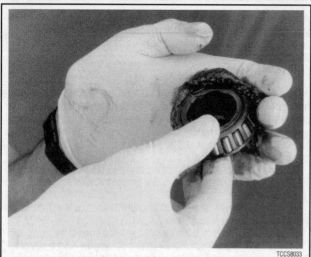

Fig. 214 Put a large dap of grease into the palm of your hand, then move the bearing in a sliding motion through the grease

- Single rear wheel vehicles—1.3–4 lbs. (0.6–1.8 kg)
- Dual rear wheel vehicles—0.9–2.2 lbs. (0.4–1 kg)
  19. Install the locknut, cotter pin and grease cap. Make certain the cap is not crooked or loose.
  20. Install the brake pads and caliper.
  21. Install the wheel and lugnuts (hand-tight). Lower the vehicle to the ground and tighten the lugnuts.

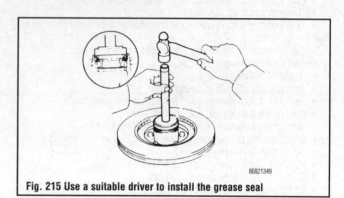

Fig. 215 Use a suitable driver to install the grease seal

TRAILER TOWING

General Recommendations

Your vehicle was primarily designed to carry passengers and cargo. It is important to remember that towing a trailer will place additional loads on your vehicles engine, drive train, steering, braking and other systems. However, if you decide to tow a trailer, using the prior equipment is a must.

Local laws may require specific equipment such as trailer brakes or fender mounted mirrors. Check your local laws.

Trailer Weight

The weight of the trailer is the most important factor. A good weight-to-horse-power ratio is about 35:1, 35 lbs. of Gross Combined Weight (GCW) for every horsepower your engine develops. Multiply the engine's rated horsepower by 35 and subtract the weight of the vehicle passengers and luggage. The number remaining is the approximate ideal maximum weight you should tow, although a numerically higher axle ratio can help compensate for heavier weight.

Hitch (Tongue) Weight

▶ See Figure 216

Calculate the hitch weight in order to select a proper hitch. The weight of the hitch is usually 9–11% of the trailer gross weight and should be measured with

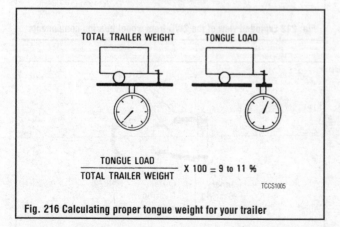

Fig. 216 Calculating proper tongue weight for your trailer

the trailer loaded. Hitches fall into various categories: those that mount on the frame and rear bumper, the bolt-on type, or the weld-on distribution type used for larger trailers. Axle mounted or clamp-on bumper hitches should never be used.

Check the gross weight rating of your trailer. Tongue weight is usually figured as 10% of gross trailer weight. Therefore, a trailer with a maximum gross weight of 2000 lbs. will have a maximum tongue weight of 200 lbs. Class I trailers fall into this category. Class II trailers are those with a gross weight rating of 2000–3000 lbs., while Class III trailers fall into the 3500–6000 lbs. category. Class IV trailers are those over 6000 lbs. and are for use with fifth wheel trucks, only.

When you've determined the hitch that you'll need, follow the manufacturer's installation instructions, exactly, especially when it comes to fastener torques. The hitch will subjected to a lot of stress and good hitches come with hardened bolts. Never substitute an inferior bolt for a hardened bolt.

Cooling

ENGINE

Aftermarket engine oil coolers are helpful for prolonging engine oil life and reducing overall engine temperatures. Both of these factors increase engine life. While not absolutely necessary in towing Class I and some Class II trailers, they are recommended for heavier Class II and all Class III towing. Engine oil cooler systems usually consist of an adapter, screwed on in place of the oil filter, a remote filter mounting and a multi-tube, finned heat exchanger, which is mounted in front of the radiator or air conditioning condenser.

Transmission

An automatic transmission is usually recommended for trailer towing. Modern automatics have proven reliable and, of course, easy to operate, in trailer towing. The increased load of a trailer, however, causes an increase in the temperature of the transmission fluid. Heat is the worst enemy of an automatic transmission. As the temperature of the fluid increases, the life of the fluid decreases.

It is essential, therefore, that you install an automatic transmission cooler and that you pay close attention to transmission fluid changes. The cooler, which consists of a multi-tube, finned heat exchanger, is usually installed in front of the radiator or air conditioning compressor, and hooked in-line with the transmission cooler tank inlet line. Follow the cooler manufacturer's installation instructions.

JUMP STARTING A DEAD BATTERY

♦ See Figure 217

Whenever a vehicle is jump started, precautions must be followed in order to prevent the possibility of personal injury. Remember that batteries contain a small amount of explosive hydrogen gas which is a by-product of battery charging. Sparks should always be avoided when working around batteries, especially when attaching jumper cables. To minimize the possibility of accidental sparks, follow the procedure carefully.

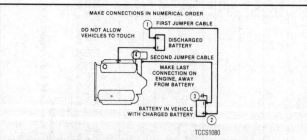

MAKE CONNECTIONS IN NUMERICAL ORDER

DO NOT ALLOW VEHICLES TO TOUCH

① FIRST JUMPER CABLE

DISCHARGED BATTERY

SECOND JUMPER CABLE

MAKE LAST CONNECTION ON ENGINE, AWAY FROM BATTERY

BATTERY IN VEHICLE WITH CHARGED BATTERY

TCCS1080

Fig. 217 Connect the jumper cables to the batteries and engine in the order shown

✳✳ CAUTION

NEVER hook the batteries up in a series circuit or the entire electrical system will go up in smoke, including the starter!

Vehicles equipped with a diesel engine may utilize two 12 volt batteries. If so, the batteries are connected in a parallel circuit (positive terminal to positive terminal, negative terminal to negative terminal). Hooking the batteries up in parallel circuit increases battery cranking power without increasing total battery voltage output. Output remains at 12 volts. On the other hand, hooking two 12 volt batteries up in a series circuit (positive terminal to negative terminal, positive terminal to negative terminal) increases total battery output to 24 volts (12 volts plus 12 volts).

Jump Starting Precautions

- Be sure that both batteries are of the same polarity (have the same terminal, in most cases NEGATIVE grounded).
- Be sure that the vehicles are not touching or a short could occur.
- On non-sealed batteries, be sure the vent cap holes are not obstructed.
- Do not smoke or allow sparks anywhere near the batteries.
- In cold weather, make sure the battery electrolyte is not frozen. This can occur more readily in a battery that has been in a state of discharge.
- Do not allow electrolyte to contact your skin or clothing.

Jump Starting Procedure

1. Make sure that the voltages of the 2 batteries are the same. Most batteries and charging systems are of the 12 volt variety.

JACKING

Your vehicle was supplied with a jack for emergency road repairs. This jack is fine for changing a flat tire or other operations not requiring you to go beneath the vehicle. If it is used in an emergency situation, carefully follow the instructions provided either with the jack or in your owner's manual. Do not attempt to use the jack in any places other than specified by the vehicle manufacturer. Always block the diagonally opposite wheel when using a jack.

A more convenient way of jacking is the use of a garage or floor jack. Never place the jack under the radiator, engine or transmission components. Severe and expensive damage will result when the jack is raised. Additionally, never jack under the floorpan or bodywork; the metal will deform.

Whenever you plan to work under the vehicle, you must support it on jackstands or ramps. Never use cinder blocks or stacks of wood to support the vehicle, even if you're only going to be under it for a few minutes. Never crawl under the vehicle when it is supported only by the tire-changing jack or other floor jack.

2. Pull the jumping vehicle (with the good battery) into a position so the jumper cables can reach the dead battery and that vehicle's engine. Make sure that the vehicles do NOT touch.

3. Place the transmissions/transaxles of both vehicles in **Neutral** (MT) or **P** (AT), as applicable, then firmly set their parking brakes.

➡**If necessary for safety reasons, the hazard lights on both vehicles may be operated throughout the entire procedure without significantly increasing the difficulty of jumping the dead battery.**

4. Turn all lights and accessories OFF on both vehicles. Make sure the ignition switches on both vehicles are turned to the **OFF** position.

5. Cover the battery cell caps with a rag, but do not cover the terminals.

6. Make sure the terminals on both batteries are clean and free of corrosion for good electrical contact.

7. Identify the positive (+) and negative (−) terminals on both batteries.

8. Connect the first jumper cable to the positive (+) terminal of the dead battery, then connect the other end of that cable to the positive (+) terminal of the booster (good) battery.

9. Connect one end of the other jumper cable to the negative (−) terminal on the booster battery and the final cable clamp to an engine bolt head, alternator bracket or other solid, metallic point on the engine with the dead battery. Try to pick a ground on the engine that is positioned away from the battery in order to minimize the possibility of the 2 clamps touching should one loosen during the procedure. DO NOT connect this clamp to the negative (−) terminal of the bad battery.

✳✳ CAUTION

Be very careful to keep the jumper cables away from moving parts (cooling fan, belts, etc.) on both engines.

10. Check to make sure that the cables are routed away from any moving parts, then start the donor vehicle's engine. Run the engine at moderate speed for several minutes to allow the dead battery a chance to receive some initial charge.

11. With the donor vehicle's engine still running slightly above idle, try to start the vehicle with the dead battery. Crank the engine for no more than 10 seconds at a time and let the starter cool for at least 20 seconds between tries. If the vehicle does not start in 3 tries, it is likely that something else is also wrong or that the battery needs additional time to charge.

12. Once the vehicle is started, allow it to run at idle for a few seconds to make sure that it is operating properly.

13. Turn ON the headlights, heater blower and, if equipped, the rear defroster of both vehicles in order to reduce the severity of voltage spikes and subsequent risk of damage to the vehicles' electrical systems when the cables are disconnected. This step is especially important to any vehicle equipped with computer control modules.

14. Carefully disconnect the cables in the reverse order of connection. Start with the negative cable that is attached to the engine ground, then the negative cable on the donor battery. Disconnect the positive cable from the donor battery and finally, disconnect the positive cable from the formerly dead battery. Be careful when disconnecting the cables from the positive terminals not to allow the alligator clips to touch any metal on either vehicle or a short and sparks will occur.

➡**Always position a block of wood on top of the jack or stand to protect the finish when lifting or supporting the vehicle.**

Small hydraulic, screw, or scissors jacks are satisfactory for raising the vehicle. Drive-on trestles or ramps are also a handy and safe way to both raise and support the vehicle. Be careful though, some ramps may be too steep to drive your vehicle onto without scraping the front bottom panels. Never support the vehicle on any suspension member (unless specifically instructed to do so by a repair manual) or underbody panel.

The following safety points cannot be overemphasized:
- Always block the opposite wheel or wheels to keep the vehicle from rolling off the jack.
- When raising the front of the vehicle, firmly apply the parking brake.
- Always use jackstands to support the vehicle when you are working underneath. Place the stands beneath the scissors jacking brackets. Before climbing underneath, rock the vehicle a bit to make sure it is firmly supported.

MAINTENANCE SCHEDULE B
1989–95 Truck

Conditions other than those listed for SCHEDULE A.

Maintenance services beyond 60,000 miles (96,000 km) should be performed at the same intervals shown in each maintenance schedule.

System	Maintenance items		Miles × 1,000	10	15	20	30	40	45	50	60
			Km × 1,000 (22R, 22R-E engine)	16	24	32	48	64	72	80	96
			Km × 1,000 (3VZ-E engine)	12	18	24	36	48	54	60	72
			Months	12	18	24	36	48	54	60	72
ENGINE	Valve clearance*	22R, 22R-E engine									A
	Valve clearance	3VZ-E engine									A
	Drive belts[1]										—
	Engine oil and oil filter*[2]			R		R	R	R		R	R
	Engine coolant[2]						R				R
	Exhaust pipes and mountings						A				A
FUEL	Idle speed										—
	Idle speed and fast idle speed	22R-E, 3VZ-E engine[3]					A				A
	Air filter*	22R engine[4]					R				R
	Fuel lines and connections										—
	Fuel tank cap gasket						R				R
IGNITION	Spark plugs**						R				R
EVAP	Charcoal canister	Calif. only									
EXHAUST	Oxygen sensor*[5]	Fed. and Canada 22R-E, 3VZ-E engine only								R (5)	
BRAKES	Brake linings and drums	4WD		—		—	—	—		—	—
		2WD		—		—	—	—		—	—
	Brake pads and discs	4WD		—		—	—	—		—	—
		2WD		—		—	—	—		—	—
	Brake line pipes and hoses	4WD					—				—
		2WD					—				—
CHASSIS	Steering linkage	4WD					—				—
		2WD					—				—
	Ball joints and dust covers	4WD					R				R
		2WD					R				R
	Drive shaft boots	4WD					—				—
		2WD					—				—
	Automatic transmission, manual transmission, transfer (4WD), differential and steering gear housing oil[6]	4WD					R				R
		2WD					—				—
	Front wheel bearing and thrust bush grease (4WD only)	4WD					R				R
	Propeller shaft grease[7]	2WD					—				—
	Bolts and nuts on chassis and body[9]	4WD		R		R	R	R		R	R
		2WD		R		R	R	R		R	R

* For vehicles sold in California
**: For vehicles sold outside California
(1) After 60,000 miles (96,000 km) or 72 months, inspect every 10,000 miles (16,000 km) or 12 months.
(2) After 60,000 miles (96,000 km) or 72 months, replace every 30,000 miles (48,000 km) or 36 months.
(3) After 60,000 miles (96,000 km) or 72 months, adjust every 30,000 miles (48,000 km) or 36 months.
(4) Adjustment at 30,000 miles (48,000 km) or 36 months only.
(5) Replace at 80,000 miles (129,000 km) only.
(6) Inspect the steering gear housing for oil leakage only.
(7) If the propeller shaft has been immersed in water, it should be re-greased daily.

8882322

MAINTENANCE SCHEDULE A
1989–95 Truck

- Towing a trailer, using a camper or car top carrier.
- Repeat short trips less than 5 miles (8 km) and outside temperatures remain below freezing.
- Extensive idling and/or low speed driving for a long distance such as police, taxi or door-to-door delivery use.
- Operating on dusty, rough, muddy or salt spread roads.

Maintenance operations: A = Check and adjust if necessary; R = Replace, change or lubricate; I = Inspect and correct or replace if necessary

Maintenance services beyond 60,000 miles (96,000 km) should be performed at the same intervals shown in each maintenance schedule.

System	Maintenance items		Miles × 1,000	5	7.5	10	15	20	22.5	25	30	35	37.5	40	45	50	52.5	55	60
			Km × 1,000 (22R, 22R-E engine)	8	12	16	24	32	36	40	48	56	60	64	72	80	84	88	96
			Km × 1,000 (3VZ-E engine)	6	9	12	18	24	27	30	36	42	45	48	54	60	63	66	72
			Months	6	9	12	18	24	27	30	36	42	45	48	54	60	63	66	72
ENGINE	Timing belt — 3VZ-E engine										R[1]								R
	Valve clearance*	22R, 22R-E engine									A								A
	Valve clearance	3VZ-E engine									A								A
	Drive belts[2]										—								—
	Engine oil and oil filter*[3]			R		R	R	R		R	R	R		R	R	R		R	R
	Engine coolant[4]										—								R
	Exhaust pipes and mountings										A								A
FUEL	Idle speed										A								A
	Idle speed and fast idle speed — 22R-E, 3VZ-E engine[4]										A								A
	Air filter* — 22R engine[5]										R								R
	Fuel lines and connections										—								—
	Fuel tank cap gasket																		R
IGNITION	Spark plugs**										R								R
EVAP	Charcoal canister	Calif. only									R								R
EXHAUST	Oxygen sensor*	Fed. and Canada 22R-E, 3VZ-E engine only									R[6]								
BRAKES	Brake linings and drums	4WD				—					—								—
		2WD				—					—								—
	Brake pads and discs	4WD				—					—								—
		2WD				—					—								—
	Brake line pipes and hoses	4WD				—					—								—
		2WD				—					—								—
CHASSIS	Steering linkage[10]	4WD									R								R
		2WD									—								—
	Ball joints and dust covers	4WD				R					R								R
		2WD									R								R
	Drive shaft boots	4WD				—					R								R
		2WD				—					—								—
	Automatic transmission, manual transmission, transfer (4WD), differential and steering gear housing oil[7]	4WD									R					R				R
		2WD									R					R				R
	Front wheel bearing and thrust bush grease (4WD only)	4WD									R					R				R
	Propeller shaft grease[8]	2WD		R		R	R			R	R	R		R	R	R		R	—
	Bolts and nuts on chassis and body[9]	4WD				R					R					R				R
		2WD				R					R					R				—

* For vehicles sold in California
**: For vehicles sold outside California
(1) For the vehicles frequently idled for extensive periods and/or driven for long distances at low speeds such as taxis, police cars and door-to-door delivery use, it is recommended to replace at 60,000 miles (96,000 km).
(2) After 60,000 miles (96,000 km) or 72 months, inspect every 10,000 miles (16,000 km) or 12 months.
(3) After 60,000 miles (96,000 km) or 72 months, replace every 30,000 miles (48,000 km) or 36 months.
(4) After 60,000 miles (96,000 km) or 72 months, adjust every 30,000 miles (48,000 km) or 36 months.
(5) Replace at 80,000 miles (129,000 km) only.
(6) Adjustment at 30,000 miles (48,000 km) or 36 months only.
(7) Inspect the steering gear housing for oil leakage only.
(8) If the propeller shaft has been immersed in water, it should be re-greased daily.
(9) Applicable when operating mainly on dusty roads. If not, follow the schedule B.
(10) Applicable when operating mainly on rough and/or muddy roads. If not, follow the schedule B.

8882321

MAINTENANCE SCHEDULE A
1989–90 Land Cruiser

Maintenance operations: A = Check and adjust if necessary; R = Replace, change or lubricate; I = Inspect and correct or replace if necessary

Maintenance services beyond 60,000 miles (96,000 km) should continue to be performed at the same intervals shown in each maintenance schedule.

System	Service interval (Odometer reading or months, whichever comes first) / Maintenance items	Months
	Miles × 1,000: 3.75, 7.5, 11.25, 15, 18.75, 22.5, 26.25, 30, 33.75, 37.5, 41.25, 45, 48.75, 52.5, 56.25, 60	
	km × 1,000: 6, 12, 18, 24, 30, 36, 42, 48, 54, 60, 66, 72, 78, 84, 90, 96	
ENGINE	Valve clearance*	A: Every 24 months
	Drive belts[1]	I: 36 months; 72 months
	Engine oil and oil filter*	R: Every 6 months
	Engine coolant[2]	R: 36 months; 72 months
	Exhaust pipes and mountings	I: Every 24 months
FUEL	Fuel lines and connections[4]	I: Every 6 months
	Fuel tank cap gasket	R: 36 months
IGNITION	Spark plug* [3]	I: Every 24 months / R: Every 36 months
EVAP	Charcoal canister (Calif. only)	
EXHAUST	Oxygen sensor* (Except California)	R: 80,000 miles (128,000 km) only
BRAKES	Brake linings and drums[5]	I: Every 36 months
	Brake pads and discs	R: Every 72 months
	Brake line pipes and hoses	R: Every 36 months
CHASSIS	Steering linkage	I: Every 72 months
	Ball joints and dust covers	I: Every 12 months
	Automatic transmission	I: Every 12 months
	Transfer, differential and steering gear box oil[6]	R: Every 24 months
	Front wheel bearing and thrust bush grease	I: Every 24 months
	Steering knuckle and chassis grease[7]	R: Every 36 months
	Propeller shaft grease[7]	R: Every 12 months
	Bolts and nuts on chassis and body[8]	I: Every 12 months

Conditions:
- Towing a trailer, using a camper or car top carrier.
- Repeated short trips of less than 5 miles (8 km) with outside temperatures remain below freezing.
- Extensive idling and/or low speed driving for long distances such as police, taxi or door-to-door delivery use.
- Operating on dusty, rough, muddy or salt spread roads.

*: For vehicle sold in California
**: For vehicle sold outside California
(1) After 60,000 miles (96,000 km) or 72 months, inspect every 7,500 miles (12,000 km) or 12 months.
(2) After 45,000 miles (72,000 km) or 36 months, replace every 30,000 miles (48,000 km) or 24 months.
(3) Applicable when operating mainly on dusty roads. If not, follow schedule B.
(4) Includes inspection of fuel tank band and vapor vent system.
(5) Also applicable to lining drum for parking brake.
(6) Inspect the steering gear box for oil leakage only.
(7) If the propeller shaft has been immersed in water, it should be re-greased within a day.
(8) Applicable only when operating mainly on rough, muddy roads. The applicable parts are listed below. For other usage conditions, refer to SCHEDULE B.
- Front and rear suspension member cross body.
- Strut bar bracket cross body
- U bolts for leaf spring.
- Bolts and nuts on chassis and body.

86821323

MAINTENANCE SCHEDULE B
1989–90 Land Cruiser

Maintenance operations: A = Check and adjust if necessary; R = Replace, change or lubricate; I = Inspect and correct or replace if necessary

Maintenance services beyond 60,000 miles (96,000 km) should be performed at the same intervals shown in each maintenance schedule.

System	Service interval (Odometer reading or months, whichever comes first) / Maintenance items	Months
	Miles × 1,000: 7.5, 15, 22.5, 30, 37.5, 45, 52.5, 60	
	km × 1,000: 12, 24, 36, 48, 60, 72, 84, 96	
ENGINE	Valve clearance*	A: Every 24 months
	Drive belts[1]	I: 36 months, 72 months
	Engine oil and oil filter*	R: Every 12 months
	Engine coolant[2]	R: Every 36 months
	Exhaust pipes and mountings	I: Every 36 months
FUEL	Air filter*	R: Every 36 months
	Fuel lines and connections[3]	I: Every 72 months
	Fuel tank cap gasket	R: Every 36 months
IGNITION	Spark plug* **	R: Every 72 months
EVAP	Charcoal canister (California only)	R: Every 36 months
EXHAUST	Oxygen sensor* (Except California)	R: 80,000 miles (128,000 km) only
BRAKES	Brake linings and drums[4]	I: Every 24 months
	Brake pads discs (Front and rear)	I: Every 24 months
	Brake line pipes and hoses	I: Every 24 months
CHASSIS	Steering linkage	I: Every 24 months
	Ball joints and dust covers	I: Every 24 months
	Front wheel bearing and thrust bush grease	R: Every 24 months
	Steering knuckle and chassis grease[5]	R: Every 24 months
	Propeller shaft grease[5]	R: Every 24 months
	Automatic transmission	I: Every 24 months
	Transfer, differential and steering gear box oil[6]	I: Every 24 months
	Bolts and nuts on chassis and body[7]	I: Every 24 months

Conditions: Conditions other than those listed for SCHEDULE A.

*: For vehicle sold in California
**: For vehicle sold outside California
(1) After 60,000 miles (96,000 km) or 72 months, inspect every 7,500 miles (12,000 km) or 12 months.
(2) After 45,000 miles (72,000 km) or 36 months, replace every 30,000 miles (48,000 km) or 24 months.
(3) Includes inspection of fuel tank band vapor vent system.
(4) Also applicable to lining drum for parking brake.
(5) If the propeller shaft has been immersed in water, it should be re-greased within a day.
(6) Inspect the steering gear box for oil leakage only.
(7) The applicable parts are listed below.
- Front and rear suspension member cross body.
- Strut bar bracket cross body
- U bolts for leaf spring.
- Bolts for sheet installation.

86821324

MAINTENANCE SCHEDULE A
1991-92 Land Cruiser

Maintenance operations: A = Check and adjust if necessary; R = Replace, change or lubricate; I = Inspect and correct or replace if necessary

Conditions:
- Towing a trailer, using a camper or car top carrier.
- Repeat short trips less than 5 miles (8 km) and outside temperatures remain below freezing.
- Extensive idling and/or low speed driving for long distances, such as police, taxi or door-to-door delivery use.
- Operating on dusty, rough, muddy or salt spread roads.

Service interval (Odometer reading or months, whichever comes first). Maintenance services beyond 60,000 miles (96,000 km) should continue to be performed at the same intervals shown in each maintenance schedule.

System	Maintenance items	3.75 / 6	7.5 / 12	11.25 / 18	15 / 24	18.75 / 30	22.5 / 36	26.25 / 42	30 / 48	33.75 / 54	37.5 / 60	41.25 / 66	45 / 72	48.75 / 78	52.5 / 84	56.25 / 90	60 / 96	Months
ENGINE	Valve clearance*								A								A	A: Every 24 months
	Drive belts(1)												I				R	I: 36 months; 72 months
	Engine oil and oil filter*	R	R	R	R	R	R	R	R	R	R	R	R	R	R	R	R	R: Every 6 months
	Engine coolant(2)												R					R: Every 36 months
	Exhaust pipes and mountings																	I: Every 24 months
FUEL	Air filter*(3)												R					I: Every 6 months; R: Every 36 months
	Fuel lines and connections(4)																	I: Every 72 months
	Fuel tank cap gasket																R	R: Every 36 months
IGNITION	Spark plugs**								R								R	I: Every 72 months
EVAP	Charcoal canister	Calif only																
EXHAUST	Oxygen sensor*	Except California								R: 80,000 miles (128,000 km) only								
BRAKES	Brake linings and drums(5)		I		I		I		I		I		I		I		I	I: Every 6 months; R: Every 36 months
	Brake pads and discs												R					I: Every 36 months
	Brake line pipes and hoses								R								R	R: Every 72 months
CHASSIS	Steering linkage												R				R	R: Every 36 months
	Ball joints and dust covers																	I: Every 12 months
	Automatic transmission								R								R	R: Every 24 months
	Transfer and differential								R								R	R: Every 24 months
	Steering gear box oil(6)																	I: Every 24 months
	Front wheel bearing and thrust bush grease												R					R: Every 36 months
	Steering knuckle and chassis grease(7)		R		R		R		R		R		R		R		R	R: Every 12 months
	Propeller shaft grease(7)		R		R		R		R		R		R		R		R	R: Every 12 months
	Bolts and nuts on chassis and body(8)								I								I	I: Every 12 months

*: For vehicle sold in California
**: For vehicle sold outside California
(1) After 60,000 miles (96,000 km) or 72 months, inspect every 7,500 miles (12,000 km) or 12 months.
(2) After 45,000 miles (72,000 km) or 36 months, replace every 30,000 miles (48,000 km) or 24 months.
(3) Applicable when operating mainly on dusty roads. If not, follow schedule B.
(4) Includes inspection of fuel tank band and vapor vent system.
(5) Also applicable to lining drum for parking brake.
(6) Inspect the steering gear box for oil leakage only.
(7) If the propeller shaft has been immersed in water, it should be re-greased within a day. The applicable parts are listed below. For other usage conditions, refer to SCHEDULE B.
(8) Bolts for sheet installation

MAINTENANCE SCHEDULE B
1991-92 Land Cruiser

Conditions: Conditions other than those listed for SCHEDULE A.

Service interval (Odometer reading or months, whichever comes first). Maintenance services beyond 60,000 miles (96,000 km) should be performed at the same intervals shown in each maintenance schedule.

System	Maintenance items	7.5 / 12	15 / 24	22.5 / 36	30 / 48	37.5 / 60	45 / 72	52.5 / 84	60 / 96	Months
ENGINE	Valve clearance*				A				A	A: Every 24 months
	Drive belts(1)						I		R	I: 36 months; 72 months
	Engine oil and oil filter*	R	R	R	R	R	R	R	R	R: Every 12 months
	Engine coolant(2)						R			R: Every 36 months
	Exhaust pipes and mountings(3)				I					I: Every 36 months
FUEL	Air filter*						R			R: Every 36 months
	Fuel lines and connections(3)								I	I: Every 72 months
	Fuel tank cap gasket								R	R: Every 72 months
IGNITION	Spark plugs**	R: 80,000 miles (129,000 km) only								I: Every 72 months
EVAP	Charcoal canister	California only								
EXHAUST	Oxygen sensor*	Except California			R: 80,000 miles (129,000 km) only					
BRAKES	Brake linings and drums(4)						I			I: Every 24 months
	Brake pads discs (Front and rear)									I: Every 24 months
	Brake line pipes and hoses									I: Every 24 months
CHASSIS	Steering linkage									I: Every 24 months
	Ball joints and dust covers									I: Every 24 months
	Front wheel bearing and thrust bush grease						R			R: Every 24 months
	Steering knuckle and chassis grease(5)				R		R		R	R: Every 24 months
	Propeller shaft grease(5)		R		R		R		R	R: Every 24 months
	Automatic transmission									I: Every 24 months
	Transfer, differential and steering gear box oil(6)									I: Every 24 months
	Bolts and nuts on chassis and body(7)									I: Every 24 months

*: For vehicle sold in California
**: For vehicle sold outside California
(1) After 60,000 miles (96,000 km) or 72 months, inspect every 7,500 miles (12,000 km) or 12 months.
(2) After 45,000 miles (72,000 km) or 36 months, replace every 30,000 miles (48,000 km) or 24 months.
(3) Includes inspection of fuel tank band and vapor vent system.
(4) Also applicable to lining drum for parking brake.
(5) If the propeller shaft has been immersed in water, it should be re-greased within a day.
(6) Inspect the steering gear box for oil leakage only.

86621326

SCHEDULE B 1993-96 LAND CRUISER

Maintenance operations:
A = Check and adjust as necessary; I = Inspect and correct or replace as necessary; R = Replace, change or lubricate.

ENGINE COMPONENTS AND EMISSION CONTROL SYSTEMS

SERVICE INTERVAL: (Use odometer reading or months, whichever comes first.)	x 1000 miles	7.5	15	22.5	30	37.5	45	52.5	60	Months
	x 1000 km	12	24	36	48	60	72	84	96	
Engine oil and oil filter*		R	R	R	R	R	R	R	R	R: Every 12 months
Valve clearance									A	A: Every 72 months
Drive belts «See note 1.»		Inspect at 48000 km (30000 miles) or 36 months and at 96000 km (60000 miles) or 72 months.								
Engine coolant «See note 2.»		Initial change at first 72000 km (45000 miles) or 36 months.								
Exhaust pipes and mountings										I: Every 36 months
Air cleaner filter*					R				R	R: Every 36 months
Fuel lines and connections, fuel tank vapor vent system hoses and fuel tank band										I: Every 36 months
Fuel tank cap gasket									R	R: Every 72 months
Spark plugs**					R				R	R: Every 36 months

CHASSIS AND BODY

SERVICE INTERVAL: (Use odometer reading or months, whichever comes first.)	x 1000 miles	7.5	15	22.5	30	37.5	45	52.5	60	Months
	x 1000 km	12	24	36	48	60	72	84	96	
Brake linings and drums		I	I	I	I	I	I	I	I	I: Every 24 months
Brake pads and discs		I	I	I	I	I	I	I	I	I: Every 24 months
Brake lines and hoses		I	I	I	I	I	I	I	I	I: Every 24 months
Steering linkage		I	I	I	I	I	I	I	I	I: Every 24 months
Ball joints and dust covers		I	I	I	I	I	I	I	I	I: Every 24 months
Automatic transmission		I	I	I	I	I	I	I	I	I: Every 24 months
Transfer and differential		I	I	I	I	I	I	I	I	I: Every 24 months
Steering gear box		I	I	I	I	I	I	I	I	I: Every 24 months
Wheel bearing and drive shaft bushing grease					R				R	R: Every 36 months
Steering knuckle grease					R				R	R: Every 24 months
Propeller shaft grease					R				R	R: Every 24 months
Bolts and nuts on chassis and body		I	I	I	I	I	I	I	I	I: Every 24 months

Charcoal canister (California and New York only)
Oxygen sensor (except California and New York) — Replace at initial 128000 km (80000 miles).

Maintenance services indicated by * or ** are required under the terms of the Emission Control Systems Warranty. See Owner's Guide or Warranty Booklet for complete warranty information.
* : For vehicles sold in California and New York only
** : For vehicles sold outside California and New York

NOTE:
1. After 96000 km (60000 miles) or 72 months, inspect every 12000 km (7500 miles) or 12 months.
2. After 72000 km (45000 miles) or 36 months, change every 48000 km (30000 miles) or 24 months.

86821328

SCHEDULE A 1993-96 LAND CRUISER

Maintenance operations:
A = Check and adjust as necessary; I = Inspect and correct or replace as necessary; R = Replace, change or lubricate.

ENGINE COMPONENTS AND EMISSION CONTROL SYSTEMS

SERVICE INTERVAL: (Use odometer reading or months, whichever comes first.)	x 1000 miles	3.75	7.5	11.25	15	18.75	22.5	26.25	30	33.75	37.5	41.25	45	48.75	52.5	56.25	60	Months
	x 1000 km	6	12	18	24	30	36	42	48	54	60	66	72	78	84	90	96	
Engine oil and oil filter*		R	R	R	R	R	R	R	R	R	R	R	R	R	R	R	R	R: Every 6 months
Valve clearance																	A	A: Every 72 months
Drive belts «See note 1.»		Inspect at 48000 km (30000 miles) or 36 months and at 96000 km (60000 miles) or 72 months.																
Engine coolant «See note 2.»		Initial change at first 72000 km (45000 miles) or 36 months.																
Exhaust pipes and mountings																		I: Every 24 months
Air cleaner filter* «See note 3.»					I				R				I				R	I: Every 6 months / R: Every 36 months

ENGINE COMPONENTS (continued)

SERVICE INTERVAL: (Use odometer reading or months, whichever comes first.)	x 1000 miles	3.75	7.5	11.25	15	18.75	22.5	26.25	30	33.75	37.5	41.25	45	48.75	52.5	56.25	60	Months
	x 1000 km	6	12	18	24	30	36	42	48	54	60	66	72	78	84	90	96	
Fuel lines and connections, fuel tank vapor vent system hoses and fuel tank band																		I: Every 36 months
Fuel tank cap gasket																	R	R: Every 72 months
Spark plugs**									R								R	R: Every 36 months
Charcoal canister (California and New York only)																		I: Every 72 months
Oxygen sensor (except California and New York only)		Replace at initial 128000 km (80000 miles).																

CHASSIS AND BODY

	x 1000 miles	3.75	7.5	11.25	15	18.75	22.5	26.25	30	33.75	37.5	41.25	45	48.75	52.5	56.25	60	Months
	x 1000 km	6	12	18	24	30	36	42	48	54	60	66	72	78	84	90	96	
Brake linings and drums		I		I		I		I		I		I		I		I		I: Every 12 months
Brake pads and discs		I		I		I		I		I		I		I		I		I: Every 12 months
Brake lines and hoses					I						I						I	I: Every 24 months
Steering linkage					I						I						I	I: Every 24 months
Ball joints and dust covers					I						I						I	I: Every 12 months
Automatic transmission									R								R	I: Every 24 months
Transfer and differential									R								R	R: Every 24 months
Steering gear box					I						I						I	I: Every 24 months

CHASSIS AND BODY (continued)

SERVICE INTERVAL: (Use odometer reading or months, whichever comes first.)	x 1000 miles	3.75	7.5	11.25	15	18.75	22.5	26.25	30	33.75	37.5	41.25	45	48.75	52.5	56.25	60	Months
	x 1000 km	6	12	18	24	30	36	42	48	54	60	66	72	78	84	90	96	
Wheel bearing and drive shaft bushing grease									R								R	R: Every 36 months
Steering knuckle grease		R		R		R		R		R		R		R		R		R: Every 12 months
Propeller shaft grease		R		R		R		R		R		R		R		R		R: Every 12 months
Bolts and nuts on chassis and body «See note 4.»					I						I						I	I: Every 12 months

Maintenance services indicated by * or ** are required under the terms of the Emission Control Systems Warranty. See Owner's Guide or Warranty Booklet for complete warranty information.
* : For vehicles sold in California and New York only
** : For vehicles sold outside California and New York

NOTE:
1. After 96000 km (60000 miles) or 72 months, inspect every 12000 km (7500 miles) or 12 months.
2. After 72000 km (45000 miles) or 36 months, change every 48000 km (30000 miles) or 24 months.
3. Applicable when you mainly operate your vehicle on dusty roads. If not, apply Schedule B.
4. Applicable when you mainly operate your vehicle on rough and/or muddy roads. If not, apply Schedule B.

86821327

SCHEDULE A 1996 TRUCK

Maintenance operations:
A = Check and adjust as necessary;   I = Inspect and correct or replace as necessary;   R = Replace, change or lubricate.

ENGINE COMPONENTS AND EMISSION CONTROL SYSTEMS

SERVICE INTERVAL: (Use odometer reading or months, whichever comes first.)	6	12	18	24	30	36	42	48	54	60	66	72	78	84	90	96	Months
x 1000 km	6	12	18	24	30	36	42	48	54	60	66	72	78	84	90	96	
x 1000 miles	3.75	7.5	11.25	15	18.75	22.5	26.25	30	33.75	37.5	41.25	45	48.75	52.5	56.25	60	
Engine oil and oil filter*	R	R	R	R	R	R	R	R	R	R	R	R	R	R	R	R	R: Every 6 months
Timing belt (5VZ-FE engine)	If you operate your vehicle under conditions of extensive idling and/or low speed driving for long distance such as police, taxi or door-to-door delivery use, replace every 96000 km (60000 miles).															A	A: Every 72 months
Valve clearance	Inspect at 96000 km (60000 miles) or 72 months.															A	
Drive belts «See note 1.»	Initial change at first 72000 km (45000 miles) or 36 months.																I: Every 24 months
Engine coolant «See note 2.»																	
Exhaust pipes and mountings							I					I					I: Every 24 months
Fuel tank cap gasket												R				R	R: Every 36 months
Spark plugs**										R							R: Every 36 months
Charcoal canister «See note 4.»																	
Air cleaner filter* «See note 3.»																	

Maintenance operations:

CHASSIS AND BODY

SERVICE INTERVAL: (Use odometer reading or months, whichever comes first.)	6	12	18	24	30	36	42	48	54	60	66	72	78	84	90	96	Months
x 1000 km	6	12	18	24	30	36	42	48	54	60	66	72	78	84	90	96	
x 1000 miles	3.75	7.5	11.25	15	18.75	22.5	26.25	30	33.75	37.5	41.25	45	48.75	52.5	56.25	60	
Fuel lines and connections, fuel tank vapor vent system hoses and fuel tank band												I					I: Every 36 months
Fuel tank cap gasket												R				R	R: Every 72 months
Spark plugs**										R							R: Every 36 months
Brake linings and drums		I		I		I		I		I		I		I		I	I: Every 12 months
Brake pads and discs		I		I		I		I		I		I		I		I	I: Every 12 months
Brake lines and hoses				I				I				I				I	I: Every 24 months
Steering linkage		I		I		I		I		I		I		I		I	I: Every 12 months
SRS airbag «See note 5.»	Initial inspect at first 120 months.																
Ball joints and dust covers		I		I		I		I		I		I		I		I	I: Every 12 months

Maintenance operations:

SERVICE INTERVAL: (Use odometer reading or months, whichever comes first.)	6	12	18	24	30	36	42	48	54	60	66	72	78	84	90	96	Months
x 1000 km	6	12	18	24	30	36	42	48	54	60	66	72	78	84	90	96	
x 1000 miles	3.75	7.5	11.25	15	18.75	22.5	26.25	30	33.75	37.5	41.25	45	48.75	52.5	56.25	60	
Drive shaft boots (four-wheel drive models)		I		I		I		I		I		I		I		I	I: Every 12 months
Automatic transmission, manual transmission and differential				R				R				R				R	R: Every 24 months
Transfer (four-wheel drive models)				R				R				R				R	R: Every 24 months
Steering gear box				I				I				I				I	I: Every 24 months
Wheel bearing grease (two-wheel drive models)																R	R: Every 48 months
Wheel bearing and drive shaft bushing grease (four-wheel drive models)																	
Propeller shaft grease (four-wheel drive models)		R		R		R		R		R		R		R		R	R: Every 12 months
Bolts and nuts on chassis and body «See note 6.»		I		I		I		I		I		I		I		I	I: Every 12 months

Maintenance services indicated by * or ● are required under the terms of the Emission Control Systems Warranty. See Owner's Guide or Warranty Booklet for complete warranty information.
* : For vehicles sold in California, Massachusetts and New York only
● : For vehicles sold outside California, Massachusetts and New York

NOTE:
1. After 96000 km (60000 miles) or 72 months, inspect every 12000 km (7500 miles) or 12 months.
2. After 72000 km (45000 miles) or 36 months, change every 48000 km (30000 miles) or 24 months.
3. Applicable when you mainly operate your vehicle on dusty roads. If not, apply Schedule B.
4. Non-maintenance item except for California, Massachusetts and New York.
5. After 120 months, inspect every 24 months.
6. Applicable when you mainly operate your vehicle on rough and/or muddy roads.

SCHEDULE B 1996 TRUCK

Maintenance operations:
A = Check and adjust as necessary;   I = Inspect and correct or replace as necessary;   R = Replace, change or replace as necessary.

ENGINE COMPONENTS AND EMISSION CONTROL SYSTEMS

SERVICE INTERVAL: (Use odometer reading or months, whichever comes first.)	12	24	36	48	60	72	84	96	Months
x 1000 km	12	24	36	48	60	72	84	96	
x 1000 miles	7.5	15	22.5	30	37.5	45	52.5	60	
Engine oil and oil filter*	R	R	R	R	R	R	R	R	R: Every 12 months
Valve clearance	Inspect at 96000 km (60000 miles) or 72 months.							A	A: Every 72 months
Drive belts «See note 1.»	Initial change at first 72000 km (45000 miles) or 72 months.							I	I: Every 36 months
Engine coolant «See note 2.»	Initial change at first 72000 km (45000 miles) or 36 months.							R	R: Every 36 months
Exhaust pipes and mountings				I				I	I: Every 36 months
Air cleaner filter*				R				R	R: Every 36 months
Fuel lines and connections, fuel tank vapor vent system hoses and fuel tank band						R			R: Every 72 months
Fuel tank cap gasket						R			R: Every 36 months
Spark plugs**								R	R: Every 36 months
Charcoal canister «See note 3.»								I	I: Every 72 months

Maintenance operations:

CHASSIS AND BODY

SERVICE INTERVAL: (Use odometer reading or months, whichever comes first.)	12	24	36	48	60	72	84	96	Months
x 1000 km	12	24	36	48	60	72	84	96	
x 1000 miles	7.5	15	22.5	30	37.5	45	52.5	60	
Brake linings and drums		I		I		I		I	I: Every 24 months
Brake pads and discs		I		I		I		I	I: Every 24 months
Brake lines and hoses		I		I		I		I	I: Every 24 months
Steering linkage		I		I		I		I	I: Every 24 months
SRS airbag «See note 4.»	Initial inspect at first 120 months.								
Drive shaft boots (four-wheel drive models)		I		I		I		I	I: Every 24 months
Automatic transmission, manual transmission and differential		I		I		I		I	I: Every 24 months
Transfer (four-wheel drive models)		I		I		I		I	I: Every 24 months
Steering gear box		I		I		I		I	I: Every 24 months
Wheel bearing grease (two-wheel drive models)				R				R	R: Every 48 months

Maintenance operations:

SERVICE INTERVAL: (Use odometer reading or months, whichever comes first.)	12	24	36	48	60	72	84	96	Months
x 1000 km	12	24	36	48	60	72	84	96	
x 1000 miles	7.5	15	22.5	30	37.5	45	52.5	60	
Wheel bearing and drive shaft bushing grease (four-wheel drive models)				R				R	R: Every 36 months
Propeller shaft grease (four-wheel drive models)		R		R		R		R	R: Every 24 months
Bolts and nuts on chassis and body		R		R		R		R	R: Every 24 months

Maintenance services indicated by * or ● are required under the terms of the Emission Control Systems Warranty. See Owner's Guide or Warranty Booklet for complete warranty information.
* : For vehicles sold in California, Massachusetts and New York only
● : For vehicles sold outside California, Massachusetts and New York

NOTE:
1. After 96000 km (60000 miles) or 72 months, inspect every 12000 km (7500 miles) or 12 months.
2. After 72000 km (45000 miles) or 36 months, change every 48000 km (30000 miles) or 24 months.
3. Non-maintenance item except for California and New York.
4. After 120 months, inspect every 24 months.

86621329

86621330

CAPACITIES (1995–1996)

Year	Model	Engine ID/VIN	Engine Displacement Liters (cc)	Engine Oil with Filter (qts.)	Transmission 4-Spd	5-Spd	Auto.	Transfer Case (pts.)	Drive Axle Front (pts.)	Rear (pts.)	Fuel Tank (gal.)	Cooling System (qts.)
1995	Pick-up	22R-E	2.4 (2366)	4.5	-							
	Pick-up	3VZ-E	3.0 (2959)	4.8	-							
	4Runner	22R-E	2.4 (2366)	4.5	-							
	4Runner	2VZ-E	3.0 (2959)	4.8	-			·				
	Tacoma	2RZ-FE	2.4 (2438)	5.8	-							
	Tacoma	3RZ-FE	2.7 (2693)	5.8	-			2.2		2.9	15.1	
	Tacoma	5VZ-FE	3.4 (3378)		-			2.2			18.0	
	T100	3RZ-FE	2.7 (2693)	5.6	-	2.7		2.2			18.0	
	T100	5VZ-FE	3.4 (3378)		-		2.4	-	·	3.8	19.8	9.2
	Land Cruiser	1FZ-FE	4.5 (4477)	7.8	-		12.6	3.9		6.8	24.0	15.0
1996	Tacoma	2RZ-FE	2.4 (2438)	5.8	-			2.2		2.9	15.1	
	Tacoma	3RZ-FE	2.7 (2693)	5.8	-			2.2			18.0	
	Tacoma	5VZ-FE	3.4 (3378)		-			2.2			18.0	
	T100	3RZ-FE	2.7 (2693)	5.6	-	2.7		-	·	3.8	19.8	9.2
	T100	5VZ-FE	3.4 (3378)		-		2.4	-			24.0	
	Land Cruiser	1FZ-FE	4.5 (4477)	7.8	-	-	12.6			6.8	25.1	15.0

1 Except 4WD automatic transmission: 8.9
2 4WD automatic transmission: 9.6
3 Automatic transmission: 11.1
4 2WD 7.5" differential: 2.8 / 2WD 8.0" differential: 3.8 / 4WD: 4.6
5 Standard: 17.2; Optional: 19.3
6 With oil cooler: 4.4
7 2WD with A340E: 3.4 / 4WD with A340H: 9.6 / 4WD with A340F: 4.2
8 Counter gear type: 2.4 / Planetary gear type: 2.4 / A340H: 1.8
9 Standard: 3.4; ADD: 4.0
10 With rear heater: 9.2
11 Short bed: 13.7 / Long bed: 17.2 / Long bed 4WD: 19.3
12 2WD manual transmission: 11.0 / 2WD automatic transmission: 10.8 / 4WD manual transmission: 11.1 / 4WD automatic transmission: 10.9
13 4WD with W56: 5.0 / 4WD with W56: 6.4 / 2WD with G58: 8.2 / 4WD with R150F: 6.4
14 With differential lock: 5.6 / Without differential lock: 5.8
15 With standard tires: 17.2 / With optional 31x10.5 tires: 18.8
16 2WD: 5.7 / 4WD: 5.5
17 G58: 8.2 / R150F: 6.4
18 Except 3VZ-E AT (VF1A type): 2.4 / 3VZ-E AT (A340H): 1.6
19 2WD with manual transmission: 10.6 / 2WD with automatic transmission: 10.5 / 4WD with manual transmission: 10.6 / 4WD with automatic transmission: 10.8
20 2WD: 5.4 / 2WD: 3.8; 4WD: 4.6 / 4WD: 4.6
21 2WD: 5.4 / 4WD: 4.6
22 2WD: 4.4 / 4WD: 4.3
23 W59: 2WD: 5.4 / 4WD: 5.2 / R150, R150F: 2WD: 5.4 / 4WD: 4.6
24 A340H: 3.4 / A43D: 5.0
25 Without ADD: 2.32 / With ADD: 2.44
26 2WD with manual transmission: 8.5 / 2WD with automatic transmission: 8.2 / 4WD with manual transmission: 8.8 / 4WD with automatic transmission: 8.7
27 Extra long: 5.4 / All others: 5.4
28 2WD: 5.5 / 4WD: 5.0
29 With ABS: 2.8 / Without ABS: 3.6

86821332

CAPACITIES (1989–1994)

Year	Model	Engine ID/VIN	Engine Displacement Liters (cc)	Engine Oil with Filter (qts.)	Transmission 4-Spd	5-Spd	Auto.	Transfer Case (pts.)	Drive Axle Front (pts.)	Rear (pts.)	Fuel Tank (gal.)	Cooling System (qts.)
1989	Pick-up	22R	2.4 (2366)	4.5								
	Pick-up	22R-E	2.4 (2366)	4.5								
	Pick-up	3VZ-E	3.0 (2959)	4.8								
	4Runner	22R	2.4 (2366)	4.5								
	4Runner	22R-E	2.4 (2366)	4.5								
	4Runner	3VZ-E	3.0 (2959)	4.8								
	Land Cruiser	3F-E	4.0 (3956)	8.2						6.8	25.1	
1990	Pick-up	22R	2.4 (2366)	4.5								
	Pick-up	22R-E	2.4 (2366)	4.5					6.0			
	Pick-up	3VZ-E	3.0 (2959)	4.8				4.4		6.8		
	4Runner	22R-E	2.4 (2366)	4.5								
	4Runner	3VZ-E	3.0 (2959)	4.8			12.6					
	Land Cruiser	3F-E	4.0 (3956)	8.2		-			6	6.8	25.1	
1991	Pick-up	22R-E	2.4 (2366)	4.5								
	Pick-up	3VZ-E	3.0 (2959)	4.8								
	4Runner	22R-E	2.4 (2366)	4.5								
	4Runner	3VZ-E	3.0 (2959)	4.8			12.6					
	Land Cruiser	3F-E	4.0 (3956)	8.2		-		4.4	6.0	6.8	25.1	
1992	Pick-up	22R-E	2.4 (2366)	4.5								
	Pick-up	3VZ-E	3.0 (2959)	4.8								
	4Runner	22R-E	2.4 (2366)	4.5								
	4Runner	3VZ-E	3.0 (2959)	4.8								
	T100	3VZ-E	3.0 (2959)	4.8		6.4	-	2.4		6.8	25.1	
	Land Cruiser	1FZ-FE	4.5 (4477)	7.8			12.6			6.8	25.1	15.0
1993	Pick-up	22R-E	2.4 (2366)	4.5								
	Pick-up	3VZ-E	3.0 (2959)	4.8								
	4Runner	22R-E	2.4 (2366)	4.5								
	4Runner	3VZ-E	3.0 (2959)	4.8								
	T100	3VZ-E	3.0 (2959)	4.8		6.4	-	2.4		6.8	25.1	
	Land Cruiser	1FZ-FE	4.5 (4477)	7.8			12.6			6.8	25.1	15.0
1994	Pick-up	22R-E	2.4 (2366)	4.5								
	Pick-up	3VZ-E	3.0 (2959)	4.8								
	4Runner	22R-E	2.4 (2366)	4.5								
	T100	3RZ-FE	2.7 (2693)	5.6		2.7	-			3.8	19.8	9.2
	T100	3VZ-E	3.0 (2959)	4.8		6.4	-	2.4		6.8		
	Land Cruiser	1FZ-FE	4.5 (4477)	7.8		-	12.6			6.8	25.1	15.0

86821331

GASOLINE ENGINE TUNE-UP

Year	Engine ID/VIN	Engine Displacement Liters (cc)	Spark Plugs Gap (in.)	Ignition Timing (deg.) MT	Ignition Timing (deg.) AT	Fuel Pump (psi)	Idle Speed (rpm) MT	Idle Speed (rpm) AT	Valve Clearance In.	Valve Clearance Ex.
1989	22R	2.4 (2366)	0.031	-	-	2.1-4.3	700	750	0.008	0.012
	22R-E	2.4 (2366)	0.031	5B	5B	35-38	750	750	0.008	0.012
	3VZ-E	3.0 (2959)	0.031	10B	10B	38-44	800	800	0.012	0.013
	3F-E	4.0 (3956)	0.031	7B	7B	37-46	650	650	0.008	0.014
1990	22R	2.4 (2366)	0.031	-	-	2.1-4.3	750	850	0.008	0.012
	22R-E	2.4 (2366)	0.031	5B	5B	38-44	750	850	0.008	0.012
	3VZ-E	3.0 (2959)	0.031	10B	10B	38-44	800	800	0.007-0.011	0.009-0.013
	3F-E	4.0 (3956)	0.031	-	7B	37-46	650	650	0.008	0.014
1991	22R-E	2.4 (2366)	0.031	5B	5B	38-44	750	850	0.008	0.012
	3VZ-E	3.0 (2959)	0.031	10B	10B	38-44	800	800	②	③
	3F-E	4.0 (3956)	0.031	-	7B	37-46	650	650	0.008	0.014
1992	22R-E	2.4 (2366)	0.031	5B	5B	38-44	750	850	0.008	0.012
	3VZ-E	3.0 (2959)	0.031	10B	10B	38-44	800	800	②	③
	1FZ-FE	4.5 (4477)	0.031	-	3B	38-44	-	600-700	0.006-0.010	0.010-0.014
1993	22R-E	2.4 (2366)	0.031	5B	5B	38-44	750	850	0.008	0.012
	3RZ-FE	2.7 (2693)	0.031	5B	-	38-44	750	-	0.008	0.012
	3VZ-E	3.0 (2959)	0.031	10B	10B	38-44	800	800	②	③
	1FZ-FE	4.5 (4477)	0.031	-	3B	38-44	-	600-700	0.006-0.010	0.010-0.014
1994	22R-E	2.4 (2366)	0.031	5B	5B	38-44	750	850	0.008	0.012
1995	2RZ-FE	2.4 (2438)	0.031	⑥	⑥	38-44	650-750	650-750	0.006-0.010	0.010-0.014

86821210

GASOLINE ENGINE TUNE-UP

Year	Engine ID/VIN	Engine Displacement Liters (cc)	Spark Plugs Gap (in.)	Ignition Timing (deg.) MT	Ignition Timing (deg.) AT	Fuel Pump (psi)	Idle Speed (rpm) MT	Idle Speed (rpm) AT	Valve Clearance In.	Valve Clearance Ex.
1995	3RZ-FE	2.7 (2693)	0.031	⑥	⑥	38-44	750	-	0.008	0.012
	3VZ-E	3.0 (2959)	0.031	10B	10B	38-44	800	800	②	③
	5VZ-FE	3.4 (3378)	0.043	④	④	38-44	650-750	650-750	0.006-0.009	0.011-0.014
	1FZ-FE	4.5 (4477)	0.031	-	3B	38-44	-	600-700	0.006-0.010	0.010-0.014
1996	22R-E	2.4 (2366)	0.031	5B	5B	38-44	750	850	0.008	0.012
	2RZ-FE	2.4 (2438)	0.031	⑥	⑥	38-44	650-750	650-750	0.006-0.010	0.010-0.014
	3RZ-FE	2.7 (2693)	0.031	10B	⑥	38-44	750	-	0.008	0.012
	3VZ-E	3.0 (2959)	0.031	10B	10B	38-44	800	800	⑦	⑨
	5VZ-FE	3.4 (3378)	0.043	⑥	⑥	38-44	650-750	650-750	0.006-0.009	0.011-0.014
	1FZ-FE	4.5 (4477)	0.031	-	3B	38-44	-	600-700	0.006-0.010	0.010-0.014

86821PPP

NOTE: The Vehicle Emission Control Information label often reflects specification changes made during production. The label figures must be used if they differ from those in this chart.

B - Before top dead center
T - T terminal shorted

1 Intake: 0.007-0.011 (cold)
2 Exhaust: 0.009-0.013 (cold)
3 10B at idle, with terminal TE1 and E1 connected of DLC1
4 5B at idle, with terminal TE1 and E1 connected of DLC1

2

ENGINE
ELECTRICAL

ELECTRONIC DISTRIBUTOR IGNITION SYSTEM

➡**For information on understanding electricity and troubleshooting electrical circuits, please refer to Section 6 of this manual.**

General Information

In order to extract the best performance and economy from your engine it is essential that it be properly tuned at regular intervals. Although computerized engine controls and more durable components have reduced ignition maintenance, a regular tune-up will keep your Toyota's engine running smoothly and will prevent the annoying minor breakdowns and poor performance associated with an untuned engine.

Electronic ignition systems offer many advantages over the conventional breaker points ignition system. By eliminating the points, maintenance requirements are greatly reduced. An electronic ignition system is capable of producing much higher voltage which in turn aids in starting, reduces spark plug fouling and provides better emission control.

The system Toyota uses consists of a distributor with a signal generator, an ignition coil and an electronic igniter. The signal generator is used to activate the electronic components of the igniter. It is located in the distributor and consists of three main components; the signal rotor, the pick-up coil and the permanent magnet. The signal rotor (not to be confused with the distributor rotor) revolves with the distributor shaft, while the pick-up coil and the permanent magnet are stationary. As the signal rotor spins, the teeth on it pass a projection leading from the pick-up coil. When this happens, voltage is allowed to flow through the system, firing the spark plugs. There is no physical contact and no electrical arcing, hence no need to replace burnt or worn parts.

Service consists of inspection of the distributor cap, rotor and the ignition wires, replacing them as necessary. In addition, the air gap between the signal rotor and the projection on the pick-up coil should be checked periodically. The resistances of the coil and pick-up circuits should be measured periodically.

Checks and Adjustments

AIR GAP

➡**The air gap is NOT adjustable. If the gap is not within specifications, the distributor must be replaced.**

1. Remove the distributor cap. Inspect the cap for cracks, carbon tracks or a worn center contact. Replace it if necessary, transferring the wires one at a time from the old cap to the new one.
2. Pull the ignition rotor (not the signal rotor) straight up and remove it. Replace it if the contact is worn, burned or pitted. Do not file the contacts.
3. Turn the engine over (you may use a socket wrench on the front pulley bolt to do this) until the projection on the pickup coil is directly opposite the signal rotor tooth.
4. Use only a non-ferrous (paper, brass, or plastic) feeler gauge set. This means that they are non-magnetic; if you have any doubt, check the feeler with a magnet. If the magnet sticks, don't use that feeler gauge.
5. Select a non-ferrous feeler blade of 0.02 in. (0.30mm) thickness, and insert it into the pick-up air gap. The gauge should just touch either side of the gap. The permissible range is 0.008–0.016 in. (0.20-0.40mm) except for 1989–90 Land Cruisers. These vehicles require only that the gap be greater than 0.001 in. (0.2mm).

Diagnosis and Testing

PRECAUTIONS

- Do not allow the ignition switch to be **ON** for more than ten minutes if the engine will not start.
- When a tachometer is connected to the system, always connect the tachometer positive lead to the ignition coil negative terminal. Some tachometers are not compatible with this system; it is recommended that you consult with the manufacturer.
- NEVER allow the ignition coil terminals to touch ground as it could result in damage to the igniter and/or the ignition coil itself.

- Do not disconnect the battery when the engine is running.
- Make sure that the igniter is always properly grounded to the body

SPARK TEST

The spark test is used to see if the ignition system is delivering electricity through the coil wire to the distributor. Use this test as a preliminary test if the engine cranks but won't start. Its a simple test, but can give a nasty shock if not performed correctly.

1. Disconnect the end of the coil wire at the distributor.
2. Use a well-insulated tool (such as ignition wire pliers) to hold the exposed end of the wire about ½ inch (13mm) from the metal of the engine block.

❋❋ CAUTION

Do not attempt to hold the wire with your bare hand. Use as much insulation as possible between the wire and the tool holding it. Do not stand on wet concrete while testing. Do not lean on the bodywork of the car while testing. The electrical charge will pass through the easiest path to ground, make certain its not you. Make sure the metal ground point nearest the cable end is safe; don't choose components that might contain fluids or electronic components.

3. Have an assistant crank the engine by turning the ignition switch to the **START** position, but only for 1 or 2 seconds. Keep clear of moving parts under the hood, and keep clothing and hair well out of the way. If the cable end is the correct distance from solid metal, a distinct, blue-white spark should jump to the metal as the engine cranks.
4. The engine may be cranked in the 1–2 seconds bursts. Longer cranking will cause the fuel injectors to deliver fuel into the cylinders, flooding the engine or at least fouling the spark plugs.
5. If no spark is present, turn the ignition switch **OFF** and proceed to check the resistance of the of the coil wire, the voltage supply to the coil (12 volts with the ignition **ON**), the coil resistance, the resistance of the pickup coil and the air gap within the distributor.

IGNITION COIL

1. With the ignition switch **OFF**, carefully remove the high-tension cable (coil-to-distributor wire) from the coil.
2. Carefully unplug the smaller electrical connector from the coil. The 1989–90 Land Cruisers and all 4 cylinder trucks and 4Runners do not have this small connector; battery wires run to external terminals on the coil. These terminals do not need to be undone.
3. Use an ohmmeter to test primary side resistance. Connect the probes to the 2 terminal pins of the small electrical connector or the two external terminals on each side of the coil.
4. Perform the check before running the engine. All the following values are on cold engines.
Resistance should be:
- 1989–91—22R engine: 0.4–0.5 ohms
- 1989–92—22R-E engine: 0.52–0.64 ohms
- 1993–95—22R-E engine: 0.36–0.55 ohms
- 1994–96—2RZ-FE and 3RZ-FE engines: 0.36–0.55 ohms
- 1989–91—3VZ-E engine: 0.4–0.5 ohms
- 1992–95—3VZ-E engine: 0.3–0.6 ohms
- 1989–90—3F-E engine: 0.52–0.64 ohms
- 1991—3F-E engine: 0.41–0.50 ohms
- 1992–96—1FZ-FE engine: 0.3–0.6 ohms
5. Check the secondary circuit resistance with the ohmmeter. Make certain the meter is set to a much higher scale than the previous test. Connect one probe of the meter to the positive coil terminal and the other to the high-tension terminal.
6. Correct resistances for a cold coil are:
- 1989–91—22R engines: 8.5–11.5k ohms
- 1989–92—22R-E engines: 11.4–15.6k ohms
- 1993–95—22R-E engine: 9.0–15.4k ohms
- 1989–91—3VZ-E engine: 10.2–13.8k ohms

- 1992–95—3VZ-E engine: 9–15k ohms
- 1994–96—2RZ-FE and 3RZ-FE engines: 9.0–15.4k ohms
- 1989–90—3F-E engine: 11.5–15.5k ohms
- 1991—3F-E engine: 10.2–13.8k ohms
- 1992–96—1FZ-FE engine: 9–15k ohms

7. If either resistance is not within specifications, the coil must be replaced.

8. For the 1989 22R engine, measure resistance from the negative coil terminal to the body of the igniter. Resistance should be infinite (no continuity). If any continuity is found, replace the coil.

PICK-UP COIL(S)

1989–91 Models

1. Detach the distributor connector.
2. Remove the distributor cap without removing the spark plug wires.
3. Remove the rotor.
4. With the ignition **OFF**, carefully detach the connector in the harness coming from the side of the distributor.
5. For 2-pin connectors, measure the resistance between the two terminals. Resistance for all 1989–90 vehicles is 140–180 ohms. For 1991 vehicles, resistance should be 205–255 ohms.
6. The 3-pin connectors are used on the Land Cruisers. Hold the connector so that the empty section of the 4-pin housing is in the lower left. The terminal in the lower right is called G-. Test each of the other terminals to G-, looking for resistance of 140–180 ohms on 1989–91 vehicles.
7. Four-pin connectors require you to hold the connector so that the square part of the housing is at the bottom. Viewing the connector in this position, the terminal on the lower right is called G-. Use the ohmmeter to check each other terminal to G-. Correct resistance for 1989–90 vehicles is 140–180 ohms; for 1991, it's 205–255 ohms.
8. If any resistance is out of specification, the distributor housing assembly must be replaced.

1992–95 Models

22R-E, 2RZ-FE AND 3RZ-FE

1. Detach the distributor connector.
2. Remove the distributor cap without removing the spark plug wires.
3. Remove the rotor.
4. With the ignition **OFF**, carefully detach the connector in the harness coming from the side of the distributor.
5. Using an ohmmeter, check the resistance of the signal generator (pick-up coil). With the engine cold, the reading should be 185–275 ohms; hot reading is 240–325 ohms.
6. If the resistance is not within specifications, replace the distributor housing assembly.

3VZ-E AND 1FZ-FE

▶ See Figures 1, 2 and 3

1. Detach the distributor connector.
2. Remove the distributor cap without removing the spark plug wires.
3. Remove the rotor.

Pickup coil resistance	Cold (−10 ~ 50°C)	Hot (50 ~ 100°C)
G1 − G⊝	125 − 200 Ω	160 − 235 Ω
G2 − G⊝	125 − 200 Ω	160 − 235 Ω
NE − G⊝	155 − 250 Ω	190 − 290 Ω

86822GPQ

Fig. 2 Pick-up coil resistance on the 3VZ-FE engine

Distributor	Air gap		0.2 − 0.4 mm (0.008 − 0.016 in.)
	Pickup coil resistance	at cold (G1 − G⊝)	185 − 275 Ω
		(G2 − G⊝)	185 − 275 Ω
		(NE − G⊝)	185 − 275 Ω
		at hot (G1 − G⊝)	240 − 325 Ω
		(G2 − G⊝)	240 − 325 Ω
		(NE − G⊝)	240 − 325 Ω

86822GPR

Fig. 3 Pick-up coil resistance on the 1FZ-FE engine

4. Using an ohmmeter, check the resistance of the pick-up coil. Refer to the chart for specifications.
5. If the resistance is not within specifications, replace the distributor housing assembly.

Ignition Coil

REMOVAL & INSTALLATION

External Mounted

▶ See Figures 4, 5, 6, 7 and 8

1. Disconnect the negative battery cable.
2. Remove the cover on the coil, if equipped.
3. Remove the coil wire lead.
4. Tag and disconnect all electrical leads to the coil. If applicable, remove the igniter.
5. Remove the two mounting bolts and lift off the ignition coil.

To install:

6. Install the coil in position and tighten the mounting bolts.
7. Connect all wires and install the igniter.
8. Connect the negative battery cable.

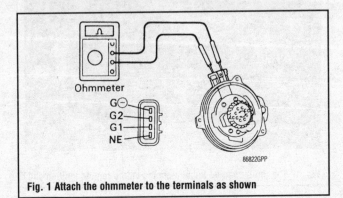

Ohmmeter

G ⊝
G2
G1
NE

86822GPP

Fig. 1 Attach the ohmmeter to the terminals as shown

86822004

Fig. 4 Remove the cover on the coil

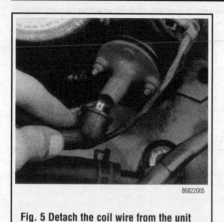

Fig. 5 Detach the coil wire from the unit

Fig. 6 Unplug the connector and remove the igniter

Fig. 7 Loosen and remove the mounting bolts securing the coil

Fig. 8 Once all leads and bolts are removed, lift the coil from the fenderwell

Internal Mounted

▶ See Figure 9

1. Disconnect the negative battery cable.
2. Remove the distributor cap and rotor.
3. Remove the ignition coil dust cover and dust proof packing.
4. Unfasten the two nuts and disconnect the three ignition coil terminals.
5. Remove the ignition coil retaining screws and the coil from the distributor.

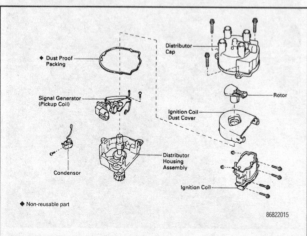

Fig. 9 Exploded view of the internal mounted ignition coil distributor

To install:

6. Before installing, remove any old packing material from the ignition coil. Apply sealer 08826-00080 or equivalent to the coil mounting surface.
7. Install the ignition coil to the distributor with the four screws.

➡When connecting the wires for the ignition coil, be sure the wires do not contact with the signal rotor or distributor housing.

8. Connect the ignition coil wires and install the two nuts.
9. Install the dust proof packing and dust cover.
10. Install the rotor and distributor cap.
11. Connect the negative battery cable.

Igniter

REMOVAL & INSTALLATION

▶ See Figures 10, 11, 12 and 13

Igniters are mounted in one of two places on your Toyota truck. On the remotely mounted coils, the igniter is secured to the coil. On the integral coil vehicles (where the coil is in the distributor) the igniter is secured to the fenderwell.

1. Disconnect the negative battery cable.
2. Separate the wiring harness connections.
3. Unbolt the igniter.
4. If mounted to the coil, loosen the nut holding the wire lead.
5. Tag and disconnect the wire lead.
6. Lift the igniter off its mount.

Fig. 10 The igniter can be found mounted on the remote ignition coil

Fig. 11 Unplug the connector and remove the retaining screws

Fig. 12 If mounted to the coil, detach the wire

Fig. 13 Check that all the connections are separated, then lift the unit from the vehicle

To install:

7. Mount the igniter to the bracket.
8. Attach the wire lead to the coil if used.
9. Connect the harness.
10. Connect the negative battery cable.

Distributor

REMOVAL

▶ **See Figures 14 thru 19**

1. Unfasten the retaining clips or remove the retaining screws and lift the distributor cap straight off. It will be easier to install the distributor if the wiring is left connected to the cap. If the wires must be removed from the cap, mark their positions to aid in installation.

➡Set cylinder No. 1 to TDC of the compression stroke. For all engines except 22R-E, set the timing mark at 0. On 22R-E, set the timing mark at 5 degrees.

2. Remove the dust cover and mark the position of the rotor relative to the distributor body; then mark the position of the body relative to the block.
3. Disconnect and tag any wires and the vacuum lines.
4. Remove the pinch-bolt and lift the distributor straight out, away from the engine. The rotor and body are marked so that they can be returned to the position from which they were removed. Do not turn or disturb the engine (unless absolutely necessary, such as for engine rebuilding), after the distributor has been removed. The relationship between the position of the distributor and the position of the engine is critical if the spark is to be sent to the proper place at the proper time.
5. Remove the distributor O-ring, discard and install a new one. Apply a light coat of engine oil on the O-ring.

Fig. 14 Remove the cap with the wires attached

Fig. 15 Mark the position of the rotor . . .

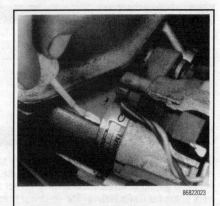

Fig. 16 . . . and the distributor base

Fig. 17 Detach the electrical connections

Fig. 18 After the pinch-bolt is removed, pull the distributor out

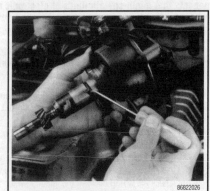

Fig. 19 Discard the O-ring. A new one should be used during installation

INSTALLATION

Timing Not Disturbed

1. Align the matchmarks on the distributor with the rotor; position the distributor to align with the mark made on the block. Insert the distributor in the block, taking care not to damage the drive gear at the bottom or to knock all the marks out of alignment.
2. Tighten the pinch-bolt to 14 ft. lbs. (19 Nm).
3. Install the distributor cap and wires. Check the ignition timing.
4. Check the idle speed.

Timing Disturbed

If the engine has been cranked, dismantled, or the timing otherwise lost, proceed as follows:

DISTRIBUTORLESS IGNITION SYSTEM

General Information

The V6 5VZ-FE engines are equipped with what is called the Toyota Direct Ignition (TDI). This system improves the timing accuracy and enhances the overall reliability of the ignition system. The TDI uses a two-cylinder simultaneous (waste spark) ignition system which fires two cylinders at the same time with one coil.

In a waste spark system, one spark plug is attached to each end of the secondary coil winding. This circuit arrangement causes one of the plugs in each cylinder pair to fire in a forward direction, and the other to fire in a reverse direction. The cylinder on the compression stroke is defined as the "event" cylinder, while the cylinder in the exhaust stroke is called the "waste" cylinder. The coils are positioned over one bank of the plugs. This arrangement has eliminated half of the normal required plug wires.

Diagnosis and Testing

SYSTEM TEST

1. Disconnect a spark plug wire from a plug.
2. Remove the spark plug.
3. Reattach the spark plug to the wire.
4. Ground the spark plug.

✳✳ CAUTION

Do not attempt to hold the wire with your bare hand. Use as much insulation as possible between the wire and the tool holding it. Do not stand on wet concrete while testing. Do not lean on the bodywork of the car while testing. The electrical charge will pass through the easiest path to ground, make certain its not you. Make sure the metal ground point nearest the cable end is safe; don't choose components that might contain fluids or electronic components.

5. Check to see if a bright blue spark occurs when the engine is being cranked.

➡To prevent any gasoline from being injected during this test, crank the engine no more than 1–2 seconds at one time.

6. If the spark does not occur, perform the following:
   a. Check the connection to the ignition coil and igniter, if it is bad, connect it securely. If this is OK then proceed.
   b. Inspect the resistance of the wires, it should be 25k ohms. Replace the wires if not to specifications. If the wires are good proceed.
   c. Check the power supply to the coil and the igniter. Turn the ignition **ON**. Check that there is battery voltage at the ignition coil positive terminal. If not, check the wiring between the ignition switch to the coil and igniter assemblies. If this is OK, proceed.
   d. Check the resistance of the ignition coils, if not within specifications, replace the coil. If this is not correct replace the coil. If it is good, continue.

1. Determine Top Dead Center (TDC) of the No. 1 cylinder's compression stroke by removing the spark plug from the No. 1 cylinder and placing your finger over the spark plug hole. Turn the engine (with a wrench on the crank pulley bolt) until the timing marks align at 0°, except on the 22R-E. On the 22R-E, align the mark at 5°. A definite compression can be felt pressing against your finger. Doing this important; the timing marks also align during the exhaust stroke, but no compression will be felt.

➡**On engines which have the spark plugs buried into the manifold, use a compression gauge.**

2. Install the rotor. Position the rotor so it is at the No. 1 cylinder firing position.
3. Slowly insert the distributor into the cylinder block.
4. Tighten the pinch-bolt to 14 ft. lbs. (19 Nm).
5. Install the distributor cap and wires. Check the ignition timing.
6. Check the idle speed.

   e. Check the resistance of the camshaft position sensor (refer to Section 4). If this is not correct, replace the sensor. If this is correct, proceed.
   f. Check the resistance of the crankshaft position sensor (refer to Section 4). If this is not correct, replace the sensor. If this is correct, proceed.
   g. If the spark is still not present, check the wiring between the ECM, distributor and igniter. If the wiring is OK, replace the igniter.

Ignition Coil Pack

TESTING

♦ See Figure 20

➡This test may be performed while the engine is cold or hot. A cold engine has a coolant temperature of less than 122°F (50°C).

1. Remove the air cleaner cap and the Mass Air Flow (MAF) sensor assembly.
2. Disconnect the wires from the coils.
3. Disconnect the coil harnesses.
4. Inspect the primary coil resistance. Using a ohmmeter, measure the resistance between the positive and negative primary terminals.
   - Cold—0.67–1.05 ohms
   - Hot—0.85–1.23 ohms
5. If the coil resistance is not as noted, replace the coil.
6. Inspect the secondary coil resistance. Using a ohmmeter, measure the resistance between the positive primary terminal and the spark plug wire.
   - Cold—9.3–16.0k ohms
   - Hot—11.7–18.8k ohms
7. If the coil resistance is not as noted, replace the coil.
8. Reinstall the coils.
9. Reconnect the coil harnesses.
10. Install the wires to the coils.
11. Install the air cleaner cap and the MAF sensor.

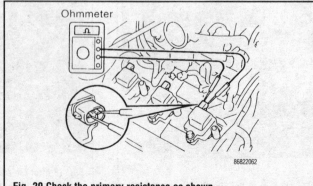

Ohmmeter

86822062

Fig. 20 Check the primary resistance as shown

REMOVAL & INSTALLATION

▶ **See Figure 21**

1. Remove the air cleaner cap and the Mass Air Flow (MAF) sensor assembly.
2. Disconnect and label the wires from the coils.
3. Unplug the connectors from the ignition coils. Remove the bolts and coils from the left cylinder head. Arrange the coils in the correct order for installation.

To install:

4. Install the coils on the cylinder heads in the reverse order of removal. Make certain the correct coil is replaced in the same place where it was removed. Tighten the mounting bolts to 69 inch lbs. (8 Nm).
5. Attach the plug wires to the coils.
6. Install the MAF sensor and the air cleaner cap.
7. Start the engine and check for proper operation.

Igniter

REMOVAL & INSTALLATION

The igniter is mounted on the fenderwell of your Toyota truck.
1. Disconnect the negative battery cable.
2. Unplug the wiring harness connections.
3. Remove the bolts retaining the igniter.
4. Lift the igniter off its mount.

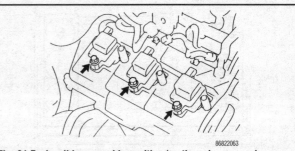

Fig. 21 Each coil is secured in position (on the valve cover above a spark plug) by a bolt

To install:
5. Mount the igniter to the bracket. Tighten the bolts until snug.
6. Connect the harness.
7. Connect the negative battery cable.

Crankshaft Position Sensor

See Section 4 for information and procedures.

Camshaft Position Sensor

See Section 4 for information and procedures.

FIRING ORDERS

▶ **See Figures 22 thru 27**

The firing order is the order in which spark is sent to each cylinder. The spark must arrive at the correct time in the combustion cycle or damage may result. For this reason, connecting the correct plug to the correct distributor terminal is critical. To avoid confusion, always replace the spark plug wires one at a time.

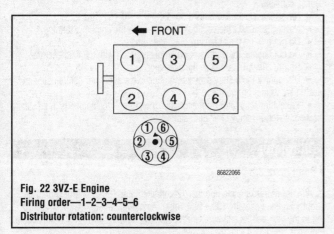

Fig. 22 3VZ-E Engine
Firing order—1–2–3–4–5–6
Distributor rotation: counterclockwise

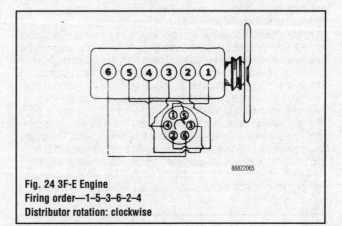

Fig. 24 3F-E Engine
Firing order—1–5–3–6–2–4
Distributor rotation: clockwise

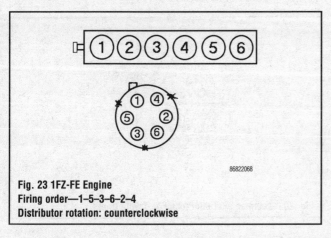

Fig. 23 1FZ-FE Engine
Firing order—1–5–3–6–2–4
Distributor rotation: counterclockwise

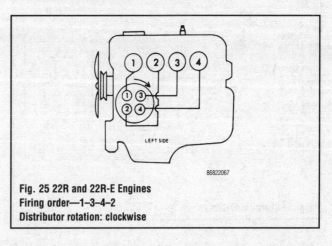

Fig. 25 22R and 22R-E Engines
Firing order—1–3–4–2
Distributor rotation: clockwise

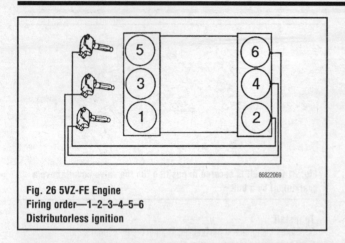

Fig. 26 5VZ-FE Engine
Firing order—1–2–3–4–5–6
Distributorless ignition

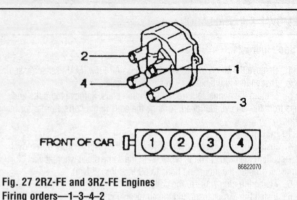

Fig. 27 2RZ-FE and 3RZ-FE Engines
Firing orders—1–3–4–2
Distributor rotation: clockwise

CHARGING SYSTEM

General Information

▶ See Figure 28

The automobile charging system provides electrical power for operation of the vehicle's ignition and all the electrical accessories. The battery serves as an electrical surge or storage tank, storing (in chemical form) the energy originally produced by the engine-driven alternator. The system also provides a means of regulating alternator output to protect the battery from being overcharged and to avoid excessive voltage to the accessories.

The vehicle's alternator is driven mechanically, through belts, by the engine crankshaft. It consists of two coils of fine wire, one stationary (the stator), and one movable (the rotor). The rotor may also be known as the armature, and consists of fine wire wrapped around an iron core which is mounted on a shaft. The electricity which flows through the two coils of wire (provided initially be the battery in some cases) creates an intense magnetic field around both the rotor and stator, and the interaction between the two fields creates voltage, allowing the alternator to power the accessories and charge the battery.

Almost all vehicles today use alternators because they are more efficient, can be rotated at higher speeds, and have fewer brush problems. In an alternator, the field rotates while all the current produced passes only through the stator windings. The brushes bear against continuous slip rings rather than a commutator. This causes the current produced to periodically reverse the direction of its flow, very similar to the power supply to your house. Diodes (electrical one-way switches) block the flow of current from traveling in the wrong direction. A series of diodes is wired together to permit the alternating flow of the stator to be converted to a pulsating, but unidirectional flow at the alternator output. This inverter circuit switches the AC current unusable by the vehicle to the standard DC or direct current. The alternator's field is wired in series with the voltage regulator.

The voltage regulator is contained within the alternator. Simply described, it consists of solid-state components whose job it is to limit the output of the

alternator to usable levels. Excess output can damage the battery and can destroy electrical components.

Alternator Precautions

To prevent damage to the alternator and regulator, the following precautionary measures must be taken when working with the electrical system.
• Never reverse the battery connections. Always check the battery polarity visually. This is to be done before any connections are made to ensure that all of the connections correspond to the battery ground polarity of the truck.
• Booster batteries must be connected properly. Make sure the positive cable of the booster battery is connected to the positive terminal of the battery which is getting the boost.
• Disconnect the battery cables before using a fast charger; the charger has a tendency to force current through the diodes in the opposite direction. This causes damage.
• Never use a fast charger as a booster for starting the truck.
• Never disconnect the voltage regulator while the engine is running.
• Do not ground the alternator output terminal.
• Do not operate the alternator on an open circuit with the field energized.
• Do not attempt to polarize the alternator.
• Disconnect the battery cables and remove the alternator before using an electric arc welder on the truck.
• Protect the alternator from excessive moisture. If the engine is to be steam cleaned, cover or remove the alternator.

Alternator

▶ See Figure 29

All current trucks use a nominal 12 volt alternator. (Exact output should be slightly higher, generally 13.5–14.4 volts). Amperage ratings may vary according to the year, model and accessories. All have a transistorized, non-adjustable regulator, integrated with the alternator.

Fig. 28 Charging system circuit

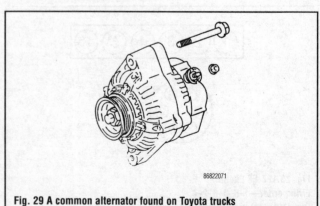

Fig. 29 A common alternator found on Toyota trucks

TESTING

▶ **See Figures 30 thru 35**

➡ **If a battery/alternator tester is available, connect the tester to the charging circuit recommended by the manufacture's instructions.**

If a tester is not available, connect a voltmeter and a ammeter to the charging circuit by doing the following:

1. Disconnect the wire from terminal (B) of the alternator and attach to the negative (-) lead of the ammeter.
2. Connect the positive (+) lead of the ammeter to terminal (B) of the alternator.
3. Connect the positive (+) lead of the voltmeter to terminal (B) of the alternator.
4. Ground the negative (-) lead of the voltmeter.
5. With the engine running from idle to 2000 rpm, check the reading on the ammeter and voltmeter. The standard amperage should be 10 A or less; standard voltage cold is 13.9–15.1 volts, and 13.5–14.3 volts hot. If the reading is greater than standard voltage, replace the regulator.

If the reading is less than the standard voltage, check the regulator and alternator:

6. With terminal (F) grounded, start the engine and check the voltmeter reading of terminal (B).
7. If the voltmeter reading is more than the standard voltage, replace the regulator.
8. If the voltmeter reading is less than standard voltage, the alternator may be faulty.
9. With the engine running at 2000 rpm turn on the high beam headlights and place the heater switch on high.
10. Check the reading on the ammeter. It should be 30 A or more.
11. If the ammeter reading is less than the standard amperage, replace the alternator.

➡ **If the battery is fully charged, the indication will sometimes be slightly less than the standard amperage.**

REMOVAL & INSTALLATION

➡ **On some models, the alternator is mounted very low on the engine. On these models, it may be necessary to remove the gravel shield and work from beneath the truck in order to gain access to the alternator. Replacing the alternator while the engine is cold is recommended. A hot engine can result in personal injury.**

22R and 22R-E Engines

▶ **See Figures 36 and 37**

1. Disconnect the negative battery terminal.
2. If the vehicle is equipped with power steering, drain the coolant from the system.

❄❄ WARNING

Housepets and small animals are attracted to the odor and taste of engine coolant (antifreeze). It is a highly poisonous mixture of chemicals; special care must be taken to protect open containers and spillage.

3. Disconnect the wiring at the back of the alternator. Lift the rubber boot, remove the nut from the external terminal and the wire.
4. If equipped with power steering, remove the engine splash shield or undercover, then the water inlet pipe.
5. If the truck has air conditioning, remove the No. 2 (lower) fan shroud.
6. Loosen the pivot bolt and remove the adjusting bolt.
7. Push the alternator towards the engine and remove the belt.
8. Hold the alternator; remove the pivot bolt then the alternator.

To install:

9. Mount the alternator on its bracket; install the pivot and adjusting bolts.
10. Install the drive belt and adjust it correctly. Tighten the pivot and adjusting bolts.

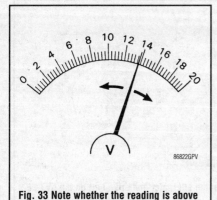

Fig. 30 Attach the ammeter and voltmeter as shown to test the charging system

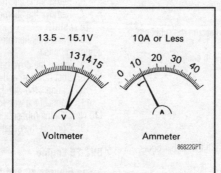

Fig. 31 With the engine running from idle to 2000 rpm, the readings should be as shown

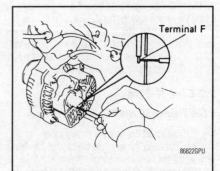

Fig. 32 With terminal (F) grounded, start the engine and check the voltmeter reading of terminal (B)

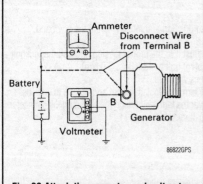

Fig. 33 Note whether the reading is above or below the standard voltage

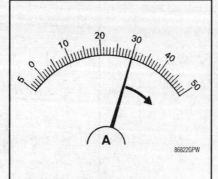

Fig. 34 The ammeter reading should be above 30 A

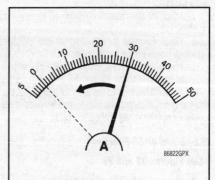

Fig. 35 If the ammeter reading is below the 30 A, replace the alternator

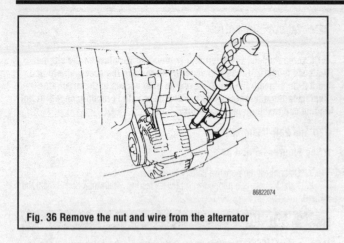

Fig. 36 Remove the nut and wire from the alternator

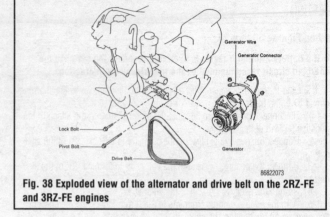

Fig. 38 Exploded view of the alternator and drive belt on the 2RZ-FE and 3RZ-FE engines

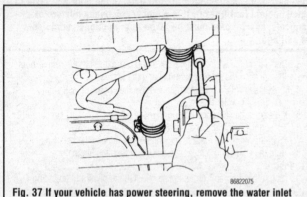

Fig. 37 If your vehicle has power steering, remove the water inlet pipe for ease of access

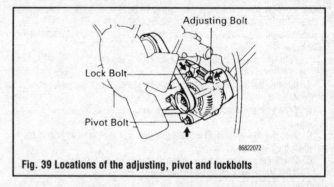

Fig. 39 Locations of the adjusting, pivot and lockbolts

11. Install the water inlet hose if it was removed. Install the fan shroud and install the engine splash shield.

12. Connect the wire to the terminal; install the nut and secure the rubber boot. Attach the wire harness connector to the alternator.

13. Close the radiator drain cock and refill the system with coolant.

14. Connect the negative battery cable.

3VZ-E and 3F-E Engines

▶ See Figures 36 and 37

1. Disconnect the negative battery cable.

2. Disconnect the wiring at the back of the alternator. Lift the rubber boot, remove the nut from the external terminal and the wire.

3. Loosen the alternator pivot bolt and loosen the upper locknut.

4. Push the alternator towards the engine and remove the belt.

5. Remove the pivot and adjusting bolts. On 3F-E, remove the nuts and bolts holding the air pump support, then remove the adjusting bar and its bolts.

6. Remove the alternator.

To install:

7. Place the alternator in position and loosely install the pivot and adjusting bolts.

8. On 3F-E, install the adjusting bar and its 2 bolts. Set the nuts and bolts just snug. Install the air pump support.

9. Install the drive belt; adjust it to the proper tension, tightening the adjusting and pivot bolts. Tighten the 3F-E adjuster bar bolts.

10. Connect the wire to the terminal and install the nut. Fit the rubber boot over the terminal. Attach the plastic wire harness connector.

11. Connect the negative battery cable.

2RZ-FE and 3RZ-FE Engines

▶ See Figures 38 and 39

1. Disconnect the negative battery cable.

2. Disconnect the wiring at the back of the alternator. Lift the rubber boot, remove the nut from the external terminal and the wire.

3. Loosen the alternator lockbolt, pivot bolt, nut and adjusting bolt.

4. Push the alternator towards the engine and remove the belt.

5. Remove the wiring harness with the clip.

6. Remove the alternator lockbolt, pivot bolt, nut.

7. Remove the alternator.

To install:

8. Place the alternator into position and loosely install the pivot and adjusting bolts.

9. Install the drive belt; adjust it to the proper tension. Tighten the lockbolt to 21 ft. lbs. (29 Nm) and the pivot bolt to 43 ft. lbs. (59 Nm).

10. Connect the wire harness with clip.

11. Connect the wire to the terminal. Install the nut and tighten to 7 ft. lbs. (10 Nm). Fit the rubber boot over the terminal. Attach the wire harness connector.

12. Connect the negative battery cable.

5VZ-FE Engine

▶ See Figures 40 and 41

1. Disconnect the negative battery cable.

2. Disconnect the wiring at the back of the alternator. Lift the rubber boot, remove the nut from the external terminal and the wire.

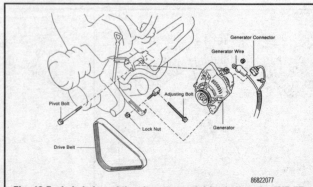

Fig. 40 Exploded view of the alternator and drive belt on the 5VZ-FE engine

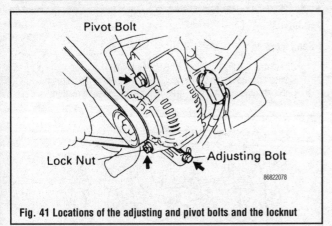

Fig. 41 Locations of the adjusting and pivot bolts and the locknut

3. Loosen the alternator locknut, pivot bolt, nut and adjusting bolt.
4. Push the alternator towards the engine and remove the belt.
5. Remove the alternator lockbolt, pivot bolt, nut.
6. Remove the alternator.

To install:

7. Place the alternator into position and loosely install the pivot and adjusting bolts.
8. Install the drive belt; adjust it to the proper tension. Tighten the locknut 25 ft. lbs. (33 Nm) and the pivot bolt 38 ft. lbs. (51 Nm).
9. Connect the wire to the terminal and install the nut and tighten. Fit the rubber boot over the terminal. Attach the wire harness connector.
10. Connect the negative battery cable.

1FZ-FE Engine

♦ **See Figures 42, 43 and 44**

1. Disconnect the negative battery terminal.
2. Remove the battery and tray.
3. Disconnect the power steering reservoir tank.
4. Loosen the lock, pivot and adjusting bolts.
5. Remove the 2 drive belts.
6. Disconnect the wiring at the back of the alternator. Lift the rubber boot, remove the nut from the external terminal and the wire.
7. Remove the wire clamp from the alternator.
8. Remove the lockbolt, nut and drive belt adjusting bar.
9. Hold the alternator; remove the pivot bolt then the alternator.

To install:

10. Mount the alternator ion the bracket with the pivot bolt. Do not tighten the bolt yet.
11. Install the drive belt adjusting bar with the bolt and nut. Tighten the bolt to 15 ft. lbs. (21Nm).
12. Temporarily install the lockbolt. Connect the alternator wiring, nut and rubber cap.
13. Connect the wire clamp to the alternator.
14. Install the drive belts.
15. Tighten the pivot bolt to 43 ft. lbs. (59 Nm) and the lockbolt to 15 ft. lbs. (21 Nm).
16. Connect the power steering reservoir tank.
17. Install the battery tray and battery.
18. Start the vehicle and check the alternator operation.

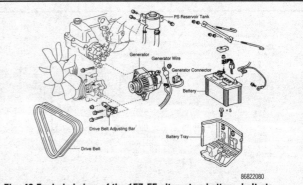

Fig. 42 Exploded view of the 1FZ-FE alternator, battery, belt etc. locations

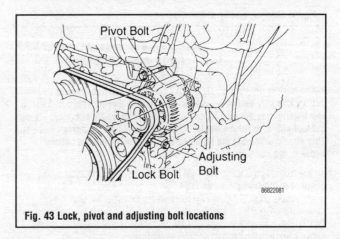

Fig. 43 Lock, pivot and adjusting bolt locations

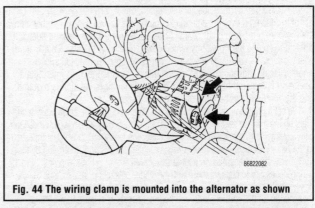

Fig. 44 The wiring clamp is mounted into the alternator as shown

Regulator

All regulators are contained inside the case of the alternator and are not replaceable as a separate unit. If the regulator fails, the entire alternator must be replaced.

STARTING SYSTEM

General Information

♦ **See Figure 45**

The battery is the first link in the chain of items which work together to provide power of the automobile engine. In most modern trucks, the battery is a lead/acid electrochemical device consisting of six two-volt subsections connected in series so the unit is capable of producing approximately 12 volts of electrical pressure.

Each subsection, or cell, consists of a series of positive and negative plates held a short distance apart in a solution of sulfuric acid and water. The two types of plates are of dissimilar metals. This causes a chemical reaction and it is this reaction which produces current flow from the battery when its positive and negative terminals are connected to an electrical appliance such as a lamp or a motor. The continued transfer of electrons would eventually convert sulfuric acid in the electrolyte to water, and make the two plates identical in chemical composition. As electrical energy is removed from the battery, its voltage output tends to drop. Thus, measuring battery voltage and battery electrolyte composition are two ways of checking the ability of the unit to supply power. During the starting of the engine, electrical energy is removed from the battery. However, if the charging circuit is in good condition and the operating conditions are nor-

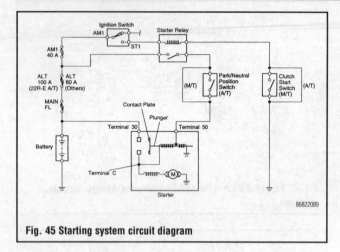

Fig. 45 Starting system circuit diagram

mal, the power removed from the battery will be replaced by the alternator which will force electrons back through the battery, reversing the normal flow, and restoring the battery to its original chemical state.

➡The battery only stores electrical energy, it does not produce it. The battery can only deliver power equal to what is stored within it. Thus, a low battery may not be able to supply enough power to crank the starter but will still operate the lights, horn and other lower amperage circuits. This should eliminate the famous line, "I know it's not the battery because the radio works . . ."

The battery and starting motor are linked by very heavy electrical cables designed to minimize resistance to the flow of current. Generally, the major power supply cable that leaves the battery goes directly to the starter, while other electrical needs are supplied by a smaller cable. During the starter operation, power flows from the battery to the starter and is grounded through the truck's frame and the battery's negative ground strap.

The starting motor is a specially designed, direct current electric motor capable of producing a very great amount of power for its size. One thing that allows the motor to produce a great deal of power is its tremendous rotating speed. It drives the engine through a tiny pinion gear (attached to the starter's armature), which drives the very large flywheel ring gear at a greatly reduced speed. Another factor allowing it to produce so much power is that only intermittent operation is required of it. Thus, little allowance for air circulation is required, and the windings can be built into a very small space.

The starter solenoid is a magnetic device which employs the small current supplied by the ignition switch. The magnetic action moves a plunger which mechanically engages the starter and electrically closes the heavy switch which connects it to the battery. The starting circuit consists of the starting signal controlled by the ignition switch, a transmission neutral safety switch or clutch pedal switch, and the wiring necessary to connect these with the starter solenoid or relay. Since the circuit has multiple switches, all must be in the ON position for the system to operate.

As soon as the engine starts, the flywheel ring gear begins turning fast enough to drive the pinion at an extremely high rate of speed. At this point, the one-way clutch begins allowing the pinion to spin faster than the starter shaft so that the starter will not operate at excessive speed. (This overrun condition is similar to riding a bicycle downhill; the rear wheel is turning faster than the chain sprocket.) When the ignition switch is released from the starter position, the solenoid is de-energized, and a spring in the solenoid assembly pulls the gear out of mesh and interrupts the current flow to the starter.

Some starters employ a separate relay, mounted away from the starter, to switch the motor and solenoid current on and off. The relay replaces the solenoid electrical switch, but does not eliminate the need for a solenoid mounted on the starter to mechanically engage the starter drive gears. The relay is used to reduce the amount of current the starting switch must carry.

Starter

TESTING

➡These tests must be performed within a 3 to 5 seconds to avoid burning out the coil.

Pull-in

▶ See Figure 46

1. Disconnect the field coil lead wire from terminal C.
2. Connect the battery to the magnetic switch shown in the diagram.
3. Check that the clutch pinion gear moves in the outward motion.

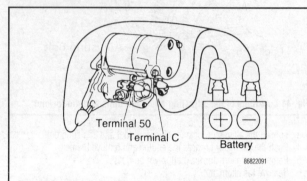

Fig. 46 Connect the battery to the magnetic switch as shown here

Hold-in

▶ See Figure 47

With the battery connected as in the picture and with the clutch pinion gear out, disconnect the negative (-) lead from terminal C. Check that the pinion gear stays in the outward position.

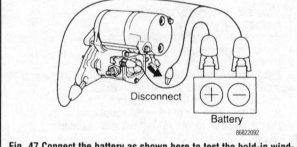

Fig. 47 Connect the battery as shown here to test the hold-in windings

Inspect Clutch Pinion Gear Return

▶ See Figure 48

Disconnect the negative lead from the magnetic switch body. Check that the pinion gear returns to the inward position.

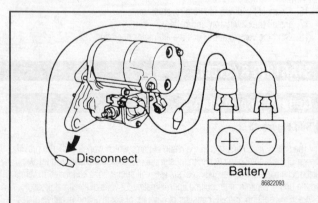

Fig. 48 Disconnect the negative lead from the starter as shown here

No-Load

♦ **See Figure 49**

1. Connect the battery ammeter to the starter as shown in the illustration.
2. Check the rotation of the starter, it must be smooth and steady with the pinion gear moving outwards. Compare the ammeter reading to the starter specifications chart.

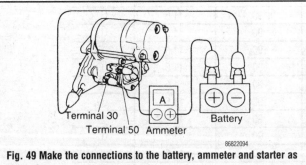

Fig. 49 Make the connections to the battery, ammeter and starter as shown here

REMOVAL & INSTALLATION

♦ **See Figure 50**

➡Replacing the starter while the engine is cold is recommended. A hot engine can result in personal injury.

1. Disconnect the negative battery cable.
2. Remove the nut and disconnect the battery cable from the magnetic switch on the starter motor.
3. Disconnect the remaining electrical connections at the starter.
4. Remove the nuts and/or bolts securing the starter to the bell housing, then pull the starter toward the front of the vehicle to remove.

To install:
5. Insert the starter into the bell housing being sure that the starter drive teeth are engaged with the flywheel teeth, not jammed against the flywheel.
6. Tighten the attaching hardware to 29 ft. lbs. (39 Nm) and replace all electrical connections.
7. Connect the positive battery cable (starter wire) to the starter.
8. Connect the negative battery cable.

RELAY REPLACEMENT

♦ **See Figure 51**

The starter relays on most of your Toyota trucks are located either in the relay block in the engine compartment or on the driver's side junction block. The cover for the relay compartments should be marked, "starter" or the relay will have a part number with a suffix of 28300-XXXXX. Simply locate the relay, pull it out, and install a new one if needed.

Battery

REMOVAL & INSTALLATION

➡Refer to Section 1 for details on battery maintenance.

1. Disconnect the negative battery cable from the terminal, then disconnect the positive cable. Special pliers and pullers are available to remove the clamps. Be careful of the many wires and fittings on and around the positive terminal.

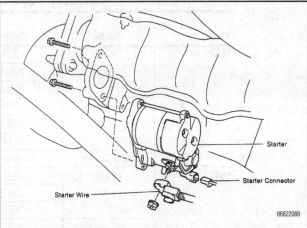

Fig. 50 Exploded view of a common starter mounting found on Toyota trucks

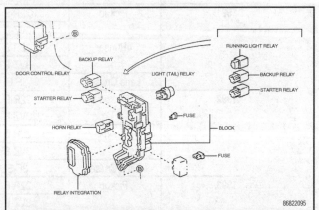

Fig. 51 The relay may be located in the driver's side junction block or in the engine compartment relay block

➡To avoid sparks, always disconnect the negative cable first and reconnect it last.

2. Unscrew and remove the battery hold down clamp.
3. Remove the battery, being careful not to spill any of the acid.

➡Spilled acid can be neutralized with a baking soda and water solution. If you somehow get acid into your eyes, flush it out with lots of clean water and get to a doctor as quickly as possible.

4. Clean the battery posts thoroughly before reinstalling or when installing a new one. A light coating of petroleum jelly or battery terminal spray protectant will help fight corrosion.
5. Clean the cable clamps using the special tools or a wire brush, both inside and out.
6. Install the battery, and the hold down clamp. Connect the positive and then the negative cable. Do not hammer them into place. The terminals should be coated with grease to prevent corrosion.

❊❊ CAUTION

Make absolutely sure that the battery is connected properly before you turn on the ignition switch. Reversed polarity can burn out an alternator in a matter of seconds.

STARTER SPECIFICATIONS

Year	Model	Engine	No-Load Test		
			Max. Amps	Max. Volts	RPM
1989	Pickup	22R	90	11.5	①
		22R-E	90	11.5	①
		3VZ-E	90	11.5	①
	4Runner	22R	90	11.5	①
		22R-E	90	11.5	①
		3VZ-E	90	11.5	①
	Land Cruiser	3F-E	90	11.5	3000
1990	Pickup	22R	90	11.5	①
		22R-E	90	11.5	①
		3VZ-E	90	11.5	①
	4Runner	22R-E	90	11.5	①
		3VZ-E	90	11.5	①
	Land Cruiser	3F-E	90	11.5	3000
1991	Pickup	22R-E	90	11.5	①
		3VZ-E	90	11.5	①
	4Runner	22R-E	90	11.5	①
		3VZ-E	90	11.5	①
	Land Cruiser	3F-E	90	11.5	3000
1992	Pickup	22R-E	90	11.5	①
		3VZ-E	90	11.5	①
	4Runner	22R-E	90	11.5	①
		3VZ-E	90	11.5	①
	Land Cruiser	3F-E	90	11.5	3000
1993	Pickup	22R-E	90	11.5	①
		3VZ-E	90	11.5	①
	4Runner	22R-E	90	11.5	①
		3VZ-E	90	11.5	①
	T100	3VZ-E	90	11.5	3500
	Land Cruiser	1FZ-FE	90	11.5	3000
1994	Pickup	22R-E	90	11.5	①
		3VZ-E	90	11.5	①
	4Runner	22R-E	90	11.5	①
		3VZ-E	90	11.5	①
	T100	3RZ-FE	90	11.5	3500
		3VZ-E	90	11.5	①
	Land Cruiser	1FZ-FE	90	11.5	3000
1995	Pickup	22R-E	90	11.5	①
		3VZ-E	90	11.5	①
	4Runner	22R-E	90	11.5	①
		3VZ-E	90	11.5	①
	Tacoma	2RZ-FE	90	11.5	3000
		3RZ-FE	90	11.5	3000
		5VZ-FE	90	11.5	3000
	T100	3RZ-FE	90	11.5	3500
		5VZ-FE	90	11.5	3500
	Land Cruiser	1FZ-FE	90	11.5	3000

86822117

STARTER SPECIFICATIONS

Year	Model	Engine	No-Load Test		
			Max. Amps	Max. Volts	RPM
1996	4Runner	3RZ-FE	90	11.5	①
		5VZ-FE	90	11.5	①
	Tacoma	2RZ-FE	90	11.5	①
		3RZ-FE	90	11.5	①
		5VZ-FE	90	11.5	①
	T100	3RZ-FE	90	11.5	3500
		5VZ-FE	90	11.5	3500
	Land Cruiser	1FZ-FE	90	11.5	3000

① 1.0 kw-3000 rpm
1.4, 1.6 kw-3500 rpm
2.0 kw-2500 rpm

86822118

SENDING UNITS AND SENSORS

Engine Coolant Temperature Sensor

OPERATION

♦ **See Figure 52**

The engine coolant temperature sensor is usually located on the thermostat housing. This sensor controls the readings of the coolant temperature gauge in your vehicle when the ignition switch is **ON**.

There is a needle that shows the temperature area of the engine. If the needle points in the red zone or higher (this is overheating), you should pull to the side of the road and turn your engine **OFF** immediately and allow it to cool.

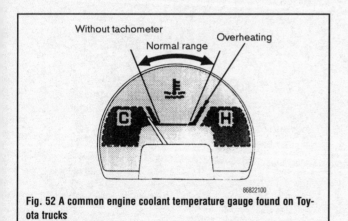

Fig. 52 A common engine coolant temperature gauge found on Toyota trucks

86822100

REMOVAL & INSTALLATION

1. Drain the coolant from the vehicle.
2. Disconnect the engine wiring protector from the brackets.
3. Disconnect the sensor wire.
4. Using a 19mm deep socket, remove the sensor and gasket.

To install:

5. Apply a thin coat of Loctite® or equivalent to the switch. By hand, carefully thread the switch into the engine. If resistance is felt at all, remove the switch and inspect the threads for any damage. Continue to install the switch. Once installed tighten to 14 ft. lbs. (20 Nm).
6. Connect the sensor wiring.
7. Install the engine wire protector to the brackets.
8. Refill the coolant. Start the engine and check for leaks.

Oil Pressure Light/Gauge Sender

OPERATION

♦ **See Figure 53**

The oil pressure sender/switch controls the gauge or idiot light operation in you vehicle while the engine is **ON**. The gauge will fluctuate during engine operation. If the pressure should stay below the normal range, you should pull to the side of the road and turn your engine **OFF** immediately. Check your oil level of the vehicle on level ground, if it is low, fill to the proper level. Start the engine and recheck after a few minutes. Oil pressure may not build up when the level is too low.

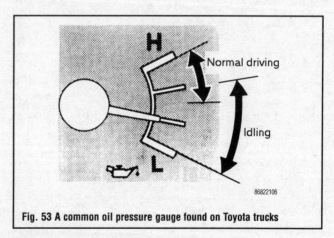

Fig. 53 A common oil pressure gauge found on Toyota trucks

86822106

REMOVAL & INSTALLATION

♦ **See Figures 54 and 55**

1. If equipped, remove the lower engine splash shield.
2. Using special tool 09816-33010 or an equivalent oil pressure switch socket, remove the switch.

To install:

3. Apply a thin coat of Loctite® or equivalent to the switch. By hand, carefully thread the switch into the engine. If resistance is felt at all, remove the switch and inspect the threads for any damage. Continue to install the switch. Once installed tighten to 11 ft. lbs. (15 Nm).
4. If removed, install the engine splash shield.
5. Start the engine and inspect for leaks.

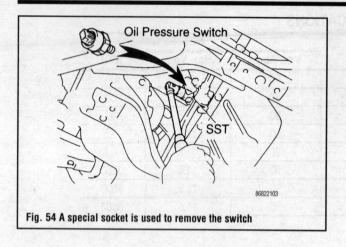

Fig. 54 A special socket is used to remove the switch

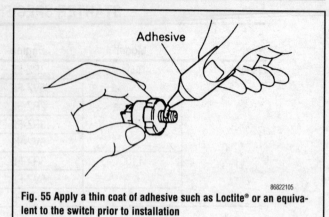

Fig. 55 Apply a thin coat of adhesive such as Loctite® or an equivalent to the switch prior to installation

3

ENGINE AND ENGINE OVERHAUL

ENGINE MECHANICAL

General Information

The piston engine is a metal block containing a series of round chambers or cylinders. These chambers may be arranged in line or in a V; hence, the description of an engine as an inline 4 or 6 or a V-6. The upper part of the engine block is usually an iron or aluminum-alloy casting. The casting forms outer walls around the cylinders with hollow areas in between, through which coolant circulates. The lower block provides a number of rigid mounting points for the crankshaft and its bearings. The lower block is referred to as the crankcase.

The crankshaft is a long, steel shaft mounted at the bottom of the engine and free to turn in its mounts. The mounting points (generally four to seven) and the bearings for the crankshaft are called main bearings. The crankshaft is the shaft which is made to turn through the function of the engine; this motion is then passed into the transmission/transaxle and on to the drive wheels.

Attached to the crankshaft are the connecting rods which run up to the pistons within the cylinders. As the air/fuel mixture explodes within the tightly sealed cylinder, the piston is forced downward. This motion is transferred through the connecting rod to the crankshaft and the shaft turns. As one piston finishes its power stroke, its next upward journey forces the burnt gasses out of the cylinder through the now-open exhaust valve. By the top of the stroke, the exhaust valve has closed and the intake valve has begun to open, allowing the fresh air/fuel charge to be sucked into the cylinder by the downward stroke of the piston. The intake valve closes, the piston once again comes back up and compresses the charge in the closed cylinder. At the top (approximately) of this stroke the spark plug fires, the charge explodes and another power stroke takes place. If you count the piston motions in between power strokes, you'll see why automotive engines are called four-stroke or four-cycle engines.

While one cylinder is performing this cycle, all the others are also contributing; but in different timing. Obviously, all the cylinders cannot fire at once or the power flow would not be steady. As any one cylinder is on its power stroke, another is on its exhaust stroke, another on intake and another on compression. These constant power pulses keep the crank turning; a large round flywheel attached to the end of the crankshaft provides a stable mass to smooth out the rotation.

At the top of the engine, the cylinder head(s) provide tight covers for the cylinders. They contain machined chambers into which the fuel charge is forced as the piston reaches the top of its travel. These combustion chambers contain at least one intake and one exhaust valve which are opened and closed through the action of the camshaft. The spark plugs are screwed into the cylinder head so that the tips of the plugs protrude into the chamber.

Since the timing of the valve action (opening and closing) is critical to the combustion process, the camshaft is driven by a belt or chain. The valves are operated either by pushrods (called overhead valves the valves are above the cam) or by the direct action of the cam pushing on the valves (overhead cam). Toyota trucks with either the 4 cylinder or V-6 engines use overhead cam (OHC) engines. The Land Cruiser 3F-E engine is a pushrod or overhead valve engine.

Lubricating oil is stored in a pan or sump at the bottom of the engine. It is force fed to all the parts of the engine by the oil pump which may be driven by either the crank or the camshaft. The oil lubricates the entire engine by travelling through passages in the block and head. Additionally, the circulation of the oil provides 25–40% of the engine cooling.

If all this seems very complicated, keep in mind that the sole purpose of any motor gas, diesel, electric, solar is to turn a shaft. The motion of the shaft is then harnessed to perform a task such as pumping water, moving the vehicle, etc. Due to the constantly changing operating conditions found in a motor vehicle, accomplishing this shaft-turning in an automotive engine requires many supporting systems such as fuel delivery, exhaust handling, lubrication, cooling, starting, etc. Operation of these systems involve principles of mechanics, vacuum, electronics, etc. Being able to identify a problem by what system is involved will allow you to begin accurate diagnosis of the symptoms and causes.

Engine Overhaul Tips

Most engine overhaul procedures are fairly standard. In addition to specific parts replacement procedures and specifications for your individual engine, this section also is a guide to acceptable rebuilding procedures. Examples of standard rebuilding practice are shown and should be used along with specific details concerning your particular engine.

Competent and accurate machine shop services will ensure maximum performance, reliability and engine life. In most instances it is more profitable for the do-it-yourself mechanic to remove, clean and inspect the component, buy the necessary parts and deliver these to a shop for actual machine work.

On the other hand, much of the rebuilding work (crankshaft, block, bearings, piston rods, and other components) is well within the scope of the do-it-yourself mechanic's tools and abilities. You will have to decide for yourself the depth of involvement you desire in an engine repair or rebuild.

TOOLS

The tools required for an engine overhaul or parts replacement will depend on the depth of your involvement. With a few exceptions, they will be the tools found in a mechanic's tool kit (see Section 1 of this manual). More in-depth work will require some or all of the following:

- a dial indicator (reading in thousandths) mounted on a universal base
- micrometers and telescope gauges
- jaw and screw-type pullers
- scraper
- valve spring compressor
- ring groove cleaner
- piston ring expander and compressor
- ridge reamer
- cylinder hone or glaze breaker
- Plastigage®
- engine stand

The use of most of these tools is illustrated in this section. Many can be rented for a one-time use from a local parts jobber or tool supply house specializing in automotive work.

Occasionally, the use of special tools is called for. See the information on Special Tools and the Safety Notice in the front of this book before substituting another tool.

INSPECTION TECHNIQUES

Procedures and specifications are given in this section for inspecting, cleaning and assessing the wear limits of most major components. Other procedures such as Magnaflux® and Zyglo® can be used to locate material flaws and stress cracks. Magnaflux® is a magnetic process applicable only to ferrous materials. The Zyglo® process coats the material with a fluorescent dye penetrant and can be used on any material. Checking for suspected surface cracks can be more readily made using spot check dye. The dye is sprayed onto the suspected area, wiped off and the area sprayed with a developer. Cracks will show up brightly.

OVERHAUL TIPS

Aluminum has become extremely popular for use in engines, due to its low weight. Observe the following precautions when handling aluminum parts:
- Never hot tank aluminum parts (the caustic hot tank solution will eat the aluminum.
- Remove all aluminum parts (identification tag, etc.) from engine parts prior to the tanking.
- Always coat threads lightly with engine oil or anti-seize compounds before installation, to prevent seizure.
- Never overtighten bolts or spark plugs especially in aluminum threads.

Stripped threads in any component can be repaired using any of several commercial repair kits (Heli-Coil®, Microdot®, Keenserts®, etc.).

When assembling the engine, any parts that will be exposed to frictional contact, the parts must be prelubed to provide lubrication at initial start-up. Any product specifically formulated for this purpose can be used, but engine oil is not recommended as a prelube in most cases.

When semi-permanent (locked, but removable) installation of bolts or nuts is desired, threads should be cleaned and coated with Loctite® or other similar, commercial non-hardening sealant.

REPAIRING DAMAGED THREADS

▶ See Figures 1, 2, 3, 4 and 5

Several methods of repairing damaged threads are available. Heli-Coil® (shown here), Keenserts® and Microdot® are among the most widely used. All involve basically the same principle—drilling out stripped threads, tapping the hole and installing a prewound insert—making welding, plugging and oversize fasteners unnecessary.

Two types of thread repair inserts are usually supplied: a standard type for most inch coarse, inch fine, metric course and metric fine thread sizes and a spark lug type to fit most spark plug port sizes. Consult the individual manufacturer's catalog to determine exact applications. Typical thread repair kits will contain a selection of prewound threaded inserts, a tap (corresponding to the outside diameter threads of the insert) and an installation tool. Spark plug inserts usually differ because they require a tap equipped with pilot threads and a combined reamer/tap section. Most manufacturers also supply blister-packed thread repair inserts separately in addition to a master kit containing a variety of taps and inserts plus installation tools.

Before attempting to repair a threaded hole, remove any snapped, broken or damaged bolts or studs. Penetrating oil can be used to free frozen threads. The offending item can be removed with locking pliers or using a screw/stud extractor. After the hole is clear, the thread can be repaired, as shown in the series of accompanying illustrations and in the kit manufacturer's instructions.

Compression Testing

▶ See Figure 6

A noticeable lack of engine power, excessive oil consumption and/or poor fuel mileage measured over an extended period are all indicators of internal engine wear. Worn piston rings, scored or worn cylinder bores, leaking head gaskets, sticking or burnt valves and worn valve seats are all possible culprits here. A check of each cylinder's compression will help you locate the problems.

As mentioned in the Tools and Equipment part of Section 1, a screw-in type compression gauge is more accurate that the type you simply hold against the spark plug hole, although it takes slightly longer to use. It's worth it to obtain a more accurate reading. Follow the procedures below.

1. Warm up the engine to normal operating temperature.
2. Remove all the spark plugs.
3. Disconnect the high tension lead (coil wire) from the ignition coil.
4. Fully open the throttle either by operating the carburetor throttle linkage by hand or by having an assistant floor the accelerator pedal.
5. Screw the compression gauge into the No.1 spark plug hole until the fitting is snug.

✳✳ WARNING

Be careful not to crossthread the plug hole. On aluminum cylinder heads use extra care, as the threads in these heads are easily ruined.

6. Ask an assistant to depress the accelerator pedal fully on both carbureted and fuel injected vehicles. Then, while you read the compression gauge, ask the assistant to crank the engine two or three times in short bursts using the ignition switch.

7. Read the compression gauge at the end of each series of cranks, and record the highest of these readings. Repeat this procedure for each of the engine's cylinders. As a general rule, new motors will have compression on the order of 150–170 psi (1034–1172 kPa). This number will decrease with age and wear. The number of pounds of pressure that your test shows is not as important as the evenness between all the cylinders. Many engines run very well with all cylinders at 105 psi (724 kPa). The lower number simply shows a gen-

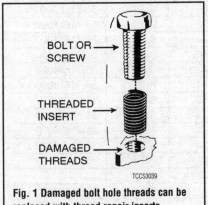

Fig. 1 Damaged bolt hole threads can be replaced with thread repair inserts

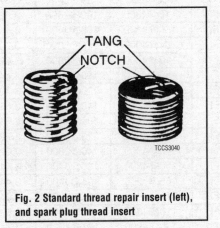

Fig. 2 Standard thread repair insert (left), and spark plug thread insert

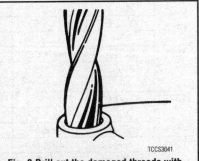

Fig. 3 Drill out the damaged threads with the specified drill. Be sure to drill completely through the hole or to the bottom of a blind hole

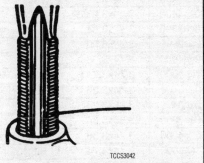

Fig. 4 Using the kit, tap the hole in order to receive the thread insert. Keep the tap well oiled and back it out frequently to avoid clogging the threads.

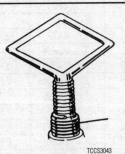

Fig. 5 Screw the threaded insert onto the installer tool until the tang engages the slot. Thread the insert into the hole until it is ¼ or ½ turn below the top surface, then remove the tool and break off the tang using a punch.

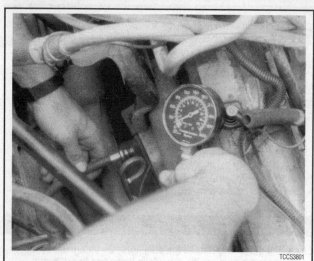

Fig. 6 A screw-in type compression gauge is more accurate and easier to use without an assistant

eral deterioration internally. This motor probably burns a little oil and may be a bit harder to start, but, based on these numbers, doesn't warrant an engine tear-down yet. Compare the highest reading of all the cylinders. Any variation of more than 10% should be considered a sign of potential trouble. For example, on a 4 cylinder engine, if your compression readings for cylinders 1 through 4 were: 135 psi (930 kPa), 125 psi (861 kPa), 90 psi (620 kPa) and 125 psi (861 kPa), it would be fair to say that cylinder number three is not working efficiently and is almost certainly the cause of your oil burning, rough idle or poor fuel mileage.

8. If a cylinder is unusually low, pour a tablespoon of clean engine oil into the cylinder through the spark plug hole and repeat the compression test. If the compression comes up after adding the oil, it appears that the cylinder's piston rings or bore are damaged or worn. (The oil sealed some of the leakage.) If the pressure remains low, the valves may not be seating properly (a valve job is needed), or the head gasket may be blown near that cylinder. If compression in any two adjacent cylinders is low, and if the addition of oil doesn't help the compression, there is leakage past the head gasket. Oil and coolant in the combustion chamber can result from this problem. There may also be evidence of water droplets on the engine oil dipstick when a head gasket has blown.

GENERAL ENGINE SPECIFICATIONS

Year	Engine ID/VIN	Engine Displacement Liters (cc)	Fuel System Type	Net Horsepower @ rpm	Net Torque @ rpm (ft. lbs.)	Bore x Stroke (in.)	Compression Ratio	Oil Pressure @ rpm
1989	22R	2.4 (2366)	2 BBL	96@4800	129@2800	3.62x3.50	9.3:1	36-71@4000
	22R-E	2.4 (2366)	EFI	116@4800	140@2800	3.62x3.350	9.3:1	36-71@4000
	3VZ-E	3.0 (2959)	EFI	150@4800	180@3400	3.44x3.23	9.0:1	36-71@4000
	3F-E	4.0 (3956)	EFI	154@4000	220@3000	3.70x3.74	8.1:1	36-71@4000
1990	22R	2.4 (2366)	2 BBL	96@4800	129@2800	3.62x3.50	9.3:1	36-71@3000
	22R-E	2.4 (2366)	EFI	116@4800	140@2800	3.62x3.50	9.3:1	36-71@3000
	3VZ-E	3.0 (2959)	EFI	150@4800	180@3400	3.44x3.23	9.0:1	36-71@4000
	3F-E	4.0 (3956)	EFI	154@4000	220@3000	3.70x3.74	8.1:1	36-71@4000
1991	22R-E	2.4 (2366)	EFI	116@4800	140@2800	3.62x3.50	9.3:1	36-71@3000
	3VZ-E	3.0 (2959)	EFI	150@4800	180@3400	3.44x3.23	9.0:1	36-71@4000
	3F-E	4.0 (3956)	EFI	154@4000	220@3000	3.70x3.74	8.1:1	36-71@4000
1992	22R-E	2.4 (2366)	EFI	116@4800	140@2800	3.62x3.50	9.3:1	36-71@3000
	3VZ-E	3.0 (2959)	EFI	150@4800	180@3400	3.44x3.23	9.0:1	36-71@4000
	3F-E	4.0 (3956)	EFI	154@4000	220@3000	3.70x3.74	8.1:1	36-71@4000
1993	22R-E	2.4 (2366)	EFI	116@4800	140@2800	3.62x3.50	9.3:1	36-71@3000
	3VZ-E	3.0 (2959)	EFI	150@4800	180@3400	3.44x3.23	9.0:1	36-71@4000
	3VZ-FE	3.0 (2952)	EFI	185@5200	195@4400	3.44x3.23	9.6:1	4.3 ①
	1FZ-FE	4.5 (4477)	EFI	212@4600	275@3200	3.94x3.74	9.0:1	36-71@3000
1994	22R-E	2.4 (2366)	EFI	116@4800	140@2800	3.62x3.50	9.3:1	36-71@3000
	3RZ-FE	2.7 (2693)	EFI	150@4800	177@4000	3.74x3.74	9.5:1	36-71@3000
	3VZ-E	3.0 (2959)	EFI	150@4800	180@3400	3.44x3.23	9.0:1	36-71@4000
	3VZ-FE	3.0 (2952)	EFI	185@5200	195@4400	3.44x3.23	9.6:1	4.3 ①
	1FZ-FE	4.5 (4477)	EFI	212@4600	275@3200	3.94x3.64	9.0:1	36-71@3000
1995	22R-E	2.4 (2366)	EFI	116@4800	140@2800	3.62x3.50	9.3:1	36-71@3000
	2RZ-FE	2.4 (2438)	EFI	142@5000	160@4000	3.74x3.38	9.5:1	36-71@3000
	3RZ-FE	2.7 (2693)	EFI	150@4800	177@4000	3.74x3.74	9.5:1	36-71@3000
	3VZ-E	3.0 (2959)	EFI	150@4800	180@3400	3.44x3.23	9.0:1	36-71@3000
	5VZ-FE	3.4 (3378)	EFI	190@4800	220@3400	3.68x3.23	9.6:1	4.3 ①
	1FZ-FE	4.5 (4477)	EFI	212@4600	275@3200	3.94x3.64	9.0:1	36-71@3000
1996	2RZ-FE	2.4 (2438)	EFI	142@5000	160@4000	3.74x3.38	9.5:1	36-71@3000
	3RZ-FE	2.7 (2693)	EFI	150@4800	177@4000	3.74x3.74	9.5:1	36-71@3000
	5VZ-FE	3.4 (3378)	EFI	190@4800	220@3400	3.68x3.23	9.6:1	4.3 ①
	1FZ-FE	4.5 (4477)	EFI	212@4600	275@3200	3.94x3.64	9.0:1	36-71@3000

1 At idle
BBL - Barrel carburetor
EFI - Electronic fuel injection

86823002

VALVE SPECIFICATIONS

Year	Engine ID/VIN	Engine Displacement Liters (cc)	Seat Angle (deg.)	Face Angle (deg.)	Spring Test Pressure (lbs. @ in.)	Spring Installed Height (in.)	Stem-to-Guide Clearance (in.) Intake	Stem-to-Guide Clearance (in.) Exhaust	Stem Diameter (in.) Intake	Stem Diameter (in.) Exhaust
1989	22R	2.4 (2366)	45①	44.5	66.1	1.594	0.0010-0.0024	0.0012-0.0026	0.3138-0.3144	0.3136-0.3142
	22R-E	2.4 (2366)	45①	44.5	66.1	1.594	0.0010-0.0024	0.0012-0.0026	0.3138-0.3144	0.3136-0.3142
	3VZ-E	3.0 (2959)	45①	44.5	54-57	1.575	0.0010-0.0024	0.0012-0.0026	0.3138-0.3144	0.3136-0.3142
	3F-E	4.0 (3956)	45	44.5	71.6	1.693	0.0012-0.0024	0.0016-0.0028	0.3140	0.3137
1990	22R	2.4 (2366)	45①	44.5	66.1	1.594	0.0010-0.0024	0.0012-0.0026	0.3138-0.3144	0.3136-0.3142
	22R-E	2.4 (2366)	45①	44.5	66.1	1.594	0.0010-0.0024	0.0012-0.0026	0.3138-0.3144	0.3136-0.3142
	3VZ-E	3.0 (2959)	45①	44.5	54-57	1.575	0.0010-0.0024	0.0012-0.0026	0.3138-0.3144	0.3136-0.3142
	3F-E	4.0 (3956)	45	44.5	71.6	1.693	0.0012-0.0024	0.0016-0.0028	0.3140	0.3137
1991	22R-E	2.4 (2366)	45①	44.5	66.1	1.594	0.0010-0.0024	0.0012-0.0026	0.3138-0.3144	0.3136-0.3142
	3VZ-E	3.0 (2959)	45①	44.5	54-57	1.575	0.0010-0.0024	0.0012-0.0026	0.3138-0.3144	0.3136-0.3142
	3F-E	4.0 (3956)	45	44.5	71.6	1.693	0.0012-0.0024	0.0016-0.0028	0.3140	0.3137
1992	22R-E	2.4 (2366)	45①	44.5	66.1	1.594	0.0010-0.0024	0.0012-0.0026	0.3138-0.3144	0.3136-0.3142
	3VZ-E	3.0 (2959)	45①	44.5	54-57	1.575	0.0010-0.0024	0.0012-0.0026	0.3138-0.3144	0.3136-0.3142
	3F-E	4.0 (3956)	45	44.5	71.6	1.693	0.0012-0.0024	0.0016-0.0028	0.3140	0.3137
1993	22R-E	2.4 (2366)	45①	44.5	66.1	1.594	0.0010-0.0024	0.0012-0.0026	0.3138-0.3144	0.3136-0.3142
	3VZ-E	3.0 (2959)	45①	44.5	54-57	1.575	0.0010-0.0024	0.0012-0.0026	0.3138-0.3144	0.3136-0.3142
	1FZ-FE	4.5 (4477)	45	44.5	53.4	1.437	0.0010-0.0024	0.0012-0.0026	0.2744-0.2750	0.2742-0.2748
1994	22R-E	2.4 (2366)	45①	44.5	66.1	1.594	0.0010-0.0024	0.0012-0.0026	0.3138-0.3144	0.3136-0.3142
	3RZ-FE	2.7 (2693)	45①	44.5	57-63	1.406	0.0010-0.0024	0.0012-0.0026	0.2350-0.2356	0.2348-0.2354
	3VZ-E	3.0 (2959)	45①	44.5	54-57	1.575	0.0010-0.0024	0.0012-0.0026	0.3138-0.3144	0.3136-0.3142
	3VZ-FE	3.0 (2952)	45	44.5	38-42	1.311	0.0010-0.0024	0.0012-0.0026	0.2350-0.2356	0.2348-0.2354
	1FZ-FE	4.5 (4477)	45	44.5	53.4	1.437	0.0010-0.0024	0.0012-0.0026	0.2744-0.2750	0.2742-0.2748

86823003

VALVE SPECIFICATIONS

Year	Engine ID/VIN	Engine Displacement Liters (cc)	Seat Angle (deg.)	Face Angle (deg.)	Spring Test Pressure (lbs. @ in.)	Spring Installed Height (in.)	Stem-to-Guide Clearance (in.) Intake	Stem-to-Guide Clearance (in.) Exhaust	Stem Diameter (in.) Intake	Stem Diameter (in.) Exhaust
1995	22R-E	2.4 (2366)	45	44.5	66.1@1.594	1.594	0.0010-0.0024	0.0012-0.0026	0.3138-0.3144	0.3136-0.3142
	2RZ-FE	2.4 (2438)	45①	44.5	40-46@1.406	1.406	0.0010-0.0024	0.0012-0.0026	0.2350-0.2356	0.2348-0.2354
	3RZ-FE	2.7 (2693)	45①	44.5	57-63@1.406	1.406	0.0010-0.0024	0.0012-0.0026	0.2350-0.2356	0.2348-0.2354
	3VZ-E	3.0 (2959)	45①	44.5	54-57@1.575	1.575	0.0010-0.0024	0.0012-0.0026	0.3138-0.3144	0.3136-0.3142
	5VZ-FE	3.4 (3378)	45	44.5	41.9-46.3@1.311	1.311	0.0010-0.0024	0.0012-0.0026	0.2350-0.2356	0.2348-0.2354
	1FZ-FE	4.5 (4477)	45	44.5	53.4@1.437	1.437	0.0010-0.0024	0.0012-0.0026	0.2744-0.2750	0.2742-0.2748
1996	2RZ-FE	2.4 (2438)	45①	44.5	40-46@1.406	1.406	0.0010-0.0024	0.0012-0.0026	0.2350-0.2356	0.2348-0.2354
	3RZ-FE	2.7 (2693)	45①	44.5	57-63@1.406	1.406	0.0010-0.0024	0.0012-0.0026	0.2350-0.2356	0.2348-0.2354
	5VZ-FE	3.4 (3378)	45	44.5	41.9-46.3@1.311	1.311	0.0010-0.0024	0.0012-0.0026	0.2350-0.2356	0.2348-0.2354
	1FZ-FE	4.5 (4477)	45	44.5	53.4@1.437	1.437	0.0010-0.0024	0.0012-0.0026	0.2744-0.2750	0.2742-0.2748

1 Blend seat with 30 and 60 degree cutters to center the 45 degree portion on valve face.

86823004

CAMSHAFT SPECIFICATIONS
All measurements given in inches.

Year	Engine ID/VIN	Engine Displacement Liters (cc)	Journal Diameter 1	2	3	4	5	Elevation In.	Ex.	Bearing Clearance	Camshaft End-Play
1994	3VZ-E	3.0 (2959)	1.0610-1.0616	1.0610-1.0616	1.0610-1.0616	1.0610-1.0616	NA	1.6598-1.6638	0.6520-1.6559	0.0014-0.0028	0.0013-0.0031
	3VZ-FE	3.0 (2952)	1.0610-1.0616	1.0610-1.0616	1.0610-1.0616	1.0610-1.0616	NA	1.9925-1.9965	1.9925-1.9965	0.0010-0.0024	0.0012-0.0031
	1FZ-FE	4.5 (4477)	1.0614-1.0620	1.0614-1.0620	1.0614-1.0620	1.0614-1.0620	NA	1.9925-1.9965	1.9925-1.9965	0.0010-0.0024	0.0031-0.0071
1995	22R-E	2.4 (2366)	1.2992	1.2992	1.2992	1.2992	NA	1.6783-1.6819	1.6807-1.6842	0.0004-0.0024	0.0031-0.0071
	2RZ-FE	2.4 (2438)	1.0614-1.0620	1.0614-1.0620	1.0614-1.0620	1.0614-1.0620	1.0614-1.0620	1.7839-1.7878	1.7740-1.7779	0.0010-0.0024	0.0016-0.0037
	3RZ-FE	2.7 (2693)	1.0614-1.0620	1.0614-1.0620	1.0614-1.0620	1.0614-1.0620	1.0614-1.0620	1.7839-1.7878	1.7740-1.7779	0.0010-0.0031	0.0016-0.0047
	5VZ-FE	3.4 (3378)	1.0610-1.0616	1.0610-1.0616	1.0610-1.0616	1.0610-1.0616	1.0610-1.0616	1.6657-1.6697	0.6520-1.6559	0.0014-0.0028	0.0013-0.0031
	1FZ-FE	4.5 (4477)	1.0614-1.0620	1.0614-1.0620	1.0614-1.0620	1.0614-1.0620	NA	1.9925-1.9965	1.9925-1.9965	0.0010-0.0024	0.0012-0.0031
1996	2RZ-FE	2.4 (2438)	1.0614-1.0620	1.0614-1.0620	1.0614-1.0620	1.0614-1.0620	1.0614-1.0620	1.7839-1.7878	1.7740-1.7779	0.0010-0.0024	0.0016-0.0037
	3RZ-FE	2.7 (2693)	1.0614-1.0620	1.0614-1.0620	1.0614-1.0620	1.0614-1.0620	1.0614-1.0620	1.7839-1.7878	1.7740-1.7779	0.0010-0.0031	0.0016-0.0047
	5VZ-FE	3.4 (3378)	1.0610-1.0616	1.0610-1.0616	1.0610-1.0616	1.0610-1.0616	1.0610-1.0616	1.6657-1.6697	0.6520-1.6559	0.0014-0.0028	0.0013-0.0031
	1FZ-FE	4.5 (4477)	1.0614-1.0620	1.0614-1.0620	1.0614-1.0620	1.0614-1.0620	NA	1.9925-1.9965	1.9925-1.9965	0.0010-0.0024	0.0012-0.0031

NA - Not Available

86823006

CAMSHAFT SPECIFICATIONS
All measurements given in inches.

Year	Engine ID/VIN	Engine Displacement Liters (cc)	Journal Diameter 1	2	3	4	5	Elevation In.	Ex.	Bearing Clearance	Camshaft End-Play
1989	22R	2.4 (2366)	1.2984-1.2992	1.2984-1.2992	1.2984-1.2992	1.2984-1.2992	NA	1.6783-1.6819	1.6807-1.6842	0.0004-0.0024	0.0031-0.0071
	22R-E	2.4 (2366)	1.2984-1.2992	1.2984-1.2992	1.2984-1.2992	1.2984-1.2992	NA	1.6783-1.6819	1.6807-1.6842	0.0004-0.0024	0.0031-0.0071
	3VZ-E	3.0 (2959)	1.0610-1.0616	1.0610-1.0616	1.0610-1.0616	1.0610-1.0616	NA	1.6598-1.6638	1.6520-1.6559	0.0014-0.0028	0.0013-0.0031
	3F-E	4.0 (3956)	1.8880-1.8888	1.8289-1.8297	1.8289-1.8297	1.7699-1.7707	NA	1.5102-1.5142	1.5059-1.5098	0.0010-0.0030	0.0079-0.0103
1990	22R	2.4 (2366)	1.2984-1.2992	1.2984-1.2992	1.2984-1.2992	1.2984-1.2992	NA	1.6783-1.6819	1.6807-1.6842	0.0004-0.0024	0.0031-0.0071
	22R-E	2.4 (2366)	1.2984-1.2992	1.2984-1.2992	1.2984-1.2992	1.2984-1.2992	NA	1.6783-1.6819	1.6807-1.6842	0.0004-0.0024	0.0031-0.0071
	3VZ-E	3.0 (2959)	1.0610-1.0616	1.0610-1.0616	1.0610-1.0616	1.0610-1.0616	NA	1.6598-1.6638	1.6520-1.6559	0.0014-0.0028	0.0013-0.0031
	3F-E	4.0 (3956)	1.8880-1.8888	1.8289-1.8297	1.8289-1.8297	1.7699-1.7707	NA	1.5102-1.5142	1.5059-1.5098	0.0010-0.0030	0.0079-0.0103
1991	22R-E	2.4 (2393)	1.2984-1.2992	1.2984-1.2992	1.2984-1.2992	1.2984-1.2992	NA	1.6783-1.6819	1.6807-1.6842	0.0004-0.0024	0.0031-0.0071
	3VZ-E	3.0 (2959)	1.0610-1.0616	1.0610-1.0616	1.0610-1.0616	1.0610-1.0616	NA	1.6598-1.6638	1.6520-1.6559	0.0010-0.0030	0.0079-0.0103
	3F-E	4.0 (3956)	1.8880-1.8888	1.8289-1.8297	1.7699-1.7707	1.7108-1.7116	NA	1.5102-1.5142	1.5059-1.5098	0.0010-0.0030	0.0079-0.0103
1992	22R-E	2.4 (2366)	1.2984-1.2992	1.2984-1.2992	1.2984-1.2992	1.2984-1.2992	NA	1.6783-1.6819	1.6807-1.6842	0.0004-0.0024	0.0031-0.0071
	3VZ-E	3.0 (2959)	1.0610-1.0616	1.0610-1.0616	1.0610-1.0616	1.0610-1.0616	NA	1.6598-1.6638	1.6520-1.6559	0.0014-0.0028	0.0013-0.0031
	3F-E	4.0 (3956)	1.8880-1.8888	1.8289-1.8297	1.7699-1.7707	1.7108-1.7116	NA	1.5102-1.5142	1.5059-1.5098	0.0010-0.0030	0.0079-0.0103
1993	22R-E	2.4 (2366)	1.2984-1.2992	1.2984-1.2992	1.2984-1.2992	1.2984-1.2992	NA	1.6783-1.6819	1.6807-1.6842	0.0004-0.0024	0.0031-0.0071
	3VZ-E	3.0 (2959)	1.0610-1.0616	1.0610-1.0616	1.0610-1.0616	1.0610-1.0616	NA	1.6598-1.6638	1.6520-1.6559	0.0014-0.0028	0.0013-0.0031
	3VZ-FE	3.0 (2952)	1.0610-1.0616	1.0610-1.0616	1.0610-1.0616	1.0610-1.0616	NA	1.6598-1.6638	1.6520-1.6559	0.0014-0.0028	0.0013-0.0031
	1FZ-FE	4.5 (4477)	1.0614-1.0620	1.0614-1.0620	1.0614-1.0620	1.0614-1.0620	NA	1.9925-1.9965	1.9925-1.9965	0.0010-0.0028	0.0012-0.0031
1994	22R-E	2.4 (2366)	1.2984-1.2992	1.2984-1.2992	1.2984-1.2992	1.2984-1.2992	NA	1.6783-1.6819	1.6807-1.6842	0.0004-0.0024	0.0031-0.0071
	3RZ-FE	2.7 (2693)	1.0614-1.0620	1.0614-1.0620	1.0614-1.0620	1.0614-1.0620	NA	1.7839-1.7878	1.7740-1.7779	0.0010-0.0031	0.0016-0.0047

86823005

CRANKSHAFT AND CONNECTING ROD SPECIFICATIONS
All measurements are given in inches.

Year	Engine ID/VIN	Engine Displacement Liters (cc)	Crankshaft Main Brg. Journal Dia.	Crankshaft Main Brg. Oil Clearance	Crankshaft Shaft End-play	Crankshaft Thrust on No.	Connecting Rod Crank Rod Clearance	Connecting Rod Oil Clearance	Connecting Rod Side Clearance
1989	22R	2.4 (2366)	2.3616-2.3622	0.0010-0.0022	0.0008-0.0087	3	2.0661-2.0866	0.0010-0.0022	0.0008-0.0087
	22R-E	2.4 (2366)	2.3616-2.3622	0.0010-0.0022	0.0008-0.0087	3	2.0661-2.0866	0.0010-0.0022	0.0008-0.0087
	3VZ-E	3.0 (2959)	2.5195-2.5197	0.0009-0.0017	0.0008-0.0098	3	2.1648-2.1654	0.0009-0.0021	0.0059-0.0130
	3F-E	4.0 (3956)	①	0.0008-0.0017	0.0024-0.0063	3	2.1252-2.1260	0.0008-0.0024	0.0043-0.0091
1990	22R	2.4 (2366)	2.3616-2.3622	0.0010-0.0022	0.0008-0.0087	3	2.0661-2.0866	0.0010-0.0022	0.0008-0.0087
	22R-E	2.4 (2366)	2.3616-2.3622	0.0010-0.0022	0.0008-0.0087	3	2.0661-2.0866	0.0010-0.0022	0.0008-0.0130
	3VZ-E	3.0 (2959)	2.5195-2.5197	0.0009-0.0017	0.0008-0.0098	3	2.1648-2.1654	0.0009-0.0021	0.0059-0.0130
	3F-E	4.0 (3956)	①	0.0008-0.0017	0.0024-0.0063	3	2.1252-2.1260	0.0008-0.0024	0.0043-0.0091
1991	22R-E	2.4 (2366)	2.3616-2.3622	0.0010-0.0022	0.0008-0.0087	3	2.0661-2.0866	0.0010-0.0022	0.0008-0.0087
	3VZ-E	3.0 (2959)	2.5195-2.5197	0.0009-0.0017	0.0008-0.0098	3	2.1648-2.1654	0.0009-0.0021	0.0059-0.0130
	3F-E	4.0 (3956)	①	0.0008-0.0017	0.0024-0.0063	3	2.1252-2.1260	0.0008-0.0024	0.0043-0.0091
1992	22R-E	2.4 (2366)	2.3616-2.3622	0.0010-0.0022	0.0008-0.0087	3	2.0661-2.0866	0.0010-0.0022	0.0008-0.0087
	3VZ-E	3.0 (2959)	2.5195-2.5197	0.0009-0.0017	0.0008-0.0098	3	2.1648-2.1654	0.0009-0.0021	0.0059-0.0130
	3F-E	4.0 (3956)	①	0.0008-0.0017	0.0024-0.0063	3	2.1252-2.1260	0.0008-0.0024	0.0043-0.0091
1993	22R-E	2.4 (2366)	2.3616-2.3622	0.0010-0.0022	0.0008-0.0087	3	2.0661-2.0866	0.0010-0.0022	0.0008-0.0087
	3VZ-E	3.0 (2959)	2.5195-2.5197	0.0009-0.0017	0.0008-0.0098	3	2.1648-2.1654	0.0009-0.0021	0.0059-0.0130
	1FZ-FE	4.5 (4477)	2.7158-2.7165	0.0016-0.0032	0.0008-0.0087	3	②	0.0013-0.0020	0.0063-0.0103
1994	22R-E	2.4 (2366)	2.3616-2.3622	0.0010-0.0022	0.0008-0.0087	3	2.0661-2.0866	0.0010-0.0022	0.0008-0.0087
	3RZ-FE	2.7 (2693)	2.3617-2.3622	0.0009-0.0019	0.0008-0.0087	3	2.0661-2.0866	0.0012-0.0023	0.0063-0.0123
	3VZ-E	3.0 (2959)	2.5195-2.5197	0.0009-0.0017	0.0008-0.0098	3	2.1648-2.1654	0.0009-0.0021	0.0059-0.0130
	1FZ-FE	4.5 (4477)	2.7158-2.7165	0.0016-0.0032	0.0008-0.0087	3	②	0.0013-0.0020	0.0063-0.0103

1 No. 1: 2.6367-2.6376
   No. 2: 2.6957-2.6967
   No. 3: 2.7548-2.7557
   No. 4: 2.8139-2.8148
2 There are five sizes of standard connecting rod bearings marked 2, 3, 4, 5 and 6 accordingly.
   Replace with one having the same number (number imprinted on outside of bearing end).
   If the number of bearing cannot be determined, select correct bearing by adding together the numbers imprinted on connecting rod and crankshaft, then selecting the bearing with the same number as the total.
   EXAMPLE: Connecting rod 3 + Crankshaft 1=4. Use bearing 4.

86823010

CRANKSHAFT AND CONNECTING ROD SPECIFICATIONS
All measurements are given in inches.

Year	Engine ID/VIN	Engine Displacement Liters (cc)	Crankshaft Main Brg. Journal Dia.	Crankshaft Main Brg. Oil Clearance	Crankshaft Shaft End-play	Crankshaft Thrust on No.	Connecting Rod Crank Rod Clearance	Connecting Rod Oil Clearance	Connecting Rod Side Clearance
1995	22R-E	2.4 (2366)	2.3616-2.3622	0.0010-0.0022	0.0008-0.0087	3	2.0661-2.0866	0.0010-0.0022	0.0008-0.0087
	2RZ-FE	2.4 (2438)	③	④	0.0008-0.0087	3	⑤	0.0012-0.0022	0.0063-0.0123
	3VZ-E	3.0 (2959)	2.5195-2.5197	0.0009-0.0017	0.0008-0.0098	3	2.1648-2.1654	0.0009-0.0021	0.0059-0.0130
	3RZ-FE	2.7 (2693)	③	④	0.0008-0.0087	3	⑤	0.0012-0.0022	0.0063-0.0123
	5VZ-FE	3.4 (3378)	2.5195-2.5197	0.0009-0.0017	0.0008-0.0087	3	2.1648-2.1654	0.0009-0.0021	0.0059-0.0130
	1FZ-FE	4.5 (4477)	2.7158-2.7165	0.0016-0.0032	0.0008-0.0087	3	②	0.0013-0.0020	0.0063-0.0103
1996	2RZ-FE	2.4 (2438)	③	④	0.0008-0.0087	3	⑤	0.0012-0.0022	0.0063-0.0123
	3VZ-E	3.0 (2959)	2.5195-2.5197	0.0009-0.0017	0.0008-0.0098	3	2.1648-2.1654	0.0009-0.0021	0.0059-0.0130
	3RZ-FE	2.7 (2693)	③	④	0.0008-0.0087	3	⑤	0.0012-0.0021	0.0063-0.0123
	5VZ-FE	3.4 (3378)	2.5195-2.5197	0.0009-0.0017	0.0008-0.0087	3	2.1648-2.1654	0.0009-0.0021	0.0059-0.0130
	1FZ-FE	4.5 (4477)	2.7158-2.7165	0.0016-0.0032	0.0008-0.0087	3	②	0.0013-0.0020	0.0063-0.0103

3 No. 3: 2.2615-2.3820
   Others: 2.3617-2.3622
4 No. 3: 0.0012-0.0022
   All others: 0.0009-0.0019
5 Mark 4: 2.2047-2.2050
   Mark 5: 2.2050-2.2052
   Mark 6: 2.2052-2.2054

86823011

PISTON AND RING SPECIFICATIONS
All measurements are given in inches.

Year	Engine ID/VIN	Engine Displacement Liters (cc)	Piston Clearance	Ring Gap			Ring Side Clearance		
				Top Compression	Bottom Compression	Oil Control	Top Compression	Bottom Compression	Oil Control
1995	22R-E	2.4 (2366)	0.0008-0.0016	0.0098-0.0185	0.0236-0.0323	0.0079-0.0224	0.0012-0.0028	0.0012-0.0028	SNUG
	2RZ-FE	2.4 (2438)	0.0022-0.0031	0.0118-0.0157	0.0157-0.0197	0.0051-0.0150	0.0008-0.0028	0.0012-0.0028	SNUG
	3RZ-FE	2.7 (2693)	0.0019-0.0024	0.0118-0.0157	0.0157-0.0194	0.0051-0.0150	0.0008-0.0028	0.0012-0.0028	0.0019-0.0024
	5VZ-FE	3.4 (3378)	0.0053-0.0060	0.0118-0.0197	0.0157-0.0236	0.0059-0.0217	0.0016-0.0031	0.0012-0.0028	0.0012-0.0028
	1FZ-FE	4.5 (4477)	0.0016-0.0024	0.0118-0.0205	0.0177-0.0264	0.0059-0.0205	0.0016-0.0031	0.0012-0.0028	SNUG
1996	2RZ-FE	2.4 (2438)	0.0022-0.0031	0.0118-0.0157	0.0157-0.0197	0.0051-0.0150	0.0008-0.0028	0.0012-0.0028	SNUG
	3RZ-FE	2.7 (2693)	0.0019-0.0024	0.0118-0.0157	0.0157-0.0194	0.0051-0.0150	0.0008-0.0028	0.0012-0.0028	0.0019-0.0024
	5VZ-FE	3.4 (3378)	0.0053-0.0060	0.0118-0.0197	0.0157-0.0236	0.0059-0.0217	0.0016-0.0031	0.0012-0.0028	0.0012-0.0028
	1FZ-FE	4.5 (4477)	0.0016-0.0024	0.0118-0.0205	0.0177-0.0264	0.0059-0.0205	0.0016-0.0031	0.0012-0.0028	SNUG

86823008

PISTON AND RING SPECIFICATIONS
All measurements are given in inches.

Year	Engine ID/VIN	Engine Displacement Liters (cc)	Piston Clearance	Ring Gap			Ring Side Clearance		
				Top Compression	Bottom Compression	Oil Control	Top Compression	Bottom Compression	Oil Control
1989	22R	2.4 (2366)	0.0008-0.0016	0.0098-0.0185	0.0236-0.0323	0.0079-0.0224	0.0012-0.0028	0.0012-0.0028	SNUG
	22R-E	2.4 (2366)	0.0008-0.0016	0.0098-0.0185	0.0236-0.0323	0.0079-0.0224	0.0012-0.0028	0.0012-0.0028	SNUG
	3VZ-E	3.0 (2959)	0.0031-0.0039	0.0090-0.0130	0.0150-0.0190	0.0060-0.0160	0.0012-0.0028	0.0012-0.0028	SNUG
	3F-E	4.0 (3956)	0.0011-0.0019	0.0079-0.0165	0.0197-0.0283	0.0079-0.0323	0.0012-0.0028	0.0020-0.0035	SNUG
1990	22R	2.4 (2366)	0.0008-0.0016	0.0098-0.0185	0.0236-0.0323	0.0079-0.0224	0.0012-0.0028	0.0012-0.0028	SNUG
	22R-E	2.4 (2366)	0.0008-0.0016	0.0098-0.0185	0.0236-0.0323	0.0079-0.0224	0.0012-0.0028	0.0012-0.0028	0.0012-0.0028
	3VZ-E	3.0 (2959)	0.0031-0.0039	0.0091-0.0327	0.0150-0.0366	0.0069-0.0354	0.0012-0.0028	0.0012-0.0028	SNUG
	3F-E	4.0 (3956)	0.0011-0.0019	0.0079-0.0165	0.0197-0.0283	0.0079-0.0323	0.0012-0.0028	0.0020-0.0035	SNUG
1991	22R-E	2.4 (2366)	0.0008-0.0016	0.0098-0.0185	0.0236-0.0323	0.0079-0.0224	0.0012-0.0028	0.0012-0.0028	SNUG
	3VZ-E	3.0 (2959)	0.0031-0.0039	0.0091-0.0327	0.0150-0.0366	0.0059-0.0354	0.0012-0.0028	0.0012-0.0028	SNUG
	3F-E	4.0 (3956)	0.0011-0.0019	0.0079-0.0165	0.0197-0.0283	0.0079-0.0323	0.0012-0.0028	0.0020-0.0035	SNUG
1992	22R-E	2.4 (2366)	0.0008-0.0016	0.0098-0.0185	0.0236-0.0323	0.0079-0.0224	0.0012-0.0028	0.0012-0.0028	SNUG
	3VZ-E	3.0 (2959)	0.0031-0.0039	0.0091-0.0327	0.0150-0.0366	0.0069-0.0354	0.0012-0.0028	0.0012-0.0028	0.0012-0.0028
	3F-E	4.0 (3956)	0.0011-0.0019	0.0079-0.0165	0.0197-0.0283	0.0079-0.0323	0.0012-0.0028	0.0020-0.0035	SNUG
1993	22R-E	2.4 (2366)	0.0008-0.0016	0.0098-0.0185	0.0236-0.0323	0.0079-0.0224	0.0012-0.0028	0.0012-0.0028	SNUG
	3VZ-E	3.0 (2959)	0.0031-0.0039	0.0091-0.0327	0.0150-0.0366	0.0059-0.0354	0.0012-0.0028	0.0012-0.0028	SNUG
	3VZ-FE	3.0 (2952)	0.0051-0.0059	0.0110-0.0197	0.0150-0.0236	0.0059-0.0234	0.0004-0.0031	0.0012-0.0028	SNUG
	1FZ-FE	4.5 (4477)	0.0016-0.0024	0.0118-0.0205	0.0177-0.0264	0.0059-0.0205	0.0016-0.0031	0.0012-0.0028	SNUG
1994	22R-E	2.4 (2366)	0.0008-0.0016	0.0098-0.0185	0.0236-0.0323	0.0079-0.0224	0.0012-0.0028	0.0012-0.0028	SNUG
	3RZ-FE	2.7 (2693)	0.0019-0.0024	0.0118-0.0157	0.0157-0.0194	0.0157-0.0194	0.0008-0.0028	0.0012-0.0028	0.0019-0.0024
	3VZ-E	3.0 (2959)	0.0031-0.0039	0.0091-0.0327	0.0150-0.0366	0.0059-0.0354	0.0012-0.0028	0.0012-0.0028	0.0012-0.0028
	3VZ-FE	3.0 (2952)	0.0051-0.0059	0.0110-0.0197	0.0150-0.0236	0.0059-0.0234	0.0004-0.0031	0.0012-0.0028	SNUG
	1FZ-FE	4.5 (4477)	0.0016-0.0024	0.0118-0.0205	0.0177-0.0264	0.0059-0.0205	0.0016-0.0031	0.0012-0.0028	SNUG

86823007

TORQUE SPECIFICATIONS
All readings in ft. lbs.

Year	Engine ID/VIN	Engine Displacement Liters (cc)	Cylinder Head Bolts	Main Bearing Bolts	Rod Bearing Bolts	Crankshaft Damper Bolts	Flywheel Bolts	Manifold Intake	Manifold Exhaust	Spark Plugs	Lug Nut
1989	22R	2.4 (2366)	53-63	69-83	40-47	120-130	73-86	13-19	26-36	11-15	76
	22R-E	2.4 (2366)	53-63	69-83	40-47	120-130	73-86	13-19	26-36	11-15	76
	3VZ-E	3.0 (2959)	③	②	①	176-186	63-67	11-15	25-33	11-15	76
	3F-E	4.0 (3956)	87-93	④	40-46	247-259	60-68	⑤	⑤	11-15	116
1990	22R	2.4 (2366)	53-63	69-83	40-47	120-130	73-86	13-19	26-36	11-15	76
	22R-E	2.4 (2366)	53-63	69-83	40-47	120-130	73-86	13-19	26-36	11-15	76
	3VZ-E	3.0 (2959)	③	②	①	176-186	63-67	11-15	25-33	11-15	76
	3F-E	4.0 (3956)	87-93	④	40-46	247-259	60-68	⑤	⑤	11-15	116
1991	22R-E	2.4 (2366)	53-63	69-83	40-47	120-130	73-86	13-19	26-36	11-15	76
	3VZ-E	3.0 (2959)	③	②	①	176-186	63-67	11-15	252-33	11-15	76
	3F-E	4.0 (3956)	87-93	④	40-46	247-259	60-68	⑤	⑤	11-15	116
1992	22R-E	2.4 (2366)	53-63	69-83	40-47	120-130	73-86	13-19	26-36	11-15	76
	3VZ-E	3.0 (2959)	③	②	①	176-186	63-67	11-15	25-33	11-15	76
	3F-E	4.0 (3956)	87-93	④	40-46	247-259	60-68	⑤	⑤	11-15	116
1993	22R-E	2.4 (2366)	53-63	69-83	40-47	120-130	73-86	13-19	26-36	11-15	76
	3VZ-E	3.0 (2959)	③	②	①	176-186	63-67	11-15	25-33	11-15	76
	3VZ-FE	3.0 (2952)	⑥	②	①	181	61	13	29	13	76
	1FZ-FE	4.5 (4477)	⑦	⑧	⑩	304	-	15	29	15	108
1994	22R-E	2.4 (2366)	53-63	69-83	40-47	120-130	73-86	13-19	26-36	11-15	76
	3RZ-FE	2.7 (2693)	⑨	⑬	⑪	192	①	22	36	14	76
	3VZ-E	3.0 (2959)	③	②	①	176-186	63-67	11-15	25-33	11-15	76
	3VZ-FE	3.0 (2952)	⑥	②	①	181	61	13	29	13	76
	1FZ-FE	4.5 (4477)	⑦	⑧	⑩	304	-	15	29	15	108
1995	22R-E	2.4 (2366)	53-63	69-83	40-47	120-130	73-86	13-19	26-36	11-15	76
	2RZ-FE	2.4 (2438)	⑨	⑬	⑪	193	65	22	36	14	76
	3RZ-FE	2.7 (2693)	⑨	⑬	⑪	⑫	⑫	22	36	14	76
	5VZ-FE	3.4 (3378)	③	②	⑫	176-186	63-67	11-15	25-33	11-15	76
	1FZ-FE	4.5 (4477)	⑦	⑧	⑩	304	-	15	29	15	108
1996	2RZ-FE	2.4 (2438)	⑨	⑬	⑪	193	65	22	36	14	76
	3RZ-FE	2.7 (2693)	⑨	⑬	⑪	⑫	⑫	22	36	14	76
	5VZ-FE	3.4 (3378)	③	②	⑫	176-186	63-67	11-15	25-33	11-15	76
	1FZ-FE	4.5 (4477)	⑦	⑧	⑩	304	-	15	29	15	108

1 Step 1: 18 ft. lbs.
  Step 2: 90 degree turn
2 Step 1: 45 ft. lbs.
  Step 2: 90 degree turn
3 Step 1: 27 ft. lbs.
  Step 2: 33 ft. lbs.
  Step 3: 90 degree turn
  Step 4: 90 degree turn
4 17mm bolts: 85 ft. lbs.
  19mm bolts: 99 ft. lbs.
5 14mm bolts: 37 ft. lbs.
  17mm bolts: 51 ft. lbs.
  Nuts: 41 ft. lbs.

6 Step 1: 25 ft. lbs.
  Step 2: 90 degree turn
  Step 3: 90 degree turn
  Recessed head bolt: 13 ft. lbs.
7 Step 1: 27 ft. lbs.
  Step 2: 90 degree turn
  Step 3: 90 degree turn
8 Step 1: 54 ft. lbs.
  Step 2: 90 degree turn

9 Step 1: 29 ft. lbs.
  Step 2: 90 degree turn
  Step 3: 90 degree turn
10 Step 1: 35 ft. lbs.
   Step 2: 90 degree turn
11 Step 1: 33 ft. lbs.
   Step 2: 90 degree turn
12 Step 1: 19 ft. lbs.
   Step 2: 90 degree turn
13 Step 1: 29 ft. lbs.
   Step 2: 90 degree turn

86823009

Engine

REMOVAL & INSTALLATION

❊❊ CAUTION

Please refer to Section 1 before discharging the compressor or disconnecting air conditioning lines. Damage to the air conditioning system or personal injury could result. Consult your local laws concerning refrigerant discharge and recycling. In many areas it may be illegal for anyone but a certified technician to service the A/C system. Always use an approved recovery station when discharging the air conditioning system.

In the process of removing the engine, you will come across a number of steps which call for the removal of a separate component or system, i.e. "disconnect the exhaust system" or "remove the radiator." In most instances, a detailed removal procedure can be found elsewhere in this manual.

It is virtually impossible to list each individual wire and hose which must be disconnected, simply because so many different model and engine combinations have been manufactured. Careful observation and common sense are the best possible additions to any repair procedure. Be absolutely sure to tag and wire or hose before it is disconnected, so that you can be assured of proper reconnection during installation.

22R Engine

1. Disconnect the negative battery cable.
2. Remove the engine undercover or splash shield.

➡ **Apply a lubricant onto any nuts or bolts that will be removed. Allow the lubricant to sit for a short while. This will aid in removal of tight and frozen nuts and bolts.**

3. Disconnect the windshield washer hose. Scribe matchmarks around the hinges, then remove the hood.
4. Drain the engine oil. Drain the engine coolant from the radiator and the cylinder block.

When draining coolant, keep in mind that cats and dogs are attracted by ethylene glycol antifreeze, and are quite likely to drink any that is left in an uncovered container or in puddles on the ground. This will prove fatal in sufficient quantity. Always drain coolant into a sealable container. Coolant may be reused unless it is contaminated or several years old.

5. Drain the automatic transmission fluid on models so equipped.
6. Disconnect the air cleaner hoses and then remove the air cleaner.
7. Remove the fan coupling with the fan.
8. Disconnect the two heater hoses at the engine.
9. Remove the radiator with the shroud attached.
10. Disconnect the accelerator cable at the carburetor.
11. Tag and disconnect the following wires:
- Vacuum switching valve (VSV) wire at the evaporative emissions (EVAP) canister
- Vacuum switching (VSV) valve wire at the air conditioning compressor
- Vacuum switch wire
- HAC wire (ex. Calif.)
- Cold mixture heater wire
- Fuel cut solenoid wire
- Water temperature sender gauge wire
- Electronic air control (EACV) valve wire (Calif.)
- Starter wire
- Oil pressure switch wire
12. Disconnect the following parts:
- Brake booster hose
- Main fuel line at the fuel inlet pipe
- Fuel return line at the fuel return pipe
- Charcoal canister hose
- HAC from the bracket (except Calif.)
- Vacuum switch, Electronic Bleed Air Control Valve (EBCV) Calif. and Vacuum Switching Valve (VSV) with its bracket
13. Remove the drive belt.
14. Remove the power steering pump from its bracket (if equipped). Move the pump to one side without disconnecting the hoses. Use a piece of stiff wire to support it. Disconnect the ground strap from the bracket.
15. On models with air conditioning, loosen the drive belt and remove the air conditioning compressor. Position it out of the way with the refrigerant lines still attached. Use several pieces of stiff wire to support the compressor.
16. Disconnect the engine ground straps at the rear and right side of the engine.
17. On trucks with a manual transmission, remove the shift lever from inside the truck.
18. On models with automatic transmission, remove the shift lever retainer.
19. Remove the rear driveshaft.
20. On vehicles with automatic transmissions, disconnect the manual shift linkage at the neutral start switch. On 4WD models with automatic transmission, disconnect the transfer shift linkage at the cross-shaft.
21. Remove the cross-shaft from the body.
22. Disconnect the speedometer cable. Be sure not to lose the felt dust protector and washers.
23. On 4WD models, remove the transfer case undercover.
24. On 4WD models, remove the stabilizer bar.
25. On 4WD models, remove the front driveshaft.
26. Remove the No. 1 frame crossmember.
27. Disconnect the oxygen sensor wiring. Disconnect the exhaust pipe at the manifold; remove the clamp from the clutch housing. Remove the exhaust pipe from the catalytic converter.
28. On models with manual transmission, remove the clutch release (slave) cylinder and its bracket from the transmission.
29. Remove the No. 1 front floor heat insulator and the brake tube heat insulator (4WD only).
30. On 2WD models, remove the four rear engine mount bolts. Raise the transmission slightly with a floor jack and then remove the four support member mounting bolts.
31. On 4WD models, remove the four rear engine mount bolts, raise the transmission slightly with a floor jack and then remove the four bolts from the side member and remove the No. 2 frame crossmember.

32. Attach an engine hoist chain to the lifting brackets on the engine. Remove the engine mount nuts and bolts and slowly lift the engine/transmission out of the truck.
33. Remove the starter. If automatic transmission, remove the oil cooler lines from the transmission. Remove the two stiffener brackets and the exhaust pipe bracket from the engine.
34. Remove the transmission.
35. If manual transmission, remove the clutch cover (pressure plate) and clutch disc.
36. Mount the engine on an engine stand, making sure the retaining bolts are tight. If an engine stand is not available, support the engine in an upright position with blocks. Never leave an engine hanging from a lift or hoist.
To install:
37. Installing the engine is essentially the reverse of the removal procedures.
38. On 2WD models, raise the transmission slightly and align the rear engine mount with the support member. Tighten the bolts to 9 ft. lbs. (13 Nm). Lower the transmission until it rests on the extension housing and then tighten the bracket mounting bolts to 19 ft. lbs. (25 Nm).
39. On 4WD models, raise the transmission slightly and tighten the No. 2 frame crossmember-to-side frame bolts to 70 ft. lbs. (95 Nm). Lower the transmission and tighten the four rear engine mount bolts to 9 ft. lbs. (13 Nm).
40. When installing the clutch release (slave) cylinder and its bracket to the manual transmission. Tighten the bracket bolts to 29 ft. lbs. (39 Nm) and the cylinder bolts to 9 ft. lbs. (13 Nm).
41. When installing the power steering pump and belt; adjust the belt correctly. Tighten the pump pulley locknut to 32 ft. lbs. (43 Nm).
42. Once the engine is installed and with the coil wire disconnected from the distributor, crank the engine with the ignition key for 5–10 seconds. This should pre-oil at least the lower part of the engine.
43. Reconnect the coil wire. Start the engine, allowing it to warm up. Check carefully for leaks.
44. Check and adjust the engine specifications as necessary.
45. Road test the vehicle. Recheck and top fluid levels as needed.

22R-E Engine

1. Disconnect the windshield washer hose. Scribe matchmarks around the hinges then remove the hood.

➡**Apply a lubricant onto any nuts or bolts that will be removed. Allow the lubricant to sit for a short while. This will aid in the removal of tight and frozen nuts and bolts.**

2. Relieve the fuel pressure.
3. Disable the air bag system.

Fuel injected engine systems remain under pressure after the engine has been turned OFF. Properly relieve the fuel pressure before disconnecting any fuel lines. Failure to do so may result in fire or personal injury.

4. Disconnect the negative battery cable.
5. Remove the engine undercover.
6. Drain the engine oil. Drain the engine coolant from the radiator and the cylinder block.

The EPA warns that prolonged contact with used engine oil may cause a number of skin disorders, including cancer! You should make every effort to minimize you exposure to used engine oil. Protective gloves should be worn when changing the oil. Wash your hands and any other exposed skin areas as soon as possible after exposure to used engine oil. Soap and water, or waterless hand cleaner should be used.

7. Drain the automatic transmission fluid on models so equipped.
8. Remove the air cleaner case and the intake air hose.
9. Remove the radiator and shroud.

10. Loosen the power steering pulley locknut and remove the power steering belt.

11. If equipped with air conditioning, remove the compressor belt.

12. Remove the alternator drive belt, fluid coupling for the fan and the fan pulley.

13. Disconnect the following wires and connectors:
- Ground strap from the left fender apron
- Alternator connector and wire
- Igniter
- Alternator wires
- Coil wire from distributor
- Ground strap at rear of engine
- Electronic Control Unit (ECU)
- Starter relay connector (manual transmission only)
- The check connector
- Air conditioning compressor connector

14. Disconnect and label these hoses: air hoses from the fuel filter and air pipe, brake booster hose, charcoal canister hose and cruise control vacuum hose if so equipped.

15. Disconnect the accelerator cable, throttle cable for the automatic transmission (if so equipped) and the cruise control cable if so equipped.

16. Remove the power steering pump from its bracket (if equipped).

17. Disconnect the ground strap from the bracket.

18. On models with air conditioning, remove the air conditioning compressor. Position it out of the way with the refrigerant lines still attached.

19. Disconnect the engine ground straps at the rear and right side of the engine.

20. On trucks with a manual transmission, remove the shift lever from inside the truck.

21. Remove the rear driveshaft.

22. On models with automatic transmission, disconnect the manual shift linkage at the neutral start switch.

23. On 4WD models with automatic transmission, disconnect the transfer shift linkage at the cross-shaft.

24. Remove the cross-shaft assembly from the body.

25. Disconnect the speedometer cable. Be sure not to lose the felt dust protector and washers.

26. Remove the transfer case undercover (4WD only).

27. Remove the stabilizer bar (4WD only).

28. Remove the front driveshaft (4WD only).

29. Remove the No. 1 frame crossmember.

30. Disconnect the oxygen sensor wiring. Disconnect the exhaust pipe at the manifold; remove the clamp from the clutch housing. Remove the exhaust pipe from the catalytic converter.

31. On models with manual transmission, remove the clutch release (slave) cylinder and its bracket from the transmission.

32. Remove the No. 1 front floor heat insulator and the brake tube heat insulator (4WD only).

33. On 2WD models, remove the four rear engine mount bolts. Raise the transmission slightly with a floor jack and then remove the support member mounting bolts.

34. On 4WD models, remove the four rear engine mount bolts, raise the transmission slightly with a floor jack and then remove the four bolts from the side member and remove the No. 2 frame crossmember.

35. Attach an engine hoist chain to the lifting brackets on the engine. Remove the engine mount nuts and bolts and slowly lift the engine/transmission out of the truck.

36. Remove the starter. If automatic transmission, remove the oil cooler lines from the transmission. Remove the two stiffener brackets and the exhaust pipe bracket from the engine.

37. Remove the transmission.

38. If manual transmission, remove the clutch cover (pressure plate) and clutch disc.

39. Mount the engine on an engine stand, making sure the retaining bolts are tight. If an engine stand is not available, support the engine in an upright position with blocks. Never leave an engine hanging from a lift or hoist.

To install:

40. Installation of the engine ie essentially the reverse of removal.

41. On 2WD models, raise the transmission slightly and align the rear engine mount with the support member and tighten the bolts to 9 ft. lbs. (13 Nm). Lower the transmission until it rests on the extension housing and then tighten the bracket mounting bolts to 19 ft. lbs. (25 Nm).

42. On 4WD models, Raise the transmission slightly and tighten the No. 2 frame crossmember-to-side frame bolts to 70 ft. lbs. (95 Nm). Lower the transmission and tighten the four rear engine mount bolts to 9 ft. lbs. (13 Nm).

43. When installing the clutch release (slave) cylinder and its bracket to the manual transmission. Tighten the bracket bolts to 29 ft. lbs. (39 Nm) and the cylinder bolts to 9 ft. lbs. (13 Nm).

44. Double check all installation items, paying particular attention to loose hoses or hanging wires, untightened nuts, poor routing of hoses and wires (too tight or rubbing) and tools left in the engine area.

45. With the coil wire disconnected from the distributor, crank the engine with the ignition key for 5–10 seconds. This should pre-oil at least the lower part of the engine.

46. Reconnect the coil wire. Start the engine, allowing it to warm up. Check carefully for leaks.

47. Check and adjust the engine specifications as necessary.

48. Road test the vehicle. Recheck and top fluid levels as needed.

3VZ-E Engine

▶ See Figures 7, 8 and 9

1. Relieve the fuel pressure.

2. Disconnect the windshield washer hose. Scribe matchmarks around the hinges, then remove the hood.

3. Disconnect the battery cables and remove the battery.

➡Apply a lubricant onto any nuts or bolts that will be removed. Allow the lubricant to sit for a short while. This will aid in the removal of tight and frozen nuts and bolts.

4. Remove the engine undercover.

5. Drain the engine oil. Drain the engine coolant from the radiator and the cylinder block.

6. Drain the automatic transmission fluid on models so equipped.

7. Disconnect the air cleaner hose and then remove the air cleaner.

8. Remove the cooling fan.

9. Remove the radiator.

10. Remove all drive belts and then remove the fluid coupling and fan pulley.

11. Tag and disconnect the following wires and connectors:
- Left side and rear ground straps
- Alternator connector and wire
- Igniter
- Oil pressure switch
- Electronic Control Unit (ECU)
- Vacuum Switching Valve (VSV)
- Starter relay (manual transmission only)
- Solenoid resistor

- Check connector
- Air conditioning compressor
12. Tag and disconnect the following hoses:
- Air hoses at the gas filter and air pipe
- Brake booster
- Cruise control vacuum hose (if equipped)
- Charcoal canister hose at the canister
- Vacuum Switching Valve (VSV)
- Oil cooler—(automatic transmissions)

13. Disconnect the accelerator cable, throttle control cable (auto. trans.) and cruise control cable if equipped.

14. Unbolt the power steering pump and position it out of the way with the hydraulic lines still connected.

15. Remove the air conditioning compressor if equipped. Position is out of the way with the lines still attached.

16. On manual transmissions, disconnect the clutch release (slave) cylinder hose.

17. Disconnect the two heater hoses from the engine.

18. Disconnect and plug the fuel inlet and outlet lines.

19. Remove the shift levers (manual transmission only).

20. Remove the rear driveshaft.

21. Disconnect the manual shift linkage (automatic transmission only).

22. Disconnect the speedometer cable—don't lose the felt dust protector and washers.

23. Remove the transfer case undercover.

24. Remove the stabilizer bar.

25. Remove the front driveshaft.

26. Remove the No.1 frame crossmember.

27. Remove the front exhaust pipe.

28. Remove the front floor heat insulator and the brake tube heat insulator.

29. Remove the four rear engine mount bolts. Raise the transmission slightly with a floor jack and remove the bolts securing the bracket and frame crossmember.

30. Attach an engine hoist chain to the lifting brackets on the engine. Remove the engine mount nuts and bolts and slowly lift the engine/transmission out of the truck.

31. Remove the starter. If automatic transmission, remove the oil cooler lines from the transmission. Remove the two stiffener brackets and the exhaust pipe bracket from the engine.

32. Remove the transmission.

33. If manual transmission, remove the clutch cover (pressure plate) and clutch disc.

34. Mount the engine on an engine stand, making sure the retaining bolts are tight. If an engine stand is not available, support the engine in an upright position with blocks. Never leave an engine hanging from a lift or hoist.

To install:

35. Installation of the engine is essentially the reverse of removal.

36. For 2WD vehicles, install the rear engine mount and bracket. Install the bolts holding the bracket to the support and tighten them to 9 ft. lbs. (13 Nm). Lower the transmission, resting it on the extension housing. Install the bracket to the mount; tighten the bolts to 19 ft. lbs. (25 Nm)

37. On 4WD vehicles, raise the transmission slightly and tighten the No. 2

frame crossmember-to-side frame bolts to 70 ft. lbs. (95 Nm). Lower the transmission and tighten the four rear engine mount bolts to 9 ft. lbs. (13 Nm).

38. Install the power steering pump and connect the ground strap. Tighten the pump pulley locknut to 32 ft. lbs. (43 Nm).

39. Double check all installation items, paying particular attention to loose hoses or hanging wires, untightened nuts, poor routing of hoses and wires (too tight or rubbing) and tools left in the engine area.

40. With the coil wire disconnected from the distributor, crank the engine with the ignition key for 5–10 seconds. This should pre-oil at least the lower part of the engine.

41. Reconnect the coil wire. Start the engine, allowing it to warm up. Check carefully for leaks.

42. Check and adjust the engine specifications as necessary.

43. Road test the vehicle. Recheck and top fluid levels as needed.

5VZ-FE Engine

2WD MODELS

♦ See Figures 10, 11 and 12

➡ Apply a lubricant onto any nuts or bolts that will be removed. Allow the lubricant to sit for a short while. This will aid in the removal of tight and frozen nuts and bolts.

1. Remove the hood.
2. Relieve the fuel pressure.
3. Disable the air bag system. Refer to Section 6.

❉ CAUTION

Work must be started only after 90 seconds from the time the ignition switch is turned to the LOCK position and the negative battery cable is disconnected.

4. Disconnect the battery and remove it from the vehicle.
5. Raise and safely support the vehicle.
6. Remove the engine under covers.
7. Drain the engine coolant.
8. Drain the engine oil.

❉ CAUTION

The EPA warns that prolonged contact with used engine oil may cause a number of skin disorders, including cancer! You should make every effort to minimize you exposure to used engine oil. Protective gloves should be worn when changing the oil. Wash your hands and any other exposed skin areas as soon as possible after exposure to used engine oil. Soap and water, or waterless hand cleaner should be used.

9. Remove the radiator.
10. Remove the power steering drive belt:
   a. Stretch the belt and loosen the fan pulley mounting bolts.

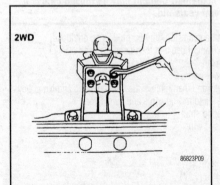

Fig. 7 On 2WD vehicles, first remove the 4 rear engine mount bolts . . .

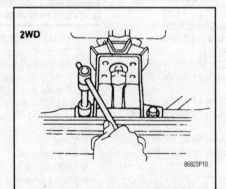

Fig. 8 . . . then remove the remaining bracket bolts

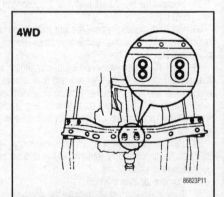

Fig. 9 On 4WD models, remove these crossmember bolts

b. Loosen the lock, pivot and adjusting bolts. Remove the drive belt from the engine.

11. If equipped with A/C, remove the A/C drive belt by loosening the idle pulley nut and adjusting bolt.

12. Loosen the lock, pivot and adjusting bolts, then remove the alternator belt.

13. Remove the fan with the fluid coupling and the fan pulleys.

14. Disconnect the power steering pump from the engine and set aside. Do not disconnect the lines from the pump.

15. If equipped with A/C, disconnect the compressor from the engine and set aside. Do not disconnect the lines from the compressor.

16. Remove the air cleaner lid, Mass Air Flow (MAF) meter and resonator.

17. Remove the air cleaner case and filter assembly.

18. Disconnect the following cables:
- If the vehicle has cruise control, disconnect the detent cable
- Accelerator
- Throttle cable on automatics

19. Disconnect the heater hoses.

20. Disconnect the following hoses:
- Brake booster vacuum
- Evaporative emissions (EVAP)
- Fuel return
- Fuel inlet

21. Disconnect the starter wire and connectors as follows:
   a. Remove the bolt retaining the ground strap.
   b. Disconnect the positive battery cable.
   c. Disconnect the starter wires and connectors.

22. Disconnect the alternator harness.

23. Remove the engine harness as follows:
   a. Remove the four screws to the right front door scuff plate.
   b. Remove clip retaining the cowl panel side trim.
   c. Disconnect the Electronic Control Unit (ECU) wiring.
   d. Detach the two connectors from the cowl wire.
   e. Disconnect the igniter.
   f. Detach the ground strap.
   g. Disconnect the six engine wire clamps.
   h. Pull the engine wire from the compartment.

24. If equipped with a manual transmission, remove the shift lever as follows:
   a. Remove the shift lever knob.
   b. Remove the four screws and the shift lever boot.
   c. Remove the six bolts to retrieve the shift lever assembly and gasket.

25. Remove the stabilizer bar.

26. Remove the driveshaft from the transmission.

27. Disconnect the speedometer cable.

28. Remove the front exhaust pipe as follows:
   a. Disconnect the heated oxygen sensor.
   b. Remove the two bolts and retainer holding the front pipe to the catalytic converter.
   c. Remove the three bolts and the support bracket.
   d. Loosen and remove the three nuts holding the front pipe and gaskets.

29. If equipped with with a manual transmission, remove the slave cylinder.

30. On automatic transmissions, remove the cross-shaft.

31. Place a jack under the transmission.

32. Remove the eight bolts retaining the transmission rear mounting bracket.

33. If equipped with A/C, remove the bolt and disconnect the wire.

34. Attach the engine hoist chain to the two engine hangers.

35. Remove the four bolts and nuts holding the front mounting insulators to the frame.

36. Lift the engine and transmission out of the vehicle.

To install:

37. Installation of the engine is essentially the reverse of the removal procedures, please note the following.

38. Raise and support the transmission slightly to install the mounting bracket. Tighten the bolts to the frame to 43 ft. lbs. (58 Nm) along with the bolts to the mounting bracket insulator to 13 ft. lbs. (18 Nm).

39. Tighten the engine mounting nuts and bolts to 28 ft. lbs. (38 Nm).

40. On manual transmissions, install the slave cylinder, tighten the mounting bolts to 9 ft. lbs. (12 Nm).

41. Attach the front exhaust pipe:
   a. Install the gaskets and attach the pipe to the engine with new mounting nuts. Tighten the nuts to 36 ft. lbs. (44 Nm).
   b. Install the support bracket and tighten the bolts to 33 ft. lbs. (44 Nm).
   c. Connect the exhaust pipe with a new gasket to the catalytic converter. Tighten the bolts to 35 ft.. lbs. (48 Nm).

42. Install the fan, fluid couplings and pulleys. Tighten to 48 inch lbs. (5 Nm).

43. Install the alternator and A/C drive belts.

44. Install the power steering pump, pulley and drive belt.

45. Install the radiator.

46. Fill the engine with oil.

47. Fill the cooling system.

48. Attach the engine splash shield.

49. Start the engine and check for leaks.

50. Top off any fluid levels.

4WD MODELS

▶ **See Figures 10 and 12**

1. Disable the air bag system.
2. Remove the transmission from the vehicle.
3. Remove the hood.

➡**Apply a lubricant onto any nuts or bolts that will be removed. Allow the lubricant to sit for a short while. This will aid in the removal of tight and frozen nuts and bolts.**

4. Disconnect the battery and remove it from the vehicle.

✷✷ CAUTION

Work must be started after 90 seconds from the time the ignition switch is turned to the LOCK position and the negative battery cable is disconnected.

5. Raise and support the vehicle safely.
6. Remove the engine splash shields.

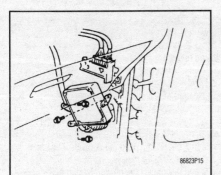

Fig. 10 The Electronic Control Unit (ECU) can be found behind the passenger's side kick panel

Fig. 11 Remove the clamps securing the engine wiring harness at the points shown

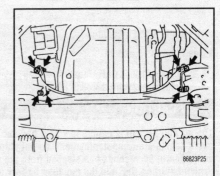

Fig. 12 Remove these four bolts and nuts that secure the engine mounting insulators to the frame

7. Drain the engine coolant.
8. Drain the engine oil.
9. Remove the radiator.
10. Remove the Power Steering (PS) drive belt:
   a. Loosen the fan pulley mounting bolts.
   b. Loosen the lock, pivot and adjusting bolts. Remove the drive belt.
11. On A/C vehicles, loosen the idle pulley nut and adjusting bolt to remove the A/C drive belt.
12. Loosen the lock, pivot and adjusting bolts. Remove the drive belt.
13. Remove the fan, fluid coupling and pulleys.
14. Disconnect the PS pump from the engine and set aside. Do not disconnect the lines from the pump.
15. On A/C vehicles, disconnect the compressor from the engine and set aside. Do not disconnect the lines from the compressor.
16. Remove the air cleaner lid, Mass Air Flow (MAF) meter and resonator.
17. Remove the air cleaner assembly.
18. Disconnect the following cables:
• Cruise control actuator and bracket
• Accelerator cable
• On automatics, remove the throttle cable
19. Disconnect the heater hoses.
20. Disconnect the following hoses:
• Brake booster vacuum hose
• Evaporative emissions (EVAP) hose
• Fuel return
• Fuel inlet
21. Disconnect the starter wire:
   a. Remove the ground strap.
   b. Remove the nuts and disconnect the positive battery cable.
   c. Disconnect the three starter wire clamps.
22. Disconnect the alternator harness.
23. Detach the engine harness:
   a. Remove the right front door scuff plate.
   b. Remove the cowl panel side trim.
   c. Disconnect the Electronic Control Unit (ECU).
   d. Disconnect the wires leading to the cowl.
   e. Remove the wiring to the igniter.
   f. Disconnect the ground strap.
   g. Remove the wire clamps.
   h. Pull out the engine wire from the cabin.
24. On A/C vehicles, remove the bolt and disconnect the compressor wire clamp.
25. Attach the engine hoist chain to the two engine hangers.
26. Remove the bolts and nuts holding the engine front mounting insulators to the frame.
27. Lift the engine out of the vehicle.

To install:

28. Install the engine assembly in the vehicle. Attach the engine mounts to the body mountings. Install the bolts and nuts but do not tighten them. Installation of the remaining components is essentially the reverse of removal. Please note the following steps.
29. Tighten the engine mounting nuts and bolts to 28 ft. lbs. (38 Nm).
30. Install the fan, coupling and pulleys. Tighten the nuts to 48 inch lbs. (5 Nm).
31. Start the engine and check for leaks.
32. Top off any fluid levels.

2RZ-FE and 3RZ-FE Engines

▶ See Figures 13 and 14

1. Turn the ignition **OFF**. Disable the air bag system, if equipped.
2. If not already done, disconnect the battery cables, negative first.

❋❋ CAUTION

The air bag system is equipped with a back-up power source. To avoid possible air bag deployment, do not start working on the vehicle until 90 seconds has elapsed from the time the ignition switch is turned OFF and the negative battery terminal is disconnected.

3. Matchmark the hood hinges, then remove the hood.
4. Relieve the fuel system pressure.

➡ Apply a lubricant onto any nuts or bolts that will be removed. Allow the lubricant to sit for a short while. This will aid in the removal of tight and frozen nuts and bolts.

5. Remove the engine splash shield.
6. Drain the engine oil, transmission fluid and cooling system.

❋❋ CAUTION

The EPA warns that prolonged contact with used engine oil may cause a number of skin disorders, including cancer! You should make every effort to minimize you exposure to used engine oil. Protective gloves should be worn when changing the oil. Wash your hands and any other exposed skin areas as soon as possible after exposure to used engine oil. Soap and water, or waterless hand cleaner should be used.

7. To remove the radiator:
   a. Remove the parking lights from the grille.
   b. Remove the two fillers.
   c. Remove the grille from the vehicle.
   d. On some California models you may need to remove the bolts retaining the air pipe.
   e. Disconnect the upper radiator hose and the reservoir hose.
   f. Remove the No. 2 fan shroud if equipped.
   g. On automatics, disconnect the oil cooler hoses.
   h. Remove the lower radiator hose.
   i. Remove the four bolts retaining the radiator, and pull the radiator from the vehicle.
8. If equipped with power steering, loosen the lock bolt and adjusting bolt to the idler pulley, then remove the drive belt.
9. On A/C vehicles, loosen the idler pulley nut and adjusting bolt. Remove the drive belt.
10. Remove the alternator drive belt, fan, fan clutch, water pump pulley, and fan shroud as follows:
   a. Loosen the water pump pulley mounting bolts.
   b. Loosen the lock, pivot and adjusting bolts for the alternator and remove the alternator drive belt.
   c. Remove the water pump pulley mount nuts.
   d. Remove the fan (with clutch) and the water pump pulley.
11. On manual transmissions, disconnect the accelerator cable from the throttle body.
12. On automatics, disconnect the accelerator and throttle cables from the throttle body.
13. For vehicles with cruise control, remove the actuator cover and disconnect the cable.
14. Remove the air cleaner top, Mass Air Flow (MAF) meter and resonator. Remove the air cleaner assembly.
15. Remove the intake air connector.
16. On vehicles with A/C, disconnect the compressor and bracket:
   a. Disconnect the A/C compressor wiring.
   b. Remove the mounting bolts and mounting bracket. Do not disconnect the A/C pressure lines from the compressor. Suspend the compressor away from the engine.
   c. Remove the compressor bracket.
17. Detach the alternator wiring.
18. Remove the heater hoses at the cowl.
19. Disconnect the following hoses:
• Brake booster vacuum
• Evaporative emissions (EVAP)
• On vehicles with power steering, disconnect the two hoses
• Fuel return
• Fuel inlet
20. Remove the power steering pump:
   a. Using special service tool 09960–10010 or an equivalent spanner wrench, remove the nut and power steering pulley.
   b. Remove the two bolts and disconnect the power steering pump.
21. Disconnect the following engine wiring:
   a. Remove the igniter connection

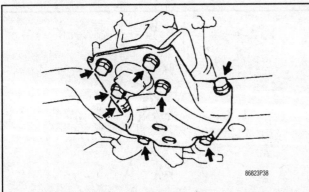

Fig. 13 Remove the bolts retaining the rear engine mounting bracket

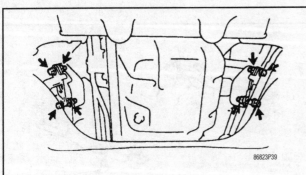

Fig. 14 The left and right engine mount retainers must be removed

   b. Disconnect the ground strap from the cowl top panel.

   c. Disconnect the two engine wire clamps.

   d. Remove the nuts holding the engine wire retainer to the cowl panel and pull out the engine wire from the vehicles cabin.

22. Disconnect the front exhaust pipe from the exhaust manifold and catalytic converter as follows:

   a. Disconnect the two heated oxygen sensors connectors.

   b. Remove the two bolts and retainer holding the front exhaust pipe to the catalytic converter.

   c. Loosen the clamp bolt and disconnect the clamp from the support bracket.

   d. Remove the two bolts to release the support bracket.

   e. Disconnect the three nuts, front exhaust pipe and gaskets from the exhaust manifold.

23. On manual transmissions, remove the shift lever assembly as follows:

   a. Remove the shift lever knob.

   b. Remove the shift lever boot.

   c. Loosen the retaining bolts for the shift lever assembly and baffle, then remove.

24. Remove the driveshaft from the vehicle.

25. Disconnect the speedometer cable from the transmission.

26. On manual transmissions, remove the slave cylinder.

27. On automatics, remove the cross-shaft.

28. Disconnect the wires at the starter.

29. Position a jack and wooden block under the transmission, then remove the rear engine mounting bracket.

30. Attach a suitable engine hoist to the engine hangers.

31. Remove the nuts and bolts from the engine mounts.

➡**Make sure the engine/transmission assembly is clear of all the wiring and hoses.**

32. Carefully lift the engine/transmission out of the vehicle.

To install:

33. Attach the engine hoist to the engine hangers. Carefully lower the engine/transmission into the vehicle. Keep the engine level, while aligning the engine mounts.

34. Installation of the remaining components is in the reverse order of removal. Please note the following steps.

35. Position a jack and wooden block under the transmission, then install the rear engine mounting bracket. Tighten the bolts to the frame to 19 ft. lbs. (58 Nm) and the bolts to the mount to 13 ft. lbs (18 Nm).

36. Remove the jack and engine hoist. Tighten the engine mounts to 28 ft. lbs. (38 Nm).

37. Install the front exhaust pipe to the manifold:

   a. Install the two gaskets and the front pipe to the manifold. Install the nuts and bolts, tighten the nuts to 46 ft. lbs. (62 Nm).

   b. Install the support bracket with the two bolts. Tighten the brackets to 29 ft. lbs. (39 Nm).

   c. Connect the clamp and tighten to 14 ft. lb.s (19 Nm).

   d. Install a new gasket on the front pipe and attach to the catalytic converter. Tighten the bolts to 29 ft. lbs. (39 Nm).

38. Install the power steering pump to the bracket, tighten the bolts to 43 ft. lbs. (58 Nm).

39. With special service tool 09960–10010 or an equivalent spanner wrench, install the power steering pulley, tighten the nut to 32 ft. lbs. (43 Nm).

40. On vehicles with A/C install the compressor:

   a. Attach the compressor bracket, tighten to 32 ft. lbs. (44 Nm).

   b. Attach the A/C compressor to the bracket, tighten the bolts 18 ft. lbs. (25 Nm).

41. Install the intake air connector, tighten the bolts to 13 ft. lbs. (18 Nm).

42. Install the water pump pulley, fan shroud, fan with fan clutch, and alternator drive belt. Stretch the belt tight and tighten the fan nuts to 16 ft. lbs. (21 Nm).

43. Install the radiator to the vehicle with the tabs on the supports through the radiator service holes. Install the bolts and tighten them to 9 ft. lbs. (12 Nm).

44. Road test the vehicle and recheck the fluids.

3F-E Engine

▶ **See Figures 15, 16 and 17**

1. Disconnect the negative battery cable.

2. Drain the engine coolant. Drain the engine oil.

✴✴ CAUTION

When draining coolant, keep in mind that cats and dogs are attracted by ethylene glycol antifreeze, and are quite likely to drink any that is left in an uncovered container or in puddles on the ground. This will prove fatal in sufficient quantity. Always drain coolant into a sealable container. Coolant may be reused unless it is contaminated or several years old.

3. Scribe matchmarks around the hood hinges, then remove the hood.

4. Relieve the fuel pressure.

5. Remove the battery and its tray.

➡**Apply a lubricant onto any nuts or bolts that will be removed. Allow the lubricant to sit for a short while. This will aid in the removal of tight and frozen nuts and bolts.**

6. Disconnect the accelerator and throttle cables.

7. Disconnect the air flow meter connector and clamp. Disconnect the ISC hose, the air pump hose, distributor hose, PCV hose, the 3 hoses from the rear of the intake chamber, and the 2 vacuum hoses from the charcoal canister. Disconnect the air intake hose clamp.

8. Remove the air intake hose, air flow meter and air cleaner cap. Remove the air cleaner element. Remove the three bolts and the air cleaner case.

9. Remove the coolant reservoir tank; remove the radiator.

10. Tag and detached the following wires and connectors:

- Oil pressure sensor
- High tension cord at the coil
- Neutral start switch and transfer near the starter
- Front differential lock
- Starter wire and connector
- Starter ground strap
- Oxygen sensor
- Alternator wire and connector
- Cooling fan

11. Disconnect the following hoses:

- Heater
- Fuel
- Transfer Case
- Brake booster
- Distributor

12. Remove the glove box. Unplug the four connectors from the ECU. Carefully pull the wiring harness through the firewall.

13. Loosen the power steering pump pulley on its shaft. Loosen the idler pulley nut and bolt; remove the drive belt. Remove the pulley mounting nut, drive pulley and woodruff key. Unbolt the power steering pump and remove it from the bracket.

14. At the air conditioning compressor, disconnect the wiring lead. Loosen the idler pulley and adjuster, then remove the belt. Remove the 4 compressor mounting bolts. Position the compressor out of the way with the hoses still attached. Use stiff wire to support the unit; do not allow the compressor to hang by the hoses.

15. Raise the vehicle, making sure it is well supported. Remove the transfer case undercover.

16. Remove the front and rear driveshafts.

17. Disconnect the speedometer cable.

18. Disconnect the ground strap from the body.

19. Detach the two vacuum hoses from the diaphragm cylinder for the transfer case.

20. Remove the transfer case shift lever.

   a. Remove the clip and pin, disconnect the shift rod from the transfer case.

   b. Remove the nut, then disconnect the washers and the shift lever with the shift rod.

21. Disconnect the transmission control rod.

22. Disconnect the exhaust pipe at the manifold.

23. Position a floor jack under the transmission with a broad piece of wood to protect the pan. Remove the eight bolts and two nuts that attach the frame crossmember and then remove the crossmember.

24. Attach an engine hoist chain to the lifting brackets on the engine. Remove the engine mount nuts and bolts, then slowly lift the engine/transmission out of the truck.

25. Remove the oil cooler lines from the transmission.

26. Remove the transmission from the engine.

27. Mount the engine on an engine stand, making sure the retaining bolts are tight. If an engine stand is not available, support the engine in an upright position with blocks. Never leave an engine hanging from a lift or hoist.

To install:

28. Install the transmission and oil cooler lines to the engine. Slowly lower the engine assembly into the engine compartment.

29. Installation of the remaining components is in the reverse order of removal. Please note the following steps.

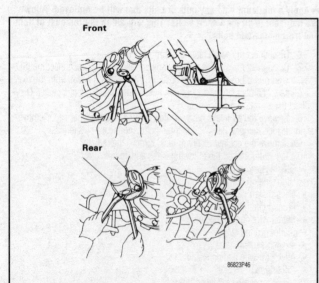

Fig. 15 Locations of the front and rear driveshafts bolts

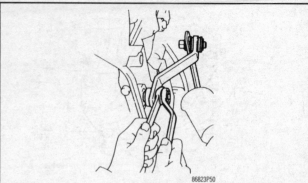

Fig. 16 Remove the two nuts securing the control rod on the transmission

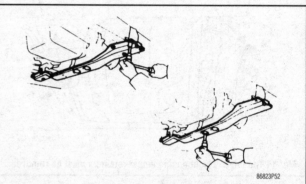

Fig. 17 Remove the outer crossmember bolts first, then the inner ones

30. Raise the transmission slightly and tighten the frame crossmember-to-chassis bolts to 29 ft. lbs. (39 Nm). Tighten the two nuts to 43 ft. lbs. (59 Nm).

31. Install the exhaust pipe with a new gasket and nuts. Tighten the nuts to 46 ft. lbs. (62 Nm).

32. Install the front and rear driveshafts. Tighten the nuts to 65 ft. lbs. (88 Nm).

33. Install the power steering pump and tighten the pulley nut to 35 ft. lbs. (47 Nm).

34. Double check all installation items, paying particular attention to loose hoses or hanging wires, untightened nuts, poor routing of hoses and wires (too tight or rubbing) and tools left in the engine area.

35. With the coil wire disconnected from the distributor, crank the engine with the ignition key for 5–10 seconds. This should pre-oil at least the lower part of the engine.

36. Reconnect the coil wire. Start the engine, allowing it to warm up. Check carefully for leaks.

37. Check and adjust the engine specifications as necessary.

38. Road test the vehicle. Recheck and top fluid levels as needed.

1FZ-FE Engine

◆ **See Figures 18, 19, 20 and 21**

1. Disable the air bag system.
2. Release the fuel pressure.
3. Disconnect the battery cables, negative first. Remove the battery and tray assembly.

✳✳ CAUTION

The air bag system is equipped with a back-up power source. To avoid possible air bag deployment, do not start working on the vehicle until 90 seconds has elapsed from the time the ignition switch is turned OFF and the negative battery terminal is disconnected.

4. Raise and safely support the vehicle.
5. Drain the engine coolant, transmission fluid and engine oil.

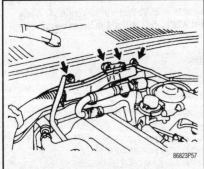

Fig. 18 Remove the bolts securing the engine ground wire and heater control valve

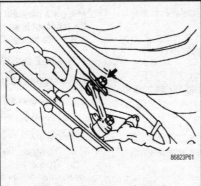

Fig. 19 Remove this nut to release the control rod

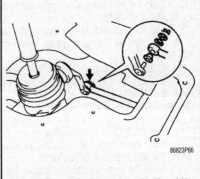

Fig. 20 Assemble the washers for the shift rod linkage as shown

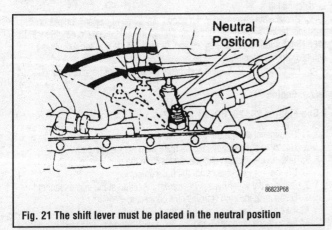

Fig. 21 The shift lever must be placed in the neutral position

⁂ CAUTION

The EPA warns that prolonged contact with used engine oil may cause a number of skin disorders, including cancer! You should make every effort to minimize you exposure to used engine oil. Protective gloves should be worn when changing the oil. Wash your hands and any other exposed skin areas as soon as possible after exposure to used engine oil. Soap and water, or waterless hand cleaner should be used.

6. Scribe the hood hinges and remove the hood.
7. Remove the radiator grille and radiator.
8. Disconnect the oil cooler hose from the oil cooler pipe.
9. Remove the air cleaner hose, cap and case assembly.
10. Disconnect the cruise control actuator cable from the throttle body.
11. Disconnect the accelerator actuator cable from the throttle body.
12. Disconnect the heater hoses.
13. Disconnect the ground wire and heater control valve from the cowl panel.
14. Remove the brake booster vacuum hose.
15. Disconnect the evaporative emissions (EVAP) hoses and fuel hoses.

⁂ CAUTION

Fuel injected engines remain under constant pressure after the ignition has been turned OFF. Properly relieve the pressure before disconnecting any fuel lines. Failure to do so may result in personal injury.

16. Disconnect the following wires and connectors:
• The two heated oxygen sensors
• DCL1 clamp
• Oil pressure gauge connections
• Alternator wire and connector
• Intake manifold from the fender apron connection
• High tension cord leading to the ignition coil

• Ground strap from the No. 1 engine hanger
• Ground strap from the air intake chamber
• Ground cable from the cylinder block
17. Loosen the idler pulley nut and adjusting nut to remove the A/C drive belt.
18. Disconnect the A/C compressor, then remove the bracket.
19. Remove the radiator pipe as follows:
  a. Remove the two nuts retaining the radiator pipe to the No. 1 oil pan.
  b. Disconnect the No. 2 radiator hose from the water inlet, then remove the radiator pipe.
20. Remove the union bolt and the two gaskets, then disconnect the pressure hose from the power steering reservoir pump.
21. Disconnect the return hose from the power steering reservoir tank.
22. Remove the glove compartment door.
23. Remove the speaker panel.
24. Disconnect the A/C amplifier.
25. Disconnect the wire leading to the Electronic Control Unit (ECU) and cowl wire.
26. Pull the engine wire out from the cabin of the vehicle.
27. Remove the stabilizer bar.
28. Matchmark the flanges of the front and rear driveshafts, then remove the shafts.
29. Remove the transfer shift lever.
  a. Remove the nut retaining the transmission control rod.
  b. Remove the shift knob.
  c. Lift up the console slightly in order to disconnect the wire.
  d. Remove the shifter console.
  e. Remove the center console box.
  f. Disconnect the wires, then remove the transfer shift lever boot and lever assembly.
  g. Pull out the retaining pin, then disconnect the shift rod.
  h. Remove the hose clamp, then the transfer shift lever.
30. Remove the front exhaust pipe.
  a. Disconnect the heated oxygen sensor.
  b. Remove the nuts and bolts holding the exhaust to the catalyst.
  c. Loosen the clamp then disconnect the from the No. 1 support bracket.
  d. Remove the No. 1 support bracket.
  e. Remove the front exhaust pipe.
31. Detach the ground strap from the heat insulator.
32. Place a jack under the transmission. Put a wooden block between the jack and the transmission oil pan to prevent damages to the pan.
33. Remove the frame crossmember.
34. Attach the engine hoist chain to the two engine hangers.
35. Remove the nuts holding the engine front mounting insulators to the frame.
36. Lift the engine with the transmission out of the vehicle slowly and carefully. Make certain the engine is clear of all wiring and hoses.
To install:
37. Attach the engine hoist chain to the engine hangers, then lower the engine and transmission into the vehicle.
38. Installation of the remaining components is in the reverse order of removal. Please note the following steps.
39. Install the front crossmember, then tighten the bolts to 45 ft. lbs. (61 Nm).
40. Tighten the nuts holding the crossmember to the engine rear mounting insulator to 54 ft. lbs. (74 Nm).

41. Tighten the nuts holding the engine front mounting insulators to the frame to 54 ft. lbs. (74 Nm).

42. Install the front exhaust pipe, then tighten the nuts to 46 ft. lbs. (63 Nm).

   a. Install the No. 1 support bracket, then tighten the bolts to 17 ft. lbs. (24 Nm).

   b. Connect the clamp and tighten to 14 ft. lbs. (19 Nm).

   c. Connect the front exhaust pipe to the rear catalyst, tighten the bolts to 34 ft. lbs. (46 Nm).

43. Install the transfer shift lever and hose clamp, tighten the bolts to 13 ft. lbs. (18 Nm).

   a. Connect the shift rod and install the pin.

   b. Install the transfer shift lever boot, transmission shift lever assembly and tighten the bolts to 4 ft. lbs. (5 Nm).

   c. Attach the connectors to the shift lever assembly.

   d. Install the center console box.

   e. Connect the pattern select switch.

   f. Install the shifter console and shift lever knob.

   g. Shift the lever into the neutral (N) position.

   h. Fully turn the control shaft lever back and return two notches. It will now be in the neutral (N) position.

   i. Connect the transmission control rod and tighten to 9 ft. lbs. (13 Nm).

44. Install the front and rear driveshafts.

   a. On the differential side, align the matchmarks on the flanges, then tighten the front shaft to 54 ft. lbs. (74 Nm), and the rear shaft to 65 ft. lbs. (88 Nm).

   b. On the transfer side, align the matchmarks on the flanges, then tighten the front shaft to 54 ft. lbs. (74 Nm), and the rear shaft to 65 ft. lbs. (88 Nm).

45. Connect the stabilizer bar brackets to 13 ft. lbs. (18 Nm).

46. Lower the vehicle so that it is resting on its suspension, tighten the bolts holding the stabilizer bar to the axle housing to 19 ft. lbs. (25 Nm).

47. Connect the power steering pressure hose with the union bolt and tighten to 42 ft. lbs. (56 Nm).

48. Install the radiator pipe as follows:

   a. Connect the No. 2 radiator hose to the water inlet.

   b. Install the two nuts holding the radiator pipe to the No. 1 oil pan. Tighten the nuts to 15 ft. lbs. (21 Nm).

49. Install the A/C bracket and tighten the bolts to 27 ft. lbs. (37 Nm).

50. Install the A/C compressor and tighten to bolts to 18 ft. lbs. (25 Nm).

51. Connect the fuel inlet hose to the fuel filter and tighten the union bolt to 22 ft. lbs. (39 Nm). Connect the fuel return line.

52. Fill the engine with oil and the transmission with proper amount and type of fluid.

53. Fill the cooling system with coolant.

54. Start the engine and check for leaks.

Rocker Arm (Valve) Cover

REMOVAL & INSTALLATION

Check the hoses, wiring, bolt threads and any other part that could be damaged or worn replace if needed.

➡Apply a lubricant onto any nuts or bolts that will be removed. Allow the lubricant to sit for a short while. This will aid in the removal of tight and frozen nuts and bolts.

22R, 22R-E and 3F-E Engines

▶ See Figures 22 thru 27

1. Disconnect the negative battery cable.
2. On 22R and 22R-E engines, remove the ground strap from the valve cover.
3. On 3F-E, remove the air cleaner hose.
4. Disconnect the PCV hose(s) from the cam cover.
5. Remove the nuts and washers. Lift the cover off the cylinder head. Cover the oil return hole in the head to prevent dirt or objects from falling in. Remove the gasket.

To install:

6. Replace the rubber half moon seals in the cylinder head if they seem to be deteriorated.
7. Replace the cover gasket if it shows any signs of damage, breaks or cracking. Install new nut gaskets, then tighten the nuts evenly to 43–48 inch lbs (4–5 Nm) for four cylinder engines and 78 inch lbs. (9 Nm) for six cylinder engines. Do not overtighten.
8. Reconnect the PCV hose and install the air cleaner hose or ground strap.

3VZ-E Engine

▶ See Figures 28, 29, 30, 31 and 32

1. Disconnect the throttle position sensor.
2. Unsecure the canister vacuum hose from the throttle body.
3. Take off the vacuum and fuel hoses from the pressure regulator.
4. Disconnect PCV hose from the hose union.
5. Remove the water bypass hose from the union at the intake manifold.
6. Disconnect the cold start injector connector.
7. Take off the vacuum hose from the fuel filter.
8. Remove the union bolt and the two gaskets and the cold start injector tube.
9. On California vehicles, disconnect the EGR temperature sensor connector.
10. Disconnect the EGR vacuum hoses from the air pipe and EGR vacuum modulator.
11. Remove the nut, bolt and intake manifold stay (support).
12. Disconnect the 2 water bypass hoses from the EGR valve.
13. Remove the 5 nuts and remove the EGR valve with the pipes and 2 gaskets.
14. Remove the 6 bolts and 2 nuts holding the intake manifold; remove the gasket.
15. Disconnect and label the following wires:
- Knock sensor connector
- Water temperature switch connector
- Start injection timer
- Water temperature sensor connector
- Water temperature sender gauge connector
- Right side engine ground strap
- Injector connector

Fig. 22 Remove the bolt securing the ground strap to the valve cover

Fig. 23 It may be necessary to use a deep socket to remove the retaining nuts

Fig. 24 Be careful not to lose the seal washers

Fig. 25 Once the retaining nuts are removed, lift the cover from the head. . .

Fig. 26 . . . then pull the rubber gasket off the head mating area

Fig. 27 Inspect the half moon seals, replace if deteriorated

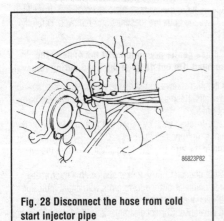

Fig. 28 Disconnect the hose from cold start injector pipe

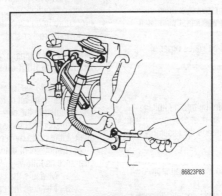

Fig. 29 Remove the bolts securing the EGR assembly to the engine

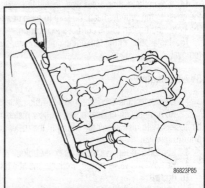

Fig. 30 Remove the two bolts retaining the engine harness in place

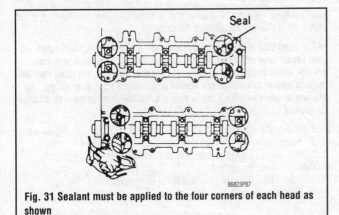

Fig. 31 Sealant must be applied to the four corners of each head as shown

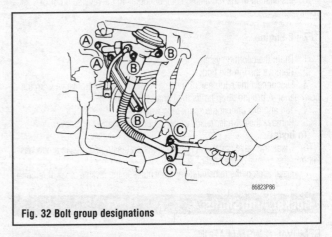

Fig. 32 Bolt group designations

16. Remove the two bolts holding the engine wire harness and remove the harness.

17. Remove the bolts holding the valve cover(s); remove the cover(s) and gaskets.

To install:

18. Examine the gasket closely for any sign of cracking or deformation. Replace the gasket if necessary.

19. Installation of the remaining components is in the reverse order of removal. Please note the following steps.

20. Place the valve cover and gaskets in place on the head and install the small retaining bolts. Tighten them evenly only to 5–6 ft. lbs. (7–8 Nm), do not overtighten.

21. Position a new gasket on the intake manifold. Install the manifold, tightening the six bolts and two nuts to 13 ft. lbs. (18 Nm).

22. Install the intake manifold support; tighten the groups of bolts to the proper torque—refer to the diagram for designations. Group (A) bolts should be

tightened to 9 ft. lbs. (12 Nm), group (B) to 13 ft. lbs. (18 Nm) and group (C) bolts to 22 ft. lbs. (29 Nm).

23. Use a new gasket and connect the cold start injector tube. Tighten the union bolt to 13 ft. lbs. (18 Nm).

2RZ-FE and 3RZ-FE Engines

1. Drain the coolant from the engine.

2. On manual transmissions, disconnect the accelerator cable from the throttle body.

3. On automatics, disconnect the accelerator and throttle cables from the throttle body.

4. On cruise control, remove the actuator cover, then disconnect the cruise control cable from the actuator.

5. Disconnect the 3 wire clamps for the engine harness.

6. Unsecure the Mass Air Flow (MAF) meter and Intake Air Temperature (IAT) sensor connections.

7. On the 3RZ-FE engine, disconnect the air hose from the air cleaner cap, then loosen the air cleaner hose clamp.

8. Loosen the 4 clips, and remove the the cleaner cap, Mass Air Flow (MAF) meter and resonator sensor.

9. Remove the air intake sensor by disconnecting the air hose for the Idle Air Control (IAC).

   a. Remove the vacuum sensing hose and wire clamp for the engine wire.

   b. Loosen the hose clamp, and remove the bolts retaining the air intake connector.

10. Remove the PCV hoses.

11. Tag and disconnect the spark plug wires.

12. Disconnect these wires:
- A/C connector if equipped
- Oil pressure sensor wire
- Engine coolant temperature sensor
- Distributor wire
- 4 engine wire clamps and engine wire

13. Remove the valve cover bolts, seal washers, cover and gasket.

To install:

14. Clean the mating areas well. Install the gasket on the cover, then install the cover to the cylinder head. Be sure to have new seal washers, then install the bolts and tighten to the correct specification.

15. Installation of the remaining components is in the reverse order of removal.

5VZ-FE Engine

1. Disconnect the accelerator and throttle cables from the throttle linkage.

2. Remove the air cleaner cover, air flow meter, and air duct assembly. It will be necessary to disconnect the wiring to the meter and unbolt the assembly, prior to removal.

3. Loosen and remove the valve cover bolts.

4. Remove the valve covers, work on one at a time to avoid confusion.

To install:

5. Clean the mating areas on the cover and cylinder head. Install the gasket to the head and seal washers. Attach the cover with the mounting bolts. Tighten until snug.

6. Installation of the remaining components is in the reverse order of removal.

1FZ-FE Engine

1. Drain the cooling system.

2. Remove the throttle body.

3. Disconnect the engine wiring harness and heater control valve from the cowl panel. A ground strap holds the harness in place.

4. Tag and disconnect the spark plug wires.

5. Remove the cylinder head (valve) cover bolts, gasket and cover.

To install:

6. Clean the valve cover and mating areas on the engine. Install a new gasket into the groove of the cover.

7. Installation of the remaining components is in the reverse order of removal.

Rocker Arm/Shafts

REMOVAL & INSTALLATION

➡️**Only the 22R, 22R-E and 3F-E engines use rocker arms. Other engines operate the valves through the direct action of the camshaft.**

22R and 22R-E Engines

♦ **See Figures 33, 34, 35, 36 and 37**

✳️✳️ WARNING

The rocker shaft assembly is held by the head bolts. Removal of the head to replace the head gasket is REQUIRED any time the head bolts are loosened. Refer to cylinder head removal and installation later in this section. NEVER attempt to re-compress the same gasket after the head bolts are loosened.

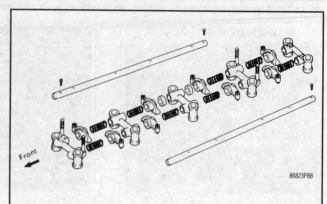

Fig. 33 Exploded view of the 22R and 22R-E rocker arms and shafts

1. Remove the valve cover. The head bolts must be loosened in the correct order to avoid head damage. Refer to the cylinder head removal and installation procedure in this section. Once the bolts are loosened and removed, carefully remove the rocker shaft assembly.

➡️**It may be necessary to use gentle prying force at the ends of the assembly to lift it off the head. Do not damage the head during removal.**

2. Remove the three retaining screws and slide the rocker supports, springs and rocker arms off of the shafts. Keep all parts in order; the shafts must be reassembled in the correct order.

3. Inspect the rocker arm-to-shaft wear by wiggling the arm laterally on the shaft. Little or no movement should be felt.

4. If movement is found in the arm-to-shaft wiggle test, measure the oil clearance as follows:

   a. Measure the outside diameter of the rocker shaft with a micrometer.

   b. Measure the inside diameter of the rocker arms with a dial indicator.

   c. The difference between the rocker arm inner diameter and the shaft outer diameter is the oil clearance. Correct clearance is 0.0004–0.0020 inches (0.01–0.05 mm). Maximum allowable clearance is 0.0031 inches (0.08 mm).

5. Inspect the face of each rocker; the surface should be smooth, shiny and evenly shaped. Any sign of metal wear or scratching requires replacement of the rocker and a detailed inspection of the camshaft.

➡️**The condition of the rocker surfaces and matching camshaft lobes are your report card for oil and filter changes. If you have been less than prompt about performing routine oil changes, chances are good that the rockers and/or cam lobes are scored or damaged from poor oiling. The amount of sludge in the head is also a good indicator of poor oil change frequency.**

6. Reassemble the rocker shafts, arms and towers in their exact original order.

7. Remove the head, install a new gasket. See cylinder head removal and installation. Install the rocker assembly and tighten the bolts gradually and in sequence to 58 ft. lbs. (78 Nm).

8. Adjust the valve clearance. Install the valve cover.

9. Test drive the vehicle.

3F-E Engine

♦ **See Figures 38, 39 and 40**

1. Remove the valve cover.

2. The rocker assembly is held by 8 bolts and 4 nuts. These must be loosened evenly in several passes. Do not loosen any one fastener all at once. Loosen the nuts and bolts in the correct order.

3. Inspect the rocker arm-to-shaft wear by wiggling the arm laterally on the shaft. Little or no movement should be felt.

4. If movement is found in the arm-to-shaft wiggle test, disassemble the shaft and keep every piece in exact order. Measure the oil clearance as follows:

   a. Measure the outside diameter of the rocker shaft with a micrometer.

   b. Measure the inside diameter of the rocker arms with a dial indicator.

   c. The difference between the rocker arm inner diameter and the shaft outer diameter is the oil clearance. Correct clearance is 0.0004–0.0020 inches (0.01–0.05 mm). Maximum allowable clearance is 0.0031 inches (0.08 mm).

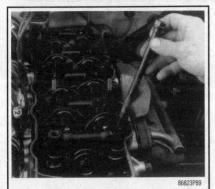

Fig. 34 After the head bolts are removed . . .

Fig. 35 . . . the rocker shaft assembly can be lifted from the head

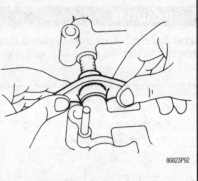

Fig. 36 Wiggle the rocker arm on the shaft to check for excessive free-play

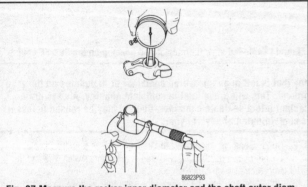

Fig. 37 Measure the rocker inner diameter and the shaft outer diameter to determine the oil clearance

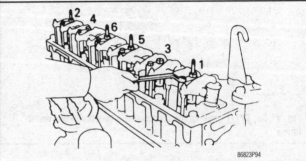

Fig. 39 Loosen the rocker assembly bolts in several passes in this order

5. Inspect the face of each rocker; the surface should be smooth, shiny and evenly shaped. Any sign of metal wear or scratching requires replacement of the rocker and a detailed inspection of the camshaft.

6. Remove the cylinder head.

➡The condition of the rocker surfaces and matching camshaft lobes are your report card for oil and filter changes. If you have been less than prompt about performing routine oil changes, chances are good that the rockers and/or cam lobes are scored or damaged from poor oiling. The amount of sludge in the head is also a good indicator of poor oil change frequency.

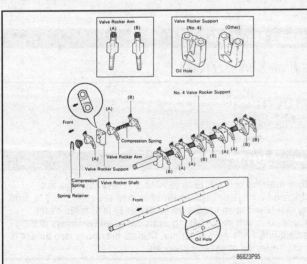

Fig. 38 Exploded view of the rocker shaft assembly on the 3F-E engine

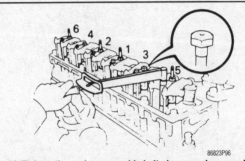

Fig. 40 Tighten the rocker assembly bolts in several passes following this order

To install:

7. Install the cylinder head with a new gasket. Reassemble the rocker shafts, arms and towers in their exact original order.

8. Place the rocker shaft assembly on the cylinder head. Align the adjusting screws with the heads of the pushrods.

9. Install the retaining bolts. Tighten them in the correct order and in several passes to the correct torque. The 12 mm bolt heads should be tightened to 15–17 ft. lbs. (21–24 Nm). The 14 mm bolt head and the nut should be tightened to 25 ft. lbs. (33 Nm).

10. Adjust the valve clearance. Reinstall the valve cover.

Thermostat

REMOVAL & INSTALLATION

◆ **See Figures 41 thru 48**

1. Drain the cooling system.

When draining coolant, keep in mind that cats and dogs are attracted by ethylene glycol antifreeze, and are quite likely to drink

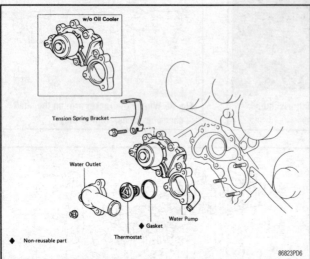

Fig. 41 On, 3VZ-E engines the thermostat is located in the water-pump

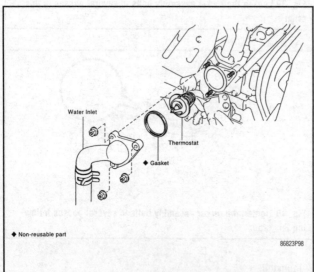

Fig. 42 Exploded view thermostat and housing on the 1FZ-FE engine

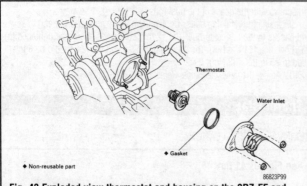

Fig. 43 Exploded view thermostat and housing on the 2RZ-FE and 3RZ-FE engines

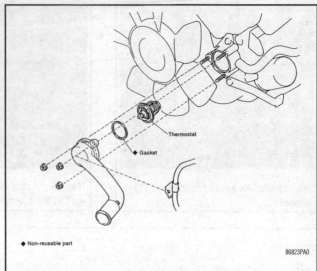

Fig. 44 Exploded view thermostat and housing on the 5VZ-FE engine

any that is left in an uncovered container or in puddles on the ground. This will prove fatal in sufficient quantity. Always drain coolant into a sealable container. Coolant may be reused unless it is contaminated or several years old.

2. On 3F-E engines, disconnect the 2 vacuum hoses from the Bimetal Vacuum Switching Valve (BVSV), the cold start injector time switch wire and the water temperature switch connector.

3. Unbolt the housing, then remove the water inlet with the hose attached. If there is not enough room to pull both housing and hose out, you may want to disconnect the hose first. On 3F-E engines, remove the clamp bolt for the ISC water bypass pipe. Make certain to remove the gasket with the water inlet. Some of it will probably stick to the housing and require scraping.

4. Remove the thermostat.

To install:

5. When installing a new thermostat always use a new gasket. Be sure that the thermostat is positioned with the spring down or into the housing. On 3VZ-E engines, the jiggle valve (looks like a small rivet in the edge of the thermostat) must be at the top or 12 o'clock position.

6. Install the water inlet. Install the bolts finger-tight until the inlet is positioned exactly. Tighten the bolts to 10–14 ft. lbs. (13–19 Nm). Overtightening will break the ear off the housing and spoil your afternoon—don't do it.

7. Installation of the remaining components is in the reverse order of removal.

Intake Manifold

REMOVAL & INSTALLATION

22R Engines

1. Disconnect the negative battery cable first.
2. Drain the cooling system.

When draining coolant, keep in mind that cats and dogs are attracted by ethylene glycol antifreeze, and are quite likely to drink any that is left in an uncovered container or in puddles on the ground. This will prove fatal in sufficient quantity. Always drain coolant into a sealable container. Coolant may be reused unless it is contaminated or several years old.

3. Remove the air cleaner assembly, complete with hoses, from the carburetor.

4. Label and disconnect the vacuum lines from the EGR valve and carburetor. Disconnect the EGR pipe from the manifold; remove the EGR valve with the pipe attached.

Fig. 45 An extension is helpful for reaching the housing retainers

Fig. 46 You may need to maneuver the housing through other hoses in the way

Fig. 47 Lift the thermostat and gasket from the inlet

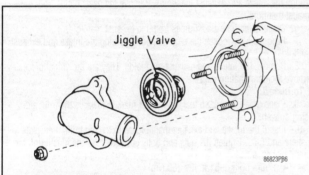

Fig. 48 If used, the thermostat must be installed with the jiggle valve at the top

5. Remove the fuel lines, electrical leads, accelerator linkage, and water hose from the carburetor.

6. Remove the water bypass hose from the manifold.

7. Unbolt and remove the intake manifold, complete with carburetor and EGR valve.

8. Cover the cylinder head ports with clean shop cloths to keep anything from falling into the cylinder head or block.

To install:

9. When installing the manifold, always replace the gasket with a new one. Tighten the mounting nuts and bolts to 14 ft. lbs. (19 Nm). Tighten the bolts in several stages working from the inside bolts outward.

22R-E Engines

▶ See Figures 49 and 50

1. Disconnect the negative battery cable.
2. Drain the cooling system.

✷✷ CAUTION

When draining coolant, keep in mind that cats and dogs are attracted by ethylene glycol antifreeze, and are quite likely to drink any that is left in an uncovered container or in puddles on the ground. This will prove fatal in sufficient quantity. Always drain coolant into a sealable container. Coolant may be reused unless it is contaminated or several years old.

3. Disconnect the air intake hose from both the air cleaner assembly on one end and the air intake chamber on the other.

4. Tag and disconnect all vacuum lines attached to the intake chamber and manifold.

5. Tag and disconnect the wires to the cold start injector, throttle position sensor, and the water hoses from the throttle body.

6. Remove the EGR valve from the intake chamber.

7. Tag and disconnect the actuator cable, accelerator cable and automatic transmission throttle cable (if equipped) from the cable bracket on the intake chamber.

8. Unbolt the air intake chamber from the intake manifold and remove the chamber with the throttle body attached.

9. Disconnect the fuel hose from the fuel delivery pipe.

10. Disconnect the air valve hose from the intake manifold.

11. Make sure all hoses, lines and wires are tagged for later installation and disconnected from the intake manifold. Unbolt the air intake chamber from the intake manifold.

12. Unbolt the manifold from the cylinder head, removing the delivery pipe and injection nozzles with the manifold.

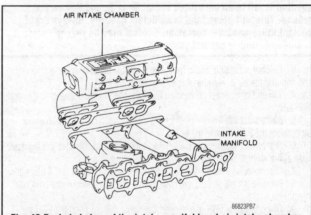

Fig. 49 Exploded view of the intake manifold and air intake chamber on the 22R-E engine

Fig. 50 With the air intake chamber removed, the manifold can now be accessed for removal

To install:

13. Position a new gasket on the cylinder head. Install the intake manifold with the fuel delivery pipe and injection nozzles.

14. Install the bolts and nuts. Tighten them evenly to 14 ft. lbs. (19 Nm).

15. Install the EGR valve. Clean the bolt threads, removing any carbon or oil deposits. For the bolt closest to the front of the engine, apply sealant to the 2 or 3 threads closest to the end of the bolt. Install the bolts and nuts.

16. Install the air intake chamber with the throttle body. Use a new gasket. Tighten the bolts to 14 ft. lbs. (19 Nm).

17. Installation of the remaining components is in the reverse order of removal.

3VZ-E Engines

▶ **See Figure 51**

1. Disconnect the negative battery cable.

※※ WARNING

The air bag system is equipped with a back-up power source. To avoid possible air bag deployment, do not start working on the vehicle until 90 seconds has elapsed from the time the ignition switch is turned OFF and the negative battery terminal is disconnected.

2. Drain the cooling system.

※※ CAUTION

When draining coolant, keep in mind that cats and dogs are attracted by ethylene glycol antifreeze, and are quite likely to drink any that is left in an uncovered container or in puddles on the ground. This will prove fatal in sufficient quantity. Always drain coolant into a sealable container. Coolant may be reused unless it is contaminated or several years old.

3. Disconnect the air intake hose from both the air cleaner assembly on one end and the air intake chamber on the other.

4. Tag and disconnect all vacuum lines attached to the intake chamber and manifold.

5. Disconnect the throttle position sensor connector at the air chamber. Disconnect the PCV hose at the union.

6. Disconnect the water bypass hose at the manifold. Remove the bypass hose at the water bypass pipe.

7. Disconnect the cold start injector and the vacuum hose at the fuel filter.

8. Remove the union bolt and two gaskets and then remove the cold start injector tube.

9. Disconnect the EGR gas temperature sensor (Calif.) and the EGR vacuum hoses from the air pipe and the vacuum modulator.

10. Remove the EGR valve.

11. Disconnect the air hose at the reed valve.

12. Remove the air intake chamber and then remove the engine wire harness (held with 2 bolts).

13. Remove the four union bolts and then remove fuel pipes.

14. Remove the timing belt covers.

15. Remove the fuel delivery pipes with their injectors.

16. Remove the water bypass outlet. Progressively loosen the retaining bolts and remove the intake manifold.

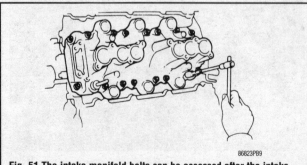

Fig. 51 The intake manifold bolts can be accessed after the intake chamber is removed

To install:

17. Install the intake manifold with new gaskets and tighten the mounting bolts to 11 to 15 ft. lbs. (13 to 20 Nm).

18. Install the water bypass outlet and tighten the two bolts to 13 ft. lbs. (18 Nm).

19. Install the timing belt covers and tighten the bolts to 74 inch lbs. (8.3 Nm).

20. Install the fuel pipes and tighten the union bolts to 22 ft. lbs. (29 Nm).

21. Installation of the remaining components is in the reverse order of removal.

3F-E Engine

▶ **See Figures 52 and 53**

1. Drain the engine coolant.

2. Disconnect the negative battery cable.

3. Disconnect the accelerator and throttle control cable.

4. Remove the air intake hose, air flow meter and air cleaner.

5. Disconnect or reposition all the wires and hoses running near or through the intake and exhaust manifolds. Remove the heat shields from the exhaust manifold.

6. Disconnect the exhaust pipes from the exhaust manifold.

7. Progressively loosen the retaining bolts holding the intake and exhaust manifold

8. Remove the intake and exhaust manifolds. They are individual pieces. Remove the combination gasket.

To install:

9. Place a new gasket so that the FRONT mark is towards the front side, facing outward.

10. Install the intake and exhaust manifolds with the ten bolts, four plate washers and nuts. Tighten the nuts and bolts correctly. Use the illustration for correct placement:

- (A) 17mm bolts—51 ft. lbs. (69 Nm)
- (B) 14 mm bolts—37 ft. lbs. (50 Nm)
- (C) nuts—41 ft. lbs. (56 Nm)

11. Installation of the remaining components is in the reverse order of removal.

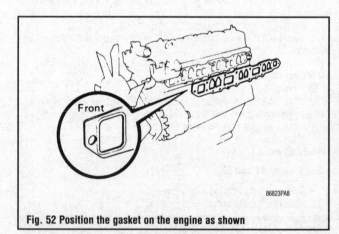

Fig. 52 Position the gasket on the engine as shown

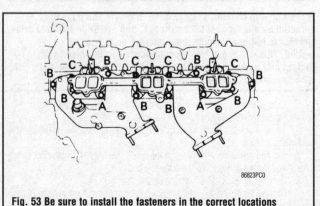

Fig. 53 Be sure to install the fasteners in the correct locations

3VZ-FE Engine

▶ See Figure 54

1. Relieve the fuel pressure.

> **⁜ CAUTION**
>
> **Fuel injected engine systems remain under pressure after the engine has been turned OFF. Properly relieve the fuel pressure before disconnecting any fuel lines. Failure to do so may result in fire or personal injury.**

2. Disconnect the negative battery cable.

> **⁜ WARNING**
>
> **The air bag system is equipped with a back-up power source. To avoid possible air bag deployment, do not start working on the vehicle until 90 seconds has elapsed from the time the ignition switch is turned OFF and the negative battery terminal is disconnected.**

3. Drain the engine coolant.
4. Remove the air cleaner hose.
5. Disconnect the following cables:
- On cruise control vehicles, the actuator cable with bracket
- Accelerator cable
- On automatics, the throttle cable
6. If equipped with an EGR valve, remove the EGR pipe and gaskets.
7. Remove the intake chamber stay as follows:
   a. Remove the oil filler tube and the throttle cable clamp.
   b. Remove the intake chamber stay.
8. Remove the following connectors:
- Vacuum Switching Valve (VSV) connector for the fuel pressure control
- Throttle positioning sensor
- Idle Air Control (IAC) valve

- Disconnect the EGR gas temperature connector
- Disconnect the VSV connector for the EGR valve
9. Disconnect the following hoses:
- PCV
- Water bypass
- Air assist from the intake air connector
- Two vacuum sensing hoses from the VSV
- EVAP
- Air hose from the power steering
- On A/C vehicles, disconnect the air hose from the idle up valve
10. Remove the bolts, nuts then the intake chamber assembly.
11. Remove the air intake connector as follows:
   a. Disconnect the engine wire from the intake connector.
   b. Disconnect the two fuel return hoses.
   c. Disconnect the brake booster vacuum hose from the intake air connector.
   d. Remove the bolt and disconnect the ground strap from the intake air connector.
   e. Disconnect the Data Link Connector No. 1 (DLC1) from the bracket of the intake connector.
   f. On A/C vehicles, disconnect the idle up valve.
   g. Remove the air intake connector from the engine.
12. Disconnect the spark plug wires.
13. Remove the engine splash shield from under the engine.
14. Disconnect the upper radiator hose.
15. Remove the power steering, A/C and alternator belts.
16. Remove the fan shroud.
17. Remove the fan shroud with the coupling and pulleys.
18. Disconnect the power steering pump from the engine, then set aside. Do not disconnect the lines from the pump.
19. On A/C vehicles, remove the compressor. Do not disconnect the lines from the compressor, then set aside. Unbolt the bracket from the compressor.
20. Remove the power steering adjusting strut. Then remove the fan bracket.
21. Set the No. 1 cylinder at TDC of the compression stroke as follows:
   a. Turn the crankshaft pulley and align it's groove with the timing mark **0** of the timing belt cover.
   b. Check the timing marks of the camshaft timing pulleys and the timing belt cover are aligned. If not, turn the crankshaft pulley one revolution (360°).

➡**If reusing the timing belt, make sure that you can still read the installation marks. If not, place new marks on the belt to match the timing marks of the camshaft timing pulleys.**

22. Remove the timing belt tensioner by alternately loosening the bolts.
23. Remove the camshaft timing pulleys.
24. Remove the crankshaft pulley.
25. Remove the timing belt idler pulley.
26. Disconnect and remove the fuel pressure regulator.
27. Disconnect the heater hose.
28. Disconnect the engine wiring from the intake manifold:
- Oil pressure sensor
- Crankshaft position sensor
- Injectors
- ECT sender gauge
- ECT sensor
- Knock sensor
- Camshaft position sensor
- Disconnect the engine wire clamps
- Remove the bolts, then disconnect the engine wire from the cylinder head
29. Remove the camshaft position sensor.
30. Unbolt the No. 3 (rear) timing belt cover.
31. Disconnect the fuel inlet hose, then remove the bo,to securing the intake manifold stay.
32. Unbolt the intake manifold assembly.

To install:

33. Installation is in the reverse order of removal. Please note the following steps.
34. Install new gaskets on the intake and tighten down to 13 ft. lbs. (18 Nm).
35. Install the intake stay, tighten to 14 ft. lbs. (18 Nm).
36. Attach the timing cover, tighten the bolts to 80 inch lbs. (9 Nm).

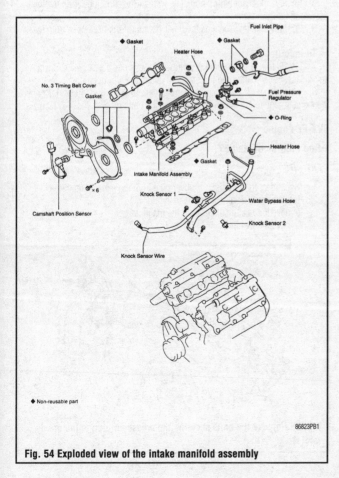

Fig. 54 Exploded view of the intake manifold assembly

37. Install the camshaft position sensor, tighten to 71 inch lbs. (8 Nm).

38. Install the timing belt idler, gears, belt and tensioner as detailed later in this section.

39. Install the timing belt cover. Tighten the bolts to 80 inch lbs. (9 Nm).

40. Install the intake manifold, then tighten the bolts and nuts to 14 ft. lbs. (18 Nm).

41. Install the air intake chamber assembly, tighten the bolts to 14 ft. lbs. (18 Nm).

42. Install new gaskets on the EGR pipe. Tighten the clamp nuts to 71 inch lbs. (8 Nm), and the EGR pipe nuts to 14 ft. lbs. (18 Nm).

1FZ-FE Engine

▶ **See Figure 55**

1. Relieve the fuel pressure.

✳✳ CAUTION

Fuel injected engine systems remain under pressure after the engine has been turned OFF. Properly receive the fuel pressure before disconnecting any fuel lines. Failure to do so may result in fire or personal injury.

2. Disconnect the negative battery cable.

✳✳ WARNING

The air bag system is equipped with a back-up power source. To avoid possible air bag deployment, do not start working on the vehicle until 90 seconds has elapsed from the time the ignition switch is turned OFF and the negative battery terminal is disconnected.

3. Drain the engine coolant.

4. Remove the air cleaner hose and cap.

5. Disconnect the following cables:
- On cruise control vehicles, the actuator cable from the throttle body
- Accelerator cable
- Throttle cable (auto transmission)

6. Disconnect the engine ground strap from the engine hanger and the ground strap from the air intake chamber.

7. Detach the connection on the intake manifold from the left fender apron.

8. Disconnect the brake booster and EVAP hoses.

9. Separate the fuel inlet and return lines from the fuel rail.

10. Remove the intake manifold stay.

11. Disconnect the following electrical connections:
- Engine Coolant Temperature (ECT) sender gauge
- ECT cut switch
- ECT sensor
- Knock sensor
- Crankshaft position sensor

12. Remove the bolt that secures the engine harness to the cylinder block.

13. Disconnect the PCV hose from the PCV valve.

14. Remove the bolt that secures the engine harness to the intake manifold.

15. Disconnect the following:
- Engine wire clamps
- EGR temperature sensor connector

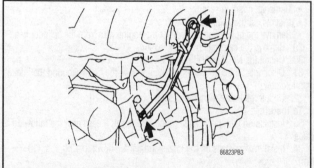

Fig. 55 Remove the two bolts securing the intake manifold stay

- The fuel injector connections
- The connector for the emission control valve set

16. Remove the water bypass pipe.

17. Remove the air intake chamber and the intake manifold assembly.
   a. Disconnect the vacuum hoses at the Thermal Vacuum (TVV) Valve.
   b. Remove the bolts securing the intake manifold and gasket.

To install:

18. Install the air intake chamber and intake manifold assembly:
   a. Place a new gasket so that the rear mark is toward the rear side.
   b. Tighten the intake manifold bolts to 15 ft. lbs. (21 Nm).
   c. Connect the vacuum hoses to the TVV.

19. Attach the intake manifold stay and tighten the bolts to 26 ft. lbs. (36 Nm).

20. Install the heater hose to the cylinder head, then connect the pipe to the intake manifold. Tighten the bolts to 15 ft. lbs. (21 Nm).

21. Installation of the remaining components is in the reverse order of removal.

Exhaust Manifold

REMOVAL & INSTALLATION

✳✳ CAUTION

When working on or around a manifold, make ceratin all surfaces are cool to the touch before beginning to work.

22R, 22R-E, and 3VZ-E Engines

1. Remove the three exhaust pipe flange bolts and disconnect the exhaust pipe from the manifold.

2. Tag and disconnect the spark plug leads. Position the spark plug wires out of the way. It is best to tie them so they don't get in your way.

3. Remove the air cleaner tube from the heat stove on carbureted engines, and remove the outer part of the heat stove.

4. Use a 14mm wrench to remove the manifold retaining nuts.

5. Remove the manifold(s), complete with air injection tubes and the inner portion of the heat stove.

6. For 22R engines, separate the inner portion of the heat stove from the manifold.

To install:

7. Installation is the reverse of removal. Always use a new gasket. Install any disassembled parts onto the manifold and place it in position.

8. Install the nuts and bolts, tightening them to 30 ft. lbs. (40 Nm). Connect the exhaust pipe to the manifold, tightening the bolts to 30 ft. lbs. (40 Nm)

5VZ-FE Engine

▶ **See Figures 56 and 57**

LEFT SIDE

1. Disconnect the exhaust crossover pipe from the exhaust manifold.

2. Remove the EGR pipe from the manifold.

3. Remove the heat insulator from the manifold.

4. Remove the manifold.

Fig. 56 Remove the bolts securing the crossover pipe to the manifold(s)

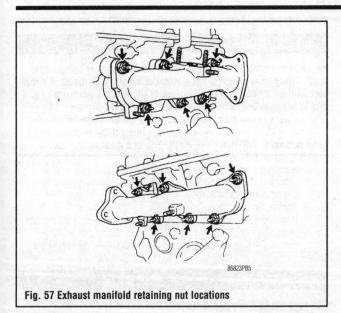

Fig. 57 Exhaust manifold retaining nut locations

To install:

5. Install the exhaust manifold to the engine with a new gasket. Tighten the nuts to 30 ft. lbs. (40 Nm). Tighten the heat insulator nuts to 71 inch lbs. (8 Nm), the EGR pipe nuts to 14 ft. lbs. (18 Nm), the clamp nuts to 71 inch lbs. (8 Nm) and the crossover pipe nuts to 33 ft. lbs. (45 Nm).

RIGHT SIDE

1. Disconnect the exhaust crossover pipe from the exhaust manifold.
2. Remove the exhaust manifold heat insulator.
3. Unfasten the six nuts securing the manifold from the engine.

To install:

4. Install the exhaust manifold with the new gasket, tighten the six nuts to 30 ft. lbs. (40 Nm). Tighten the nuts to 71 inch lbs. (8 Nm) and the crossover pipe nuts to 33 ft. lbs. (45 Nm).

2RZ-FE and 3RZ-FE Engines

▸ See Figures 58 and 59

1. Raise and safely support the vehicle.
2. Disconnect the front exhaust pipe from the manifold and mounting bracket.
3. Lower the vehicle.
4. Remove the heat insulator.
5. Remove the exhaust manifold and gasket.

To install:

6. Install the exhaust manifold and gasket to the engine, tighten the nuts to 36 ft. lbs. (49 Nm). Tighten the insulator bolts and nuts to 48 inch lbs. (5 Nm). Tighten the front pipe nuts to 46 ft. lbs. (62 Nm), the support bracket to 29 ft. lbs. (39 Nm) and the clamp nut to 14 ft. lbs. (19 Nm).

Fig. 58 The exhaust pipe must be disconnected from the manifold and its mounting bracket

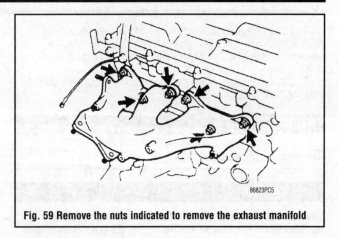

Fig. 59 Remove the nuts indicated to remove the exhaust manifold

1FZ-FE Engine

▸ See Figures 60 and 61

1. Disconnect the negative battery cable.
2. Raise and safely support the vehicle.
3. Working from underneath the vehicle, disconnect the oxygen sensor wire.
4. Remove the nuts and bolts holding the front header pipe to the catalytic converter.
5. Loosen the pipe clamp.
6. Remove the pipe bracket.
7. Unbolt the front header pipe and remove the gasket.
8. Lower the vehicle, remove the exhaust manifold heat shield insulators.
9. Remove the nuts from the exhaust manifolds, then the gaskets.

To install:

10. Install the exhaust manifolds to the engine with new gaskets. Uniformly tighten the nuts in several passes. Tighten the nuts to 29 ft. lbs. (39 Nm).

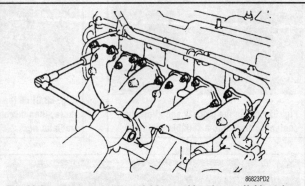

Fig. 60 An extension can be useful for reaching the manifold retaining nuts

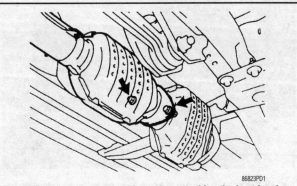

Fig. 61 Be sure to use new gaskets when attaching the front header pipe to the catalyst

11. Install the exhaust manifold heat insulators, tighten down to 14 ft. lbs. (19 Nm). Tighten the header pipe nuts to 46 ft. lbs. (63 Nm), the support bracket to 17 ft. lbs. (24 Nm) and the clamp nut to 14 ft. lbs. (19 Nm).

12. Connect the header pipe to the catalyst using a new gasket to prevent leaks. Tighten to 34 ft. lbs. (46 Nm).

Radiator

REMOVAL & INSTALLATION

❉❉ CAUTION

When draining coolant, keep in mind that cats and dogs are attracted by ethylene glycol antifreeze, and are quite likely to drink any that is left in an uncovered container or in puddles on the ground. This will prove fatal in sufficient quantity. Always drain coolant into a sealable container. Coolant may be reused unless it is contaminated or several years old.

22R and 22R-E Engines

▶ See Figures 62, 63, 64, 65 and 66

1. Discharge the A/C system using an approved recovery/recycling machine. Refer to Section 1.

2. Drain the cooling system.

❉❉ CAUTION

Never remove the pressure cap while the engine is running or personal injury from scalding hot coolant or steam may result. If possible, wait until the engine has cooled to remove the pressure cap. If this is not possible, wrap a thick cloth around the pressure cap and turn it slowly to the stop. Step back while the pressure is released from the cooling system. When you are sure all the pressure has been released, still using the cloth, turn and remove the cap.

3. Remove the engine undercover.

4. Disconnect the reservoir hose and remove the reservoir tank.

5. Disconnect the radiator hoses.

6. Remove the fan shrouds.

7. Disconnect the oil cooler lines (automatic transmission). Plug the lines immediately to prevent oil loss.

8. Remove the nuts and bolts holding the radiator. Carefully remove the radiator.

❉❉ WARNING

Use care not to damage the radiator fins or the cooling fan.

To install:

9. Installation is the reverse of removal. If replacing with a used radiator or your old one, install a new O-ring onto the petcock.

10. Install the radiator and tighten the mounting bolts. Make certain all the rubber mounts and bushings are present and correctly placed.

Fig. 62 Loosen the petcock and allow the coolant to drain into a suitable container

Fig. 63 Slide the clamp up the overflow hose, then disconnect the hose from the radiator neck

Fig. 64 Be careful when disconnecting the lower hose, as some coolant may be left in it

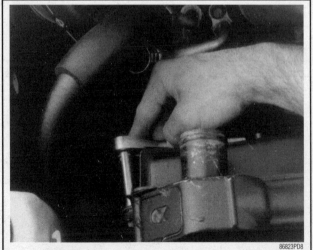

Fig. 65 It may be necessary to use a deep socket or an extension to access the retainers

Fig. 66 Be careful not to damage the fins when lifting the radiator out

3F-E Engine

▶ See Figures 67 and 68

1. Discharge the air conditioning system using a recovery/recycling machine. Refer to Section 1.
2. Drain the cooling system.

※※ CAUTION

Never remove the pressure cap while the engine is running or personal injury from scalding hot coolant or steam may result. If possible, wait until the engine has cooled to remove the pressure cap. If this is not possible, wrap a thick cloth around the pressure cap and turn it slowly to the stop. Step back while the pressure is released from the cooling system. When you are sure all the pressure has been released, still using the cloth, turn and remove the cap.

3. On some models you will need to remove the front turn signal lamp assembly, radiator grille and headlamp doors.
4. Remove the hood lock brace, hood lock, grille center support and grille. Some models need the engine under cover removed.
5. On 1989–90 models you need to remove the battery and battery case.
6. Disconnect the A/C discharge hose and liquid tube.

➡ **Make sure you cap the open fittings immediately to keep moisture from entering the system.**

7. Remove the condenser.
8. Disconnect the coolant reservoir hose and tank.
9. Remove the radiator hoses.
10. Remove the power steering and air conditioning belts.
11. Remove the fan shroud, fan and the fluid coupling. The alternator belt will need removal also.
12. Disconnect the A/T oil cooler hoses. You will need a suitable container to catch oil which will leak out on removal.
13. Remove the radiator mounting bolts, nuts and then the radiator.
To install:
14. Installation is the reverse of the removal procedure. If replacing with a used radiator or your old one, install a new O-ring onto the petcock.

15. Install the radiator and tighten the (A) mounting bolts to 13 ft. lbs. (18 Nm) and (B) bolts to 9 ft. lbs. (12 Nm). Make certain all the rubber mounts and bushings are present and correctly placed.

2RZ-FE and 3RZ-FE Engines

▶ See Figures 69, 70 and 71

1. Drain the engine cooling system. Have a large enough suitable container to collect the fluid.
2. Remove the grille, filler and clearance lamps.
3. On the 3RZ-FE California models, disconnect the air pipe.
4. Disconnect the upper radiator hose.
5. Unsecure the radiator reservoir hose.
6. On some T100 models you will need to remove the power steering belt. With A/C vehicles, loosen the idler pulley nut and adjusting bolt, then remove the belt. Remove the alternator belt, fan with coupling, water pump pulley and fan shroud.
7. Remove the lower radiator hose, remember there will still be some fluid left in the lower hose. Have a catch pan available.
8. Remove the fan shroud.
9. On A/T vehicles, disconnect the oil cooler hoses. Plug the openings so oil does not drip out.
10. Remove the 4 mounting bolts for the radiator. When lifting the radiator from the vehicle, you will have some fluid still in the tank. Watch for spillage, it's slippery.
To install:
11. Installation is in the reverse order of the removal procedure. After the radiator has been removed there may be supports on the old unit. Remove them and install on the new radiator.
12. If replacing with a used radiator or your old one, install a new O-ring onto the petcock.
13. With the supports on the new unit, install into the vehicle. Insert the tabs of the support through the radiator service holes. Tighten the radiator down to 9 ft. lbs. (12 Nm).

5VZ-FE Engine

1. Drain the engine cooling system, discard of the O-ring.
2. Remove the front bumper filler.

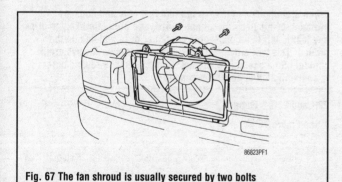

Fig. 67 The fan shroud is usually secured by two bolts

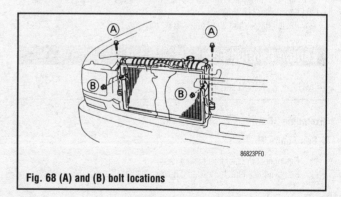

Fig. 68 (A) and (B) bolt locations

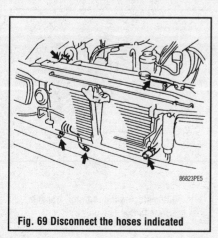

Fig. 69 Disconnect the hoses indicated

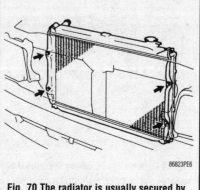

Fig. 70 The radiator is usually secured by four bolts

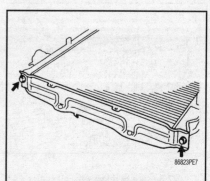

Fig. 71 If the radiator is being replaced, the support brackets should be transferred to the new radiator

3. Take off the clearance lamps and grille assemblies.
4. Disconnect the upper, reserve and lower radiator hoses.
5. Unbolt the fan shroud.
6. On A/T vehicles, disconnect and plug the oil cooler lines.
7. Unbolt the radiator and lift out of the vehicle.

To install:

8. Installation is in the reverse order of the removal procedures. If replacing with a used radiator or your old one, install a new O-ring onto the petcock.
9. Install the radiator and tighten the mounting bolts to 9 ft. lbs. (12 Nm). Make certain all the rubber mounts and bushings are present and correctly placed.

1FZ-FE Engine

▶ See Figures 72, 73 and 74

1. Drain the engine cooling system.
2. Disconnect the battery cables, negative first.
3. Remove the hold-down clamp and battery.
4. Unbolt and remove the ground strap.
5. Remove the battery tray.
6. Remove the grille.
7. Disconnect the radiator bypass hose.
8. Disconnect the upper radiator and reservoir hoses.
9. Loosen the water pump pulley mounting nuts.
10. Loosen the lock, pivot and adjusting bolts of the alternator, then remove the drive belts.
11. Disconnect the oil cooler hose from the clamp on the fan shroud.
12. Unbolt and remove the fan shroud.
13. Remove the water pump pulley mounting nuts.
14. Pull out the fan with the coupling, water pump pulley and fan shroud.
15. Disconnect the A/T oil cooler hoses.
16. Unclamp, then remove the lower radiator hose.
17. Unbolt and remove the radiator, remember there will be some fluid still in the unit.

To install:

18. Installation is in the reverse order of the removal procedure. Remove the radiator brackets from the old unit if replacing with a new or used one.
19. Install the brackets to the new unit, tighten to 9 ft. lbs. (12 Nm)and the install a new O-ring on the petcock.
20. Install the radiator, tighten the mounting nuts to 9 ft. lbs. (12 Nm) and bolts to 13 ft. lbs. (18 Nm).
21. Install the shroud and tighten to 43 inch lbs. (5 Nm).

Engine Cooling Fan

REMOVAL & INSTALLATION

Excluding 3F-E

▶ See Figure 75

1. Disconnect the negative battery cable.
2. Remove the alternator, A/C and power steering (if applicable) belts.

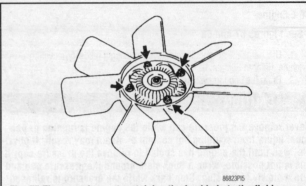

Fig. 75 There are four nuts retaining the fan blade to the fluid coupling

3. Loosen the bolts holding the water pump pulley (not the pump).
4. Remove the nuts, then remove the fan clutch (fluid coupling), fan and pulley.

To install:

5. Installation is in the reverse order of the removal procedure. Tighten the nuts to 16 ft. lbs. (21 Nm).

3F-E Engine

1. With the ignition **OFF**, unplug the connector at the fan.
2. Remove the fan shroud.
3. Remove the air intake silencer and support.
4. Remove the 3 bolts and remove the cooling fan assembly.
5. Installation is in the reverse order of the removal procedure.

Water Pump

REMOVAL & INSTALLATION

✳✳ CAUTION

When draining coolant, keep in mind that cats and dogs are attracted by ethylene glycol antifreeze, and are quite likely to drink any that is left in an uncovered container or in puddles on the ground. This will prove fatal in sufficient quantity. Always drain coolant into a sealable container. Coolant may be reused unless it is contaminated or several years old.

22R and 22R-E Engines

▶ See Figure 76

1. Drain the cooling system.
2. Unfasten the fan shroud securing bolts and remove the fan shroud, if so equipped.

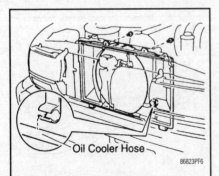

Fig. 72 Be sure to release the oil cooler hose from the clamp before removing the fan shroud

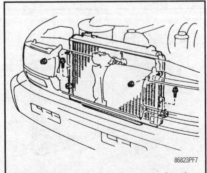

Fig. 73 Remove the fasteners securing the radiator, the remove the radiator from the vehicle

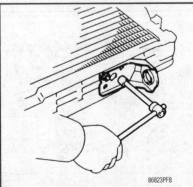

Fig. 74 If replacing the radiator, transfer the brackets to the new unit

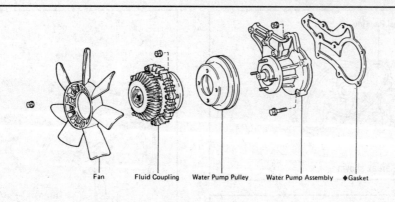

♦ Non-reusable part

86823PJ5

Fig. 76 Exploded view of the water pump mounting—22R and 22R-E engines

3. Loosen the alternator adjusting bolt and remove the drive belt.

4. Remove the air conditioning and/or power steering pump drive belt.

5. Detach the bypass hose from the water pump.

6. Unfasten the water pump retaining bolts and remove the water pump and fan assembly, using care not to damage the radiator with the fan.

To install:

7. Install the water pump and tighten the mounting bolts to 27 ft. lbs. (37 Nm). Always use a new gasket between the pump body and its mounting. Installation of the remaining components is in the reverse order of the removal procedure.

3F-E Engine

♦ **See Figure 77**

1. Drain the coolant from the engine and radiator.

2. Loosen the idler pulley and adjusting bolts to remove the power steering belt.

3. Disconnect the coolant reservoir hose and the radiator inlet hose.

4. Remove the bolts holding the fan shroud.

5. Remove the alternator drive belt, fan, fluid coupling, and water pump pulley.

6. Remove the alternator.

7. Remove the air conditioning belt.

8. Disconnect the four hoses from the water pump.

9. Remove bolt, two nuts and power steering idler pulley with the bracket assembly.

10. Remove the bolt and nut, alternator support bracket, water pump and gasket.

To install:

11. Install a new gasket, the water pump and alternator support bracket with the bolt and nut.

12. Install the power steering idler pulley with the bracket assembly.

13. Tighten the water pump nuts and bolt to 27 ft. lbs. (37 Nm).

14. Installation of the remaining components is in the reverse order of the removal procedure.

3VZ-E and 5VZ-FE Engines

♦ **See Figures 78 and 79**

> **❊❊ WARNING**
>
> This procedure requires the removal of the timing belt. This is an extensive procedure requiring precision work, particularly at reassembly. Refer to Timing Belt—Removal and Installation in this section. Do not attempt to remove the belt without consulting the procedure; extreme engine damage may occur.

1. Drain the cooling system.

2. Remove the timing belt.

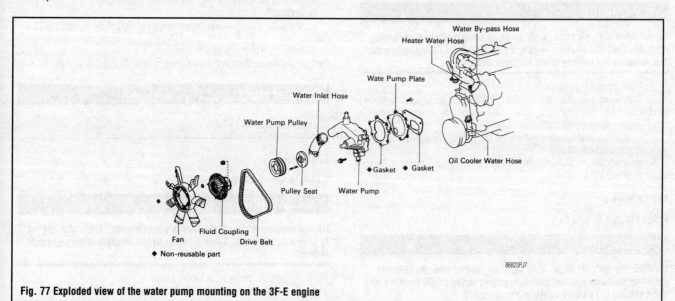

♦ Non-reusable part

86823PJ7

Fig. 77 Exploded view of the water pump mounting on the 3F-E engine

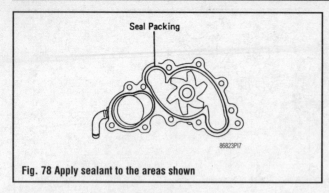

Fig. 78 Apply sealant to the areas shown

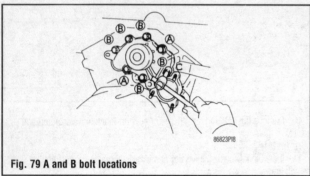

Fig. 79 A and B bolt locations

3. Remove the idler pulley.
4. Remove the thermostat.
5. Disconnect the oil cooler hose from the pump.
6. Remove the seven bolts and the tension spring bracket and then remove the water pump.
To install:
7. Apply sealer to the pump-to-block mating surface and then tighten the bolts marked **A** in the illustration to 13 ft. lbs. (18 Nm). Tighten those marked **B** to 14 ft. lbs. (20 Nm).

➡**The parts must be installed within 5 minutes of application of the sealant. Otherwise you must remove and reapply the sealer.**

8. Installation of the remaining components is in the reverse order of the removal procedure.

2RZ-FE and 3RZ-FE Engines

1. Disconnect the negative battery cable.

> ✳✳ **CAUTION**
>
> On vehicles with air bags, work must be started after 90 seconds from the time the ignition switch is turned to the LOCK position and the negative (–) battery cable is disconnected.

2. Drain the cooling system.
3. If equipped with an air conditioning compressor or power steering pump drive belts, it may be necessary to loosen the adjusting bolt, remove the drive belt(s) and move the component(s) out of the way.
4. Remove the fluid coupling with the fan and water pump pulley.
5. Remove the water pump.
6. Installation is in the reverse order of the removal procedure. Bleed the cooling system.

1FZ-FE Engine

▶ See Figures 80 and 81

> ✳✳ **CAUTION**
>
> On vehicles with air bags, work must be started after 90 seconds from the time the ignition switch is turned to the LOCK position and the negative (–) battery cable is disconnected.

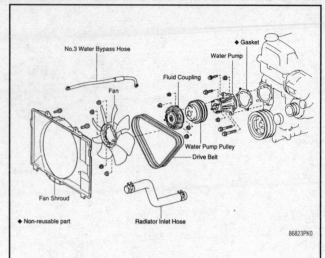

Fig. 80 Exploded view of the water pump mounting on the 1FZ-FE engine

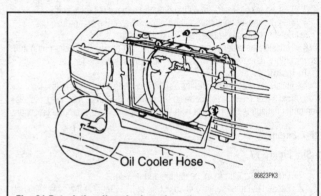

Fig. 81 Detach the oil cooler hose from the clamp on the fan shroud

1. Disconnect the negative battery cable.
2. Drain the engine coolant.
3. Disconnect the water bypass and radiator inlet hoses.
4. Remove the drive belts, fan assembly, and the fan shroud.
5. Disconnect the oil cooler hose from the clamp on the fan shroud. Remove the bolts holding the fan shroud to the radiator.
6. Remove the 4 bolts, 2 nuts, water pump, and the gasket.
To install:
7. Install the water pump using a new gasket. Tighten the fasteners to 15 ft. lbs. (21 Nm).
8. Installation of the remaining components is in the reverse order of the removal procedure. Tighten the fan shroud bolts to 43 inch lbs. (5 Nm).

Cylinder Head

REMOVAL & INSTALLATION

22R Engine

▶ See Figures 82 thru 92

> ✳✳ **WARNING**
>
> Do not perform this operation on a warm engine. Allow the engine to cool a minimum of 3 hours, more in hot weather. Overnight cold is best.

1. Disconnect the negative battery cable.
2. Drain the cooling system (both radiator and block). If the coolant is to be reused, place a large, clean container underneath the drains.

✳✳ CAUTION

When draining coolant, keep in mind that cats and dogs are attracted by ethylene glycol antifreeze, and are quite likely to drink any that is left in an uncovered container or in puddles on the ground. This will prove fatal in sufficient quantity. Always drain coolant into a sealable container. Coolant may be reused unless it is contaminated or several years old.

3. Remove the three exhaust pipe flange nuts and separate the pipe from the manifold.

4. Remove the air cleaner assembly, complete with hoses, from the carburetor.

➡ **Cover the carburetor with a clean shop cloth so that nothing can fall into it.**

5. Drain the engine oil.

✳✳ CAUTION

Used motor oil may cause skin cancer if repeatedly left in contact with the skin for prolonged periods. Although this is unlikely unless you handle oil on a daily basis, it is wise to thoroughly wash your hands with soap and water immediately after handling used motor oil.

6. Disconnect the radiator inlet hose from the water outlet.

7. If equipped with an air pump, disconnect the air hose from the check valve.

8. Disconnect the two heater hoses from the engine.

9. Disconnect the accelerator cable at the carburetor.

10. Label and disconnect these wires:
- EGR temperature sensor (Calif. vehicles only)
- Water temperature gauge sender
- Cold mixture heater wire
- Temperature sensor
- Fuel cut solenoid valve
- EACV (Calif. only)

11. Label and disconnect these hoses and lines:
- Charcoal canister
- Brake booster
- Main fuel line from fuel inlet
- Fuel return line
- HAC from the bracket
- Electronic Bleed Air Control Valve (EBCV) Calif. vehicles
- Vacuum Switching Valve (VSV) with bracket

12. Remove the rear ground strap from the engine.

13. Remove the distributor.

14. Remove the spark plugs.

15. Disconnect the power steering bracket from the head.

16. Remove the valve cover and gasket.

17. Turn the engine clockwise (use the crank pulley bolt) until the piston is at TDC on its compression stroke. Matchmark the timing sprocket to the cam chain, and remove the semi-circular plug. Remove the cam sprocket bolt. Slide the distributor drive gear and spacer off the cam. Leave the chain on the sprocket; either wire the cam sprocket in place just off the end of the cam or rest the sprocket on the vibration damper. Tie a string or wire to the chain to keep it from falling in the engine.

18. Remove the timing chain cover bolt at the front of the head. This must be done before the head bolts are removed.

19. Remove the cylinder head bolts in the correct order. Make two or three passes to loosen the bolts. Do not loosen any one bolt all at once; improper removal could cause head damage.

20. Using prybars applied evenly at the front and the rear of the valve rocker assembly, gently pry the assembly off its mounting dowels.

21. Lift the head off its dowels. If it is difficult to lift off, pry gently with small tools. Do NOT damage the mating surfaces of the head or block. Do NOT attempt to slide the head off the block; it must be lifted.

22. Support the head on a workbench.

23. Remove the manifolds and EGR valve from the head on the workbench. Keep the head covered and the manifold ports plugged with rags when not working on it.

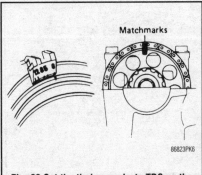

Fig. 82 Set the timing marks to TDC on the No.1 cylinder; matchmark the chain and sprocket

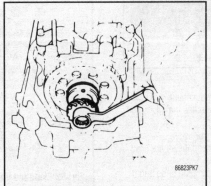

Fig. 83 Using a 19mm wrench, remove the cam sprocket bolt

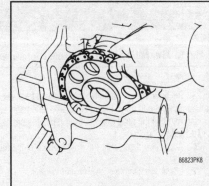

Fig. 84 Do not allow the chain to slip off while removing the sprocket

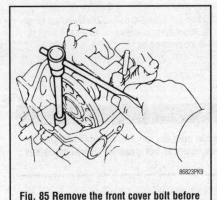

Fig. 85 Remove the front cover bolt before loosening the cylinder head bolts

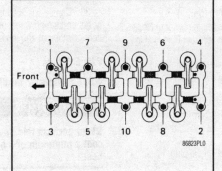

Fig. 86 Loosen the head bolts in this order, in three passes

Fig. 87 Lift the old gasket from the block

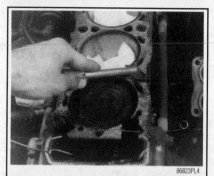

Fig. 88 Scrape and clean the area well, insert paper into the cylinder walls to keep dirt from entering

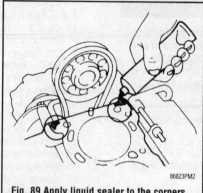

Fig. 89 Apply liquid sealer to the corners of the block

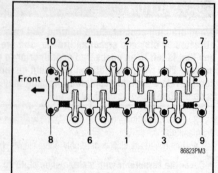

Fig. 90 Tighten the head bolts in this order, in three passes

Fig. 91 Always tighten the bolts with a torque wrench

24. Re-drain the oil pan after the head has been removed; coolant will have spilled into the pan during head removal. The coolant will pollute the oil.

25. Remove the head gasket, never reuse the old one.

To install:

26. Apply liquid sealer to the front corners of the block and install the head gasket.

27. Lower the head over the locating dowels. Do not attempt to slide it into place.

28. Install the rocker arm assembly over its positioning dowels.

29. Install the cylinder head bolts. Tighten the bolts sequentially in three passes and in the correct order, coming to final torque on the last pass. Correct tightness is 58 ft. lbs. (78 Nm)

30. Install the chain cover bolt. Tighten it to 9 ft. lbs. (13 Nm).

31. Hold the chain and sprocket; double check that the engine is set to No. 1 TDC. If necessary, manually rotate the engine so that the sprocket hole is also at the top.

32. Fit the sprocket over the camshaft dowel. If the chain won't allow the sprocket to reach, rotate the crankshaft back and forth, while lifting up on the chain and sprocket.

33. Install the distributor drive gear and cam thrust plate over the chain sprocket; tighten the camshaft sprocket bolt to 58 ft. lbs. (78 Nm).

34. Set the No. 1 piston at TDC of its compression stroke and adjust the valves.

35. Apply sealant only the cylinder head side of the half-circle plug; install the plug.

36. Install the valve cover

37. Attach the power steering bracket and compressor support to the cylinder head. Tighten the bolts to 33 ft. lbs. (44 Nm)

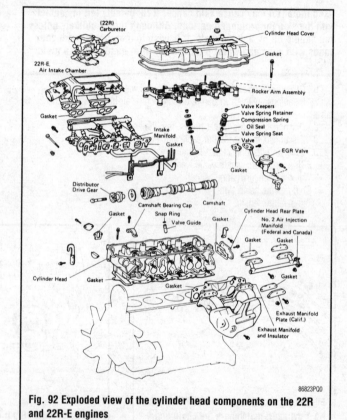

Fig. 92 Exploded view of the cylinder head components on the 22R and 22R-E engines

38. Installation of the remaining components is in the reverse order of the removal procedure.

39. Double check all installation items, paying particular attention to loose hoses or hanging wires, untightened nuts, poor routing of hoses and wires (too tight or rubbing) and tools left in the engine area.

22R-E Engine

▶ **See Figures 82 thru 92**

☀☀ CAUTION

Never perform this operation on a hot engine. Allow the engine to cool a minimum of 3 hours, more in hot weather. Overnight cold is best.

1. Disconnect the negative battery cable.

2. Drain the coolant from the radiator and the cylinder block. Drain the engine oil.

When draining coolant, keep in mind that cats and dogs are attracted by ethylene glycol antifreeze, and are quite likely to drink any that is left in an uncovered container or in puddles on the ground. This will prove fatal in sufficient quantity. Always drain coolant into a sealable container. Coolant may be reused unless it is contaminated or several years old.

3. Remove the air intake connector.

4. Disconnect the oxygen sensor wire.

5. Remove the three nuts attaching the manifold to the exhaust pipe and then separate the two.

6. Remove the oil dipstick.

7. Remove the distributor with the spark plug leads attached. Remove the spark plugs.

8. Disconnect the upper radiator hose and the two heater hoses where they attach to the engine and then position them out of the way.

9. Disconnect the actuator cable, the accelerator cable and the throttle cable for the automatic transmission at their brackets.

10. Tag and disconnect the following:
- Both PCV vacuum hoses
- Brake booster hose
- Actuator hose (if equipped with cruise control)
- Air control valve hoses (with power steering)
- Vacuum Switching Valve (VSV) hoses (with air conditioning)
- Evaporative emissions (EVAP) hoses
- EGR valve hose
- fuel pressure hose
- Reed valve hose
- Pressure regulator hose
- Vacuum hoses from throttle body
- Water bypass hoses from throttle body

11. Tag and disconnect the EGR vacuum modulator hoses and then remove the modulator itself along with the bracket.

12. Disconnect the following wires:
- Cold start injector pipe and wire
- Throttle position sensor wire
- EGR temperature sensor (Calif. vehicles)

13. Disconnect the cold start injector pipe from the intake chamber. Remove the bolt holding the EGR pipe to the air chamber. Disconnect the chamber from the stay. Remove the air chamber-to-intake manifold bolts and then lift off the chamber with the throttle body, resonator and gasket.

14. Disconnect the fuel return hose.

15. Tag and disconnect the following wires:
- Knock sensor
- Oil pressure sender gauge or switch
- Starter wire
- Transmission wires
- Compressor wires (with air conditioning)
- Injector wires
- Water temperature sender gauge wire
- Overdrive temperature switch wire (auto. trans.)
- Igniter wire
- Vacuum Switching Valve (VSV) wires
- Cold start injector time switch wire
- Water temperature sensor wire

16. Disconnect the bypass hose at the intake manifold.

17. If equipped with power steering, remove the pump and position it out of the way without disconnecting the hydraulic lines.

18. Remove the power steering bracket from the cylinder head.

19. Remove the valve cover.

➡**Be sure to cover the oil return hole in the cylinder head with a clean rag to prevent anything from falling in.**

20. Turn the engine clockwise (use the crank pulley bolt) until the No. 1 piston is at TDC on its compression stroke. Matchmark the timing sprocket to the cam chain, and remove the semi-circular plug. Remove the cam sprocket bolt. Slide the distributor drive gear and spacer off the cam. Leave the chain on the sprocket; either wire the cam sprocket in place just off the end of the cam or rest the sprocket on the vibration damper.

21. Remove the timing chain cover bolt at the front of the head. This must be done before the head bolts are removed.

22. Remove the cylinder head bolts in the correct order. Make two or three passes to loosen the bolts. Do not loosen any one bolt all at once; improper removal could cause head damage.

➡**Refer to the bolt removal diagram for the 22R engine. All illustrations for the 22R are valid for the 22R-E.**

23. Using prybars applied evenly at the front and the rear of the valve rocker assembly, gently pry the assembly off its mounting dowels.

24. Lift the head off its dowels. If it is difficult to lift off, pry gently with small tools. Do NOT damage the mating surfaces of the head or block. Do NOT attempt to slide the head off the block; it must be lifted.

25. Support the head on a workbench.

26. Remove the manifolds and EGR valve from the head on the workbench. Keep the head covered and the manifold ports plugged with rags when not working on it.

27. Re-drain the oil pan after the head has been removed; coolant will have spilled into the pan during head removal. The coolant will pollute the oil.

To install:

28. Apply liquid sealer to the front corners of the block and install the head gasket.

29. Lower the head over the locating dowels. Do not attempt to slide it into place.

30. Install the rocker arm assembly over its positioning dowels.

31. Install the cylinder head bolts. Tighten the bolts sequentially in three passes and in the correct order, coming to final torque on the last pass. Correct tightness is 58 ft. lbs. (78 Nm)

➡**Refer to the bolt tightening diagram for the 22R engine. All illustrations for the 22R are valid for the 22R-E.**

32. Install the chain cover bolt. Tighten it to 9 ft. lbs. (13 Nm).

33. Hold the chain and sprocket; double check that the engine is set to No. 1 TDC. If necessary, manually rotate the engine so that the sprocket hole is also at the top.

34. Fit the sprocket over the camshaft dowel. If the chain won't allow the sprocket to reach, rotate the crankshaft back and forth, while lifting up on the chain and sprocket.

35. Install the distributor drive gear and cam thrust plate over the chain sprocket; tighten the camshaft sprocket bolt to 58 ft. lbs. (78 Nm).

36. Set the No. 1 piston at TDC of its compression stroke and adjust the valves.

37. Apply sealant only the cylinder head side of the half-circle plug; install the plug.

38. Install the valve cover.

39. Attach the power steering bracket to the head; tighten the 4 bolts to 33 ft. lbs. (44 Nm).

40. Installation of the remaining components is in the reverse order of the removal procedure.

41. Double check all installation items, paying particular attention to loose hoses or hanging wires, untightened nuts, poor routing of hoses and wires (too tight or rubbing) and tools left in the engine area.

3F-E Engine

◆ **See Figures 93 thru 98**

1. Drain the engine coolant.

When draining coolant, keep in mind that cats and dogs are attracted by ethylene glycol antifreeze, and are quite likely to drink any that is left in an uncovered container or in puddles on the ground. This will prove fatal in sufficient quantity. Always drain coolant into a sealable container. Coolant may be reused unless it is contaminated or several years old.

2. Disconnect the negative battery cable.

3. Scribe matchmarks around the hood hinges and then remove the hood.

4. Disconnect the accelerator and throttle cables.

5. Remove the hoses and parts as follows:

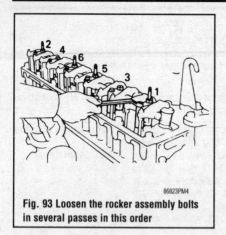

Fig. 93 Loosen the rocker assembly bolts in several passes in this order

Fig. 94 Remove the pushrods and keep them in order

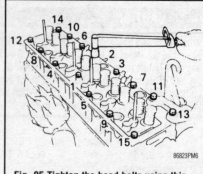

Fig. 95 Tighten the head bolts using this sequence in several passes to prevent head damage

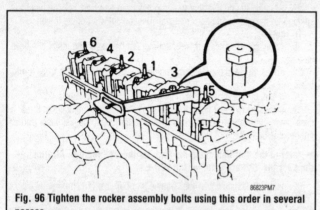

Fig. 96 Tighten the rocker assembly bolts using this order in several passes

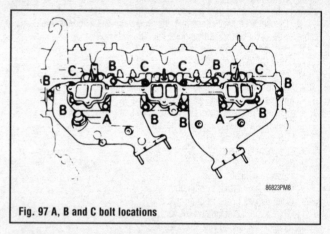

Fig. 97 A, B and C bolt locations

- Air intake
- Air flow meter
- Air cleaner cap
- ISC hose
- Air pump hose
- Distributor hose
- PCV hose
- Remove the 3 hoses from the rear of the intake chamber
- The 2 hoses from the charcoal canister vacuum valve

6. Unbolt the power steering pump and position it out of the way without disconnecting the hydraulic lines.

7. Unbolt the air conditioning compressor and position it out of the way without disconnecting the refrigerant lines. Use stiff wire or string to support the compressor; do not allow it to hang by the hoses.

8. Remove the power steering pump and air conditioning compressor brackets.

9. Disconnect the high tension cords from the spark plugs and the coil.

10. Disconnect and remove the hoses from the water outlet, water pump, oil cooler and heater water pipe. Unbolt the oil cooler water pipe from the cylinder head.

11. Disconnect the upper radiator hose.

12. Disconnect and plug the fuel lines.

13. Raise the vehicle, making certain it is well supported. Disconnect the exhaust pipe at the manifold.

14. Remove the air pump.

15. Remove the fuel delivery pipe along with the fuel injectors.

16. Remove the air injection manifold.

17. Remove the intake and exhaust manifolds.

18. Disconnect the water bypass hose at the water outlet and then remove the outlet assembly.

19. Remove the spark plugs.

20. Disconnect the water bypass hose from the water outlet. Remove the water outlet assembly.

21. Loosen the eight bolts and four nuts that attach the rocker shaft assembly in several stages and in the correct order. Remove the rocker shaft.

22. Remove the twelve pushrods, be sure they are kept in the order in which they were removed; they MUST be installed in the original locations. An upside-down box with holes punched in the bottom will make a good holder; number the holes to correspond with the pushrod position.

23. Loosen the cylinder head bolts in several stages and in the opposite order of the tightening sequence. Remove the air pump bracket and engine hanger.

❊❊ CAUTION

Removing the head bolts in the improper order could cause damage to the aluminum cylinder head.

24. Lift the cylinder head off of its mounting dowels, Do NOT attempt to slide the head off; it must be lifted. Support the head on a workbench.

To install:

25. Lift the old gasket from the block, then scrape and clean the area well. Insert paper into the cylinder walls to keep dirt from entering.

26. Install the cylinder head on the cylinder block using a new gasket.

27. Lightly coat the threads of the cylinder head bolts with engine oil and then install them into the head. Tighten them in several stages, in the order shown.

28. Install the pushrods in the order that they were removed.

29. Position the rocker shaft assembly on the cylinder head and align the rocker arm adjusting screws with the heads of the pushrods. Tighten the mounting bolts with a 12mm head to 17 ft. lbs. (24 Nm); tighten the bolts with a 14mm head to 25 ft. lbs. (33 Nm).

30. Adjust the valve clearance and install the spark plugs.

31. Install the valve cover and tighten the cap nuts to 78 inch lbs. (9 Nm).

32. Install the water outlet and connect the bypass hose. Tighten the bolts to 18 ft. lbs. (25 Nm).

33. Install the intake and exhaust manifolds using a new gasket. Make sure the front mark on the gasket is towards the front of the engine. Tighten the following:
- 17mm bolts (A)—51 ft. lbs. (69 Nm)
- 14mm bolts (B)—37 ft. lbs. (50 Nm)
- Nuts (C)—41 ft. lbs. (56 Nm)

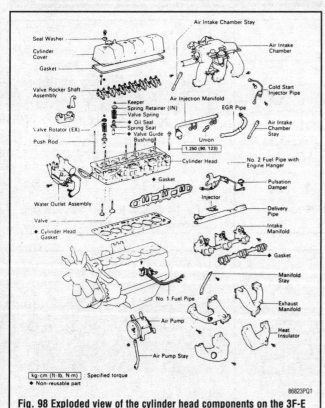

Fig. 98 Exploded view of the cylinder head components on the 3F-E engine

34. Install the heat insulators and the manifold stay. Tighten the manifold heat shield bolts to 9 ft. lbs. (12 Nm) and the manifold support bolts to 22 ft. lbs. (29 Nm).

35. Install the air injection manifold and tighten the union nuts and clamp bolts to 15 ft. lbs. (21 Nm).

36. Installation of the remaining components is in the reverse order of the removal procedure.

37. Double check all installation items, paying particular attention to loose hoses or hanging wires, untightened nuts, poor routing of hoses and wires (too tight or rubbing) and tools left in the engine area.

38. Road test the vehicle for general performance and driveability.

3VZ-E Engine

▶ **See Figures 99 thru 116**

1. Disconnect the negative battery cable.
2. Remove the air cleaner hose and case.
3. Drain the engine coolant.

✳✳ CAUTION

When draining coolant, keep in mind that cats and dogs are attracted by ethylene glycol antifreeze, and are quite likely to drink any that is left in an uncovered container or in puddles on the ground. This will prove fatal in sufficient quantity. Always drain coolant into a sealable container. Coolant may be reused unless it is contaminated or several years old.

4. Remove the radiator.
5. Unbolt the power steering pump and position it out of the way with the hoses still attached.
6. Remove the alternator drive belt; remove the fluid coupling and fan pulley.
7. Disconnect the following wires and connectors:
- Ground strap from left fender apron
- Alternator connector and wire
- Igniter connector

- Oil pressure switch connector
- Ground strap at rear of engine
- Electronic Control Unit (ECU) connectors
- Vacuum Switching Valve (VSV) connectors
- Starter relay connector (manual trans. only)
- Solenoid resistor connector
- The check connector
- Compressor connector

8. Disconnect the following hoses:
- Power steering air hoses
- Brake booster hose
- Cruise control vacuum hose
- Charcoal canister hose at the canister
- Vacuum Switching Valve (VSV) vacuum hose

9. Disconnect the accelerator, throttle and cruise control cables.

10. Disconnect the clutch release (slave) cylinder hose (manual transmission only).

11. Disconnect the two heater hoses and the two fuel lines.

12. Remove the left side scuff plate (inside the left door). Disconnect the oxygen sensor. Disconnect the exhaust pipe at the manifold and at the catalytic converter. Remove the front exhaust pipe.

13. Remove the timing belt. Follow the procedure given later in this section. This is a complicated operation requiring accurate work.

14. Remove the distributor with the spark plug leads attached; position it out of the way.

15. Remove the air intake chamber as follows:
   a. Disconnect the throttle position sensor connector
   b. Disconnect the canister vacuum hose from the throttle body.
   c. Disconnect the vacuum and fuel hoses from the pressure regulator.
   d. Remove the PCV hose from the hose union.
   e. Disconnect the water bypass hose from the intake manifold hose union.
   f. Remove the water bypass hose from the water bypass pipe.
   g. Disconnect the cold start injector connector.
   h. Disconnect the vacuum hose from the fuel filter.
   i. Remove the union bolt and remove the cold start injector tube.
   j. On California vehicles, disconnect the EGR temperature sensor.

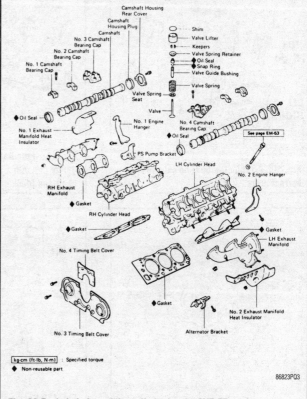

Fig. 99 Exploded view of the cylinder head—3VZ-FE engine

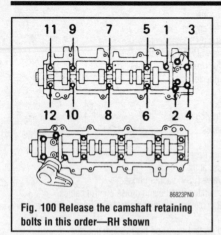

Fig. 100 Release the camshaft retaining bolts in this order—RH shown

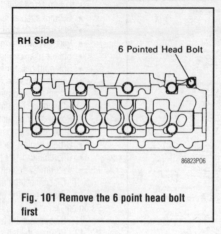

Fig. 101 Remove the 6 point head bolt first

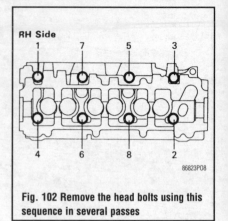

Fig. 102 Remove the head bolts using this sequence in several passes

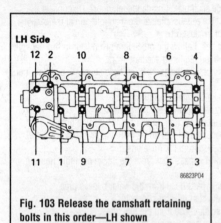

Fig. 103 Release the camshaft retaining bolts in this order—LH shown

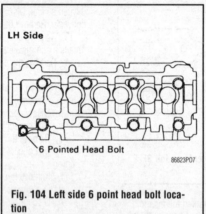

Fig. 104 Left side 6 point head bolt location

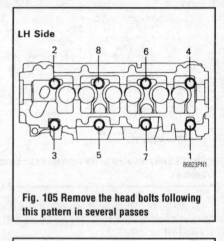

Fig. 105 Remove the head bolts following this pattern in several passes

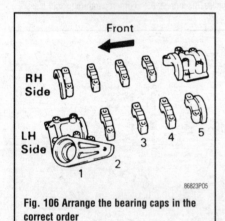

Fig. 106 Arrange the bearing caps in the correct order

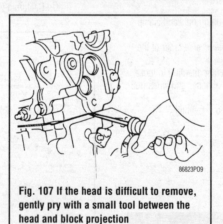

Fig. 107 If the head is difficult to remove, gently pry with a small tool between the head and block projection

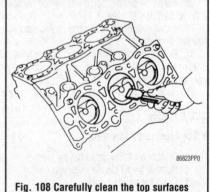

Fig. 108 Carefully clean the top surfaces of the pistons . . .

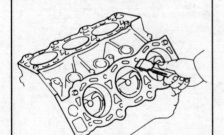

Fig. 109 . . . and the cylinder block

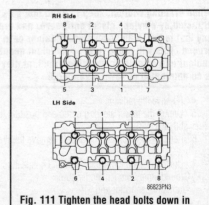

Fig. 110 Position the new gaskets on the cylinder heads as shown

Fig. 111 Tighten the head bolts down in several passes in this order

k. Disconnect the EGR vacuum hoses from the air pipe and EGR vacuum modulator.

l. Remove the intake chamber support. Remove the two water hoses from the EGR valve if so equipped.

m. Remove the EGR valve with its pipes.

n. Remove the six bolts and two nuts; remove the intake air chamber.

16. Disconnect the following, then when all are released, remove the two bolts and the engine wire harness:

- Knock sensor
- Water temperature sensor
- Cold start injector time switch connector
- Water temperature sensor
- Water temperature sender gauge connector
- Right ground strap
- Injector connectors

17. Remove the fuel pipes.

18. Remove the timing belt cover.

19. Remove the idler pulley and the timing belt cover.

20. Remove the right side fuel delivery pipe (fuel rail) with the injectors.

21. Remove the left fuel delivery pipe and the injectors.

22. Disconnect the hose and remove the water bypass outlet.

23. Remove the intake manifold.

24. Remove the knock sensor wire.

25. Remove the exhaust crossover pipe.

Right side:

26. Remove the reed valve with the air injection manifold.

27. Remove the water bypass pipe mounting bolt.

28. Remove the alternator.

29. Remove the right valve cover.

30. Remove the two bolts and camshaft housing rear cover. Loosen each cam bearing cap bolt a little at a time and in the correct order. When all are loose, remove the bolts and remove the camshaft along with the oil seal and housing plug.

31. Remove the 6-point head bolt from the head first.

32. Loosen the cylinder head bolts in several stages, in the correct order.

✴✴ WARNING

Removing the head bolts in the improper order could cause damage to the cylinder head.

33. Lift the cylinder head off of its mounting dowels; do NOT attempt to slide it off. If the head is difficult to remove, gently pry with a small tool between the head and block projection. Take great care not to damage the mating faces of the block and/or head. Support the head on wooden blocks on the workbench.

34. Remove the exhaust manifold from the cylinder head.

Left side:

35. Remove the alternator.

36. Remove the oil dipstick. Remove the bolt and remove the dipstick guide tube.

37. Remove the left valve cover.

38. Remove the two bolts and camshaft housing rear cover. Loosen each cam bearing cap bolt a little at a time and in the correct order. When all are loose, remove the bolts and remove the camshaft along with the oil seal and housing plug.

39. Remove the 6 point cylinder head bolt.

40. Loosen the cylinder head bolts in several stages, in the correct order.

✴✴ WARNING

Removing the head bolts in the improper order could cause damage to the cylinder head.

41. Lift the cylinder head off of its mounting dowels; do NOT attempt to slide it off. If the head is difficult to remove, gently pry with a small tool between the head and block projection. Take great care not to damage the mating faces of the block and/or head. Support the head on wooden blocks on the workbench.

42. Remove the exhaust manifold from the cylinder head.

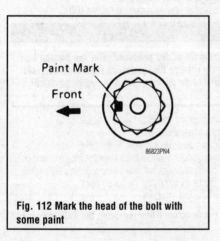

Fig. 112 Mark the head of the bolt with some paint

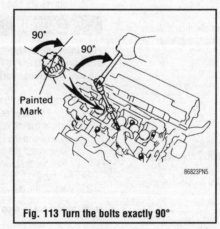

Fig. 113 Turn the bolts exactly 90°

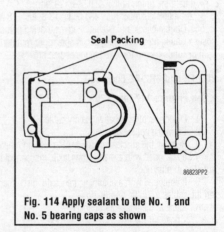

Fig. 114 Apply sealant to the No. 1 and No. 5 bearing caps as shown

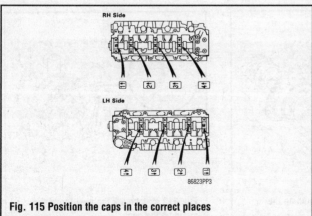

Fig. 115 Position the caps in the correct places

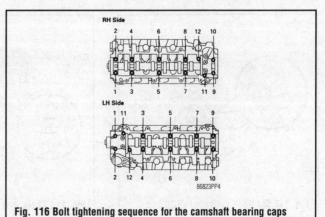

Fig. 116 Bolt tightening sequence for the camshaft bearing caps

To install:

43. Lift the old gasket from the block, then scrape and clean the area well. Insert paper into the cylinder walls to keep dirt from entering.

44. Install the cylinder head on the cylinder block using a new gasket.

45. Lightly coat the threads of the cylinder head bolts with engine oil and then install them into the head. Do not install bolt 6-pointed bolt at this time. Tighten the bolts in several stages, in the order shown to 33 ft. lbs. (44 Nm). After the initial tightening, mark the front side of the top of the bolt with paint. Tighten the bolts an additional 90° (¼ turn) and check that the mark is now facing the side of the head. Tighten the bolts an additional 90° and check that the mark is now facing the rear of the head.

46. Install the 6-pointed head bolts. Tighten to 30 ft. lbs. (41 Nm).

47. Install the camshaft. Install a new oil seal and install the camshaft housing plug.

48. Apply sealant to the Nos. 1 and 5 cam bearing caps. Place the caps on each journal with the arrows correctly aligned. On the right head, they should point to the front; on the left head, to the rear.

49. Temporarily tighten the bearing cap bolts just snug, a little at a time and in the correct order.

50. Use a seal driver of the correct diameter to install the oil seal. Use the same tool to install the camshaft housing plug.

51. Tighten the bearing caps to the correct torque. Make several passes and proceed in the proper order. Correct torque is 12 ft. lbs. (16 Nm).

52. Check the valve clearance. Intake 0.007–0.011 in. (0.18–0.28mm) and exhaust is 0.009–0.013 in. (0.22–0.32mm).

53. Install the valve covers.

54. Install of the remaining components is in the reverse order of the removal procedure. Please note the following torque specifications:
- Injection manifold bolts to the cylinder head: 27 ft. lbs. (37 Nm);
- Injection manifold bolts to the exhaust manifold: 22 ft. lbs. (29 Nm);
- Crossover pipe bolts: 29 ft. lbs. (39 Nm);
- Intake manifold bolts: 13 ft. lbs. (18 Nm);
- Water bypass outlet bolts: 13 ft. lbs. (18 Nm);
- Fuel delivery pipes and injectors: 9 ft. lbs. (13 Nm);
- Timing belt cover bolts: 74 inch lbs. (8 Nm);
- Fuel pipe union bolts: 22 ft. lbs. (29 Nm);
- Air intake chamber nuts and bolts: 13 ft. lbs. (18 Nm).

55. Double check all installation items, paying particular attention to loose hoses or hanging wires, untightened nuts, poor routing of hoses and wires (too tight or rubbing) and tools left in the engine area.

2RZ-FE and 3RZ-FE Engines

▶ See Figures 117 thru 125

1. Disconnect the negative battery cable.
2. Drain the engine coolant.
3. Remove the air cleaner cap, MAF meter, and the resonator.
4. If equipped with a manual transaxle, disconnect the accelerator cable from the throttle body.
5. If equipped with a automatic transaxle, disconnect the accelerator and throttle cables from the throttle body.
6. Remove the intake air connector:
- Air hose for the IAC

- Vacuum sensing hose
- Wire clamp for the engine
7. Remove the oil dipstick guide.
8. If equipped with A/C, remove the air conditioning idle-up valve.
9. Remove the power steering drive belt.
10. Remove the idler pulley by removing the three bolts.
11. Remove the power steering drive belt, pump and bracket.
12. Remove the PCV hoses.
13. Remove the distributor connector, hold-down bolts, and the distributor.
14. Remove the water housing as follows:
    a. Disconnect the radiator inlet hose.
    b. Disconnect the ECT sender gauge connector.
    c. Remove the two bolts, water housing, and the gasket.
15. Remove the throttle body.
16. Disconnect the following harnesses:
- If equipped with A/C, the A/C compressor wiring
- Oil pressure sensor
- ECT sensor
- EGR gas temperature sensor
- EGR Vacuum Switching Valve (VSV)
17. Disconnect the engine wire as follows:
    a. Remove the two bolts and disconnect the engine wire from the intake chamber.
    b. Disconnect the five engine wire clamps and engine wire.
    c. Detach the following connectors:
- Knock sensor
- Crankshaft position sensor
- Fuel pressure control VSV
    d. Disconnect the DLC1 from the bracket.
    e. Disconnect the two engine wire clamps.
    f. Remove the bolt, then disconnect the engine wire from the engine.
18. Disconnect the fuel injectors.
19. Remove the cylinder head rear cover by disconnecting the heater bypass hose, then remove the three bolts.
20. Remove the EGR valve and vacuum modulator.
21. Remove the intake chamber stay by removing the two bolts.

❈❈ CAUTION

Fuel injection systems remain under pressure after the engine has been turned OFF. Properly relieve fuel pressure before disconnecting any fuel lines. Failure to do so may result in fire or personal injury.

22. Unbolt the fuel return pipe and separate from the hoses.
23. Remove the intake chamber as follows:
    a. Disconnect the vacuum hose from the gas filter.
    b. Disconnect the brake booster vacuum hose from the intake chamber.
    c. Remove the three bolts, two nuts, air intake chamber, and the gasket.
24. Remove the fuel inlet tube by removing the union bolts.
25. Remove the delivery pipe and injectors.

➡Be careful not to drop the injectors when removing the delivery pipe.

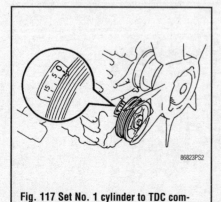

Fig. 117 Set No. 1 cylinder to TDC compression stroke

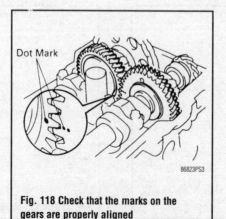

Fig. 118 Check that the marks on the gears are properly aligned

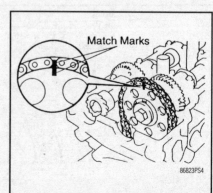

Fig. 119 Matchmark the position of the chain on the gear

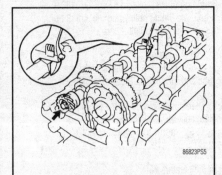

Fig. 120 Hold the hexagon head portion of the camshaft with a wrench and remove the bolt securing the gear

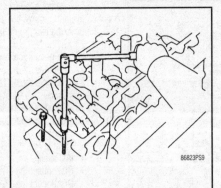

Fig. 121 Remove the 2 bolts at the front of the head first

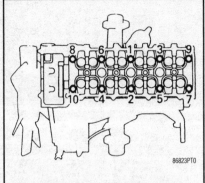

Fig. 122 Remove the cylinder head bolts using this order, in several passes

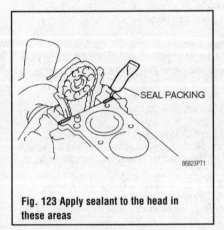

Fig. 123 Apply sealant to the head in these areas

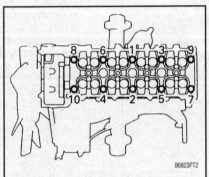

Fig. 124 Tighten the head bolts down uniformly using this sequence in several passes

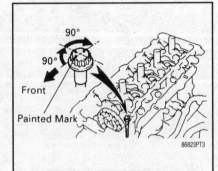

Fig. 125 Mark the front of the bolt with paint and retighten bolts 90 degrees in the proper sequence

26. Remove the intake manifold by removing the three bolts and two nuts.
27. Disconnect the front exhaust pipe from the exhaust manifold.
28. Remove the heat insulator by loosening the two bolts and two nuts.
29. Remove the exhaust manifold and gasket.
30. Remove the engine hangers.
31. Remove the valve cover and gasket.
32. Remove the spark plug wires and plugs from the engine.
33. Set No. 1 cylinder to TDC compression stroke. The groove on the crankshaft pulley should align with the **0** mark on the timing chain cover and the timing marks (one and two dots) of the camshaft gears should form a straight line in respect to the cylinder head surface. If not, turn the crankshaft 1 revolution (360 degrees).
34. Remove the chain tensioner and gasket by removing the two nuts.
35. Remove the camshaft timing gear.

➡Since the thrust clearance of the camshaft is small, the camshaft must be kept level while it is being removed. If the camshaft is not kept level, the portion of the cylinder head receiving the shaft thrust may crack or be damaged, causing the camshaft to seize or break.

36. Remove exhaust camshaft.

✳✳ WARNING

When removing the camshaft, make sure that the torsional spring force of the sub-gear has been eliminated by the above operation.

37. Remove the intake camshaft.

➡Do not pry on or attempt to force the camshaft with a tool or other object.

38. Remove the 2 bolts in the front of the head before the other head bolts are removed. Uniformly loosen and remove the remaining head bolts, in several passes, in the sequence shown.
39. Lift the cylinder head from the block and place the head on wooden blocks on a bench.

To install:

40. Before installing, thoroughly clean the gasket mating surfaces and check for warpage.
41. Apply sealant (08826–00080 or equivalent) to the 2 locations, as shown. Place a new head gasket on the block and install the cylinder head.
42. Lightly coat the cylinder head bolts with engine oil. Install the bolts and tighten in several passes in the sequence shown:
   a. Tighten all bolts to 29 ft. lbs. (39 Nm).
   b. Mark the front of the bolt with paint and retighten bolts 90 degrees in the proper sequence.
   c. Retighten an additional 90 degrees. Check that the painted mark is now facing rearward.
43. Install and tighten the 2 front mounting bolts to 15 ft. lbs.(21 Nm).

➡If any of the bolts break, deform or do not meet the torque specification, replace them.

44. Installation of the remaining components is in the reverse order of the removal procedure. Please note the following.
45. Set No. 1 cylinder to TDC compression stroke: the crankshaft pulley groove will align with the **0** mark on timing cover and camshafts timing marks with one dot and two dots will be in a straight line on the cylinder head surface.
46. Install the timing gearas detailed later in this section.

5VZ-FE Engine

▶ **See Figures 126, 127 and 128**

1. Disconnect the negative battery cable.
2. Relieve the fuel system pressure.

✳✳ CAUTION

Fuel injection systems remain under pressure after the engine has been turned OFF. Properly relieve fuel pressure before disconnecting any fuel lines. Failure to do so may result in fire or personal injury.

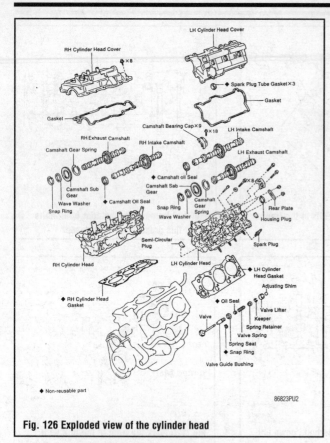

RH Cylinder Head Cover
LH Cylinder Head Cover
×8
Spark Plug Tube Gasket×3
Gasket
Gasket
Camshaft Bearing Cap×9
×18
LH Intake Camshaft
RH Exhaust Camshaft
RH Intake Camshaft
Camshaft Gear Spring
LH Exhaust Camshaft
Camshaft oil Seal
Camshaft Sub Gear
Camshaft Sab Gear
×8
Wave Washer
Camshaft Oil Seal
Snap Ring
Snap Ring
Camshaft Gear Spring
Wave Washer
Rear Plate
Housing Plug
Semi-Circular Plug
Spark Plug
RH Cylinder Head
LH Cylinder Head
LH Cylinder Head Gasket
Adjusting Shim
RH Cylinder Head Gasket
Oil Seal
Valve Lifter
Valve
Keeper
Spring Retainer
Valve Spring
Spring Seat
Snap Ring
Valve Guide Bushing

◆ Non-reusable part

86823PU2

Fig. 126 Exploded view of the cylinder head

3. Remove the engine under cover.
4. Drain the cooling system.
5. Remove the front exhaust pipe as follows.
6. Disconnect the air cleaner cap, MAF meter and the resonator.
7. Disconnect the following cables:
- If equipped with cruise control, disconnect the actuator cable with the bracket
- Accelerator cable
- With A/T, Throttle cable
8. Disconnect the heater hose.
9. Disconnect the upper radiator hose from the engine.
10. Remove the power steering drive belt.
11. Remove the A/C drive belt by loosening the idle pulley nut and adjusting bolt.
12. Loosen the lock bolt, pivot bolt and adjusting bolt and the alternator drive belt.
13. Remove the fan shroud by removing the two clips.
14. Remove the fan with the fluid coupling and fan pulleys.
15. Disconnect the power steering pump from the engine and set aside. Do not disconnect the lines from the pump.
16. If equipped with A/C, remove the compressor from the engine and set aside. Do not disconnect the lines from the compressor.
17. If equipped with A/C, remove the A/C bracket.
18. Remove the spark plug wires with the ignition coils.
19. Remove the spark plugs.
20. Remove the timing belt cover.
21. Remove the fan bracket.
22. Set the No. 1 cylinder at TDC of the compression stroke.
    a. Turn the crankshaft pulley and align its groove with the timing mark **0** of the No. 1 timing belt cover.
    b. Check that the timing marks of the camshaft timing pulleys and the No. 3 timing belt cover are aligned. If not, turn the crankshaft pulley one revolution (360°).

➡**If re-using the timing belt, make sure that you can still read the installation marks. If not, place new installation marks on the timing belt to match the timing marks of the camshaft timing pulleys.**

23. Remove the timing belt tensioner by alternately loosening the two bolts.

24. Remove the camshaft timing pulleys. Using SST 09960–10010 or equivalent, remove the pulley bolt, the timing pulley and the straight pin. Remove the two timing pulleys with the timing belt.
25. Remove the bolt and the idler pulley.
26. Remove the alternator from the engine.
27. Remove the nuts and remove the EGR pipe and two gaskets.
28. Remove the oil filler tube and throttle cable clamp by removing the bolt and two nuts.
29. Remove the intake chamber stay by removing the two bolts.
30. Remove the following connectors:
    Vacuum Switching Valve (VSV) for the fuel pressure control
- Throttle Position Sensor (TPS)
- IAC valve
- Disconnect the EGR gas temperature wire
- Disconnect the VSV wire for the EGR valve
31. Disconnect the following hoses:
- PCV hoses
- Water bypass
- Air assist hose from the intake air wire
- Two vacuum sensing hoses from the VSV
- EVAP
- Air hose from the power steering
- If equipped with A/C, disconnect the air hose from the A/C idle up valve.
32. Remove the four bolts, two nuts and remove the air intake chamber assembly from the engine.
33. Remove the intake air connector as follows:
    a. Disconnect the engine wire from the intake air connector by removing the bolt.
    b. Disconnect the two fuel return hoses.
    c. Disconnect the brake booster vacuum hose from the intake air connector.
    d. Remove the bolt and disconnect the ground strap from the intake air harness.
    e. Disconnect the DLC1 from the bracket of the intake air harness.
    f. If equipped with A/C, disconnect idle up valve.
    g. Remove the intake air connector from the engine by removing the three bolts and two nuts.
34. Disconnect the engine wire from the intake manifold as follows:
    a. Disconnect the following wiring:
- Oil pressure sensor
- Crankshaft position sensor
- Six injectors
- ECT sender gauge
- ECT sensor
- Knock sensor
- Camshaft position sensor
    b. Disconnect the three engine wire clamps.
    c. Remove the three bolts, then disconnect the engine wire from the cylinder head.
35. Remove the camshaft position sensor.
36. Remove the (rear) timing belt cover by removing the six bolts.
37. Remove the fuel pressure regulator.
38. Remove the intake manifold assembly.
39. Remove the power steering pump bracket.
40. Remove the oil dipstick and guide.
41. Remove the exhaust crossover pipe and gaskets by removing the six nuts.
42. Remove the left hand exhaust manifold by unbolting the heat insulator and nuts for the exhaust manifold.
43. Remove the right hand exhaust manifold by unbolting the heat insulator and six nuts for the exhaust manifold.
44. Remove the eight bolts, seal washers, valve cover and gasket. Remove both valve covers.
45. Remove the semi-circular plugs.
46. Remove the right side exhaust camshaft.

✲✲ WARNING

Do not pry on or attempt to force the camshaft with a tool or other object.

47. Remove the right side intake camshaft.
48. Remove the left side exhaust camshaft.
49. Remove the left side intake camshaft.

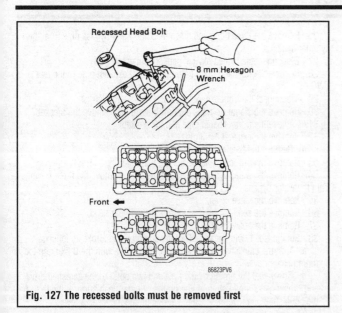

Fig. 127 The recessed bolts must be removed first

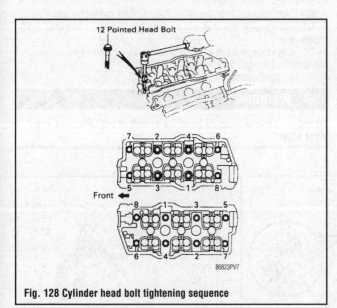

Fig. 128 Cylinder head bolt tightening sequence

50. Remove the valve lifters and shims from the cylinder head. Arrange the valve lifters and shims in correct order.

51. Using a 8mm hexagon wrench, remove the cylinder head (recessed head) bolt on each cylinder head, then repeat for the other side.

52. Uniformly loosen and remove the eight cylinder head (12 pointed head) bolts on each cylinder head. Loosen the bolts in several passes using the reverse of the tightening sequence.

53. Remove the 16 cylinder head bolts and plate washers.

54. Lift the cylinder head from the dowels on the cylinder block.

To install:

55. Place two new cylinder head gaskets in position on the cylinder block.

56. Place the two cylinder heads on the dowels of the cylinder block.

57. Apply a light coat of engine oil on the threads and under the heads of the cylinder head bolts.

58. Install and uniformly tighten the cylinder head bolts on each cylinder as follows:

   a. In several passes and in the sequence shown, tighten the cylinder bolts to 25 ft. lbs. (34 Nm).

   b. Mark the front of the cylinder head bolt with paint.

   c. Retighten the cylinder head bolts by 90° in order.

   d. Check that the painted mark is now at a 90° angle to the front.

59. Apply a light coat of engine oil on the threads and under the heads of the recessed cylinder head bolts. Using a 8mm hexagon wrench, install the

cylinder head bolt on each cylinder head, then repeat for the other side. Tighten the bolts to 13 ft. lbs. (18 Nm).

60. Install the valve lifters and shims. Check that the valve lifter rotates smoothly by hand.

61. Installation of the remaining components is in the reverse order of the removal procedure.

1FZ-FE Engine

♦ **See Figures 129 thru 135**

1. Release the fuel system pressure.

2. Disconnect the battery cables and remove the battery and the battery tray.

❊❊ **CAUTION**

On vehicles with air bags, work must be started after 90 seconds from the time the ignition switch is turned to the LOCK position and the negative (-) battery cable is disconnected.

3. Drain the engine coolant.

4. Remove the air cleaner hose and cap.

5. Disconnect the cruise control actuator cable from the throttle body.

6. Disconnect the accelerator cable from the throttle body.

7. Disconnect the throttle cable from the throttle body.

8. Disconnect the engine ground strap from the engine hanger and the ground strap from the air intake chamber.

9. Unplug the connector on the intake manifold from the left fender apron.

10. Disconnect the brake booster vacuum hose.

11. Disconnect the EVAP hose and disconnect the fuel return hose.

❊❊ **CAUTION**

Fuel injection systems remain under pressure after the engine has been turned *OFF*. Properly relieve fuel pressure before disconnecting any fuel lines. Failure to do so may result in fire or personal injury.

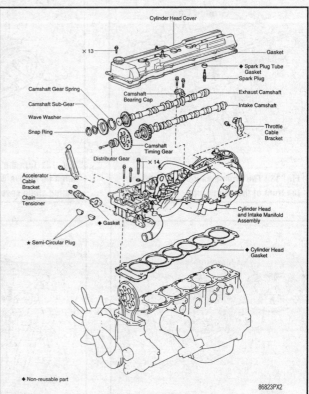

Fig. 129 Exploded view of the cylinder head components—1FZ-FE engine

12. Disconnect the heater hoses.
13. Disconnect the engine wire and heater valve from the cowl panel.
14. Remove the cylinder head covers
15. Remove the distributor.
16. Disconnect the P/S reservoir tank.
17. Disconnect the radiator inlet hose and the water bypass hose.
18. Remove the alternator.
19. Remove the throttle body.
20. Remove the oil dipsticks and guides for the engine and transmission.
21. Remove the intake manifold stay.
22. Disconnect the fuel inlet hose from the fuel filter.
23. Unplug the following connectors:
    a. ECT sender gauge connector, the ECT cut switch connector, and the ECT sensor connector
    b. Knock sensor connector
    c. Crankshaft position sensor connector
24. Remove the bolt and disconnect the engine wire harness from the cylinder block.
25. Disconnect the following:
    a. Oil level sensor connector
    b. Two connectors from the transmission
    c. Starter connector
    d. Disconnect the two heated oxygen sensor connectors
    e. Disconnect the Park/Neutral Position (PNP) switch connector
    f. Remove the two bolts and disconnect the engine wire from the intake manifold and the cylinder block
    g. Disconnect the PCV hose from the PCV valve
    h. Remove the bolt holding the engine wire to the intake manifold
    i. Unplug the connector for the emission control valve set assembly and the three injector connectors
    j. Disconnect the engine wire harness clamp
    k. Disconnect the EGR gas temperature sensor connector
    l. Disconnect the clamp of the No. 6 injector wire from the bracket
    m. Disconnect the engine wire harness from the cylinder head and the intake manifold

26. Remove the three bolts and disconnect the water bypass pipe from the cylinder head.
27. Disconnect the heated oxygen sensor connector.
28. Remove the nuts and bolts holding the front exhaust pipe to the rear TWC.
29. Disconnect the front exhaust pipe and remove the gasket.
30. Remove the clamp from the support bracket and remove the bracket.
31. Remove the front exhaust pipe and the gaskets.
32. Remove the exhaust manifolds.
    a. Remove the heat insulators.
33. Remove the ground cable, heater pipe and gasket.
34. Remove the water bypass outlet and the pipe. Remove the three O–rings from the water bypass outlet and the pipe.
35. Remove the valve cover.
36. Remove the semi–circular plug from the cylinder head.
37. Remove the spark plugs.
38. Set the No. 1 cylinder to TDC of the compression stroke as follows:
    a. Turn the crankshaft pulley and align its groove with the **0** mark on the timing chain cover.
    b. Check that the timing marks (one and two dots) of the camshaft drive and driven gears are in straight line on the cylinder head surface. If not, turn the crankshaft one revolution (360°) and align the marks as above.
39. Remove the chain tensioner.
40. Place matchmarks on the camshaft timing gear and the timing chain.
41. Hold the intake camshaft with a wrench and remove the bolt and the distributor gear.
42. Remove the camshaft timing gear and chain from the intake camshaft and leave on the slipper and the damper.
43. Remove the camshafts.

✳✳ WARNING

Do not pry on or attempt to force the camshaft with a tool or any other object.

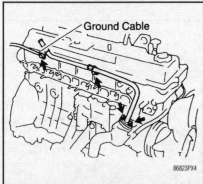

Fig. 130 The ground cable is secured to the front of the heater pipe

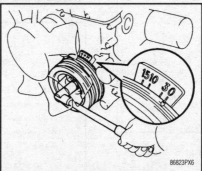

Fig. 131 Turn the crankshaft pulley and align its groove with the 0 mark on the timing chain cover

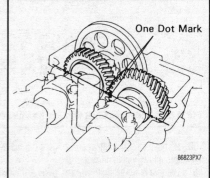

Fig. 132 Check that the marks on the camshaft pulley are aligned

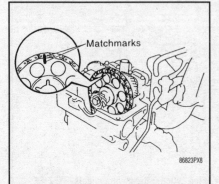

Fig. 133 Matchmark the position of the chain on the gear

Fig. 134 Remove these two bolts prior to the rest of the head bolts

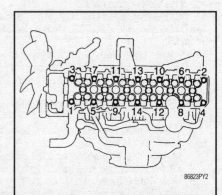

Fig. 135 Uniformly loosen and remove the 14 head bolts in several passes

44. Remove the two bolts in the front of the head first.
45. Loosen and remove the 14 cylinder head bolts in sequence using several passes.

✸✸ WARNING

Cylinder head warpage or cracking could result from removing bolts in incorrect order.

46. Lift the cylinder head from the dowels on the cylinder block and place the cylinder head on wooden blocks on the bench. If the cylinder head is difficult to lift off, pry between the cylinder head and the cylinder block with a flat prying tool.
47. Remove the alternator bracket.
48. Remove the two nuts, the water outlet, and the gasket.
49. Loosen the union nut and remove the EGR pipe and gasket.
50. Remove the heater inlet pipe and hose.
51. Remove the air intake chamber and the intake manifold assembly.
52. Remove the water bypass hose.
53. Remove the engine hanger brackets.
54. Remove the two engine wire clamp brackets.
55. Remove the accelerator cable bracket and the throttle cable bracket.
56. Remove the valve lifters and shims. Arrange the valve lifters and shims in correct order for reinstallation.
To install:
57. Install the valve lifters and shims. Check to make sure that the valve lifter rotates smoothly by hand.
58. Install the accelerator cable bracket and the throttle cable bracket.
59. Install the engine wire clamp brackets.
60. Install the engine hangers. Tighten to 30 ft. lbs. (41 Nm).
61. Install the air intake chamber and intake manifold assembly.
62. Install the heater hose to the cylinder head, and connect the pipe to the intake manifold. Tighten the bolts to 15 ft. lbs. (21 Nm).
63. Temporarily install the union nut to the EGR valve. Install the EGR pipe to the cylinder head. Tighten the bolts to 15 ft. lbs. (21 Nm). Tighten the union nut to 58 ft. lbs. (78 Nm).
64. Install a new gasket and the water outlet. Tighten the nuts to 15 ft. lbs. (21 Nm).
65. Install the alternator bracket and tighten the bolts to 32 ft. lbs. (43 Nm).
66. Apply sealant on the end of the engine block by the timing belt.
67. Install a new cylinder head gasket on the cylinder block.
68. Install the cylinder head.
69. Install the cylinder head bolts as follows:
   a. The cylinder head bolts are tightened in three progressive steps. Apply a light coat of engine oil on the threads and under the heads of the cylinder head bolts.
   b. Install the 14 cylinder head bolts and tighten progressively in sequence to 29 ft. lbs. (39 Nm). Use the reverse of the loosening sequence.
   c. Mark the front of the cylinder head bolt head with paint.
   d. Retighten the cylinder head bolts by 90° in numerical order.
   e. Retighten the cylinder head bolts an additional 90° so that the painted mark is now facing to the rear.

✸✸ WARNING

Do not combine steps D and E in one pass; the above steps must be followed exactly and in order to prevent cylinder head damage or pre-mature gasket failure.

   f. Install and tighten the two bolts at the front of the head to 15 ft. lbs. (21 Nm).
70. Installation of the remaining components is in the reverse order of the removal procedure. Installation of the timing gear is detailed later in this section.

Oil Pan

REMOVAL & INSTALLATION

22R, 22R-E, and 3VZ-E Engines

▶ See Figures 136, 137, 138, 139 and 140

1. Elevate the front of the vehicle and support it securely on stands.
2. Remove the undercover from the engine.
3. Drain the oil and discard of the drain plug gasket.

➡If you haven't replaced the oil filter lately, you may want to for a complete oil change.

✸✸ CAUTION

Used motor oil may cause skin cancer if repeatedly left in contact with the skin for prolonged periods. Although this is unlikely unless you handle oil on a daily basis, it is wise to thoroughly wash your hands with soap and water immediately after handling used motor oil.

4. If the truck or 4Runner is 4WD, remove the front differential.
5. For 22R and 22R-E engines, remove the engine mounting bolts. Position a floor jack under the transmission and raise the engine/transmission assembly slightly.
6. Remove the oil pan bolts and remove the oil pan. If the pan is difficult to remove do NOT pry with screwdrivers or similar tools; the flange of the pan will be deformed. Tap the bottom or side of the pan with a wooden or plastic mallet to dislodge the pan from the block. If this is still not sufficient, make a wooden wedge and gently tap it into the seam.
7. Use a razor blade to scrape the cylinder block and oil pan mating surfaces clean of any old sealing material. All traces of the old gasket and sealer must be removed.
8. Clean the pan thoroughly, using solvent if necessary. If the pan has a magnet in it, examine the catchings closely; excess metal is a sign of excess engine wear. A few slivers are normal. Use good judgment in relating the amount of debris to the amount of time the pan has been in place.

➡Do not use a solvent which will affect the paint on the pan. Flecks of paint remaining in the pan can plug the oil pick-up or dissolve in the oil, polluting it.

Fig. 136 It may be necessary to use an extension to reach the bolts retaining the oil pan

Fig. 137 Be careful when lowering the pan, some oil may be left in it

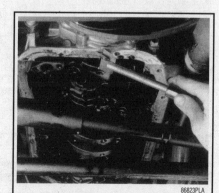
Fig. 138 Scrape the block of any old sealant left behind

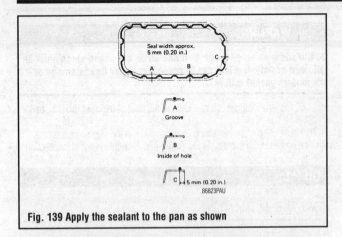

Fig. 139 Apply the sealant to the pan as shown

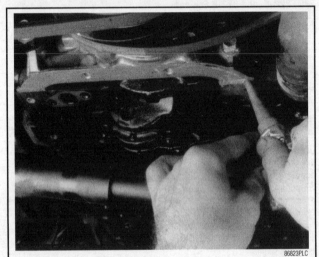

Fig. 140 For 22R and 22R-E engines, apply sealant to the 4 corners of the cylinder block

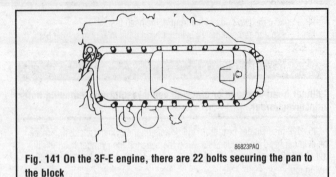

Fig. 141 On the 3F-E engine, there are 22 bolts securing the pan to the block

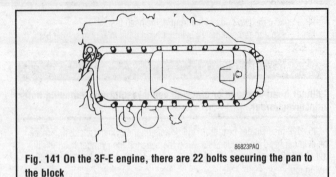

Fig. 142 Apply sealant to the 4 corners of the block

To install:

9. On 22R, 22R-E and 3VZ-E engines, the pan does not use a pre-made gasket; all sealing is accomplished by correct application of sealant from a tube. Cut the tip of the tube to produce a bead about 2/10 inch (5mm) only. Place the continuous bead of sealant in the groove, looping to the inside of the bolt holes and keeping about 2/10 inch (5mm) from the edge of the pan.

10. For 22R, and 22R-E, apply sealant to the 4 points on the cylinder block at the chain cover and rear oil seal retainer.

11. Install the pan to the block within about 5 minutes of applying the sealer and install the retaining bolts and nuts finger-tight.

12. Tighten the nuts and bolts, working from the center of the pan in a circular pattern to the ends. For 22R and 22R-E, tighten to 9 ft. lbs. (13 Nm); for 3VZ-E engines, 52 inch lbs. (6 Nm).

13. Installation of the remaining components is in the reverse order of the removal procedure.

3F-E Engine

♦ See Figures 141 and 142

1. Elevate the front of the vehicle and support it securely on stands.
2. Remove the undercover from the engine.
3. Drain the oil and discard of the drain plug gasket.

➡If you haven't replaced the oil filter lately, you may want to for a complete oil change.

⁕⁕ CAUTION

Used motor oil may cause skin cancer if repeatedly left in contact with the skin for prolonged periods. Although this is unlikely unless you handle oil on a daily basis, it is wise to thoroughly wash your hands with soap and water immediately after handling used motor oil.

4. Remove the oil pan bolts and remove the oil pan. If the pan is difficult to remove do NOT pry with screwdrivers or similar tools; the flange of the pan will be deformed. Tap the bottom or side of the pan with a wooden or plastic mallet to dislodge the pan from the block. If this is still not sufficient, make a wooden wedge and gently tap it into the seam.

5. Use a razor blade or equivalent to scrape the cylinder block and oil pan mating surfaces of any old sealing material. All traces of the old gasket and sealer must be removed.

6. Clean the pan thoroughly, using solvent if necessary. If the pan has a magnet in it, examine the catchings closely; excess metal is a sign of excess engine wear. A few slivers are normal. Use good judgment in relating the amount of debris to the amount of time the pan has been in place.

➡Do not use a solvent which will affect the paint on the pan. Flecks of paint remaining in the pan can plug the oil pick-up or dissolve in the oil, polluting it.

To install:

7. Install a new gasket on the pan without sealant.

8. Apply sealant to the 4 points on the cylinder block at the chain cover and rear oil seal retainer.

9. Install the pan to the block within about 5 minutes of applying the sealer and install the retaining bolts and nuts finger-tight.

10. Tighten the nuts and bolts. Working from the center of the pan in a circular pattern to the ends to 69 inch lbs. (8 Nm).

11. Installation of the remaining components is in the reverse order of the removal procedure.

5VZ-FE Engine

♦ See Figures 143 and 144

1. Disconnect the negative battery cable.

⁕⁕ CAUTION

On vehicles with air bags, work must be started after 90 seconds from the time the ignition switch is turned to the LOCK position and the negative (−) battery cable is disconnected.

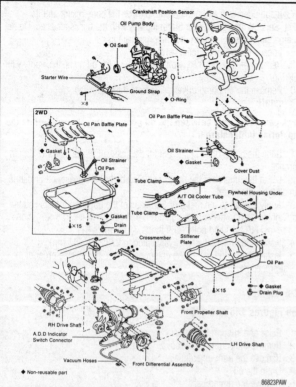

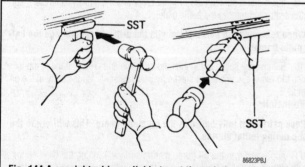

Fig. 143 Exploded view of the oil pan and related components on the 5VZ-FE engine

Fig. 144 A special tool is available to cut through the sealant of the oil pan

2. Raise and safely support the vehicle.
3. Remove the crankshaft timing pulley.
4. Drain the engine oil.
5. If equipped with 4WD, remove the front differential.
6. On A/T remove the oil cooler tube and clamp.
7. Remove the stiffener plate.
8. Unbolt the flywheel housing dust cover.
9. Disconnect the starter wire clamp.
10. Remove the crankshaft position sensor.
11. Unbolt and lower the oil pan.
12. Using SST 09032–00100 or equivalent and a brass bar, separate the oil pan from the cylinder block.

To install:

13. Clean the pan thoroughly, using solvent if necessary. If the pan has a magnet in it, examine the catchings closely; excess metal is a sign of excess engine wear. A few slivers are normal. Use good judgment in relating the amount of debris to the amount of time the pan has been in place.
14. Install the crankshaft position sensor.
15. Install the oil pan baffle plate.
16. Apply seal packing to the oil pan and install the pan to the cylinder block. Tighten the nuts and bolts to 66 inch lbs. (8 Nm). If parts are not assem-

bled within 5 minutes of applying time, the effectiveness of the seal packing is lost and must be removed and reapplied.

17. Installation of the remaining components is in the reverse order of the removal procedure.

2RZ-FE and 3RZ-FE Engines

1. Disconnect the negative battery cable from the battery.
2. Raise and support the vehicle safely.
3. Remove the engine under cover.
4. Drain the engine oil.
5. Remove the 16 mounting bolts and two nuts to the oil pan.
6. Remove the oil pan from the engine.

➡**Be careful not to damage the oil pan flanges of the oil pan and cylinder block.**

To install:

7. Apply sealant 08826–00080 or equivalent, to the oil pan.

➡**Parts must be assembled within 5 minutes of application. Otherwise the material must be removed and reapplied.**

8. Install the oil pan and mounting bolts. Tighten the bolts and nuts to 9 ft. lbs. (13 Nm).
9. Installation of the remaining components is in the reverse order of the removal procedure.

1FZ-FE Engine

▶ **See Figure 145**

1. Disconnect the negative battery cable.

❋❋ CAUTION

On vehicles with air bags, work must be started after 90 seconds from the time the ignition switch is turned to the LOCK position and the negative (−) battery cable is disconnected.

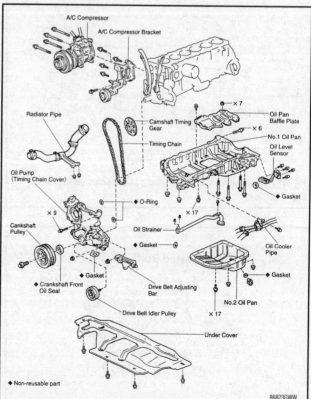

Fig. 145 Exploded view of the oil pan and related components on the 1FZ-FE engine

2. Raise and safely support the vehicle.
3. Drain the engine oil.
4. Remove the engine undercover.
5. Disconnect the oil cooler pipe bracket from the No. 1 oil pan.
6. Remove the oil level sensor.
7. Remove the bolts holding the No. 1 oil pan to the transmission housing.
8. Remove the No. 2 oil pan.
9. Remove the No. 1 oil pan.

To install:
10. Install the No. 1 oil pan.
   a. Apply seal packing to the No. 1 oil pan.
   b. Install the oil pan and tighten the 14 mm bolts to 32 ft. lbs. (44 Nm) and the 12 mm bolts to 14 ft. lbs. (20 Nm).
11. Apply seal packing to the No. 2 oil pan and tighten the bolt to 6 ft. lbs. (8 Nm) and the nuts to 7 ft. lbs. (9 Nm).
12. Install the bolts holding the No. 1 oil pan to the transmission housing and tighten to 53 ft. lbs. (72 Nm).
13. Installation of the remaining components is in the reverse order of the removal procedure.

Oil Pump

REMOVAL & INSTALLATION

22R and 22R-E Engines

▶ **See Figures 146 and 147**

1. Raise and support the vehicle.
2. Drain the oil from the engine.
3. Remove the oil pan.
4. Remove the four bolts holding the oil strainer.
5. Remove the drive belts.
6. Remove the crankshaft pulley.

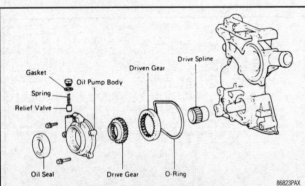

Fig. 146 Exploded view of the oil pump components on 22R and 22R-E engines

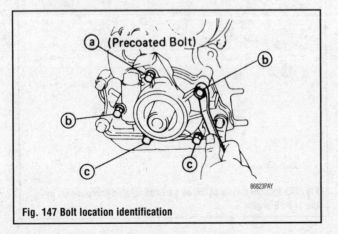

Fig. 147 Bolt location identification

7. If equipped with air conditioning, remove the compressor and its bracket. Support the compressor out of the way with the hoses attached. Do not discharge the system or loosen the compressor lines.
8. Loosen the oil pump relief valve bolt.
9. Remove the five retaining bolts and remove the oil pump assembly with the O-ring.
10. Remove the oil pump drive spline from the crankshaft.

To install:

➡ **Pack petroleum jelly between the oil pump gears. This will prime the pump during initial startup.**

11. Slide the pump drive spline onto the crankshaft. Place a new O-ring in the groove.
12. Clean the threads of the upper retaining bolt of all grease, oil or debris. Clean all the bolts holes.
13. Apply sealant to the first 2 or 3 threads of the upper bolt. Install all the bolts. Tighten each to the correct tightness. Bolt (a): 18 ft. lbs. (25 Nm); bolts (b) 14 ft. lbs. (19 Nm) and bolts (c) 9 ft. lbs. (13 Nm).
14. Install and tighten the relief valve plug to 27 ft. lbs. (37 Nm).
15. Installation of the remaining components is in the reverse order of the removal procedure.

3VZ-E Engine

▶ **See Figures 148 and 149**

1. Raise and support the vehicle.
2. Remove the engine under cover.
3. Remove the front differential.
4. Drain the oil.
5. Remove the timing belt.
6. Remove the crankshaft timing pulley.
7. Raise the engine slightly and remove the oil pan.
8. Remove the two bolts and nuts; remove the oil strainer and gasket.
9. Insert a drift between the cylinder block and the oil pan baffle plate, cut off the sealer and remove the baffle plate.

➡ **When removing the baffle plate with the drift, do not damage the baffle plate flange.**

10. Remove the seven bolts, then the oil pump and O-ring. If the pump is difficult to remove, use a small plastic mallet to carefully tap the body of the oil pump.

To install:

➡ **Pack petroleum jelly between the oil pump gears. This will prime the pump during initial startup.**

11. Apply sealer to the oil pump mating surface, running the bead on the inside of the bolt holes. Position a new O-ring in the groove in the cylinder block and install the pump so that the spline teeth of the drive gear engage the large teeth on the crankshaft. Tighten the mounting bolts to 14 ft. lbs. (20 Nm).
12. Installation of the remaining components is in the reverse order of the removal procedure. Tighten the oil strainer bolts to 61 inch lbs. (7 Nm).

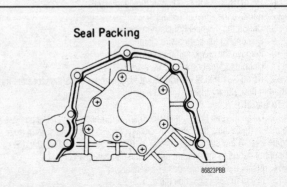

Fig. 148 Do not apply too much sealant on the pump surface, it can block the oil passage

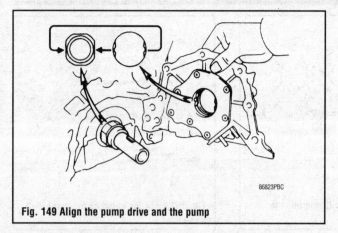

Fig. 149 Align the pump drive and the pump

2RZ-FE and 3RZ-FE Engines

➡The oil pump assembly is mounted in the timing chain cover. To properly service the oil pump, the timing chain cover must be removed from the cylinder block.

1. Disconnect the negative battery cable.
2. Drain the engine oil and cooling system.
3. Raise and safely support the vehicle.
4. Remove the engine under cover.
5. If equipped with 4WD, remove the front differential and halfshaft assembly.
6. For the California vehicles with 3RZ-FE engine, remove the two bolts and disconnect the air pipe.
7. Disconnect the upper hose from the radiator.
8. Remove the oil dipstick guide by removing the bolt.
9. If equipped with power steering, remove the drive belt by loosening the lock bolt and adjusting bolt.
10. Remove the fan shrouds.
11. If equipped with A/C, loosen the idler pulley nut and adjusting bolt, then remove the drive belt from the engine.
12. Remove the alternator drive belt, fan (with fan clutch), water pump pulley, and the fan shroud:
    a. Loosen the water pump pulley mounting nuts.
    b. Loosen the lock, pivot and adjusting bolts for the alternator and remove the alternator drive belt from the engine.
    c. Remove the four water pump pulley mounting nuts.
    d. Remove the fan (with fan clutch) and the water pump pulley.
13. Remove the cylinder head from the engine.
14. If equipped with A/C, disconnect the compressor and bracket:
    a. Remove the four mounting bolts and disconnect the compressor from the bracket. Do not disconnect the A/C pressure lines from the compressor. Suspend the compressor away from the engine.
    b. Remove the A/C compressor bracket by removing the four bolts.
15. Remove the alternator, adjusting bar and bracket:
    a. Loosen the lock bolt, pivot bolt, nut and adjusting bolt at the alternator.
    b. After loosening the adjusting bolt, remove the alternator drive belt from the engine.
    c. Disconnect the alternator wiring.
    d. Remove the nut, then disconnect the alternator wire.
    e. Disconnect the wire harness with the clip.
    f. Remove the lock bolt, pivot bolt, nut, and the alternator from the engine.
    g. Unbolt and remove the adjusting bar.
    h. Unbolt and remove the bracket.
16. Unbolt and remove the crankshaft position sensor.
17. If equipped with 2WD, remove the stiffener plates by removing the eight bolts.
18. Remove the flywheel housing undercover and dust seal.
19. Remove the oil pan by removing the 16 mounting bolts and 2 nuts.

➡Be careful not to damage the flanges of the oil pan and cylinder block.

20. Remove the two bolts, two nuts, oil strainer, and gasket.

21. Remove the crankshaft pulley:
    a. If equipped with A/C, remove the No. 2 and No. 3 crankshaft pulleys by removing the four bolts.
    b. Using SST 09213–54015 and 09330–00021 or equivalents, remove the crankshaft pulley bolts.
    c. Remove the crankshaft pulley.
22. Remove the timing chain cover as follows:
    a. Remove the two water bypass pipe mounting nuts.
    b. Remove the two timing chain cover mounting bolts.
    c. Remove the nine mounting bolts and two mounting nuts from the timing chain cover.
    d. Using a rubber hammer, loosen the chain cover, timing chain cover and three gaskets.
23. Disassemble the oil pump from the front cover, pump cover, drive rotor, driven rotor and O-ring.
24. Remove the relief valve as follows:
    a. Using snapring pliers, remove the snapring for the relief valve.
    b. Remove the retainer, spring(s) and relief valve from the front cover.

To install:
25. Install the relief valve.

➡Pack petroleum jelly between the oil pump gears. This will prime the pump during initial startup.

26. Install the drive and driven rotors.
27. Install the timing chain cover.
28. Install the 2 rear timing chain cover mounting bolts and water bypass pipe mounting nuts. Tighten to 13 ft. lbs. (18 Nm).
29. Align the pulley set key with the key groove of the pulley and slide on the pulley. Install and tighten the pulley bolt to 193 ft. lbs. (260 Nm).
30. If equipped with A/C, install the crankshaft pulleys with the four bolts. Tighten the bolts to 18 ft. lbs. (25 Nm).
31. Install the oil strainer, then tighten the fasteners to 13 ft. lbs. (18 Nm).
32. Clean the oil pan and cylinder block mating surfaces. Apply seal packing to the oil pan. Install the oil pan and tighten the fasteners to 9 ft. lbs. (13 Nm).
33. Installation of the remaining components is in the reverse order of the removal procedure.

5VZ-FE Engine

▶ See Figures 150, 151, 152 and 153

1. Disconnect the negative battery cable.

✳✳ CAUTION

On vehicles with air bags, work must be started after 90 seconds from the time the ignition switch is turned to the LOCK position and the negative (–) battery cable is disconnected.

2. Raise and safely support the vehicle.
3. Remove the crankshaft timing pulley.
4. Drain the engine oil.
5. If equipped with 4WD, remove the front differential.
6. On A/T remove the oil cooler tube and clamp.
7. Remove the stiffener plate.
8. Unbolt the flywheel housing dust cover.
9. Disconnect the starter wire clamp.
10. Remove the crankshaft position sensor.
11. Unbolt and lower the oil pan.
12. Remove the oil strainer and gasket.
13. Unbolt the oil pan baffle plate.
14. Remove the 8 bolts, ground strap and oil pump from the engine. Using a plastic head hammer, carefully tap the pump body. Remove the O-ring from the block.

To install:
15. Clean the pan thoroughly, using solvent if necessary. If the pan has a magnet in it, examine the catchings closely; excess metal is a sign of excess engine wear. A few slivers are normal. Use good judgment in relating the amount of debris to the amount of time the pan has been in place.
16. Apply multi-purpose grease to the oil seal lip. Using a special service tool 09306–37010 or equivalent, along with a hammer, tap in the oil seal until it's surface is flush with the pump body edge.

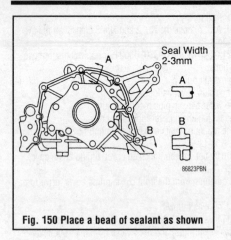

Fig. 150 Place a bead of sealant as shown

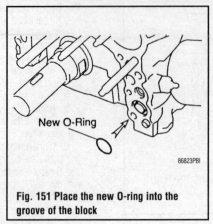

Fig. 151 Place the new O-ring into the groove of the block

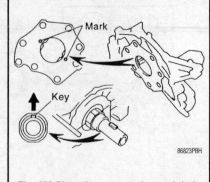

Fig. 152 Place the pump on the crankshaft as shown

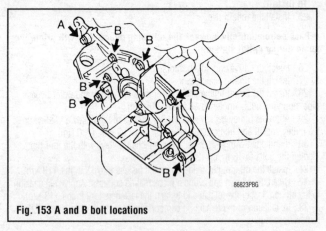

Fig. 153 A and B bolt locations

➡Pack petroleum jelly between the oil pump gears. This will prime the pump during initial startup.

17. Remove any old sealant from the surface of the pump, apply new sealant to the oil pump, place the new O-ring into the groove of the block. Place the pump on the crankshaft with the spline teeth of the drive rotor engaged with the large teeth of the crankshaft.

18. Tighten fasteners A to 15 ft. lbs. (20 Nm) and B to 31 ft. lbs. (42 Nm).

19. Install the crankshaft position sensor.

20. Install the oil pan baffle plate.

21. Apply seal packing to the oil pan and install the pan to the cylinder block. Tighten the nuts and bolts to 66 inch lbs. (8 Nm). If parts are not assembled within 5 minutes of applying time, the effectiveness of the seal packing is lost and must be removed and reapplied.

22. Installation of the remaining components is in the reverse order of the removal procedure.

3F-E Engines

▶ See Figures 154, 155 and 156

1. Raise and support the vehicle.
2. Remove the oil pan.
3. Remove the oil outlet pipe.
4. Remove the oil pump retaining bolt(s) and nuts; remove the strainer and gasket.
5. Remove the retaining bolt and remove the oil pump.
6. Remove the oil pump cover and inspect the parts for nicks, scoring, grooving, etc.

To install:

➡Pack petroleum jelly between the oil pump gears. This will prime the pump during initial startup.

7. Install the oil pump so that the slot in the oil pump shaft is in alignment with the protrusion on the governor shaft of the distributor. Tighten the mounting bolts just snug.

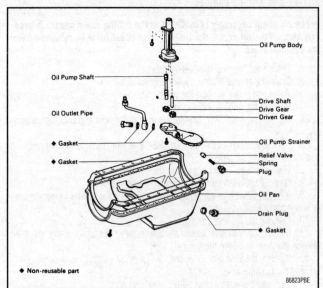

Fig. 154 Exploded view of the components for the oil pump on the 3F-E engine

◆ Non-reusable part

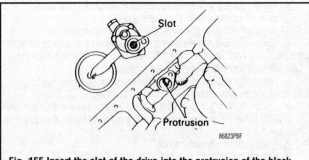

Fig. 155 Insert the slot of the drive into the protrusion of the block

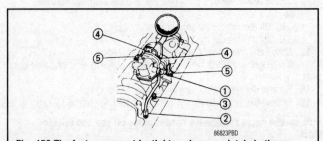

Fig. 156 The fasteners must be tightened appropriately in the correct order

8. Use new gaskets. Install the outlet pipe and strainer, again tightening the bolts just snug.

9. Tighten the nuts and bolts in the order shown. When tightening the oil strainer nuts (4) press the strainer onto the cylinder block so there is no gap between the strainer and block. Use the illustration and tighten the nuts and bolts: (1) and (2), 33 ft. lbs. (44 Nm); (3), 13 ft. lbs. (18 Nm); (4), 10 ft. lbs. (13 Nm) and (5), 13 ft. lbs. (18 Nm).

10. Install the oil pan.

11. Lower the vehicle, fill the engine with oil and check for leaks.

1FZ-FE Engine

▶ **See Figures 157 thru 162**

1. Drain the engine oil.
2. Remove the engine under cover.
3. Drain the engine coolant.

☀☀ CAUTION

When draining coolant, keep in mind that cats and dogs are attracted by ethylene glycol antifreeze, and are quite likely to drink any that is left in an uncovered container or in puddles on the ground. This will prove fatal in sufficient quantity. Always drain coolant into a sealable container. Coolant may be reused unless it is contaminated or several years old.

4. Remove the radiator.

5. Disconnect the A/C compressor and bracket. Loosen the pulley nut and adjusting bolt, then remove the belt. Put the compressor aside but do not disconnect any of the lines.

6. Disconnect the radiator hose from the inlet, then unbolt and remove the radiator pipe.

7. Remove the water pump.

8. Remove the cylinder head.

9. Disconnect the oil cooler bracket from the No. 1 oil pan.

10. Remove the oil level sensor and gasket.

11. Unbolt the No. 1 oil pan to the transmission housing.

12. Remove the No. 2 oil pan, this is the pan with 17 mounting bolts. Insert a blade or SST 09032–00100, between the No. 1 and No. 2 oil pans, remove the sealer and lower the No. 2 pan.

13. Remove the 21 bolts retaining the No. 1 oil pan. Pry portions A between the block and the No. 1 pan.

14. Remove the oil pan baffle plate.

15. Remove the oil strainer.

16. Unbolt the crankshaft pulley, then remove the drive belt idler pulley.

17. Unbolt the oil pump (timing chain cover), timing chain and camshaft gear.

To install:

18. Turn the crankshaft until the set key on the crankshaft faces downwards.

19. Install the timing chain and camshaft gear.

20. Attach the oil pump (timing chain cover).

21. Install the drive belt idler pulley then the crankshaft pulley.

22. Install the oil strainer to the block with a new gasket, tighten to 14 ft. lbs. (20 Nm).

23. Install the oil baffle plate, tighten to 78 inch lbs. (9 Nm).

24. Remove any old sealant from the No. 1 oil pan. Clean all the components with a non-residue solvent. Apply sealant to the No. 1 pan as shown. Install the pan within 5 minutes of application.

25. Install a new gasket into position. Pour approximately 0.9 cu. in. 15 cm3 of engine oil into the oil pump hole. Install the No. 1 pan, tighten the 14mm heads to 31 ft. lbs. (44 Nm) and the 12mm heads to 14 ft. lbs. (20 Nm).

26. Apply sealant to the No. 2 pan, assemble within 5 minutes. Tighten the bolts to 69 inch lbs. (8 Nm) and the nuts to 78 inch lbs. (9 Nm).

27. Attach the bolts retaining the No. 1 pan to the transmission housing, tighten to 53 ft. lbs. (73 Nm).

28. Installation of the remaining components is in the reverse order of the removal procedure.

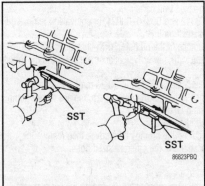

Fig. 157 A special tool is available to cut through the sealant of the oil pan

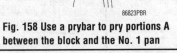
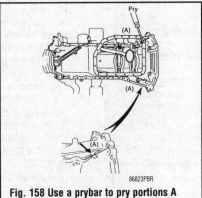

Fig. 158 Use a prybar to pry portions A between the block and the No. 1 pan

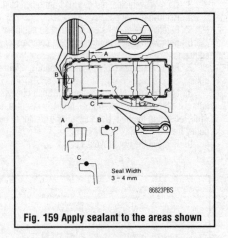

Fig. 159 Apply sealant to the areas shown

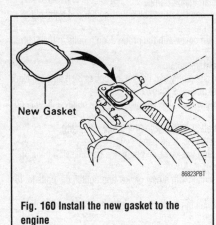

Fig. 160 Install the new gasket to the engine

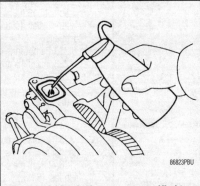

Fig. 161 Add only the amount specified to the oil pump hole

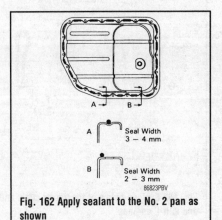

Fig. 162 Apply sealant to the No. 2 pan as shown

Timing Belt Cover and Seal

REMOVAL & INSTALLATION

Refer to the timing belt procedure in this section to remove the timing belt cover. There is no replaceable seal.

Timing Chain Cover and Seal

REMOVAL & INSTALLATION

22R and 22R-E Engines

♦ See Figure 163

✳✳ WARNING

This procedure requires tools which may not be in your tool box. Pulley extractors and a counter-holding bar are required. Do not attempt the procedure without these tools available.

1. Remove the cylinder head.
2. Drain the cooling system and remove the radiator.

✳✳ CAUTION

When draining coolant, keep in mind that cats and dogs are attracted by ethylene glycol antifreeze, and are quite likely to drink any that is left in an uncovered container or in puddles on the ground. This will prove fatal in sufficient quantity. Always drain coolant into a sealable container. Coolant may be reused unless it is contaminated or several years old.

3. On 4WD vehicles, remove the front differential.
4. Remove the oil pan.
5. Remove the power steering belt if so equipped.
6. Remove the air conditioning belt, compressor and bracket if so equipped.
7. Remove the fan clutch (fluid coupling), with the fan and water pump pulley.
8. Matchmark and remove the crankshaft pulley(s). This bolt will be very tight. Use the counter-holding bar, properly installed, to hold the crankshaft while loosening the bolt. Use the pulley extractor or gear puller to remove the pulley from the shaft.
9. Remove the water bypass pipe.

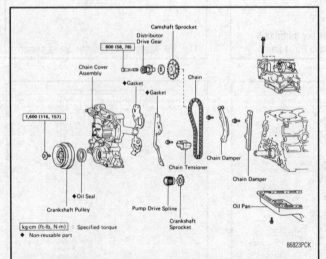

Fig. 163 Exploded view of the timing chain and cover on the 22R and 22R-E engines

10. Remove the fan belt adjusting bar. If necessary, remove the lower power steering bracket for access.
11. Disconnect the heater water outlet pipe.
12. Unbolt the timing chain cover assembly. Be careful to loosen only the correct bolts.

To install:

13. Install new gaskets over the dowels on the timing chain cover. Slide the cover assembly over the dowels and pump spline. Insert the retaining bolts and tighten them. Tighten the 8mm bolts to 9 ft. lbs. (13 Nm) and the 10mm bolts to 29 ft. lbs. (39 Nm).
14. Installation of the remaining components is in the reverse order of the removal procedure..
15. When installing the crankshaft pulley and bolt, use the counter-holding tool to hold the pulley and tighten the bolt to 116 ft. lbs. (157 Nm).

1FZ-FE Engine

1. Disconnect the negative battery cable.

✳✳ CAUTION

On vehicles with air bags, work must be started after 90 seconds from the time the ignition switch is turned to the LOCK position and the negative (−) battery cable is disconnected.

2. Raise and safely support the vehicle.
3. Drain the engine oil and the engine coolant.
4. Remove the engine under cover.
5. Remove the radiator.
6. Disconnect and remove the A/C compressor and the bracket.
7. Remove the radiator pipe by disconnecting the radiator hose from the water inlet.
8. Remove the water pump and the gasket.
9. Remove the cylinder head.
10. Disconnect the oil cooler pipe bracket from the No. 1 oil pan.
11. Remove the oil level sensor.
12. Remove the bolts holding the No. 1 oil pan to the transmission housing.
13. Remove the No. 2 and No. 1 oil pans.
14. Using SST 09213–58012 and 09330–00021, or equivalent, remove pulley bolt. Remove the crankshaft pulley.
15. Check the thrust clearance of the oil pump driveshaft gear.
   a. Using a dial indicator with a lever type attachment, measure the thrust clearance.
   b. Maximum thrust clearance is 0.0118 in. (0.30 mm).
   c. If the thrust clearance is greater than maximum, replace the oil pump driveshaft gear and/or timing chain cover.
16. Remove the drive belt idler pulley.
17. Remove the bolts securing the timing chain cover, then remove the cover. Be sure to note the locations of the different sized bolts.

To install:

18. Clean the gasket mating areas well.
19. Install the timing chain cover. Apply packing seal to the cover before installation.
   a. Engage the gear of the oil pump drive rotor with the gear of the oil pump drive gear, and install the oil pump.
   b. Install the oil pump and the drive belt adjusting bar and tighten the bolts to 15 ft. lbs. (21 Nm).
20. Install the timing chain cover with the proper length bolts in their correct locations.
21. Remove the cord from the chain.
22. Install the drive belt idler pulley and tighten the bolt to 32 ft. lbs. (43 Nm).
23. Install the crankshaft pulley as follows:
   a. Align the pulley set key with the key groove of the pulley and slide on the pulley.
   b. Install the pulley bolt and tighten to 304 ft. lbs. (412 Nm).
24. Install the No. 1 and the No. 2 oil pans.
25. Install the bolts holding the No. 1 oil pan to the transmission housing. Tighten to 53 ft. lbs. (72 Nm).
26. Install the oil level sensor with a new gasket and tighten the bolts to 48 inch lbs. (5 Nm).
27. Installation of the remaining components is in the reverse order of the removal procedure.

2VZ-FE and 3RZ-FE Engines

1. Disconnect the negative battery cable.

✳✳ CAUTION

On vehicles with air bags, work must be started after 90 seconds from the time the ignition switch is turned to the LOCK position and the negative (–) battery cable is disconnected.

2. Raise and safely support the vehicle.
3. Drain the engine coolant and the engine oil.
4. Remove the cylinder head.
5. Remove the radiator.
6. On 4WD vehicles, remove the front differential.
7. Remove the oil pan.
8. If equipped with P/S, remove the power steering belt.
9. If equipped with A/C, remove the A/C belt, compressor, and the bracket.
10. Remove the fluid coupling with the fan and the water pump pulley.
   a. Loosen the water pump pulley bolts. Loosen the belt adjusting bolt and the pivot bolt of the alternator and remove the drive belt.
   b. Remove the set nuts, the fluid coupling with the fan, and the water pump pulley.
11. Remove the crankshaft pulley.
12. Remove the water bypass pipe.
13. Remove the fan belt adjusting bar. With P/S, remove the lower P/S bracket.
14. Disconnect the heater water outlet pipe.
15. Remove the chain cover assembly. Remove the bolts shown by the arrows.

To install:

16. Clean the gasket mating areas well.
17. Install the timing chain cover assembly.
   a. Remove the old cover gaskets and install new gaskets.
   b. Slide the cover assembly over the dowels and the pump spline. Tighten the 8 mm bolts to 9 ft. lbs. (13 Nm) and the 10 mm bolts to 29 ft. lbs. (13 Nm).
18. Install the fan belt adjusting bar to the chain cover and the cylinder head. Tighten to 9 ft. lbs. (13 Nm).
19. Installation of the remaining components is in the reverse order of the removal procedure. Tighten the crankshaft pulley bolt to 116 ft. lbs. (157 Nm).

SEAL REPLACEMENT

Cover Removed

▶ See Figures 164 and 165

1. Unbolt the timing chain cover assembly. Be careful to loosen only the correct bolts.
2. Pry out the seal from the cover with a flat-bladded tool.
3. It is a good idea to remove the oil pump from the timing cover and replace the O-ring.

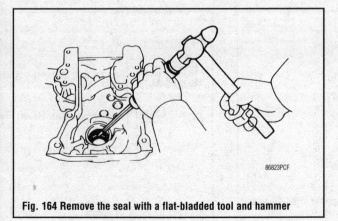

Fig. 164 Remove the seal with a flat-bladded tool and hammer

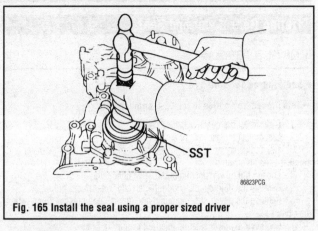

Fig. 165 Install the seal using a proper sized driver

To install:

4. Clean and inspect the timing cover area. Install new gaskets around the dowel areas and pump spline.
5. Apply multi-purpose grease to the new oil seal lip.
6. Tap the seal into place with SST 09223–50010/60010 or equivalent, and a hammer. Do this until the seal surface is flush with the cover edge.
7. Install the cover, tighten the bolts as specified for your engine.
8. If the oil pump was removed, install a new O-ring behind the pump prior to installation.

Cover Not Removed

▶ See Figures 166 and 167

1. Unbolt and remove the oil pump.
2. Using a knife, carefully cut off the oil seal lip. With a flat-bladded tool, (preferably with tape around it) pry the seal from the cover.

To install:

3. Apply multi-purpose grease to the new oil seal lip.
4. Tap the seal into place with SST 09223–50010/60011 or equivalent, and a hammer. Do this until the seal surface is flush with the cover edge.
5. Install the oil pump with a new O-ring.

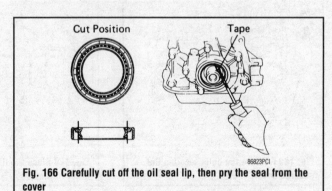

Fig. 166 Carefully cut off the oil seal lip, then pry the seal from the cover

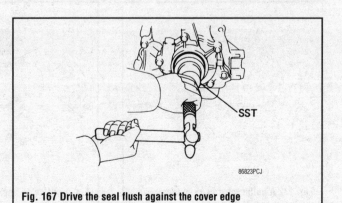

Fig. 167 Drive the seal flush against the cover edge

Timing Gear Cover and Seal

REMOVAL & INSTALLATION

◆ **See Figures 168 thru 175**

➥ **This procedure applies to the 3F-E engines only.**

1. Disconnect the negative battery cable.
2. Disconnect the accelerator and throttle cables.
3. Tag and disconnect, then remove the air intake hose, air flow meter and air cleaner as an assembly.
4. Loosen the power steering pump drive pulley nut.
5. Remove the fluid coupling with the fan and water pump pulley.
6. Remove the power steering pump and the A/C compressor along with their brackets.
7. Remove the power steering idler pulley and it's bracket.
8. Remove the distributor.
9. Remove the valve rocker shaft assembly.
10. Remove the ten bolts, two nuts, pushrod cover and gasket off the pushrod cover.
11. Remove the valve lifters, place them in the exact order as removed.
12. Remove the power steering pulley from the crankshaft pulley.
13. Using SST 09213–58011 and SST 09330–00021 or equivalent, along with a 46mm socket, remove the pulley mount bolt for the crankshaft.
14. Remove the pulley using SST 09213–60017 or an equivalent puller.
15. Remove the oil cooler pipe with the hose.
16. Remove the timing gear cover and gasket.

To install:

17. Clean all the gasket mating areas. With multi-purpose grease, install a new seal into the timing cover. A seal installer will be needed for this job such as 09223–50010.
18. There are three sizes of timing gear cover bolts. Apply adhesive to the two A bolts. Install a new gasket and then position the cover. Finger tighten all

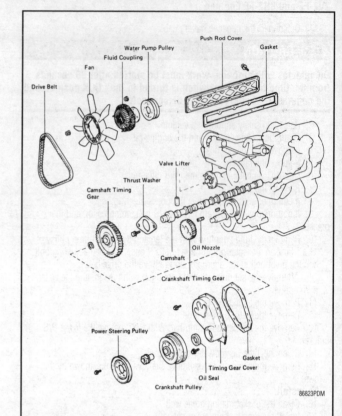

Fig. 168 Exploded view of the timing gear assembly on the 3F-E engine

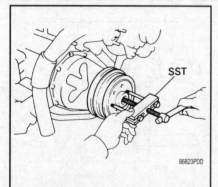

Fig. 169 Remove the bolts securing the pushrod cover

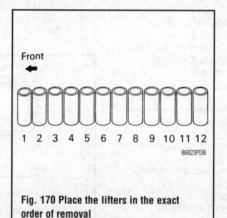

Fig. 170 Place the lifters in the exact order of removal

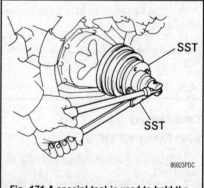

Fig. 171 A special tool is used to hold the crankshaft pulley while loosening the bolt

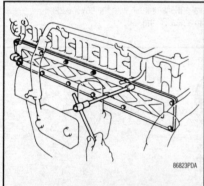

Fig. 172 A puller must be used to remove the crankshaft pulley

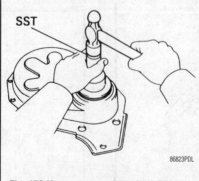

Fig. 173 Use an appropriate sized driver to seat the seal

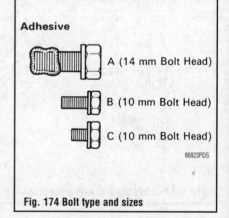

Fig. 174 Bolt type and sizes

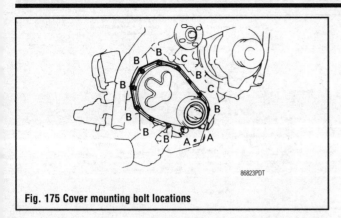

Fig. 175 Cover mounting bolt locations

the bolts. Align the crankshaft pulley set key with the groove of the pulley; gently tap the pulley onto the crankshaft. Tighten the cover bolts marked **A** to 18 ft. lbs. (25 Nm); and the **B** and **C** to 43 inch lbs. (5 Nm).

19. Installation of the remaining components is in the reverse order of the removal procedure. Tighten the pulley bolt to 253 ft. lbs. (343 Nm). Use of the counter-holding tool is required.

Timing Gears

REMOVAL & INSTALLATION

◗ **See Figures 176 thru 182**

➡ **This procedure applies to 3F-E engines only.**

1. Remove the timing gear cover and gasket.
2. Using a flat-bladded tool and a hammer, tap out the old oil seal from the cover.

3. Check the timing gear backlash. Using a dial indicator, measure the backlash at several places while turning the camshaft clockwise and counterclockwise.
   - Standard backlash:0.0039–0.0072 in. (0.100–0.183mm)
   - Maximum backlash:0.0098 in. (0.25mm)
4. If the backlash is greater than the maximum, replace the camshaft and crankshaft timing gears.
5. Align the large (camshaft) gear so that the holes allow access to the camshaft retaining bolts. Remove the bolts and very carefully remove the cam with the gear attached. Use great care not to damage the bearing surfaces on the way out.
6. Tap out the gear set key.
7. With a puller or SST 09213–60017, remove the timing gear.
8. If necessary, remove the oil nozzle.
9. Use snaring pliers to remove the snaring holding the cam gear. Mount the camshaft in a press and, using the proper size tip on the press, press the cam out of the gear.

To install:

10. Clean all the gasket mating areas. With multi-purpose grease, install a new seal into the timing cover. A seal installer will be needed for this job such as 09223–50010.
11. Install the timing gear set key to the camshaft. Assemble the cam, thrust plate and timing gear. Use a press with the proper installation tool. Align the timing gear set key with the groove in the gear and press in the camshaft.
12. Install the snaring.
13. Use a thickness gauge to measure the thrust clearance between the thrust plate and camshaft. Correct clearance is 0.200mm–0.290mm (0.0079–0.0114 in.). Maximum allowance is 0.33mm (0.0130 in.).
14. Put the timing gear on the crankshaft with the timing gear facing forwards. Align the timing gear set key with the groove of the timing gear. Use a small mallet to tap the gear into place. Use a plastic mallet to tap in the pulley set key.
15. Turn the crankshaft timing gear clockwise to place the key groove upwards.
16. Insert the camshaft into the block. Take great care not to damage the camshaft bearing surfaces on the way in.

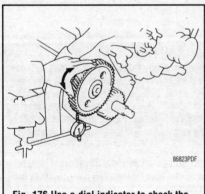

Fig. 176 Use a dial indicator to check the timing gear backlash

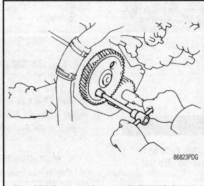

Fig. 177 Remove the camshaft retaining bolts through the access holes

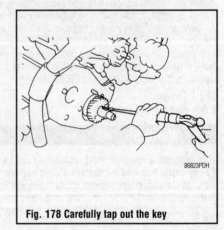

Fig. 178 Carefully tap out the key

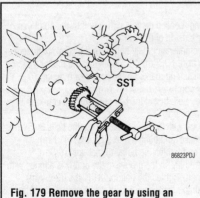

Fig. 179 Remove the gear by using an appropriate puller

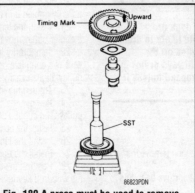

Fig. 180 A press must be used to remove and install the camshaft gear

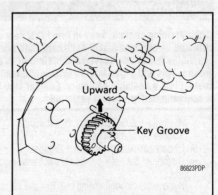

Fig. 181 Turn the crankshaft timing gear clockwise to place the key groove upwards

Fig. 182 Align the timing marks of the crankshaft and camshaft

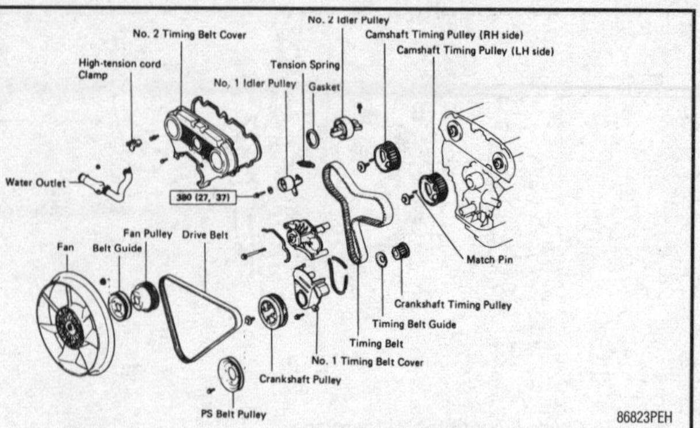

Fig. 183 Exploded view of the 3VZ-E engine timing belt routing

17. Align the timing marks of the crankshaft and camshaft. Note that the one mark is on a tooth, the other in a valley. The tooth and valley must align exactly. Cylinder No. 6 should be at TDC compression if everything is set properly.

18. Install the two bolts holding the thrust washer to the block; tighten to 9 ft. lbs. (12 Nm).

19. Recheck the timing gear backlash.

20. Install the oil nozzle if it was removed. Make certain the oil hole points towards the gear mesh area. Use a hammer and small chisel to stake the threads of the oil nozzle.

21. Install the timing gear cover with a new seal and gasket.

Timing Belt

REMOVAL & INSTALLATION

** CAUTION **

The timing system is complex. Severe engine damage will occur if you make any mistakes. Do not attempt this procedure unless you are highly experienced with this type of repair. If you are at all unsure of your abilities, consult an expert. Double-check all your work and be sure everything is correct before you attempt to start the engine.

If you operate your vehicle under conditions of extensive idling and or low speed driving for long periods of time; such as a taxi, police or delivery service, replace the belt every 60,000 miles (96,000 km).

3VZ-E Engine
◆ See Figures 183 thru 192

※ WARNING

This procedure requires tools which may not be in your tool box. Pulley extractors and a counter-holding bar are required. Do not attempt the procedure without these tools available.

1. Disconnect the negative battery cable and drain the coolant.
2. Remove the radiator and shroud.

※ CAUTION

When draining coolant, keep in mind that cats and dogs are attracted by ethylene glycol antifreeze, and are quite likely to drink any that is left in an uncovered container or in puddles on the ground. This will prove fatal in sufficient quantity. Always drain coolant into a sealable container. Coolant may be reused unless it is contaminated or several years old.

3. Remove the power steering belt and pump.
4. Remove the spark plugs.
5. Disconnect the air hoses at the air pipe.
6. Disconnect the water bypass hose at the air pipe and then remove the water outlet.
7. Remove the air conditioning belt. Remove the alternator drive belt, fluid coupling, guide and fan pulley.
8. Disconnect the high tension cords and their clamps at the upper timing

belt cover and then remove the cover and its three gaskets.

9. Rotate the crankshaft pulley until the groove on its lip is aligned with the 0 on the lower timing belt cover. This should set the No. 1 cylinder at TDC of its compression stroke. The matchmarks on the camshaft timing pulleys must be in alignment with those on the upper rear timing cover. If not, rotate the engine 360° (one complete revolution).

10. Remove the power steering pulley. Use the counter-holding tool to aid in removal of the crank pulley bolt. Remove the crankshaft pulley using a puller.

11. Remove the fan pulley bracket, then remove the lower timing belt cover.

12. If the belt is to be reused after removal, mark an arrow on the belt showing the direction of rotation. Make matchmarks on the side of the belt and the flanges of each pulley. These marks will be important at reassembly.

13. Remove the timing belt guide and remove the tension spring.

14. Loosen the idler pulley bolt and shift the idler to the left as far as it will go. Temporarily tighten the bolt and release tension from the belt. Slide the belt off the pulleys.

15. Protect the belt at all times. Do not bend, twist or turn the belt inside out. Do not allow the belt to contact oil, grease or water.

16. Examine the belt closely for cracking, loose teeth, separation or wear on the edges. If the belt is not virtually perfect, replace it.

17. Check the idler pulleys for smoothness of rotation. Inspect the tension spring; correct length is 2.15 in. (54.6mm).

To install:

** CAUTION **

Before starting the engine, carefully rotate the crankshaft by hand through at least two full revolutions (use a socket and breaker bar on the crankshaft pulley centerbolt). If you feel any resistance, STOP! There is something wrong - most likely, valves are contacting the pistons. You must find the problem before proceeding. Check your work and see if any updated repair information is available.

18. Make absolutely certain the marks on the camshaft pulleys are aligned with the marks on the rear belt cover. Check the lower belt pulley; align its mark with the mark on the oil pump housing.

19. Install the timing belt onto the cam pulleys, idler pulley, water pump pulley and crank pulley. If reusing the original belt, take great care to align all the marks made before removal.

20. Loosen the idler pulley bolt; move it as far to the right as it will go. Temporarily tighten it in position.

21. Install the tension spring. Loosen the idle pulley bolt to the point that the idler pulley moves with the tension of the spring.

22. Temporarily install the crankshaft pulley bolt. Turn the crankshaft clockwise two complete revolutions from TDC to TDC. Check that the mark on each pulley aligns exactly with the matching marks on the covers or housing.

23. Tighten the idler pulley bolt to 27 ft. lbs. (37 Nm).

24. Remove the crank pulley bolt. Install the belt guide on the crankshaft pulley. The cupped side faces outward.

25. Install the bottom belt cover with the two gaskets. tighten the bolts to (5 Nm).

26. Install the fan pulley bracket and tighten it to 30 ft. lbs. (41 Nm).

27. Install the upper cover and tighten the bolts 48 inch lbs. (5 Nm).

28. Position the crankshaft pulley so the groove in the pulley is aligned with the woodruff key in the crankshaft. Tighten the bolt to 181 ft. lbs. (245 Nm).

29. Installation of the remaining components is in the reverse order of the removal procedure.

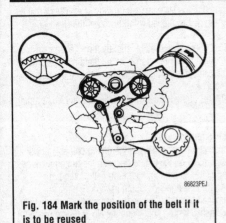

Fig. 184 Mark the position of the belt if it is to be reused

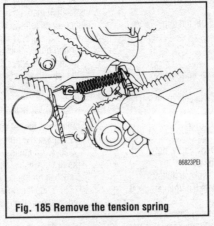

Fig. 185 Remove the tension spring

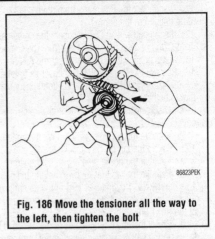

Fig. 186 Move the tensioner all the way to the left, then tighten the bolt

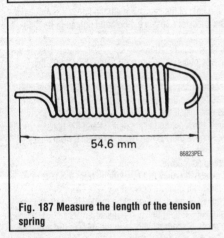

Fig. 187 Measure the length of the tension spring

54.6 mm

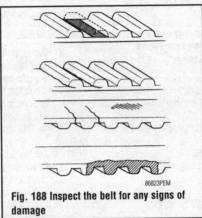

Fig. 188 Inspect the belt for any signs of damage

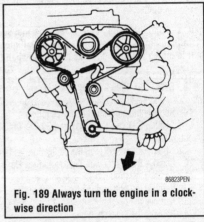

Fig. 189 Always turn the engine in a clockwise direction

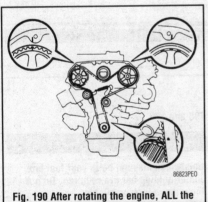

Fig. 190 After rotating the engine, ALL the timing marks must align perfectly

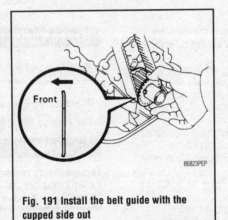

Fig. 191 Install the belt guide with the cupped side out

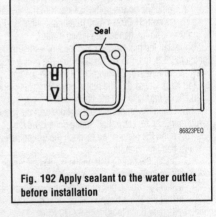

Fig. 192 Apply sealant to the water outlet before installation

5VZ-FE Engine

1. Disconnect the negative battery cable.

❊❊ CAUTION

On vehicles with air bags, work must be started after 90 seconds from the time the ignition switch is turned to the LOCK position and the negative (–) battery cable is disconnected.

2. Raise and safely support the vehicle.
3. Remove the engine under cover.
4. Drain the engine coolant.

❊❊ CAUTION

When draining coolant, keep in mind that cats and dogs are attracted by ethylene glycol antifreeze, and are quite likely to drink any that is left in an uncovered container or in puddles on the ground. This will prove fatal in sufficient quantity. Always drain coolant into a sealable container. Coolant may be reused unless it is contaminated or several years old.

5. Disconnect the upper radiator hose from the engine.
6. Remove the Power Steering (P/S) drive belt as follows:
   a. Loosen the fan pulley mounting nuts.
   b. Loosen the lock bolt, pivot bolt, and the adjusting bolt. Remove the drive belt from the engine.
7. Remove the A/C drive belt by loosening the idler pulley nut and the adjusting bolt.
8. Loosen the lock bolt, pivot bolt, and the adjusting bolt and the alternator drive belt.
9. Remove the fan shroud by removing the two clips.
10. Remove the fan with the fluid coupling and fan pulleys.

11. Disconnect the P/S pump from the engine and set aside. Do not disconnect the lines from the pump.

12. If equipped with A/C, disconnect the compressor from the engine and set aside. Do not disconnect the lines from the compressor.

13. If equipped with A/C, disconnect the A/C bracket.

14. Remove the No. 2 timing belt cover as follows:

a. Disconnect the camshaft position sensor connector from the No. 2 timing belt cover.

b. Disconnect the three spark plug wire clamps from the No. 2 timing belt cover.

c. Remove the six bolts and remove the timing belt cover.

15. Remove the fan bracket as follows:

a. Remove the power steering adjusting strut by removing the nut.

b. Remove the fan bracket by removing the bolt and nut.

16. Set the No. 1 cylinder at TDC of the compression stroke as follows:

a. Turn the crankshaft pulley and align its groove with the timing mark **0** of the No. 1 timing belt cover.

b. Check that the timing marks of the camshaft timing pulleys and the No. 3 timing belt cover are aligned. If not, turn the crankshaft pulley one revolution (360°).

➡ **If re–using the timing belt, make sure that you can still read the installation marks. If not, place new installation marks on the timing belt to match the timing marks of the camshaft timing pulleys.**

17. Remove the timing belt tensioner by alternately loosening the two bolts.

18. Using SST 09960–10010 or equivalent, remove the pulley bolt, the timing pulley and the knock pin. Remove the two timing pulleys with the timing belt.

19. Remove the crankshaft pulley as follows:

a. Using SST 09213–54015 and 09330–00021, or equivalent, loosen the pulley bolt.

b. Remove the SST tool, the pulley bolt, and the pulley.

20. Remove the starter wire bracket and the No. 1 timing belt cover.

21. Remove the timing belt guide and remove the timing belt.

22. Remove the bolt and the No. 2 idler pulley.

23. Remove the pivot bolt, the No. 1 idler pulley, and the plate washer.

24. Remove the crankshaft gear.

To install:

25. Install the crankshaft timing gear as follows:

a. Align the timing pulley set key with the key groove of the gear.

b. Using SST 09214–60010, or equivalent, and a hammer, tap in the timing gear with the flange side facing inward.

26. Install the plate washer and the No. 1 idler pulley with the pivot bolt and tighten to 26 ft. lbs. (35 Nm). Check that the pulley bracket moves smoothly.

27. Install the No. 2 timing belt idler with the bolt. Tighten the bolt to 30 ft. lbs. (40 Nm). Check that the pulley bracket moves smoothly.

28. Temporarily install the timing belt.

a. Using the crankshaft pulley bolt, turn the crankshaft and align the timing marks of the crankshaft timing pulley and the oil pump body.

b. Align the installation mark on the timing belt with the dot mark of the crankshaft timing pulley.

c. Install the timing belt on the crankshaft timing pulley, No. 1 idler, and the water pump pulleys.

29. Install the timing belt guide with the cup side facing outward.

30. Install the No. 1 timing belt cover and starter wire bracket. Tighten the timing belt cover bolts to 80 inch lbs. (9 Nm).

31. Install the crankshaft pulley as follows:

a. Align the pulley set key with the key groove of the crankshaft pulley.

b. Install the pulley bolt and tighten it to 184 ft. lbs. (250 Nm).

32. Install the left camshaft timing pulley as follows:

a. Install the knock pin to the camshaft.

b. Align the knock pin hole of the camshaft with the knock pin groove of the timing pulley.

c. Slide the timing pulley on the camshaft with the flange side facing outward. Tighten the pulley bolt to 81 ft. lbs. (110 Nm).

33. Set the No. 1 cylinder to TDC of the compression stroke as follows:

a. Turn the crankshaft pulley, and align its groove with the timing mark **0** of the No. 1 timing belt cover.

b. Turn the camshaft, align the knock pin hole of the camshaft with the timing mark of the No. 3 timing belt cover.

c. Turn the camshaft timing pulley, align the timing marks of the camshaft timing pulley and the No. 3 timing belt cover.

34. Connect the timing belt to the left camshaft timing pulley. Check that the installation mark on the timing belt is aligned with the end of the No. 1 timing belt cover as follows:

a. Using SST 09960–01000 or equivalent, slightly turn the left camshaft timing pulley clockwise. Align the installation mark on the timing belt with the timing mark of the camshaft timing pulley, and hang the timing belt on the left camshaft timing pulley.

b. Align the timing marks of the left camshaft pulley and the No. 3 timing belt cover.

c. Check that the timing belt has tension between the crankshaft timing pulley and the left camshaft timing pulley.

35. Install the right camshaft timing pulley and the timing belt as follows:

a. Align the installation mark on the timing belt with the timing mark of the right camshaft timing pulley, and hang the timing belt on the right camshaft timing pulley with the flange side facing inward.

b. Slide the right camshaft timing pulley on the camshaft. Align the timing marks on the right camshaft timing pulley and the No. 3 timing belt cover.

c. Align the knock pin hole of the camshaft with the knock pin groove of the pulley and install the knock pin. Install the bolt and tighten to 81 ft. lbs. (110 Nm).

36. Set the timing belt tensioner as follows:

a. Using a press, slowly press in the pushrod using 220–2,205 lbs. (981–9,807 N) of force.

b. Align the holes of the pushrod and housing, pass a 1.5 mm hexagon wrench through the holes to keep the setting position of the pushrod.

c. Release the press and install the dust boot to the tensioner.

37. Install the timing belt tensioner and alternately tighten the bolts to 20 ft. lbs. (28 Nm). Using pliers, remove the 1.5 mm hexagon wrench from the belt tensioner.

38. Check the valve timing as follows:

a. Slowly turn the crankshaft pulley two revolutions from the TDC to TDC. Always turn the crankshaft pulley clockwise.

b. Check that each pulley aligns with the timing marks. If the timing marks do not align, remove the timing belt and reinstall it.

39. Installation of the remaining components is in the reverse order of the removal procedure.

Timing Chain

REMOVAL & INSTALLATION

22R and 22R-E Engines

◗ **See Figures 193, 194, 195 and 196**

❊❊ WARNING

This procedure requires tools which may not be in your tool box. Pulley extractors and a counter-holding bar are required. Do not attempt the procedure without these tools available.

1. Remove the cylinder head.

2. Drain the cooling system and remove the radiator.

❊❊ CAUTION

When draining coolant, keep in mind that cats and dogs are attracted by ethylene glycol antifreeze, and are quite likely to drink any that is left in an uncovered container or in puddles on the ground. This will prove fatal in sufficient quantity. Always drain coolant into a sealable container. Coolant may be reused unless it is contaminated or several years old.

3. On 4WD vehicles, remove the front differential.

4. Remove the oil pan.

5. Remove the power steering belt if so equipped.

6. Remove the air conditioning belt, compressor and bracket if so equipped.

7. Remove the fan clutch (fluid coupling), with the fan and water pump pulley.

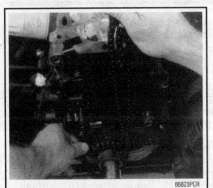

Fig. 193 The chain can be removed after the front cover is removed

86823PCR

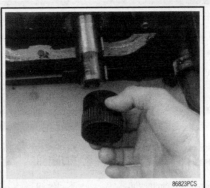

Fig. 194 The drive spline comes off first . . .

86823PCS

Fig. 195 . . . then the crankshaft sprocket

86823PCT

Fig. 196 Using a scraper, clean the gasket areas thoroughly

86823PCU

8. Matchmark and remove the crankshaft pulley(s). This bolt will be very tight. Use the counter-holding bar, properly installed, to hold the crankshaft while loosening the bolt. Use the pulley extractor or gear puller to remove the pulley from the shaft.

9. Remove the water bypass pipe.

10. Remove the fan belt adjusting bar. If necessary, remove the lower power steering bracket for access.

11. Disconnect the heater water outlet pipe.

12. Unbolt the timing chain cover assembly. Be careful to loosen only the correct bolts.

13. Remove the chain from the crankshaft sprocket. Remove the camshaft sprocket and the chain upwards.

14. Remove the oil pump drive splines and the crankshaft sprocket.

To install:

15. Remove any gasket material left on the cylinder block.

16. Turn the crankshaft until the shaft key is on top.

17. Slide the sprocket over the key on the crankshaft. Place the timing chain on the sprocket with the single bright link aligned with the timing mark on the sprocket.

18. Place the timing chain on the cam sprocket so that the link BETWEEN the two bright links is aligned with the timing mark on the sprocket. Make certain the chain is positioned within the dampers (guides).

19. Turn the camshaft sprocket counter-clockwise (viewed from the front) to take slack out of the chain. Rest or support the camshaft sprocket and chain at the top of the block.

20. Install the oil pump drive over the crankshaft key. It may be necessary to use a seal or bearing installing tool.

21. Install new gaskets over the dowels on the timing chain cover. Slide the cover assembly over the dowels and pump spline. Insert the retaining bolts and

tighten them. Tighten the 8mm bolts to 9 ft. lbs. (13 Nm) and the 10mm bolts to 29 ft. lbs. (39 Nm).

22. Installation of the remaining components is in the reverse order of the removal procedure. Tighten the crankshaft pulley bolt to 116 ft. lbs. (157 Nm).

2RZ-FE and 3RZ-FE Engines

◆ **See Figures 197 thru 208**

1. Disconnect the negative battery cable.

❊❊ CAUTION

On vehicles with air bags, work must be started after 90 seconds from the time the ignition switch is turned to the LOCK position and the negative (−) battery cable is disconnected.

2. Raise and safely support the vehicle.

3. Remove the engine under cover.

4. Drain the engine coolant and the engine oil.

5. On 4WD vehicles, remove the front differential.

6. Remove the alternator belt, fan with coupling and the water pump pulley.

a. Loosen the water pump pulley bolts. Loosen the belt adjusting bolt and the pivot bolt of the alternator and remove the drive belt.

b. Remove the set nuts, the fluid coupling with the fan, and the water pump pulley.

7. Remove the cylinder head.

8. If equipped with A/C, remove the A/C belt, compressor, and the bracket.

9. Remove the alternator adjusting bar and bracket.

10. Unbolt and remove the crankshaft position sensor and O-ring.

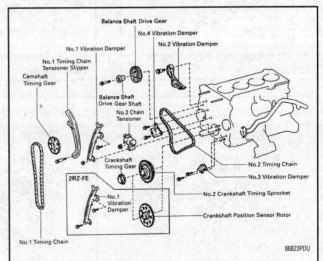

Fig. 197 Exploded view of the 2RZ-FE and 3RZ-FE engine timing chain components

11. On 2WD vehicles, remove the stiffener plates.
12. Unbolt the flywheel housing under cover and dust seal.
13. Remove the oil pan.
14. Unsecure the oil strainer and gasket.
15. Remove the crankshaft pulley, a pulley bolt remover will be needed.
16. Remove the water bypass pipe.
17. Remove the chain cover assembly. Remove the bolts shown by the arrows.
18. Remove the No. 1 timing chain and camshaft gear.
19. Remove the crankshaft timing gear.
20. Remove the No. 1 timing chain tensioner slipper and No. 1 vibration damper. On the 2RZ-FE, remove the two bolts and the damper. On the 3RZ-FE, remove the bolt, nut and No. 1 damper.
21. On the 2RZ-FE remove the crankshaft position sensor rotor and the timing chain oil jet.
22. On the 3RZ-FE engine, remove the No. 2 and No. 3 vibration dampers and the No. 2 chain tensioner as follows:

a. Install a pin to the No. 2 tensioner and lock the plunger.
b. Remove the bolt and the No. 2 damper.
c. Remove the 2 bolts and the No. 3 damper.
d. Unsecure the nut and the No. 2 tensioner.
23. Remove the balance shaft driven gear, shaft, No. 2 timing chain and the No. 2 crankshaft sprocket.
a. Unbolt the balance shaft driven gear.
b. Remove the balance shaft gear with the shaft.
c. Remove the No. 2 timing chain with the No. 2 crankshaft timing sprocket.
24. Remove the No. 4 vibration damper.

To install:
25. Install the No. 4 vibration dampener.
26. Install the No. 2 timing chain, No. 2 crankshaft timing sprocket, balance shaft drive gear and shaft as follows:
a. Install the No. 2 chain by matching the marked links with the timing marks on the crankshaft sprocket and balance shaft timing sprocket.

Fig. 198 The crankshaft position sensor is located below the water pump

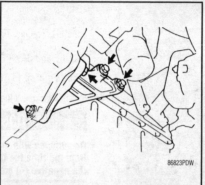

Fig. 199 Remove the bolts securing the stiffener plates

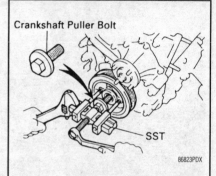

Fig. 200 The crankshaft pulley must be removed with a special puller

Fig. 201 Remove the bolts securing the front cover

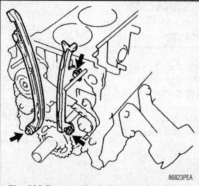

Fig. 202 Remove the bolts securing the chain dampers

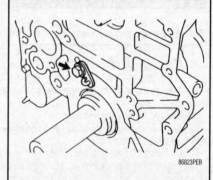

Fig. 203 The oil jet is secured by a single bolt

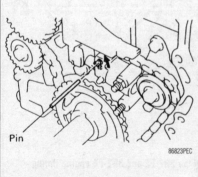

Fig. 204 Lock the plunger with an appropriate pin

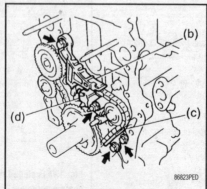

Fig. 205 Fastener locations for the tensioner and damper

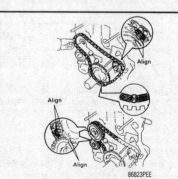

Fig. 206 Align the small timing mark of the balance shaft drive gear with the timing mark of the balance shaft timing gear

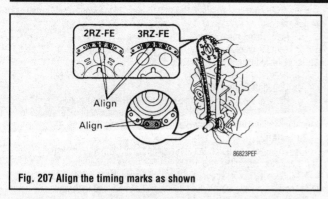

Fig. 207 Align the timing marks as shown

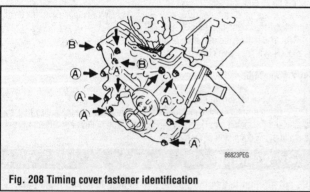

Fig. 208 Timing cover fastener identification

b. Fit the other mark link of the No. 2 chain onto the sprocket behind the large timing mark of the balance shaft gear.

c. Insert the balance shaft gear shaft through the balance shaft drive gear so that it fits into the thrust plate hole. Align the small timing mark of the balance shaft drive gear with the timing mark of the balance shaft timing gear.

d. Install the bolt to the balance shaft gear and tighten to 18 ft. lbs. (25 Nm).

e. Check each timing mark is matched with the corresponding mark link.

27. Install the No. 2, No. 3 vibration dampers and the No. 2 chain tensioner.

➡**Assemble the chain tensioner with the pin installed, then remove the pin after assembly.**

a. Install the No. 2 chain tensioner with the nut, tighten to 13 ft. lbs. (18 Nm).

b. Install the No. 3 damper with the bolts, tighten to 13 ft. lbs. (18 Nm).

c. Install the No. 2 damper, tighten to 20 ft. lbs. (27 Nm).

d. Remove the pin from the No. 2 chain tensioner and free the plunger.

28. On the 2RZ-FE engines, install the oil jet and crankshaft position sensor rotor. Make sure the front mark of the cavity of the rotor is facing forward.

29. Install the No. 1 timing chain tensioner slipper and the No. 1 vibration damper.

a. Install the No. 1 damper, tighten to 22 ft. lbs. (29 Nm).

b. Install the slipper, tighten to 20 ft. lbs. (27 Nm).

c. Check that the slipper moves smoothly.

30. Install the crankshaft timing gear.

31. Install the No. 1 timing chain and camshaft timing gear.

a. Align the timing mark between the mark link of the No. 1 timing chain, and install the No. 1 timing chain to the gear.

b. Align the timing mark of the crankshaft timing gear with the mark of the No. 1 timing chain, then install the No. 1 timing chain.

c. Tie the No. 1 chain with a wire or cord, make sure it does not come loose.

32. Install the timing chain cover assembly.

a. Remove the old cover gaskets and install new gaskets.

b. Slide the cover assembly over the dowels and the pump spline. Tighten the following:
• 12mm **A** bolts—14 ft. lbs. (20 Nm)
• 12mm **B** bolts—18 ft. lbs. (25 Nm)
• 14mm bolts—32 ft. lbs. (44 Nm)
• 14mm nut—14 ft. lbs. (20 Nm)

c. Attach the water bypass pipe.

d. Remove the cord or wire from the chain.

33. Install the crankshaft pulley, tighten the bolt to 193 ft. lbs. (260 Nm). On A/C vehicles, install the crankshaft pulleys with bolts and tighten to 18 ft. lbs. (25 Nm).

34. Installation of the remaining components is in the reverse order of the removal procedure.

1FZ-FE Engine

▶ **See Figure 209**

1. Disconnect the negative battery cable.

※※ CAUTION

On vehicles with air bags, work must be started after 90 seconds from the time the ignition switch is turned to the LOCK position and the negative (–) battery cable is disconnected.

2. Raise and safely support the vehicle.
3. Drain the engine oil and the engine coolant.
4. Remove the engine under cover.
5. Remove the radiator.
6. Disconnect and remove the A/C compressor and the bracket.
7. Remove the radiator pipe. Disconnect the radiator hose from the water inlet.
8. Remove the water pump and the gasket.
9. Remove the cylinder head.
10. Disconnect the oil cooler pipe bracket from the No. 1 oil pan.
11. Remove the oil level sensor.
12. Remove the bolts holding the No. 1 oil pan to the transmission housing.
13. Remove the No. 2 and No. 1 oil pans.
14. Using SST 09213–58012 and 09330–00021, or equivalent, remove the pulley bolt. Remove the crankshaft pulley.
15. Check the thrust clearance of the oil pump driveshaft gear as follows:

a. Using a dial indicator with a lever type attachment, measure the thrust clearance.

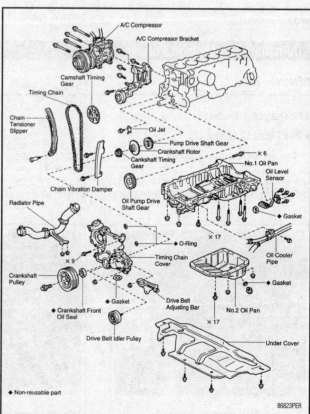

Fig. 209 Exploded view of the timing chain assembly on the 1FZ-FE engine

b. Maximum thrust clearance is 0.0118 in. (0.30 mm).

c. If the thrust clearance is greater than maximum, replace the oil pump driveshaft gear and/or timing chain cover.

16. Remove the drive belt idler pulley.

17. Remove the timing chain cover.

18. Remove the timing chain and the camshaft timing gear.

19. Pull out the crankshaft timing gear.

20. Remove the chain tensioner slipper and the vibration damper.

21. Remove the oil jet.

22. Remove the crankshaft rotor.

23. Remove the oil pump driveshaft gear.

24. Remove the pump driveshaft gear.

To install:

25. Clean the mating areas well.

26. Set the crankshaft. Turn the crankshaft until the set key on the crankshaft is facing downward (6 o'clock).

27. Install the pump driveshaft gear.

28. Apply a light coat of engine oil on the shaft portion of the oil pump driveshaft gear, and install the gear.

29. Install the crankshaft rotor.

30. Install the oil jet and tighten the bolt to 14 ft. lbs. (20 Nm).

31. Install the vibration damper and tighten the bolts to 14 ft. lbs. (20 Nm).

32. Install the chain tensioner slipper and tighten the bolt to 51 ft. lbs. (69 Nm). Make sure that the slipper moves smoothly.

33. Install the crankshaft timing gear.

34. Install the timing chain and the camshaft timing gear as follows:

a. Install the timing chain on the camshaft timing gear with the bright link aligned with the timing mark on the camshaft timing gear.

b. Install the timing chain on the crankshaft timing gear with the other bright link aligned with the timing mark on the crankshaft timing gear.

c. Tie the timing chain with a cord to make sure that it doesn't come loose.

35. Install the timing chain cover.

36. Remove the cord from the chain.

37. Install the drive belt idler pulley and tighten the bolt to 32 ft. lbs. (43 Nm).

38. Install the crankshaft pulley. Align the pulley set key with the key groove of the pulley and slide on the pulley. Install the pulley bolt and tighten to 304 ft. lbs. (412 Nm).

39. Installation of the remaining components is in the reverse order of the removal procedure.

Camshaft Sprockets or Pulleys

REMOVAL & INSTALLATION

22R and 22R-E Engines

▶ **See Figures 210 thru 215**

The cam sprocket is removed from the camshaft as part of the rocker assembly removal. If the sprocket is to be removed from the chain, support the chain with a piece of wood, keeping the chain engaged to the lower sprocket. If the chain comes off the bottom sprocket, the timing chain covers must be removed

and the chain correctly installed. Refer to Timing Chain Removal and Installation in this section.

1. Remove the valve cover.

2. Turn the crankshaft until the No. 1 cylinder position is set at TDC compression. Matchmark the sprocket and chain.

3. Remove the halfmoon plug from the front of the cylinder head.

4. Unbolt the cam sprocket.

5. Remove the distributor drive gear and camshaft thrust plate.

6. Remove the cam sprocket. Using a piece of wire, support the chain to keep it in position for sprocket installation.

To install:

7. While holding up on the sprocket and chain, turn the crankshaft until No. 1 and No. 4 cylinders reach TDC.

8. Place the sprocket over the camshaft dowel. If the chain does not seem long enough on replacement, turn the crankshaft back and forth while pulling up on the chain and sprocket.

9. Place the distributor drive gear and camshaft thrust plate over the chain sprocket. Tighten the bolt to 58 ft. lbs. (78 Nm).

10. Install the halfmoon plug on the cylinder head.

11. Install the valve cover.

3VZ-E Engines

1. Remove the timing belt.

2. Use a properly installed counter-holding tool, remove the pulley bolt, the pulley and the match pin. Note that the left and right pulleys are different; label them to avoid confusion at reassembly.

✳✳ WARNING

Do not attempt to use the timing belt to hold tension on the pulley while loosening or tightening the retaining bolt.

To install:

3. Align the camshaft match pin hole with the marks on the upper rear timing belt cover. This places the pin holes straight up.

4. Install the timing pulleys, observing correct placement of the left and right pulleys. Install the bolts, but do not install the match pin. Check that the bolt head is not touching the pulley (Its not completely tight yet). Align the timing mark on each pulley with the marks on the No. 3 timing cover.

5. Install the timing belt, following the procedures given in this section. After the engine has been rotated two full revolutions and the timing marks checked, tighten the idler pulley bolt.

6. Remove the camshaft timing pulley bolts. Align the pulley match pin hole with the hole in the camshaft; install the match pin. Reinstall the camshaft pulley bolts.

7. Using a properly installed counter-holding tool, tighten the pulley bolts to 80 ft. lbs. (108 Nm). Take great care not to nick or damage the timing belt.

8. Continue with the re-installation of the timing belt.

Camshaft and Bearings

➡**These engines are not equipped with replaceable camshaft bearings.**

Fig. 210 Matchmark the chain and sprocket

Fig. 211 Remove the halfmoon plug located at the front of the cylinder head

Fig. 212 Loosen and remove the drive gear bolt

Fig. 213 Pull the bolt and gear off the cam

Fig. 214 Lift the chain up, then slide the gear off the cam

Fig. 215 Secure the chain in place with a length of wire

REMOVAL & INSTALLATION

3F-E Engines

To service the camshaft on these engines, please refer to the Timing Gear Removal and Installation procedure.

22R, 22R-E and 3VZ-E Engines

1. Perform the Cylinder Head Removal procedure (for your engine) far enough to gain access to the camshaft bearing cap bolts. If you are going to remove the head anyway, remove the cam after removing the cylinder head.

2. Prior to removing the camshaft, measure its end-play with a feeler gauge. Consult the Camshaft Specifications Chart for the correct end-play. If the end-play is beyond this, replace the head.

3. Remove the bearing cap bolts. Remove the caps. Keep them in order, or mark them.

4. Measure the bearing oil clearance by placing a piece of Plastigage® on each journal. Replace the caps and tighten their bolts to 13-16 ft. lbs. (18-22 Nm).

5. Remove the caps and measure each piece of Plastigage®. If the clearance is greater than 2 in. (0.1mm), replace the head and cam.

6. Lift the camshaft out of the head.

To install:

7. Coat all of the camshaft bearing journals with engine oil.

8. Lay the camshaft in the head.

9. Install the bearing caps in numerical order with their arrows pointing forward (toward the front of the engine) on 4 cylinder engines. On the 3VZ-E, the right bank arrows point forward with No. 1 cap at the front of the engine. The left bank arrows point to the rear with the No. 1 cap at the rear.

10. Install the cap bolts and tighten them in three passes and in the correct order, to 13-16 ft. lbs. (18-22 Nm).

11. Complete the cylinder head installation procedure and/or valve rocker installation.

2RZ-FE and 3RZ-FE Engines

♦ **See Figures 216, 217, 218, 219 and 220**

1. Remove the timing chain from the engine.

2. Remove the exhaust camshaft by bringing the service bolt hole of the driven sub-gear upwards. Turn the hexagon wrench head portion of the exhaust camshaft with a wrench.

3. Secure the exhaust camshaft sub-gear to the main gear with a service bolt. The thread diameter should be 0.23 in. (6mm) with a thread pitch of 0.04 in. (1.0mm) and a bolt length of 0.63–0.79 in. (16–20mm).

➡ **When you remove the camshaft, be sure that the torsional spring force of the sub-gear has been eliminated by the above operation.**

4. Uniformly loosen and remove the exhaust bearing cap bolts (10 of them), in several passes. Use the reverse order of the tightening sequence. Remove the 5 bearing caps and the camshaft. Do the same for the intake camshafts.

➡ **If the camshaft is not being lifted out straight and level, reinstall the No. 3 cap with the 2 bolts. Alternately loosen then remove the bearing cap bolts with the camshaft pulled up. Do not pry on or force the camshaft.**

5. Inspect the camshafts for runout. Inspect the cam lobes and journals. The bearings are part of the cam and should be inspected for flaking or scoring. If the bearings are damaged, replace the caps and the cylinder head as a set. The camshaft journal oil and thrust clearances should be checked.

To install:

➡ **When installing the camshafts; since the thrust clearance of the shafts is small, the cam must be kept level while it is being installed. If it is not kept level, the portion of the head receiving the shaft thrust may crack or be damaged. This can cause the camshaft to seize or break.**

6. Install the intake camshaft as follows:
   a. Apply multi-purpose grease to the thrust portion of the intake camshaft.
   b. Position the intake camshaft with the pin facing upward.
   c. Install the bearing caps in their proper locations. Apply a light coat of engine oil to the threads and install the cap bolts. Uniformly tighten the cap bolts in the sequence shown to 12 ft. lbs. (16 Nm).

7. Install the exhaust camshaft as follows:
   a. Apply engine oil to the thrust portion of the intake camshaft.
   b. Engage the exhaust camshaft gear to the intake camshaft gear by matching the timing marks (one and two dots) on each other.
   c. Roll down the exhaust camshaft onto the bearing journals while engaging the gears with each other. Install the bearing caps in their proper locations.
   d. Apply a light coat of engine oil to the threads and install the cap bolts. Uniformly tighten the cap bolts in the sequence shown to 12 ft. lbs. (16 Nm).
   e. Remove the service bolt from the driven sub-gear. Check that the intake and exhaust camshafts turn smoothly.

8. Set No. 1 cylinder to TDC compression stroke. The crankshaft pulley groove aligns with the **0** mark on timing cover and camshaft timing marks with one dot and two dots will be in a straight line on the cylinder head surface.

9. Install the timing gear. Place the gear over the straight pin of the intake camshaft.
   a. Hold the intake camshaft with a wrench. Install and tighten the bolt to 54 ft. lbs. (74 Nm).
   b. Hold the exhaust camshaft and install the bolt and distributor gear. Tighten the bolt to 34 ft. lbs. (46 Nm).

10. Install the chain tensioner, using a new gasket (mark toward the front) as follows:
   a. Release the ratchet pawl, fully push in the plunger and apply the hook to the pin so that the plunger cannot spring out.
   b. Turn the crankshaft pulley clockwise to provide some slack for the chain on the tensioner side.
   c. Push the tensioner by hand until it touches the head installation surface, then install the 2 nuts. Tighten the nuts to 13 ft. lbs. (18 Nm). Check that the hook of the tensioner is not released.
   d. Turn the crankshaft to the left so that the hook of the chain tensioner is released from the pin of the plunger, allowing the plunger to spring out and the slipper to be pushed into the chain.

11. Installation of the remaining components is in the reverse order of the removal procedure.

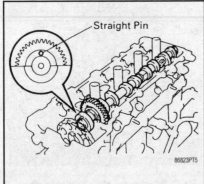

Fig. 216 Install the camshaft with the pin facing upwards

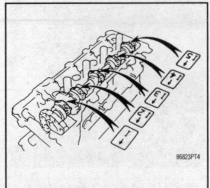

Fig. 217 Intake camshaft bearing cap locations

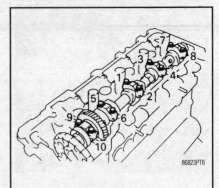

Fig. 218 Tighten the intake bearing caps following this order

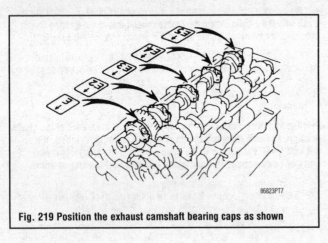

Fig. 219 Position the exhaust camshaft bearing caps as shown

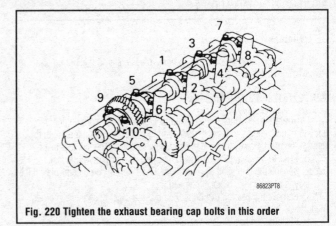

Fig. 220 Tighten the exhaust bearing cap bolts in this order

5VZ-FE Engine

▶ **See Figures 221 thru 234**

1. Remove both valve covers.
2. Remove the semi-circular plugs.
3. Remove the right exhaust camshaft as follows:

   a. Bring the service bolt hole of the driven sub-gear upward by turning the hexagon head portion of the exhaust camshaft with a wrench.

   b. Align the timing mark (2 dot marks) of the camshaft drive and driven gears by turning the camshaft with a wrench.

   c. Secure the exhaust camshaft sub-gear to the driven gear with a service bolt 0.2 in. (6mm) diameter, 0.6 in. (16–20mm) bolt length and 0.4 in. (1.0mm) in thread pitch.

➡ **When removing the camshaft, make sure that the torsional spring force of the sub-gear has been eliminated by the above operation.**

   d. Uniformly loosen and remove the bearing cap bolts in several passes, in the sequence shown.

   e. Remove the bearing caps and camshaft. Make a note of the bearing cap positions for proper installation.

✳✳ WARNING

Do not pry on or attempt to force the camshaft with a tool or other object.

4. Remove the right hand intake camshaft as follows:

   a. Uniformly loosen and remove the bearing cap bolts in several passes, in the sequence shown.

   b. Remove the bearing caps, oil seal and camshaft. Make a note of the bearing cap positions for proper installation.

5. Remove the left exhaust camshaft as follows:

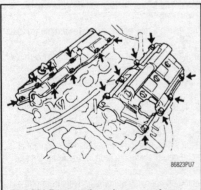

Fig. 221 Remove the valve covers from the engine

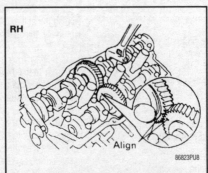

Fig. 222 Align the gears by turning the camshaft with a wrench on the flats provided—not the lobes

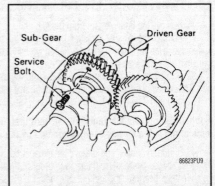

Fig. 223 Use an appropriate size bolt to release the spring force of the sub-gear

a. Align the timing mark (1 dot mark) of the camshaft drive and driven gears by turning the camshaft with a wrench.

b. Secure the exhaust camshaft sub-gear to the driven gear with a service bolt 0.2 in. (6mm) diameter, 0.6 in. (16–20mm) bolt length and 0.4 in. (1.0mm) in thread pitch.

➡ **When removing the camshaft, make sure that the torsional spring force of the sub-gear has been eliminated by the above operation.**

c. Uniformly loosen and remove the bearing cap bolts in several passes, in the sequence shown.

d. Remove the bearing caps and camshaft. Make a note of the bearing cap positions for proper installation.

✳✷ WARNING

Do not pry on or attempt to force the camshaft out.

6. Remove the left hand intake camshaft as follows:

a. Uniformly loosen and remove the bearing cap bolts in several passes, in the sequence shown.

b. Remove the bearing caps, oil seal and camshaft. Make a note of the bearing cap positions for proper installation.

To install:

7. Install the right intake camshaft as follows:

a. Apply engine oil to the thrust portion of the intake camshaft.

b. Position the intake camshaft at a 90° angle of the timing mar (2 dot marks) on the cylinder head.

c. Install the bearing caps in their proper locations. Apply a light coat of engine oil to the threads and install the cap bolts.

d. Apply a light coat of engine oil on the threads and under the heads of the bearing cap bolts.

e. Uniformly tighten the cap bolts in the sequence shown to 12 ft. lbs. (16 Nm).

8. Install the right exhaust camshaft as follows:

a. Apply engine oil to the thrust portion of the intake camshaft.

b. Align the timing marks (2 dot marks) of the camshaft drive and driven gears.

c. Roll down the exhaust camshaft onto the bearing journals while engaging the gears with each other. Install the bearing caps in their proper locations.

d. Apply a light coat of engine oil to the threads and install the cap bolts.

e. Apply a light coat of engine oil on the threads and under the heads of the bearing cap bolts.

f. Uniformly tighten the cap bolts in the sequence shown to 12 ft. lbs. (16 Nm).

g. Remove the service bolt from the driven sub-gear. Check that the intake and exhaust camshafts turns smoothly.

h. Align the timing marks (2 dot mark) of the camshaft drive and driven gears by turning the camshaft with a wrench.

9. Install the left intake camshaft as follows:

a. Apply engine oil to the thrust portion of the intake camshaft.

b. Position the intake camshaft at a 90° angle of the timing mar (1 dot marks) on the cylinder head.

c. Install the bearing caps in their proper locations. Apply a light coat of engine oil to the threads and install the cap bolts.

d. Apply a light coat of engine oil on the threads and under the heads of the bearing cap bolts.

e. Uniformly tighten the cap bolts in the sequence shown to 12 ft. lbs. (16 Nm).

10. Install the left exhaust camshaft as follows:

a. Apply engine oil to the thrust portion of the intake camshaft.

b. Align the timing marks (1 dot marks) of the camshaft drive and driven gears.

c. Roll down the exhaust camshaft onto the bearing journals while engaging the gears with each other. Install the bearing caps in their proper locations.

d. Apply a light coat of engine oil to the threads and install the cap bolts.

e. Apply a light coat of engine oil on the threads and under the heads of the bearing cap bolts.

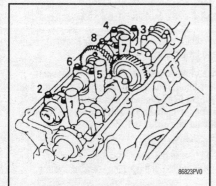

Fig. 224 Right hand exhaust camshaft bearing cap loosening sequence

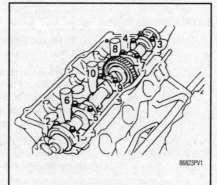

Fig. 225 Loosen the right hand intake camshaft caps as shown

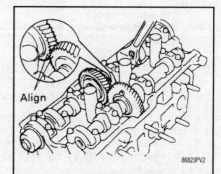

Fig. 226 When aligning the gears, turn the camshaft with the flats provided. Do not place a wrench on the lobes

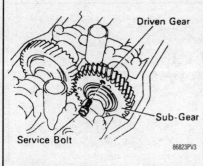

Fig. 227 The spring force of the sub-gear can be overcome by using an appropriate sized bolt

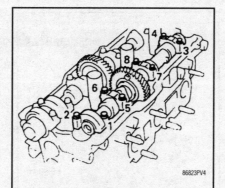

Fig. 228 Left hand exhaust camshaft bearing cap loosening sequence

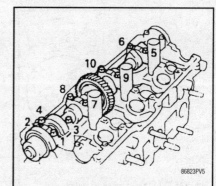

Fig. 229 The left hand intake camshaft caps must be loosened as shown

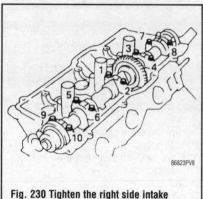

Fig. 230 Tighten the right side intake camshaft caps as shown

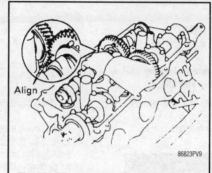

Fig. 231 When installing the exhaust camshaft, make sure the timing marks align

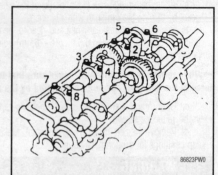

Fig. 232 Use this sequence to tighten the right hand exhaust camshaft bearing cap bolts

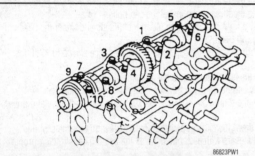

Fig. 233 Tightening sequence for the left hand intake camshaft cap bolts

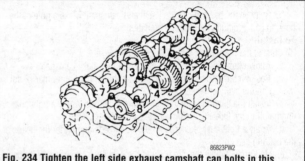

Fig. 234 Tighten the left side exhaust camshaft cap bolts in this sequence

f. Uniformly tighten the cap bolts in the sequence shown to 12 ft. lbs. (16 Nm).

g. Remove the service bolt.

11. Installation of the remaining components is in the reverse order of the removal procedure.

1FZ-FE Engine

♦ **See Figures 235, 236, 237 and 238**

1. Remove the valve cover.
2. Remove the semi–circular plug from the cylinder head.
3. Remove the spark plugs.
4. Set the No. 1 cylinder to TDC of the compression stroke as follows:

a. Turn the crankshaft pulley and align its groove with the **0** mark on the timing chain cover.

b. Check that the timing marks (one and two dots) of the camshaft drive and driven gears are in straight line on the cylinder head surface. If not, turn the crankshaft one revolution (360°) and align the marks.

5. Remove the chain tensioner.

6. Place matchmarks on the camshaft timing gear and the timing chain and remove the camshaft timing gear.

a. Hold the intake camshaft with a wrench and remove the bolt and the distributor gear.

b. Remove the camshaft timing gear and chain from the intake camshaft and leave on the slipper and the damper.

➡Since the thrust clearance of the camshaft is small, the camshaft must be kept level while it is being removed. If the camshaft is not kept level, the portion of the cylinder head receiving the shaft thrust may crack or be damaged, causing the camshaft to seize or break. To avoid this, the following steps should be carried out.

7. Remove the exhaust camshaft as follows:

a. Bring the service bolt hole of the driven sub–gear upward by turning the hexagon wrench head portion of the exhaust camshaft with a wrench.

b. Secure the exhaust camshaft sub–gear to the main gear with a service bolt. When removing the camshaft, make sure that the torsional spring force of the sub–gear has been eliminated by the above operation.

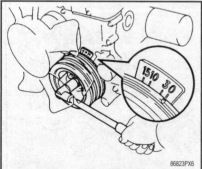

Fig. 235 Turn the crankshaft pulley and align its groove with the 0 mark on the timing chain cover

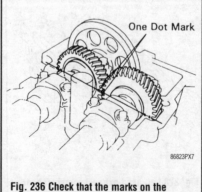

Fig. 236 Check that the marks on the camshaft pulley are aligned

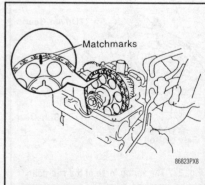

Fig. 237 Matchmark the position of the chain on the gear

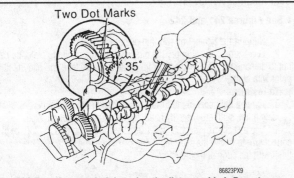

Fig. 238 Turn the camshaft by using the flats provided. Do not use the lobes

c. Set the timing mark (two dot marks) of the camshaft driven gear at approximately a 35° angle by turning the hexagon wrench head portion of the intake camshaft with a wrench.

d. Lightly push the camshaft towards the rear without applying excessive force.

e. Loosen and remove the No. 1 bearing cap bolts, alternately loosening the left and right bolts uniformly.

f. Loosen and remove the No. 2, No. 3, No. 5 and the No. 7 bearing cap bolts, alternately loosening the left and right bolts uniformly in several passes, in sequence.

➡**Do not remove the No. 4 and No. 6 bearing cap bolts at this stage.**

g. Remove the four bearing caps.

h. Alternately and uniformly loosen and remove the No. 4 and the No. 6 bearing cap bolts.

i. If the camshaft is not being lifted out straight and level, retighten the four No. 4 and No. 6 bearing cap bolts. Then reverse the order of the above steps from (g) to (e) and repeat steps from (c) to (h) once again.

j. Remove the two bearing caps and exhaust camshaft. Do not pry on or attempt to force the camshaft with a tool or any other object.

8. Remove the intake camshaft as follows:

a. Set the timing mark (two dot marks) of the camshaft drive gear at approximately a 25° angle by turning the hexagon wrench head portion of the intake camshaft with a wrench.

➡**This angle allows the No. 1 and the No. 4 cylinder cam lobes of the intake camshaft to push their valve lifters evenly.**

b. Lightly push the intake camshaft towards the front without applying excessive force.

c. Loosen and remove the No. 1 bearing cap bolts, alternately loosening the left and the right bolts uniformly.

d. Loosen and remove the No. 3, No. 4, No. 6 and the No. 7 bearing cap bolts, alternately loosening the left and right bolts uniformly in several passes in sequence.

➡**Do not remove the No. 2 and No. 5 bearing cap bolts at this stage.**

e. Remove the four bearing caps.

f. Alternately and uniformly loosen and remove the No. 2 and the No. 5 bearing cap bolts.

g. If the camshaft is not being lifted out straight and level, retighten the four No. 2 and No. 5 bearing cap bolts. Then reverse the order of the above steps from (e) to (c) and repeat steps from (a) to (f) once again.

h. Remove the two bearing caps and the exhaust camshaft.

To install:

9. Install the intake camshaft as follows:

➡**Since the thrust clearance of the camshaft is small, the camshaft must be kept level while it is being installed. If the camshaft is not kept level, the portion of the cylinder head receiving the shaft thrust may crack or be damaged, causing the camshaft to seize or break. To avoid this, the following steps should be carried out.**

a. Apply engine oil to the thrust portion of the intake camshaft.

b. Lightly place the intake camshaft on top of the cylinder head so that the No. 1 and the No. 4 cylinder cam lobes face downward.

c. Lightly push the camshaft towards the front without applying excessive force. Place the No. 2 and the No. 5 bearing caps in their proper location.

d. Temporarily tighten these bearing cap bolts uniformly and alternately in several passes until the bearing caps are snug with the cylinder head.

e. Place the No. 3, No. 4, No. 6, and the No. 7 bearing caps in their proper location. Temporarily tighten these bearing cap bolts, alternately tightening the left and right bolts uniformly.

f. Place the No. 1 bearing cap in its proper location. Check that there is no gap between the cylinder head and the contact surface of the bearing cap.

g. Uniformly tighten the 14 bearing cap bolts in several passes to 12 ft. lbs. (16 Nm).

10. Install the exhaust camshaft as follows:

a. Set the timing mark (two dot marks) of the camshaft drive gear at approximately 35° angle by turning the hexagon wrench head portion of the intake camshaft with a wrench.

b. Apply engine oil to the thrust portion of the exhaust camshaft. Engage the exhaust camshaft gear to the intake camshaft gear by matching the timing marks (two dot marks) on each gear.

c. Roll down the exhaust camshaft onto the bearing journals while engaging the gears with each other. Lightly push the intake camshaft towards the front without applying excessive force.

d. Install the No. 4 and the No. 6 bearing caps in their proper location. Temporarily tighten these bearing cap bolts, alternately tightening the left and right bolts uniformly.

e. Place the No. 2, No. 3, No. 5, and the No. 7 bearing caps in their proper location. Temporarily tighten these bearing cap bolts, alternately tightening the left and right bolts uniformly.

f. Tighten the 14 bearing cap bolts in several passes to 12 ft. lbs. (16 Nm).

g. Bring the service bolt installed in the driven sub–gear upward by turning the hexagon wrench head portion of the camshaft with a wrench. Remove the service bolt.

h. Check that the intake and the exhaust camshafts turn smoothly.

11. Set the No. 1 cylinder to TDC of the compression stroke as follows:

a. Turn the crankshaft pulley, and align its groove with the timing mark **0** of the timing chain cover. Turn the camshaft so that the timing marks with one and two dots will be in straight line on the cylinder head surface.

12. Install the camshaft timing gear as follows:

a. Check that the matchmarks on the camshaft timing gear and the timing chain are aligned. Place the gear over the straight pin of the intake camshaft.

b. Align the straight pin of the distributor gear with the straight pin groove of the intake camshaft gear.

c. Hold the intake camshaft with a wrench, install and tighten the bolt to 54 ft. lbs. (74 Nm).

13. Install the chain tensioner. Push the tensioner by hand until it touches the head installation surface, then install and tighten the two nuts to 15 ft. lbs. (21 Nm).

14. Check the valve timing as follows:

a. Turn the crankshaft pulley, and align its groove with the timing mark **0** of the timing chain cover. Always turn the crankshaft clockwise.

b. Check that the timing marks (one and two dots) of the camshaft drive and driven gears are in straight line on the cylinder head surface. If not, turn the crankshaft one revolution (360°) and align the marks.

15. Installation of the remaining components is in the reverse order of the removal procedure. Be sure to check the valve clearance.

INSPECTION

A dial indicator, micrometer and inside micrometer are all needed to properly measure the camshaft and camshaft housing. If these instruments are available, proceed; if they are not available, have the parts checked at a reputable machine shop. Camshaft specifications are included in the Camshaft Specifications chart in this section.

1. Using the micrometer, measure the height of each cam lobe. If a lobe height is less than the minimum specified, the lobe is worn and the cam must be replaced.

2. Place the cam in V-blocks and measure its run-out at the center journal with a dial indicator. Replace the cam if run-out exceeds:

- 22R and 22R-E—0.0078 in. (0.20mm)
- 3F-E—0.0118 in. (0.30mm)
- All others—0.0024 in. (0.06mm)

3. Using the micrometer, measure journal diameter, jot down the readings and compare the readings with those listed in the Camshaft Specifications chart. Measure the housing bore inside diameter with the inside micrometer, and jot the measurements down. Subtract the journal diameter measurement from the housing bore measurement. If the clearance is greater than the maximum listed under Bearing Clearance in the chart, replace the camshaft and/or the housing.

Rear Main Oil Seal

REMOVAL & INSTALLATION

Seal Retainer On Engine

♦ **See Figures 239 and 240**

1. Remove the transmission.
2. Remove the clutch cover assembly and flywheel (manual trans.) or the flexplate (auto. trans.).
3. Use a small, sharp knife to cut off the lip of the oil seal. Take great care not to score any metal with the knife.
4. Use a small prybar to pry the old seal from the retaining plate. Be careful not to damage the plate. Protect the tip of the tool with tape and pad the fulcrum point with cloth.
5. Inspect the crankshaft and seal lip contact surfaces for any sign of damage.

To install:

6. Apply a light coat of multi-purpose grease to the lip of a new oil seal. Loosely fit the seal into place by hand, making sure it is not crooked.
7. Use a seal driver such as (SST 09223–15030 and 09950–70010) of the correct size to install the seal. Tap it into place until the surface of the seal is flush with the edge of the housing.

➡Use the correct tools. Homemade substitutes may install the seal crooked, resulting in oil leaks and premature seal failure.

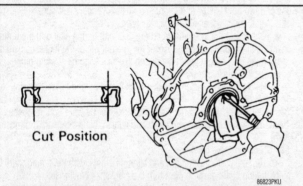

Fig. 239 Cut off the oil seal lip, then pry the seal out of the retaining plate

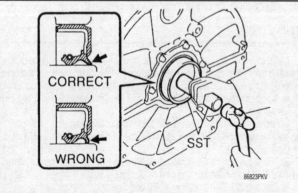

Fig. 240 Tap the new seal into place

Seal Retainer Removed

♦ **See Figures 241 and 242**

1. Support the retainer on two thin pieces of wood.
2. Use a small prybar to pry the old seal from the retaining plate. Be careful not to damage the plate. Protect the tip of the tool with tape and pad the fulcrum point with cloth.

To install:

3. Apply a light coat of multi-purpose grease to the lip of a new oil seal. Loosely fit the seal into place by hand, making sure it is not crooked.
4. Use a seal driver such as (SST 09223–15030 and 09950–70010) of the correct size to install the seal. Tap it into place until the surface of the seal is flush with the edge of the housing.

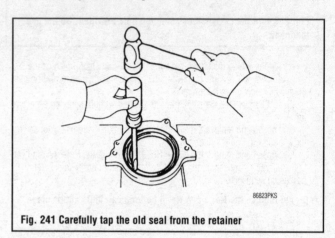

Fig. 241 Carefully tap the old seal from the retainer

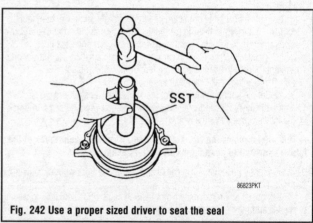

Fig. 242 Use a proper sized driver to seat the seal

Flywheel and Ring Gear

REMOVAL & INSTALLATION

♦ **See Figures 243 and 244**

1. If the engine is installed in the vehicle, remove the transmission.
2. Remove the clutch assembly, if equipped with manual transmission.
3. Make matchmarks on the flywheel and crankshaft end. Loosen the bolts holding the flywheel or driveplate a little at a time and in a crisscross pattern. Remove the flywheel; it is heavy. Protect the flywheel and ring-gear from damage or impact. Keep your feet clear of accidental dropping.

To install:

4. Thoroughly clean the flywheel bolts. Coat the first 3 or 4 threads of each bolt with sealant. Install the flywheel, aligning the previously made marks. Install the bolts finger-tight.
5. Tighten the bolts in a crisscross pattern and in several passes to the correct tightness:

- 22R/22R-E manual trans—80 ft. lbs. (108 Nm)
- 3VZ-E and 2RZ-FE manual trans—65 ft. lbs. (88 Nm)

Fig. 243 If a flywheel holding tool is not available, an air gun may be used to loosen the bolts

Fig. 244 Always tighten the bolts with a torque wrench. Note the flywheel holding tool

- 3RZ-FE manual trans—1st pass; 19 ft. lbs. (26 Nm), 2nd pass turn 90°
- 22R, 22R-E, 3VZ-E and 5VZ-FE auto. trans.—61 ft. lbs. (83 Nm)
- 3F-E auto trans.—64 ft. lbs. (87 Nm)

- 2RZ-FE and 3RZ-FE auto trans.—54 ft. lbs. (74 Nm)
- 1FZ-FE auto trans.—74 ft. lbs. (100 Nm)
6. Reinstall the transmission.

ENGINE RECONDITIONING

Determining Engine Condition

Anything that generates heat and/or friction will eventually burn or wear out (for example, a light bulb generates heat, therefore its life span is limited). With this in mind, a running engine generates tremendous amounts of both; friction is encountered by the moving and rotating parts inside the engine and heat is created by friction and combustion of the fuel. However, the engine has systems designed to help reduce the effects of heat and friction and provide added longevity. The oiling system reduces the amount of friction encountered by the moving parts inside the engine, while the cooling system reduces heat created by friction and combustion. If either system is not maintained, a break-down will be inevitable. Therefore, you can see how regular maintenance can affect the service life of your vehicle. If you do not drain, flush and refill your cooling system at the proper intervals, deposits will begin to accumulate in the radiator, thereby reducing the amount of heat it can extract from the coolant. The same applies to your oil and filter; if it is not changed often enough it becomes laden with contaminates and is unable to properly lubricate the engine. This increases friction and wear.

There are a number of methods for evaluating the condition of your engine. A compression test can reveal the condition of your pistons, piston rings, cylinder bores, head gasket(s), valves and valve seats. An oil pressure test can warn you of possible engine bearing, or oil pump failures. Excessive oil consumption, evidence of oil in the engine air intake area and/or bluish smoke from the tailpipe may indicate worn piston rings, worn valve guides and/or valve seals. As a general rule, an engine that uses no more than one quart of oil every 1000 miles is in good condition. Engines that use one quart of oil or more in less than 1000 miles should first be checked for oil leaks. If any oil leaks are present, have them fixed before determining how much oil is consumed by the engine, especially if blue smoke is not visible at the tailpipe.

COMPRESSION TEST

▶ **See Figure 245**

A noticeable lack of engine power, excessive oil consumption and/or poor fuel mileage measured over an extended period are all indicators of internal engine wear. Worn piston rings, scored or worn cylinder bores, blown head gaskets, sticking or burnt valves, and worn valve seats are all possible culprits. A check of each cylinder's compression will help locate the problem.

➡A screw-in type compression gauge is more accurate than the type you simply hold against the spark plug hole. Although it takes slightly longer to use, it's worth the effort to obtain a more accurate reading.

1. Make sure that the proper amount and viscosity of engine oil is in the crankcase, then ensure the battery is fully charged.
2. Warm-up the engine to normal operating temperature, then shut the engine **OFF**.
3. Disable the ignition system.
4. Label and disconnect all of the spark plug wires from the plugs.
5. Thoroughly clean the cylinder head area around the spark plug ports, then remove the spark plugs.
6. Set the throttle plate to the fully open (wide-open throttle) position. You can block the accelerator linkage open for this, or you can have an assistant fully depress the accelerator pedal.
7. Install a screw-in type compression gauge into the No. 1 spark plug hole until the fitting is snug.

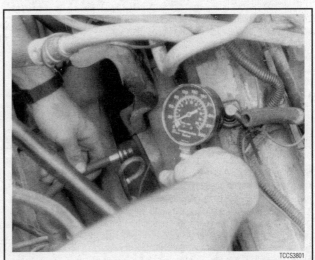

Fig. 245 A screw-in type compression gauge is more accurate and easier to use without an assistant

Be careful not to crossthread the spark plug hole.

8. According to the tool manufacturer's instructions, connect a remote starting switch to the starting circuit.

9. With the ignition switch in the **OFF** position, use the remote starting switch to crank the engine through at least five compression strokes (approximately 5 seconds of cranking) and record the highest reading on the gauge.

10. Repeat the test on each cylinder, cranking the engine approximately the same number of compression strokes and/or time as the first.

11. Compare the highest readings from each cylinder to that of the others. The indicated compression pressures are considered within specifications if the lowest reading cylinder is within 75 percent of the pressure recorded for the highest reading cylinder. For example, if your highest reading cylinder pressure was 150 psi (1034 kPa), then 75 percent of that would be 113 psi (779 kPa). So the lowest reading cylinder should be no less than 113 psi (779 kPa).

12. If a cylinder exhibits an unusually low compression reading, pour a tablespoon of clean engine oil into the cylinder through the spark plug hole and repeat the compression test. If the compression rises after adding oil, it means that the cylinder's piston rings and/or cylinder bore are damaged or worn. If the pressure remains low, the valves may not be seating properly (a valve job is needed), or the head gasket may be blown near that cylinder. If compression in any two adjacent cylinders is low, and if the addition of oil doesn't help raise compression, there is leakage past the head gasket. Oil and coolant in the combustion chamber, combined with blue or constant white smoke from the tailpipe, are symptoms of this problem. However, don't be alarmed by the normal white smoke emitted from the tailpipe during engine warm-up or from cold weather driving. There may be evidence of water droplets on the engine dipstick and/or oil droplets in the cooling system if a head gasket is blown.

OIL PRESSURE TEST

Check for proper oil pressure at the sending unit passage with an externally mounted mechanical oil pressure gauge (as opposed to relying on a factory installed dash-mounted gauge). A tachometer may also be needed, as some specifications may require running the engine at a specific rpm.

1. With the engine cold, locate and remove the oil pressure sending unit.

2. Following the manufacturer's instructions, connect a mechanical oil pressure gauge and, if necessary, a tachometer to the engine.

3. Start the engine and allow it to idle.

4. Check the oil pressure reading when cold and record the number. You may need to run the engine at a specified rpm, so check the specifications.

5. Run the engine until normal operating temperature is reached (upper radiator hose will feel warm).

6. Check the oil pressure reading again with the engine hot and record the number. Turn the engine **OFF**.

7. Compare your hot oil pressure reading to specification. If the reading is low, check the cold pressure reading against the chart. If the cold pressure is well above the specification, and the hot reading was lower than the specification, you may have the wrong viscosity oil in the engine. Change the oil, making sure to use the proper grade and quantity, then repeat the test.

Low oil pressure readings could be attributed to internal component wear, pump related problems, a low oil level, or oil viscosity that is too low. High oil pressure readings could be caused by an overfilled crankcase, too high of an oil viscosity or a faulty pressure relief valve.

Buy or Rebuild?

Now if you have determined that your engine is worn out, you must make some decisions. The question of whether or not an engine is worth rebuilding is largely a subjective matter and one of personal worth. Is the engine a popular one, or is it an obsolete model? Are parts available? Will it get acceptable gas mileage once it is rebuilt? Is the car it's being put into worth keeping? Would it be less expensive to buy a new engine, have your engine rebuilt by a pro, rebuild it yourself or buy a used engine from a salvage yard? Or would it be simpler and less expensive to buy another car? If you have considered all these matters, and have still decided to rebuild the engine, then it is time to decide how you will rebuild it.

➡The editors at Chilton feel that most engine machining should be performed by a professional machine shop. Think of it as an assurance that the job has been done right the first time. There are many expensive and specialized tools required to perform such tasks as boring and honing an engine block or having a valve job done on a cylinder head. Even inspecting the parts requires expensive micrometers and gauges to properly measure wear and clearances. A machine shop can deliver to you clean, and ready to assemble parts, saving you time and aggravation. Your maximum savings will come from performing the removal, disassembly, assembly and installation of the engine and purchasing or renting only the tools required to perform these tasks.

A complete rebuild or overhaul of an engine involves replacing all of the moving parts (pistons, rods, crankshaft, camshaft, etc.) with new ones and machining the non-moving wearing surfaces of the block and heads. Unfortunately, this may not be cost effective. For instance, your crankshaft may have been damaged or worn, but it can be machined undersize for a minimal fee.

So although you can replace everything inside the engine, it is usually wiser to replace only those parts which are really needed, and, if possible, repair the more expensive ones. Later in this section, we will break the engine down into its two main components: the cylinder head and the engine block. We will discuss each component, and the recommended parts to replace during a rebuild on each.

Engine Overhaul Tips

Most engine overhaul procedures are fairly standard. In addition to specific parts replacement procedures and specifications for your individual engine, this section is also a guide to acceptable rebuilding procedures. Examples of standard rebuilding practice are given and should be used along with specific details concerning your particular engine.

Competent and accurate machine shop services will ensure maximum performance, reliability and engine life. In most instances it is more profitable for the do-it-yourself mechanic to remove, clean and inspect the component, buy the necessary parts and deliver these to a shop for actual machine work.

Much of the assembly work (crankshaft, bearings, piston rods, and other components) is well within the scope of the do-it-yourself mechanic's tools and abilities. You will have to decide for yourself the depth of involvement you desire in an engine repair or rebuild.

TOOLS

The tools required for an engine overhaul or parts replacement will depend on the depth of your involvement. With a few exceptions, they will be the tools found in a mechanic's tool kit (see Section 1 of this manual). More in-depth work will require some or all of the following:

- A dial indicator (reading in thousandths) mounted on a universal base
- Micrometers and telescope gauges
- Jaw and screw-type pullers
- Scraper
- Valve spring compressor
- Ring groove cleaner
- Piston ring expander and compressor
- Ridge reamer
- Cylinder hone or glaze breaker
- Plastigage®
- Engine stand

The use of most of these tools is illustrated in this section. Many can be rented for a one-time use from a local parts jobber or tool supply house specializing in automotive work.

Occasionally, the use of special tools is called for. See the information on Special Tools and the Safety Notice in the front of this book before substituting another tool.

OVERHAUL TIPS

Aluminum has become extremely popular for use in engines, due to its low weight. Observe the following precautions when handling aluminum parts:

- Never hot tank aluminum parts (the caustic hot tank solution will eat the aluminum.)
- Remove all aluminum parts (identification tag, etc.) from engine parts prior to the tanking.
- Always coat threads lightly with engine oil or anti-seize compounds before installation, to prevent seizure.

• Never overtighten bolts or spark plugs especially in aluminum threads. When assembling the engine, any parts that will be exposed to frictional contact must be prelubed to provide lubrication at initial start-up. Any product specifically formulated for this purpose can be used, but engine oil is not recommended as a prelube in most cases.

When semi-permanent (locked, but removable) installation of bolts or nuts is desired, threads should be cleaned and coated with Loctite® or another similar, commercial non-hardening sealant.

CLEANING

▶ **See Figures 246, 247, 248 and 249**

Before the engine and its components are inspected, they must be thoroughly cleaned. You will need to remove any engine varnish, oil sludge and/or carbon deposits from all of the components to insure an accurate inspection. A crack in the engine block or cylinder head can easily become overlooked if hidden by a layer of sludge or carbon.

Most of the cleaning process can be carried out with common hand tools and readily available solvents or solutions. Carbon deposits can be chipped away using a hammer and a hard wooden chisel. Old gasket material and varnish or sludge can usually be removed using a scraper and/or cleaning solvent. Extremely stubborn deposits may require the use of a power drill with a wire brush. If using a wire brush, use extreme care around any critical machined surfaces (such as the gasket surfaces, bearing saddles, cylinder bores, etc.). USE OF A WIRE BRUSH IS NOT RECOMMENDED ON ANY ALUMINUM COMPONENTS. Always follow any safety recommendations given by the manufacturer of the tool and/or solvent.

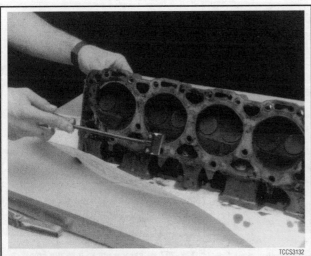

Fig. 246 Use a gasket scraper to remove the old gasket material from the mating surfaces

✳✳ CAUTION

Always wear eye protection during any cleaning process involving scraping, chipping or spraying of solvents.

An alternative to the mess and hassle of cleaning the parts yourself is to drop them off at a local garage or machine shop. They should have the necessary equipment to properly clean all of the parts for a nominal fee.

Remove any oil galley plugs, freeze plugs and/or pressed-in bearings and carefully wash and degrease all of the engine components including the fasteners and bolts. Small parts such as the valves, springs, etc., should be placed in a metal basket and allowed to soak. Use pipe cleaner type brushes, and clean all passageways in the components.

Use a ring expander and remove the rings from the pistons. Clean the piston ring grooves with a special tool or a piece of broken ring. Scrape the carbon off of the top of the piston. You should never use a wire brush on the pistons. After preparing all of the piston assemblies in this manner, wash and degrease them again.

✳✳ WARNING

Use extreme care when cleaning around the cylinder head valve seats. A mistake or slip may cost you a new seat.

When cleaning the cylinder head, remove carbon from the combustion chamber with the valves installed. This will avoid damaging the valve seats.

Engine Preparation

To properly rebuild an engine, you must first remove it from the vehicle, then disassemble and diagnose it. Ideally you should place your engine on an engine stand. This affords you the best access to the engine components. Remove the flywheel or flexplate before installing the engine to the stand.

Now that you have the engine on a stand, and assuming that you have drained the oil and coolant from the engine, it's time to strip it of all but the necessary components. Before you start disassembling the engine, you may want to take a moment to draw some pictures, or fabricate some labels or containers to mark the locations of various components and the bolts and/or studs which fasten them. Modern day engines use a lot of little brackets and clips which hold wiring harnesses and such, and these holders are often mounted on studs and/or bolts that can be easily mixed up. The manufacturer spent a lot of time and money designing your vehicle, and they wouldn't have wasted any of it by haphazardly placing brackets, clips or fasteners on the vehicle. If it's present when you disassemble it, put it back when you assemble, you will regret not remembering that little bracket which holds a wire harness out of the path of a rotating part.

You should begin by unbolting any accessories still attached to the engine, such as the water pump, power steering pump, alternator, etc. Then, unfasten any manifolds (intake or exhaust) which were not removed during the engine removal procedure. Finally, remove any covers remaining on the engine such as the rocker arm, front or timing cover and oil pan. Some front covers may require the vibration damper and/or crank pulley to be removed beforehand. The idea is to reduce the engine to the bare necessities of cylinder head(s), valve train, engine block, crankshaft, pistons and connecting rods, plus any other 'in block' components such as oil pumps, balance shafts and auxiliary shafts.

Fig. 247 Before cleaning and inspection, use a ring expander tool to remove the piston rings

Fig. 248 Clean the piston ring grooves using a ring groove cleaner tool, or . . .

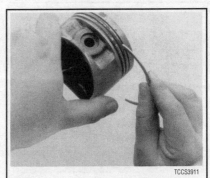

Fig. 249 . . . use a piece of an old ring to clean the grooves. Be careful, the ring can be quite sharp

Finally, remove the cylinder head(s) from the engine block and carefully place on a bench. Disassembly instructions for each component follow later in this section.

Cylinder Head

There are two basic types of cylinder heads used on today's automobiles: the Overhead Valve (OHV) and the Overhead Camshaft (OHC). The latter can also be broken down into two subgroups: the Single Overhead Camshaft (SOHC) and the Dual Overhead Camshaft (DOHC). Generally, if there is only a single camshaft on a head, it is just referred to as an OHC head. Also, an engine with an OHV cylinder head is also known as a pushrod engine.

Most cylinder heads these days are made of an aluminum alloy due to its light weight, durability and heat transfer qualities. However, cast iron was the material of choice in the past, and is still used on many vehicles. Whether made from aluminum or iron, all cylinder heads have valves and seats. Some use two valves per cylinder, while the more hi-tech engines will utilize a multi-valve configuration using 3, 4 and even 5 valves per cylinder. When the valve contacts the seat, it does so on precision machined surfaces, which seals the combustion chamber. All cylinder heads have a valve guide for each valve. The guide centers the valve to the seat and allows it to move up and down within it. The clearance between the valve and guide can be critical. Too much clearance and the engine may consume oil, lose vacuum and/or damage the seat. Too little, and the valve can stick in the guide causing the engine to run poorly if at all, and possibly causing severe damage. The last component all automotive cylinder heads have are valve springs. The spring holds the valve against its seat. It also returns the valve to this position when the valve has been opened by the valve train or camshaft. The spring is fastened to the valve by a retainer and valve locks (sometimes called keepers). Aluminum heads will also have a valve spring shim to keep the spring from wearing away the aluminum.

An ideal method of rebuilding the cylinder head would involve replacing all of the valves, guides, seats, springs, etc. with new ones. However, depending on how the engine was maintained, often this is not necessary. A major cause of valve, guide and seat wear is an improperly tuned engine. An engine that is running too rich, will often wash the lubricating oil out of the guide with gasoline, causing it to wear rapidly. Conversely, an engine which is running too lean will place higher combustion temperatures on the valves and seats allowing them to wear or even burn. Springs fall victim to the driving habits of the individual. A driver who often runs the engine rpm to the redline will wear out or break the springs faster then one that stays well below it. Unfortunately, mileage takes it toll on all of the parts. Generally, the valves, guides, springs and seats in a cylinder head can be machined and re-used, saving you money. However, if a valve is burnt, it may be wise to replace all of the valves, since they were all operating in the same environment. The same goes for any other component on the cylinder head. Think of it as an insurance policy against future problems related to that component.

Unfortunately, the only way to find out which components need replacing, is to disassemble and carefully check each piece. After the cylinder head(s) are disassembled, thoroughly clean all of the components.

DISASSEMBLY

OHV Heads

▶ See Figures 255 thru 260

Before disassembling the cylinder head, you may want to fabricate some containers to hold the various parts, as some of them can be quite small (such as keepers) and easily lost. Also keeping yourself and the components organized will aid in assembly and reduce confusion. Where possible, try to maintain a components original location; this is especially important if there is not going to be any machine work performed on the components.

1. If you haven't already removed the rocker arms and/or shafts, do so now.
2. Position the head so that the springs are easily accessed.
3. Use a valve spring compressor tool, and relieve spring tension from the retainer.

➡ **Due to engine varnish, the retainer may stick to the valve locks. A gentle tap with a hammer may help to break it loose.**

Fig. 255 When removing an OHV valve spring, use a compressor tool to relieve the tension from the retainer

Fig. 256 A small magnet will help in removal of the valve locks

Fig. 257 Be careful not to lose the small valve locks (keepers)

Fig. 258 Remove the valve seal from the valve stem—O-ring type seal shown

Fig. 259 Removing an umbrella/positive type seal

Fig. 260 Invert the cylinder head and withdraw the valve from the valve guide bore

4. Remove the valve locks from the valve tip and/or retainer. A small magnet may help in removing the locks.

5. Lift the valve spring, tool and all, off of the valve stem.

6. If equipped, remove the valve seal. If the seal is difficult to remove with the valve in place, try removing the valve first, then the seal. Follow the steps below for valve removal.

7. Position the head to allow access for withdrawing the valve.

➡Cylinder heads that have seen a lot of miles and/or abuse may have mushroomed the valve lock grove and/or tip, causing difficulty in removal of the valve. If this has happened, use a metal file to carefully remove the high spots around the lock grooves and/or tip. Only file it enough to allow removal.

8. Remove the valve from the cylinder head.

9. If equipped, remove the valve spring shim. A small magnetic tool or screwdriver will aid in removal.

10. Repeat Steps 3 though 9 until all of the valves have been removed.

OHC Heads

▶ **See Figures 261 and 262**

Whether it is a single or dual overhead camshaft cylinder head, the disassembly procedure is relatively unchanged. One aspect to pay attention to is careful labeling of the parts on the dual camshaft cylinder head. There will be an intake camshaft and followers as well as an exhaust camshaft and followers and they must be labeled as such. In some cases, the components are identical and could easily be installed incorrectly. DO NOT MIX THEM UP! Determining which is which is very simple; the intake camshaft and components are on the same side of the head as was the intake manifold. Conversely, the exhaust camshaft and components are on the same side of the head as was the exhaust manifold.

CUP TYPE CAMSHAFT FOLLOWERS

▶ **See Figures 263, 264 and 265**

Most cylinder heads with cup type camshaft followers will have the valve spring, retainer and locks recessed within the follower's bore. You will need a C-clamp style valve spring compressor tool, an OHC spring removal tool (or equivalent) and a small magnet to disassemble the head.

1. If not already removed, remove the camshaft(s) and/or followers. Mark their positions for assembly.

2. Position the cylinder head to allow use of a C-clamp style valve spring compressor tool.

➡It is preferred to position the cylinder head gasket surface facing you with the valve springs facing the opposite direction and the head laying horizontal.

3. With the OHC spring removal adapter tool positioned inside of the follower bore, compress the valve spring using the C-clamp style valve spring compressor.

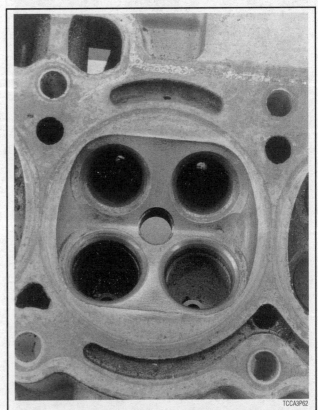

Fig. 262 Example of a multi-valve cylinder head. Note how it has 2 intake and 2 exhaust valve ports

Fig. 261 Exploded view of a valve, seal, spring, retainer and locks from an OHC cylinder head

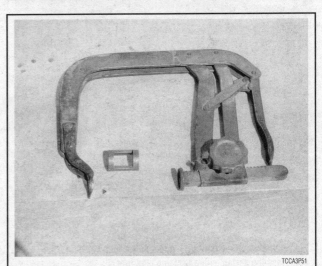

Fig. 263 C-clamp type spring compressor and an OHC spring removal tool (center) for cup type followers

4. Remove the valve locks. A small magnetic tool or screwdriver will aid in removal.

5. Release the compressor tool and remove the spring assembly.

6. Withdraw the valve from the cylinder head.

7. If equipped, remove the valve seal.

➡**Special valve seal removal tools are available. Regular or needle-nose type pliers, if used with care, will work just as well. If using ordinary pliers, be sure not to damage the follower bore. The follower and its bore are machined to close tolerances and any damage to the bore will effect this relationship.**

Fig. 264 Most cup type follower cylinder heads retain the camshaft using bolt-on bearing caps

Fig. 265 Position the OHC spring tool in the follower bore, then compress the spring with a C-clamp type tool

8. If equipped, remove the valve spring shim. A small magnetic tool or screwdriver will aid in removal.

9. Repeat Steps 3 through 8 until all of the valves have been removed.

ROCKER ARM TYPE CAMSHAFT FOLLOWERS

◆ See Figures 266 thru 274

Most cylinder heads with rocker arm-type camshaft followers are easily disassembled using a standard valve spring compressor. However, certain models may not have enough open space around the spring for the standard tool and may require you to use a C-clamp style compressor tool instead.

1. If not already removed, remove the rocker arms and/or shafts and the camshaft. If applicable, also remove the hydraulic lash adjusters. Mark their positions for assembly.

2. Position the cylinder head to allow access to the valve spring.

3. Use a valve spring compressor tool to relieve the spring tension from the retainer.

➡**Due to engine varnish, the retainer may stick to the valve locks. A gentle tap with a hammer may help to break it loose.**

4. Remove the valve locks from the valve tip and/or retainer. A small magnet may help in removing the small locks.

5. Lift the valve spring, tool and all, off of the valve stem.

6. If equipped, remove the valve seal. If the seal is difficult to remove with the valve in place, try removing the valve first, then the seal. Follow the steps below for valve removal.

7. Position the head to allow access for withdrawing the valve.

➡**Cylinder heads that have seen a lot of miles and/or abuse may have mushroomed the valve lock grove and/or tip, causing difficulty in removal of the valve. If this has happened, use a metal file to carefully remove the high spots around the lock grooves and/or tip. Only file it enough to allow removal.**

8. Remove the valve from the cylinder head.

9. If equipped, remove the valve spring shim. A small magnetic tool or screwdriver will aid in removal.

10. Repeat Steps 3 though 9 until all of the valves have been removed.

INSPECTION

Now that all of the cylinder head components are clean, it's time to inspect them for wear and/or damage. To accurately inspect them, you will need some specialized tools:

• A 0–1 in. micrometer for the valves
• A dial indicator or inside diameter gauge for the valve guides
• A spring pressure test gauge

If you do not have access to the proper tools, you may want to bring the components to a shop that does.

Valves

◆ See Figures 275 and 276

The first thing to inspect are the valve heads. Look closely at the head, margin and face for any cracks, excessive wear or burning. The margin is the best place to look for burning. It should have a squared edge with an even width all around the diameter. When a valve burns, the margin will look melted and the edges rounded. Also inspect the valve head for any signs of tulipping. This will show as a lifting of the edges or dishing in the center of the head and will usually not occur to all of the valves. All of the heads should look the same, any that seem dished more than others are probably bad. Next, inspect the valve lock grooves and valve tips. Check for any burrs around the lock grooves, especially if you had to file them to remove the valve. Valve tips should appear flat, although slight rounding with high mileage engines is normal. Slightly worn valve tips will need to be machined flat. Last, measure the valve stem diameter with the micrometer. Measure the area that rides within the guide, especially towards the tip where most of the wear occurs. Take several measurements along its length and compare them to each other. Wear should be even along the length with little to no taper. If no minimum diameter is given in the specifications, then the stem should not read more than 0.001 in. (0.025mm) below the unworn portion of the stem. Any valves that fail these inspections should be replaced.

Fig. 266 Example of the shaft mounted rocker arms on some OHC heads

Fig. 267 Another example of the rocker arm type OHC head. This model uses a follower under the camshaft

Fig. 268 Before the camshaft can be removed, all of the followers must first be removed . . .

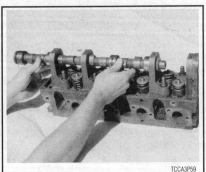

Fig. 269 . . . then the camshaft can be removed by sliding it out (shown), or unbolting a bearing cap (not shown)

Fig. 270 Compress the valve spring . . .

Fig. 271 . . . then remove the valve locks from the valve stem and spring retainer

Fig. 272 Remove the valve spring and retainer from the cylinder head

Fig. 273 Remove the valve seal from the guide. Some gentle prying or pliers may help to remove stubborn ones

Fig. 274 All aluminum and some cast iron heads will have these valve spring shims. Remove all of them as well

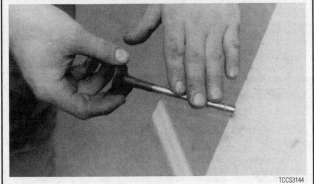

Fig. 275 Valve stems may be rolled on a flat surface to check for bends

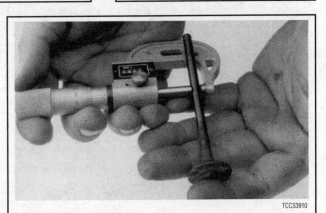

Fig. 276 Use a micrometer to check the valve stem diameter

Springs, Retainers and Valve Locks

▶ **See Figures 277 and 278**

The first thing to check is the most obvious, broken springs. Next check the free length and squareness of each spring. If applicable, insure to distinguish between intake and exhaust springs. Use a ruler and/or carpenter's square to measure the length. A carpenter's square should be used to check the springs for squareness. If a spring pressure test gauge is available, check each springs rating and compare to the specifications chart. Check the readings against the specifications given. Any springs that fail these inspections should be replaced.

The spring retainers rarely need replacing, however they should still be checked as a precaution. Inspect the spring mating surface and the valve lock retention area for any signs of excessive wear. Also check for any signs of cracking. Replace any retainers that are questionable.

Valve locks should be inspected for excessive wear on the outside contact area as well as on the inner notched surface. Any locks which appear worn or broken and its respective valve should be replaced.

Cylinder Head

There are several things to check on the cylinder head: valve guides, seats, cylinder head surface flatness, cracks and physical damage.

VALVE GUIDES

▶ **See Figure 279**

Now that you know the valves are good, you can use them to check the guides, although a new valve, if available, is preferred. Before you measure anything, look at the guides carefully and inspect them for any cracks, chips or

breakage. Also if the guide is a removable style (as in most aluminum heads), check them for any looseness or evidence of movement. All of the guides should appear to be at the same height from the spring seat. If any seem lower (or higher) from another, the guide has moved. Mount a dial indicator onto the spring side of the cylinder head. Lightly oil the valve stem and insert it into the cylinder head. Position the dial indicator against the valve stem near the tip and zero the gauge. Grasp the valve stem and wiggle towards and away from the dial indicator and observe the readings. Mount the dial indicator 90 degrees from the initial point and zero the gauge and again take a reading. Compare the two readings for an out of round condition. Check the readings against the specifications given. An Inside Diameter (I.D.) gauge designed for valve guides will give you an accurate valve guide bore measurement. If the I.D. gauge is used, compare the readings with the specifications given. Any guides that fail these inspections should be replaced or machined.

VALVE SEATS

A visual inspection of the valve seats should show a slightly worn and pitted surface where the valve face contacts the seat. Inspect the seat carefully for severe pitting or cracks. Also, a seat that is badly worn will be recessed into the cylinder head. A severely worn or recessed seat may need to be replaced. All cracked seats must be replaced. A seat concentricity gauge, if available, should be used to check the seat run-out. If run-out exceeds specifications the seat must be machined (if no specification is available given use 0.002 in. or 0.051mm).

CYLINDER HEAD SURFACE FLATNESS

▶ **See Figures 280 and 281**

After you have cleaned the gasket surface of the cylinder head of any old gasket material, check the head for flatness.

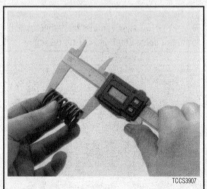

TCCS3907

Fig. 277 Use a caliper to check the valve spring free-length

TCCS3908

Fig. 278 Check the valve spring for squareness on a flat surface; a carpenter's square can be used

TCCS3142

Fig. 279 A dial gauge may be used to check valve stem-to-guide clearance; read the gauge while moving the valve stem

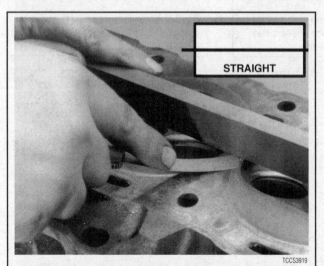

TCCS3919

Fig. 280 Check the head for flatness across the center of the head surface using a straightedge and feeler gauge

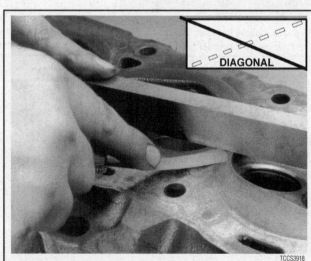

TCCS3918

Fig. 281 Checks should also be made along both diagonals of the head surface

Place a straightedge across the gasket surface. Using feeler gauges, determine the clearance at the center of the straightedge and across the cylinder head at several points. Check along the centerline and diagonally on the head surface. If the warpage exceeds 0.003 in. (0.076mm) within a 6.0 in. (15.2cm) span, or 0.006 in. (0.152mm) over the total length of the head, the cylinder head must be resurfaced. After resurfacing the heads of a V-type engine, the intake manifold flange surface should be checked, and if necessary, milled proportionally to allow for the change in its mounting position.

CRACKS AND PHYSICAL DAMAGE

Generally, cracks are limited to the combustion chamber, however, it is not uncommon for the head to crack in a spark plug hole, port, outside of the head or in the valve spring/rocker arm area. The first area to inspect is always the hottest: the exhaust seat/port area.

A visual inspection should be performed, but just because you don't see a crack does not mean it is not there. Some more reliable methods for inspecting for cracks include Magnaflux®, a magnetic process or Zyglo®, a dye penetrant. Magnaflux® is used only on ferrous metal (cast iron) heads. Zyglo® uses a spray on fluorescent mixture along with a black light to reveal the cracks. It is strongly recommended to have your cylinder head checked professionally for cracks, especially if the engine was known to have overheated and/or leaked or consumed coolant. Contact a local shop for availability and pricing of these services.

Physical damage is usually very evident. For example, a broken mounting ear from dropping the head or a bent or broken stud and/or bolt. All of these defects should be fixed or, if unrepairable, the head should be replaced.

Camshaft and Followers

Inspect the camshaft(s) and followers as described earlier in this section.

REFINISHING & REPAIRING

Many of the procedures given for refinishing and repairing the cylinder head components must be performed by a machine shop. Certain steps, if the inspected part is not worn, can be performed yourself inexpensively. However, you spent a lot of time and effort so far, why risk trying to save a couple bucks if you might have to do it all over again?

Valves

Any valves that were not replaced should be refaced and the tips ground flat. Unless you have access to a valve grinding machine, this should be done by a machine shop. If the valves are in extremely good condition, as well as the valve seats and guides, they may be lapped in without performing machine work.

It is a recommended practice to lap the valves even after machine work has been performed and/or new valves have been purchased. This insures a positive seal between the valve and seat.

LAPPING THE VALVES

➡ **Before lapping the valves to the seats, read the rest of the cylinder head section to insure that any related parts are in acceptable enough condition to continue. Also, remember that before any valve seat machining and/or lapping can be performed, the guides must be within factory recommended specifications.**

1. Invert the cylinder head.
2. Lightly lubricate the valve stems and insert them into the cylinder head in their numbered order.
3. Raise the valve from the seat and apply a small amount of fine lapping compound to the seat.
4. Moisten the suction head of a hand-lapping tool and attach it to the head of the valve.
5. Rotate the tool between the palms of both hands, changing the position of the valve on the valve seat and lifting the tool often to prevent grooving.
6. Lap the valve until a smooth, polished circle is evident on the valve and seat.
7. Remove the tool and the valve. Wipe away all traces of the grinding compound and store the valve to maintain its lapped location.

✳✳ WARNING

Do not get the valves out of order after they have been lapped. They must be put back with the same valve seat with which they were lapped.

Springs, Retainers and Valve Locks

There is no repair or refinishing possible with the springs, retainers and valve locks. If they are found to be worn or defective, they must be replaced with new (or known good) parts.

Cylinder Head

Most refinishing procedures dealing with the cylinder head must be performed by a machine shop. Read the sections below and review your inspection data to determine whether or not machining is necessary.

VALVE GUIDE

➡ **If any machining or replacements are made to the valve guides, the seats must be machined.**

Unless the valve guides need machining or replacing, the only service to perform is to thoroughly clean them of any dirt or oil residue.

There are only two types of valve guides used on automobile engines: the replaceable-type (all aluminum heads) and the cast-in integral-type (most cast iron heads). There are four recommended methods for repairing worn guides.

- Knurling
- Inserts
- Reaming oversize
- Replacing

Knurling is a process in which metal is displaced and raised, thereby reducing clearance, giving a true center, and providing oil control. It is the least expensive way of repairing the valve guides. However, it is not necessarily the best, and in some cases, a knurled valve guide will not stand up for more than a short time. It requires a special knurlizer and precision reaming tools to obtain proper clearances. It would not be cost effective to purchase these tools, unless you plan on rebuilding several of the same cylinder head.

Installing a guide insert involves machining the guide to accept a bronze insert. One style is the coil-type which is installed into a threaded guide. Another is the thin-walled insert where the guide is reamed oversize to accept a split-sleeve insert. After the insert is installed, a special tool is then run through the guide to expand the insert, locking it to the guide. The insert is then reamed to the standard size for proper valve clearance.

Reaming for oversize valves restores normal clearances and provides a true valve seat. Most cast-in type guides can be reamed to accept an valve with an oversize stem. The cost factor for this can become quite high as you will need to purchase the reamer and new, oversize stem valves for all guides which were reamed. Oversizes are generally 0.003–0.030 in. (0.076–0.762mm), with 0.015 in. (0.381mm) being the most common.

To replace cast-in type valve guides, they must be drilled out, then reamed to accept replacement guides. This must be done on a fixture which will allow centering and leveling off of the original valve seat or guide, otherwise a serious guide-to-seat misalignment may occur making it impossible to properly machine the seat.

Replaceable-type guides are pressed into the cylinder head. A hammer and a stepped drift or punch may be used to install and remove the guides. Before removing the guides, measure the protrusion on the spring side of the head and record it for installation. Use the stepped drift to hammer out the old guide from the combustion chamber side of the head. When installing, determine whether or not the guide also seals a water jacket in the head, and if it does, use the recommended sealing agent. If there is no water jacket, grease the valve guide and its bore. Use the stepped drift, and hammer the new guide into the cylinder head from the spring side of the cylinder head. A stack of washers the same thickness as the measured protrusion may help the installation process.

VALVE SEATS

➡ **Before any valve seat machining can be performed, the guides must be within factory recommended specifications. If any machining occurred or if replacements were made to the valve guides, the seats must be machined.**

If the seats are in good condition, the valves can be lapped to the seats, and the cylinder head assembled. See the valves section for instructions on lapping.

If the valve seats are worn, cracked or damaged, they must be serviced by a machine shop. The valve seat must be perfectly centered to the valve guide, which requires very accurate machining.

CYLINDER HEAD SURFACE

If the cylinder head is warped, it must be machined flat. If the warpage is extremely severe, the head may need to be replaced. In some instances, it may be possible to straighten a warped head enough to allow machining. In either case, contact a professional machine shop for service.

➥Any OHC cylinder head that shows excessive warpage should have the camshaft bearing journals align bored after the cylinder head has been resurfaced.

✳✳ WARNING

Failure to align bore the camshaft bearing journals could result in severe engine damage including but not limited to: valve and piston damage, connecting rod damage, camshaft and/or crankshaft breakage.

CRACKS AND PHYSICAL DAMAGE

Certain cracks can be repaired in both cast iron and aluminum heads. For cast iron, a tapered threaded insert is installed along the length of the crack. Aluminum can also use the tapered inserts, however welding is the preferred method. Some physical damage can be repaired through brazing or welding. Contact a machine shop to get expert advice for your particular dilemma.

ASSEMBLY

The first step for any assembly job is to have a clean area in which to work. Next, thoroughly clean all of the parts and components that are to be assembled. Finally, place all of the components onto a suitable work space and, if necessary, arrange the parts to their respective positions.

OHV Engines

1. Lightly lubricate the valve stems and insert all of the valves into the cylinder head. If possible, maintain their original locations.
2. If equipped, install any valve spring shims which were removed.
3. If equipped, install the new valve seals, keeping the following in mind:
   • If the valve seal presses over the guide, lightly lubricate the outer guide surfaces.
   • If the seal is an O-ring type, it is installed just after compressing the spring but before the valve locks.
4. Place the valve spring and retainer over the stem.
5. Position the spring compressor tool and compress the spring.
6. Assemble the valve locks to the stem.
7. Relieve the spring pressure slowly and insure that neither valve lock becomes dislodged by the retainer.
8. Remove the spring compressor tool.
9. Repeat Steps 2 through 8 until all of the springs have been installed.

OHC Engines

▶ See Figure 282

CUP TYPE CAMSHAFT FOLLOWERS

To install the springs, retainers and valve locks on heads which have these components recessed into the camshaft follower's bore, you will need a small screwdriver-type tool, some clean white grease and a lot of patience. You will also need the C-clamp style spring compressor and the OHC tool used to disassemble the head.

1. Lightly lubricate the valve stems and insert all of the valves into the cylinder head. If possible, maintain their original locations.
2. If equipped, install any valve spring shims which were removed.
3. If equipped, install the new valve seals, keeping the following in mind:
   • If the valve seal presses over the guide, lightly lubricate the outer guide surfaces.

Fig. 282 Once assembled, check the valve clearance and correct as needed

   • If the seal is an O-ring type, it is installed just after compressing the spring but before the valve locks.
4. Place the valve spring and retainer over the stem.
5. Position the spring compressor and the OHC tool, then compress the spring.
6. Using a small screwdriver as a spatula, fill the valve stem side of the lock with white grease. Use the excess grease on the screwdriver to fasten the lock to the driver.
7. Carefully install the valve lock, which is stuck to the end of the screwdriver, to the valve stem then press on it with the screwdriver until the grease squeezes out. The valve lock should now be stuck to the stem.
8. Repeat Steps 6 and 7 for the remaining valve lock.
9. Relieve the spring pressure slowly and insure that neither valve lock becomes dislodged by the retainer.
10. Remove the spring compressor tool.
11. Repeat Steps 2 through 10 until all of the springs have been installed.
12. Install the followers, camshaft(s) and any other components that were removed for disassembly.

ROCKER ARM TYPE CAMSHAFT FOLLOWERS

1. Lightly lubricate the valve stems and insert all of the valves into the cylinder head. If possible, maintain their original locations.
2. If equipped, install any valve spring shims which were removed.
3. If equipped, install the new valve seals, keeping the following in mind:
   • If the valve seal presses over the guide, lightly lubricate the outer guide surfaces.
   • If the seal is an O-ring type, it is installed just after compressing the spring but before the valve locks.
4. Place the valve spring and retainer over the stem.
5. Position the spring compressor tool and compress the spring.
6. Assemble the valve locks to the stem.
7. Relieve the spring pressure slowly and insure that neither valve lock becomes dislodged by the retainer.
8. Remove the spring compressor tool.
9. Repeat Steps 2 through 8 until all of the springs have been installed.
10. Install the camshaft(s), rockers, shafts and any other components that were removed for disassembly.

Engine Block

GENERAL INFORMATION

A thorough overhaul or rebuild of an engine block would include replacing the pistons, rings, bearings, timing belt/chain assembly and oil pump. For OHV engines also include a new camshaft and lifters. The block would then have the cylinders bored and honed oversize (or if using removable cylinder sleeves, new sleeves installed) and the crankshaft would be cut undersize to provide new wearing surfaces and perfect clearances. However, your particular engine may not have everything worn out. What if only the piston rings have worn out and the clearances on everything else are still within factory specifications? Well, you could just replace the rings and put it back together, but this would be a very rare example. Chances are, if one component in your engine is worn, other components are sure to follow, and soon. At the very least, you should always replace the rings, bearings and oil pump. This is what is commonly called a "freshen up".

Cylinder Ridge Removal

Because the top piston ring does not travel to the very top of the cylinder, a ridge is built up between the end of the travel and the top of the cylinder bore.

Pushing the piston and connecting rod assembly past the ridge can be difficult, and damage to the piston ring lands could occur. If the ridge is not removed before installing a new piston or not removed at all, piston ring breakage and piston damage may occur.

➡ It is always recommended that you remove any cylinder ridges before removing the piston and connecting rod assemblies. If you know that new pistons are going to be installed and the engine block will be bored oversize, you may be able to forego this step. However, some ridges may actually prevent the assemblies from being removed, necessitating its removal.

There are several different types of ridge reamers on the market, none of which are inexpensive. Unless a great deal of engine rebuilding is anticipated, borrow or rent a reamer.

1. Turn the crankshaft until the piston is at the bottom of its travel.
2. Cover the head of the piston with a rag.
3. Follow the tool manufacturers instructions and cut away the ridge, exercising extreme care to avoid cutting too deeply.
4. Remove the ridge reamer, the rag and as many of the cuttings as possible. Continue until all of the cylinder ridges have been removed.

DISASSEMBLY

♦ See Figures 283 and 284

The engine disassembly instructions following assume that you have the engine mounted on an engine stand. If not, it is easiest to disassemble the engine on a bench or the floor with it resting on the bell housing or transmission mounting surface. You must be able to access the connecting rod fasteners and turn the crankshaft during disassembly. Also, all engine covers (timing, front, side, oil pan, whatever) should have already been removed. Engines which are seized or locked up may not be able to be completely disassembled, and a core (salvage yard) engine should be purchased.

Pushrod Engines

If not done during the cylinder head removal, remove the pushrods and lifters, keeping them in order for assembly. Remove the timing gears and/or timing chain assembly, then remove the oil pump drive assembly and withdraw the camshaft from the engine block. Remove the oil pick-up and pump assembly. If equipped, remove any balance or auxiliary shafts. If necessary, remove the cylinder ridge from the top of the bore. See the cylinder ridge removal procedure earlier in this section.

OHC Engines

If not done during the cylinder head removal, remove the timing chain/belt and/or gear/sprocket assembly. Remove the oil pick-up and pump assembly and, if necessary, the pump drive. If equipped, remove any balance or auxiliary shafts. If necessary, remove the cylinder ridge from the top of the bore. See the cylinder ridge removal procedure earlier in this section.

All Engines

Rotate the engine over so that the crankshaft is exposed. Use a number punch or scribe and mark each connecting rod with its respective cylinder number. The cylinder closest to the front of the engine is always number 1. However, depending on the engine placement, the front of the engine could either be the flywheel or damper/pulley end. Generally the front of the engine faces the front of the vehicle. Use a number punch or scribe and also mark the main bearing caps from front to rear with the front most cap being number 1 (if there are five caps, mark them 1 through 5, front to rear).

⁑ **WARNING**

Take special care when pushing the connecting rod up from the crankshaft because the sharp threads of the rod bolts/studs will score the crankshaft journal. Insure that special plastic caps are installed over them, or cut two pieces of rubber hose to do the same.

Fig. 283 Place rubber hose over the connecting rod studs to protect the crankshaft and cylinder bores from damage

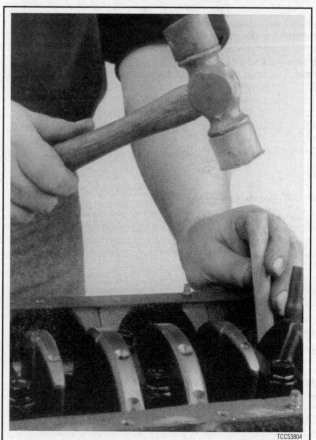

Fig. 284 Carefully tap the piston out of the bore using a wooden dowel

Again, rotate the engine, this time to position the number one cylinder bore (head surface) up. Turn the crankshaft until the number one piston is at the bottom of its travel, this should allow the maximum access to its connecting rod. Remove the number one connecting rods fasteners and cap and place two lengths of rubber hose over the rod bolts/studs to protect the crankshaft from damage. Using a sturdy wooden dowel and a hammer, push the connecting rod up about 1 in. (25mm) from the crankshaft and remove the upper bearing insert. Continue pushing or tapping the connecting rod up until the piston rings are out of the cylinder bore. Remove the piston and rod by hand, put the upper half of the bearing insert back into the rod, install the cap with its bearing insert installed, and hand-tighten the cap fasteners. If the parts are kept in order in this manner, they will not get lost and you will be able to tell which bearings came form what cylinder if any problems are discovered and diagnosis is necessary. Remove all the other piston assemblies in the same manner. On V-style engines, remove all of the pistons from one bank, then reposition the engine with the other cylinder bank head surface up, and remove that banks piston assemblies.

The only remaining component in the engine block should now be the crankshaft. Loosen the main bearing caps evenly until the fasteners can be turned by hand, then remove them and the caps. Remove the crankshaft from the engine block. Thoroughly clean all of the components.

INSPECTION

Now that the engine block and all of its components are clean, it's time to inspect them for wear and/or damage. To accurately inspect them, you will need some specialized tools:

- Two or three separate micrometers to measure the pistons and crankshaft journals
- A dial indicator
- Telescoping gauges for the cylinder bores
- A rod alignment fixture to check for bent connecting rods

If you do not have access to the proper tools, you may want to bring the components to a shop that does.

Generally, you shouldn't expect cracks in the engine block or its components unless it was known to leak, consume or mix engine fluids, it was severely overheated, or there was evidence of bad bearings and/or crankshaft damage. A visual inspection should be performed on all of the components, but just because you don't see a crack does not mean it is not there. Some more reliable methods for inspecting for cracks include Magnaflux®, a magnetic process or Zyglo®, a dye penetrant. Magnaflux® is used only on ferrous metal (cast iron). Zyglo® uses a spray on fluorescent mixture along with a black light to reveal the cracks. It is strongly recommended to have your engine block checked professionally for cracks, especially if the engine was known to have overheated and/or leaked or consumed coolant. Contact a local shop for availability and pricing of these services.

Engine Block

ENGINE BLOCK BEARING ALIGNMENT

Remove the main bearing caps and, if still installed, the main bearing inserts. Inspect all of the main bearing saddles and caps for damage, burrs or high spots. If damage is found, and it is caused from a spun main bearing, the block will need to be align-bored or, if severe enough, replacement. Any burrs or high spots should be carefully removed with a metal file.

Place a straightedge on the bearing saddles, in the engine block, along the centerline of the crankshaft. If any clearance exists between the straightedge and the saddles, the block must be align-bored.

Align-boring consists of machining the main bearing saddles and caps by means of a flycutter that runs through the bearing saddles.

DECK FLATNESS

The top of the engine block where the cylinder head mounts is called the deck. Insure that the deck surface is clean of dirt, carbon deposits and old gasket material. Place a straightedge across the surface of the deck along its centerline and, using feeler gauges, check the clearance along several points. Repeat the checking procedure with the straightedge placed along both diagonals of the deck surface. If the reading exceeds 0.003 in. (0.076mm) within a 6.0 in. (15.2cm) span, or 0.006 in. (0.152mm) over the total length of the deck, it must be machined.

CYLINDER BORES

♦ See Figure 285

The cylinder bores house the pistons and are slightly larger than the pistons themselves. A common piston-to-bore clearance is 0.0015–0.0025 in. (0.0381mm–0.0635mm). Inspect and measure the cylinder bores. The bore should be checked for out-of-roundness, taper and size. The results of this inspection will determine whether the cylinder can be used in its existing size and condition, or a rebore to the next oversize is required (or in the case of removable sleeves, have replacements installed).

The amount of cylinder wall wear is always greater at the top of the cylinder than at the bottom. This wear is known as taper. Any cylinder that has a taper of 0.0012 in. (0.305mm) or more, must be rebored. Measurements are taken at a number of positions in each cylinder: at the top, middle and bottom and at two points at each position; that is, at a point 90 degrees from the crankshaft centerline, as well as a point parallel to the crankshaft centerline. The measurements are made with either a special dial indicator or a telescopic gauge and micrometer. If the necessary precision tools to check the bore are not available, take the block to a machine shop and have them mike it. Also if you don't have the tools to check the cylinder bores, chances are you will not have the necessary devices to check the pistons, connecting rods and crankshaft. Take these components with you and save yourself an extra trip.

For our procedures, we will use a telescopic gauge and a micrometer. You will need one of each, with a measuring range which covers your cylinder bore size.

1. Position the telescopic gauge in the cylinder bore, loosen the gauges lock and allow it to expand.

➡Your first two readings will be at the top of the cylinder bore, then proceed to the middle and finally the bottom, making a total of six measurements.

2. Hold the gauge square in the bore, 90 degrees from the crankshaft centerline, and gently tighten the lock. Tilt the gauge back to remove it from the bore.
3. Measure the gauge with the micrometer and record the reading.
4. Again, hold the gauge square in the bore, this time parallel to the crankshaft centerline, and gently tighten the lock. Again, you will tilt the gauge back to remove it from the bore.
5. Measure the gauge with the micrometer and record this reading. The difference between these two readings is the out-of-round measurement of the cylinder.
6. Repeat steps 1 through 5, each time going to the next lower position, until you reach the bottom of the cylinder. Then go to the next cylinder, and continue until all of the cylinders have been measured.

The difference between these measurements will tell you all about the wear in your cylinders. The measurements which were taken 90 degrees from the crankshaft centerline will always reflect the most wear. That is because at this position

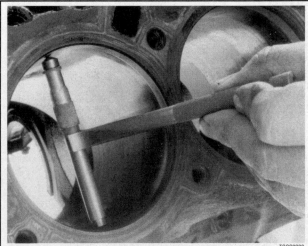

TCCS3209

Fig. 285 Use a telescoping gauge to measure the cylinder bore diameter—take several readings within the same bore

is where the engine power presses the piston against the cylinder bore the hardest. This is known as thrust wear. Take your top, 90 degree measurement and compare it to your bottom, 90 degree measurement. The difference between them is the taper. When you measure your pistons, you will compare these readings to your piston sizes and determine piston-to-wall clearance.

Crankshaft

Inspect the crankshaft for visible signs of wear or damage. All of the journals should be perfectly round and smooth. Slight scores are normal for a used crankshaft, but you should hardly feel them with your fingernail. When measuring the crankshaft with a micrometer, you will take readings at the front and rear of each journal, then turn the micrometer 90 degrees and take two more readings, front and rear. The difference between the front-to-rear readings is the journal taper and the first-to-90 degree reading is the out-of-round measurement. Generally, there should be no taper or out-of-roundness found, however, up to 0.0005 in. (0.0127mm) for either can be overlooked. Also, the readings should fall within the factory specifications for journal diameters.

If the crankshaft journals fall within specifications, it is recommended that it be polished before being returned to service. Polishing the crankshaft insures that any minor burrs or high spots are smoothed, thereby reducing the chance of scoring the new bearings.

Pistons and Connecting Rods

PISTONS

▶ See Figure 286

The piston should be visually inspected for any signs of cracking or burning (caused by hot spots or detonation), and scuffing or excessive wear on the skirts. The wrist pin attaches the piston to the connecting rod. The piston should move freely on the wrist pin, both sliding and pivoting. Grasp the connecting rod securely, or mount it in a vise, and try to rock the piston back and forth along the centerline of the wrist pin. There should not be any excessive play evident between the piston and the pin. If there are C-clips retaining the pin in the piston then you have wrist pin bushings in the rods. There should not be any excessive play between the wrist pin and the rod bushing. Normal clearance for the wrist pin is approx. 0.001–0.002 in. (0.025mm–0.051mm).

Use a micrometer and measure the diameter of the piston, perpendicular to the wrist pin, on the skirt. Compare the reading to its original cylinder measurement obtained earlier. The difference between the two readings is the piston-to-wall clearance. If the clearance is within specifications, the piston may be used as is. If the piston is out of specification, but the bore is not, you will need a new piston. If both are out of specification, you will need the cylinder rebored and oversize pistons installed. Generally if two or more pistons/bores are out of specification, it is best to rebore the entire block and purchase a complete set of oversize pistons.

CONNECTING ROD

You should have the connecting rod checked for straightness at a machine shop. If the connecting rod is bent, it will unevenly wear the bearing and piston, as well as place greater stress on these components. Any bent or twisted connecting rods must be replaced. If the rods are straight and the wrist pin clearance is within specifications, then only the bearing end of the rod need be checked. Place the connect-

ing rod into a vice, with the bearing inserts in place, install the cap to the rod and torque the fasteners to specifications. Use a telescoping gauge and carefully measure the inside diameter of the bearings. Compare this reading to the rods original crankshaft journal diameter measurement. The difference is the oil clearance. If the oil clearance is not within specifications, install new bearings in the rod and take another measurement. If the clearance is still out of specifications, and the crankshaft is not, the rod will need to be reconditioned by a machine shop.

➡You can also use Plastigage® to check the bearing clearances. The assembling section has complete instructions on its use.

Camshaft

Inspect the camshaft and lifters/followers as described earlier in this section.

Bearings

All of the engine bearings should be visually inspected for wear and/or damage. The bearing should look evenly worn all around with no deep scores or pits. If the bearing is severely worn, scored, pitted or heat blued, then the bearing, and the components that use it, should be brought to a machine shop for inspection. Full-circle bearings (used on most camshafts, auxiliary shafts, balance shafts, etc.) require specialized tools for removal and installation, and should be brought to a machine shop for service.

Oil Pump

➡The oil pump is responsible for providing constant lubrication to the whole engine and so it is recommended that a new oil pump be installed when rebuilding the engine.

Completely disassemble the oil pump and thoroughly clean all of the components. Inspect the oil pump gears and housing for wear and/or damage. Insure that the pressure relief valve operates properly and there is no binding or sticking due to varnish or debris. If all of the parts are in proper working condition, lubricate the gears and relief valve, and assemble the pump.

REFINISHING

▶ See Figure 287

Almost all engine block refinishing must be performed by a machine shop. If the cylinders are not to be rebored, then the cylinder glaze can be removed with a ball hone. When removing cylinder glaze with a ball hone, use a light or penetrating type oil to lubricate the hone. Do not allow the hone to run dry as this may cause excessive scoring of the cylinder bores and wear on the hone. If new pistons are required, they will need to be installed to the connecting rods. This should be performed by a machine shop as the pistons must be installed in the correct relationship to the rod or engine damage can occur.

Pistons and Connecting Rods

▶ See Figure 288

Only pistons with the wrist pin retained by C-clips are serviceable by the home-mechanic. Press fit pistons require special presses and/or heaters to

Fig. 286 Measure the piston's outer diameter, perpendicular to the wrist pin, with a micrometer

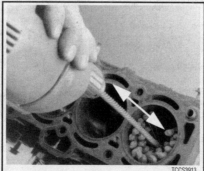

Fig. 287 Use a ball type cylinder hone to remove any glaze and provide a new surface for seating the piston rings

Fig. 288 Most pistons are marked to indicate positioning in the engine (usually a mark means the side facing the front)

remove/install the connecting rod and should only be performed by a machine shop.

All pistons will have a mark indicating the direction to the front of the engine and the must be installed into the engine in that manner. Usually it is a notch or arrow on the top of the piston, or it may be the letter F cast or stamped into the piston.

C-CLIP TYPE PISTONS

1. Note the location of the forward mark on the piston and mark the connecting rod in relation.
2. Remove the C-clips from the piston and withdraw the wrist pin.

➡**Varnish build-up or C-clip groove burrs may increase the difficulty of removing the wrist pin. If necessary, use a punch or drift to carefully tap the wrist pin out.**

3. Insure that the wrist pin bushing in the connecting rod is usable, and lubricate it with assembly lube.
4. Remove the wrist pin from the new piston and lubricate the pin bores on the piston.
5. Align the forward marks on the piston and the connecting rod and install the wrist pin.
6. The new C-clips will have a flat and a rounded side to them. Install both C-clips with the flat side facing out.
7. Repeat all of the steps for each piston being replaced.

ASSEMBLY

Before you begin assembling the engine, first give yourself a clean, dirt free work area. Next, clean every engine component again. The key to a good assembly is cleanliness.

Mount the engine block into the engine stand and wash it one last time using water and detergent (dishwashing detergent works well). While washing it, scrub the cylinder bores with a soft bristle brush and thoroughly clean all of the oil passages. Completely dry the engine and spray the entire assembly down with an anti-rust solution such as WD-40® or similar product. Take a clean lint-free rag and wipe up any excess anti-rust solution from the bores, bearing saddles, etc. Repeat the final cleaning process on the crankshaft. Replace any freeze or oil galley plugs which were removed during disassembly.

Crankshaft

▶ **See Figures 289, 290, 291, 292 and 293**

1. Remove the main bearing inserts from the block and bearing caps.
2. If the crankshaft main bearing journals have been refinished to a definite undersize, install the correct undersize bearing. Be sure that the bearing inserts and bearing bores are clean. Foreign material under inserts will distort bearing and cause failure.
3. Place the upper main bearing inserts in bores with tang in slot.

➡**The oil holes in the bearing inserts must be aligned with the oil holes in the cylinder block.**

4. Install the lower main bearing inserts in bearing caps.
5. Clean the mating surfaces of block and rear main bearing cap.
6. Carefully lower the crankshaft into place. Be careful not to damage bearing surfaces.
7. Check the clearance of each main bearing by using the following procedure:
   a. Place a piece of Plastigage® or its equivalent, on bearing surface across full width of bearing cap and about ¼ in. off center.
   b. Install cap and tighten bolts to specifications. Do not turn crankshaft while Plastigage® is in place.
   c. Remove the cap. Using the supplied Plastigage® scale, check width of Plastigage® at widest point to get maximum clearance. Difference between readings is taper of journal.
   d. If clearance exceeds specified limits, try a 0.001 in. or 0.002 in. undersize bearing in combination with the standard bearing. Bearing clearance must be within specified limits. If standard and 0.002 in. undersize bearing

does not bring clearance within desired limits, refinish crankshaft journal, then install undersize bearings.

8. After the bearings have been fitted, apply a light coat of engine oil to the journals and bearings. Install the rear main bearing cap. Install all bearing caps except the thrust bearing cap. Be sure that main bearing caps are installed in original locations. Tighten the bearing cap bolts to specifications.
9. Install the thrust bearing cap with bolts finger-tight.
10. Pry the crankshaft forward against the thrust surface of upper half of bearing.
11. Hold the crankshaft forward and pry the thrust bearing cap to the rear. This aligns the thrust surfaces of both halves of the bearing.
12. Retain the forward pressure on the crankshaft. Tighten the cap bolts to specifications.
13. Measure the crankshaft end-play as follows:
   a. Mount a dial gauge to the engine block and position the tip of the gauge to read from the crankshaft end.
   b. Carefully pry the crankshaft toward the rear of the engine and hold it there while you zero the gauge.
   c. Carefully pry the crankshaft toward the front of the engine and read the gauge.
   d. Confirm that the reading is within specifications. If not, install a new thrust bearing and repeat the procedure. If the reading is still out of specifications with a new bearing, have a machine shop inspect the thrust surfaces of the crankshaft, and if possible, repair it.
14. Install the rear main seal.
15. Rotate the crankshaft so as to position the first rod journal to the bottom of its stroke.

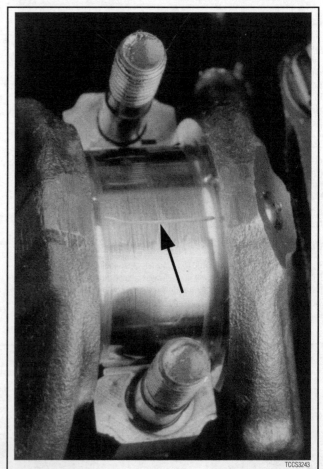

Fig. 289 Apply a strip of gauging material to the bearing journal, then install and torque the cap

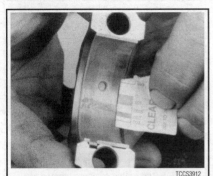

Fig. 290 After the cap is removed again, use the scale supplied with the gauging material to check the clearance

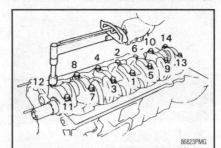

Fig. 291 All main bearing caps for Toyota truck engines should be tightened in the proper sequence, from the insie working outward—3F-E and 1FZ-FE shown, others use the same pattern, but with fewer bolts

Fig. 292 A dial gauge may be used to check crankshaft end-play

Fig. 293 Carefully pry the crankshaft back and forth while reading the dial gauge for end-play

Pistons and Connecting Rods

♦ See Figures 294, 295, 296, 297 and 298

1. Before installing the piston/connecting rod assembly, oil the pistons, piston rings and the cylinder walls with light engine oil. Install connecting rod bolt protectors or rubber hose onto the connecting rod bolts/studs. Also perform the following:

a. Select the proper ring set for the size cylinder bore.

b. Position the ring in the bore in which it is going to be used.

c. Push the ring down into the bore area where normal ring wear is not encountered.

d. Use the head of the piston to position the ring in the bore so that the ring is square with the cylinder wall. Use caution to avoid damage to the ring or cylinder bore.

e. Measure the gap between the ends of the ring with a feeler gauge. Ring gap in a worn cylinder is normally greater than specification. If the ring gap is greater than the specified limits, try an oversize ring set.

f. Check the ring side clearance of the compression rings with a feeler gauge inserted between the ring and its lower land according to specification. The gauge should slide freely around the entire ring circumference without binding. Any wear that occurs will form a step at the inner portion of the lower land. If the lower lands have high steps, the piston should be replaced.

2. Unless new pistons are installed, be sure to install the pistons in the cylinders from which they were removed. The numbers on the connecting rod and bearing cap must be on the same side when installed in the cylinder bore. If a connecting rod is ever transposed from one engine or cylinder to another, new bearings should be fitted and the connecting rod should be numbered to correspond with the new cylinder number. The notch on the piston head goes toward the front of the engine.

3. Install all of the rod bearing inserts into the rods and caps.

4. Install the rings to the pistons. Install the oil control ring first, then the second compression ring and finally the top compression ring. Use a piston ring expander tool to aid in installation and to help reduce the chance of breakage.

5. Make sure the ring gaps are properly spaced around the circumference of the piston. Fit a piston ring compressor around the piston and slide the piston and connecting rod assembly down into the cylinder bore, pushing it in with the wooden hammer handle. Push the piston down until it is only slightly below the top of the cylinder bore. Guide the connecting rod onto the crankshaft bearing journal carefully, to avoid damaging the crankshaft.

6. Check the bearing clearance of all the rod bearings, fitting them to the crankshaft bearing journals. Follow the procedure in the crankshaft installation above.

7. After the bearings have been fitted, apply a light coating of assembly oil to the journals and bearings.

8. Turn the crankshaft until the appropriate bearing journal is at the bottom of its stroke, then push the piston assembly all the way down until the connecting rod bearing seats on the crankshaft journal. Be careful not to allow the bearing cap screws to strike the crankshaft bearing journals and damage them.

9. After the piston and connecting rod assemblies have been installed, check the connecting rod side clearance on each crankshaft journal.

10. Prime and install the oil pump and the oil pump intake tube.

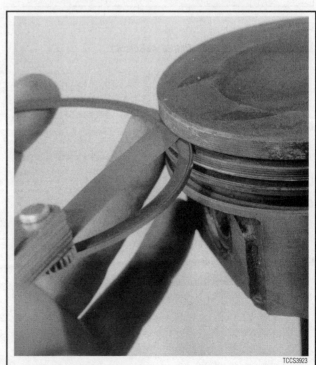

Fig. 294 Checking the piston ring-to-ring groove side clearance using the ring and a feeler gauge

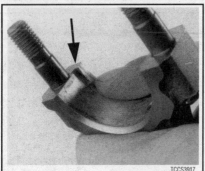

Fig. 295 The notch on the side of the bearing cap matches the tang on the bearing insert

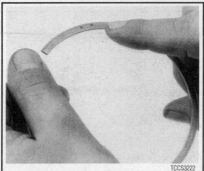

Fig. 296 Most rings are marked to show which side of the ring should face up when installed to the piston

Fig. 297 Remember to place lengths of rubber hose over the rod bolts in order to protect the crankshaft from damage . . .

Fig. 298 . . . then install the piston and rod assembly into the block using a ring compressor and the handle of a hammer

OHV Engines

CAMSHAFT, LIFTERS AND TIMING ASSEMBLY

1. Install the camshaft.
2. Install the lifters/followers into their bores.
3. If applicable, install the balance shafts.
4. Install the timing gears/chain assembly.

CYLINDER HEAD(S)

1. Install the cylinder head(s) using new gaskets.
2. Assemble the rest of the valve train (pushrods and rocker arms and/or shafts).

OHC Engines

CYLINDER HEAD(S)

1. Install the cylinder head(s) using new gaskets.
2. Install the timing sprockets/gears and the belt/chain assemblies.

Engine Covers and Components

Install the timing cover(s) and oil pan. Refer to your notes and drawings made prior to disassembly and install all of the components that were removed. Install the engine into the vehicle.

Engine Start-up and Break-in

STARTING THE ENGINE

Now that the engine is installed and every wire and hose is properly connected, go back and double check that all coolant and vacuum hoses are connected. Check that your oil drain plug is installed and properly tightened. If not already done, install a new oil filter onto the engine. Fill the crankcase with the proper amount and grade of engine oil. Fill the cooling system with a 50/50 mixture of coolant/water.

1. Connect the vehicle battery.
2. Start the engine. Keep your eye on your oil pressure indicator; if it does not indicate oil pressure within 10 seconds of starting, turn the vehicle **OFF**.

✳✳ WARNING

Damage to the engine can result if it is allowed to run with no oil pressure. Check the engine oil level to make sure that it is full. Check for any leaks and if found, repair the leaks before continuing. If there is still no indication of oil pressure, you may need to prime the system.

3. Confirm that there are no fluid leaks (oil or other).
4. Allow the engine to reach normal operating temperature (the upper radiator hose will be hot to the touch).
5. At this point any necessary checks or adjustments can be performed, such as ignition timing.
6. Install any remaining components or body panels which were removed.

BREAKING IT IN

Make the first miles on the new engine, easy ones. Vary the speed but do not accelerate hard. Most importantly, do not lug the engine, and avoid sustained high speeds until at least 100 miles. Check the engine oil and coolant levels frequently. Expect the engine to use a little oil until the rings seat. Change the oil and filter at 500 miles, 1500 miles, then every 3000 miles past that.

KEEP IT MAINTAINED

Now that you have just gone through all of that hard work, keep yourself from doing it all over again by thoroughly maintaining it. Not that you may not have maintained it before, heck you could have had one to two hundred thousand miles on it before doing this. However, you may have bought the vehicle used, and the previous owner did not keep up on maintenance. Which is why you just went through all of that hard work. See?

TORQUE SPECIFICATIONS

Component	U.S.	Metric
22R		
Cylinder head-to-cylinder block	58 ft. lbs.	78 Nm
Intake manifold-to-cylinder head	14 ft. lbs.	19 Nm
Exhaust manifold-to-cylinder head	33 ft. lbs.	44 Nm
Crankshaft bearing cap-to-cylinder block	76 ft. lbs.	103 Nm
Connecting rod cap-to-connecting rod	51 ft. lbs.	69 Nm
Crankshaft pulley-to-crankshaft	116 ft. lbs.	157 Nm
Flywheel-to-crankshaft	80 ft. lbs.	108 Nm
Camshaft bearing cap-to-cylinder head	14 ft. lbs.	20 Nm
Oil pan-to-block	9 ft. lbs.	13 Nm
22R-E		
Cylinder head-to-valve cover	52 inch lbs.	6 Nm
Cylinder head-to-camshaft bearing cap	14 ft. lbs.	20 Nm
Cylinder head-to-spark plug	13 ft. lbs.	18 Nm
Cylinder head-to-intake manifold	14 ft. lbs.	19 Nm
Cylinder head-to-exhaust manifold	33 ft. lbs.	44 Nm
Cylinder head-to-head rear cover	9 ft. lbs.	13 Nm
Block-to-chain dampner	16 ft. lbs.	22 Nm
Block-to-cylinder head	58 ft. lbs.	78 Nm
Block-to-engine mounting	29 ft. lbs.	39 Nm
Block-to-rear seal retainer	13 ft. lbs.	18 Nm
Oil cooler relief valve-to-block	51 ft. lbs.	69 Nm
Block-to-crankshaft bearing cap	76 ft. lbs.	103 Nm
Block-to-oil strainer	9 ft. lbs.	13 Nm
Block-to-oil pan	9 ft. lbs.	13 Nm
Valve clearance adjusting screw	18 ft. lbs.	25 Nm
Crankshaft pulley-to-No. 2 pulley	14 ft. lbs.	19 Nm
Air intake chamber-to-intake manifold	14 ft. lbs.	19 Nm
Connecting rod-to-connecting rod cap	51 ft. lbs.	69 Nm
Crankshaft-to-crankshaft-to-pulley	116 ft. lbs.	157 Nm
Crankshaft-to-flywheel	80 ft. lbs.	108 Nm
Crankshaft-to-drive plate	61 ft. lbs.	83 Nm
Oil pan-to-drain plug	18 ft. lbs.	25 Nm
3VZ-E		
No. 1 idler pulley-to-cylinder head	25 ft. lbs.	34 Nm
Crankshaft pulley-to-crankshaft	181 ft. lbs.	245 Nm
Camshaft pulley-to-camshaft	80 ft. lbs.	108 Nm
Fan bracket-to-block	30 ft. lbs.	41 Nm
Cooling fan-to-coupling	48 inch lbs.	5 Nm
Water outlet-to-No. 2 idler pulley	74 inch lbs.	8 Nm
Cylinder head-to-spark plug	13 ft. lbs.	18 Nm
Exhaust manifold-to-cylinder head	29 ft. lbs.	39 Nm
Cylinder head-to-block		
(1st) 12 pointed head	33 ft. lbs.	44 Nm
(2nd)	Turn 90 degrees	
(3rd)	Turn 90 degrees	

TORQUE SPECIFICATIONS

Component	U.S.	Metric
3VZ-E		
6 pointed head	30 ft. lbs.	41 Nm
	12 ft. lbs.	16 Nm
Valve cover-to-cylinder head	48 inch lbs.	5 Nm
Engine crossover pipe-to-exhaust manifold	29 ft. lbs.	39 Nm
Intake manifold-to-cylinder head	13 ft. lbs.	18 Nm
No. 3 timing belt cover-to-cylinder head	74 inch lbs.	8 Nm
No. 2 idler pulley-to-intake manifold	13 ft. lbs.	18 Nm
No. 4 timing belt cover-to-No. 3 cover	74 inch lbs.	8 Nm
Air intake chamber-to-intake manifold	13 ft. lbs.	18 Nm
Oil pump-to-block	14 ft. lbs.	20 Nm
Oil pan-to-block	52 inch lbs.	6 Nm
Water pump-to-block		
short bolt	14 ft. lbs.	20 Nm
long bolt	13 ft. lbs.	18 Nm
Crankshaft-to-flywheel	65 ft. lbs.	88 Nm
Crankshaft-to-driveplate	61 ft. lbs.	83 Nm
5VZ-FE		
No. 1 idler pulley-to-oil pump	26 ft. lbs.	35 Nm
No. 1 timing belt cover-to-oil pump	80 inch lbs.	9 Nm
Crankshaft pulley-to-crankshaft	184 ft. lbs.	250 Nm
Camshaft timing pulley-to-camshaft	81 ft. lbs.	110 Nm
No. 2 timing belt cover-to-No. 3 belt cover	80 inch lbs.	9 Nm
Camshaft bearing cap-to-cylinder head	12 ft. lbs.	16 Nm
Cylinder head rear plate-to-head	71 inch lbs.	8 Nm
Exhaust manifold-to-cylinder head	30 ft. lbs.	40 Nm
Intake manifold-to-cylinder head	13 ft. lbs.	18 Nm
No. 3 timing belt cover-to-cylinder head	80 inch lbs.	9 Nm
Camshaft position sensor-to-cylinder head	71 inch lbs.	8 Nm
Connecting rod cap-to-connecting rod		
(1st)	18 ft. lbs.	25 Nm
(2nd)	Turn 90 degrees	
Main bearing cap-to-block		
(1st)	45 ft. lbs.	61 Nm
(2nd)	Turn 90 degrees	
Rear oil seal retainer-to-block	71 inch lbs.	8 Nm
Rear end plate-to-block	66 inch lbs.	7 Nm
Driveplate-to-crankshaft	61 ft. lbs.	83 Nm
Flywheel-to-crankshaft	63 ft. lbs.	85 Nm
2RZ-FE and 3RZ-FE		
Cylinder head-to-timing chain cover	15 ft. lbs.	21 Nm
Camshaft bearing cap-to-cylinder head	12 ft. lbs.	16 Nm
Camshaft timing gear-to-intake camshaft	54 ft. lbs.	74 Nm
No. 1 chain tensioner-to-cylinder head	15 ft. lbs.	21 Nm
Cylinder head rear cover-to-head	10 ft. lbs.	13 Nm
Exhaust manifold-to-cylinder head	36 ft. lbs.	49 Nm

86823C11

86823C12

TORQUE SPECIFICATIONS

2RZ-FE and 3RZ-FE

Component		U.S.	Metric
Intake manifold-to-cylinder head		22 ft. lbs.	29 Nm
Air intake chamber-to-intake manifold		15 ft. lbs.	20 Nm
Balance shaft drive gear-to-balance shaft		18 ft. lbs.	25 Nm
Timing chain cover-to-block			
12mm head bolt	A	14 ft. lbs.	20 Nm
	B	18 ft. lbs.	25 Nm
14mm head	bolt	32 ft. lbs.	44 Nm
	nut	14 ft. lbs.	20 Nm
Crankshaft pulley-to-cranksaht		193 ft. lbs.	260 Nm
Oil pan-to-block		9 ft. lbs.	13 Nm
Stiffener plate-to-block		27 ft. lbs.	37 Nm
Connecting rod cap-to-connecting rod	(1st)	33 ft. lbs.	45 Nm
	(2nd)	Turn 90 degrees	Turn 90 degrees
Main bearing cap-to-block	(1st)	29 ft. lbs.	39 Nm
	(2nd)	Turn 90 degrees	Turn 90 degrees
No. 1 balance shaft-to-timing gear		26 ft. lbs.	36 Nm
Balance shaft-to-block		13 ft. lbs.	18 Nm
Rear oil seal retainer-to-block		10 ft. lbs.	14 Nm
Rear end plate-to-block		13 ft. lbs.	18 Nm
Flywheel-to-cranksaht	2RZ-FE	65 ft. lbs.	88 Nm
	3RZ-FE (1st)	19 ft. lbs.	27 Nm
	(2nd)	Turn 90 degrees	Turn 90 degrees
Driveplate-to-crankshaft		54 ft. lbs.	74 Nm

3F-E

Component	U.S.	Metric
Cylinder head-to-block	90 ft. lbs.	123 Nm
Valve rocker support-to-head		
12mm bolt head	17 ft. lbs.	24 Nm
14mm head bolt and nut	25 ft. lbs.	33 Nm
Manifold-to-cylinder head		
14mm bolt head	37 ft. lbs.	50 Nm
17mm bolt head	51 ft. lbs.	69 Nm
nut	41 ft. lbs.	56 Nm
Valve cover-to-cylinder head	78 inch lbs.	9 Nm
Intake manifold-to-intake stay	22 ft. lbs.	29 Nm
Block-to-intake stay	22 ft. lbs.	29 Nm
Exhaust manifold-to-exhaust pipe	46 ft. lbs.	62 Nm
Camshaft thrust washer-to-block	9 ft. lbs.	12 Nm
Crankshaft pulley-to-crankshaft	253 ft. lbs.	343 Nm
Pushrod cover-to-block	35 inch lbs.	4 Nm
Connecting rod cap-to-connecting rod	43 ft. tbs.	59 Nm
Driveplate-to-crankshaft	64 ft. lbs.	87 Nm

86823C13

TORQUE SPECIFICATIONS

Compnant		U.S.	Metric
1FZ-FE			
Air intake chamber-to-intake manifold		15 ft. lbs.	21 Nm
Cylinder head-to-block	(1st)	29 ft. lbs.	39 Nm
	(2nd)	Turn 90 degrees	Turn 90 degrees
	(3rd)	Turn 90 degrees	Turn 90 degrees
Cylinder head-to-timing chain		15 ft. lbs.	21 Nm
Camshaft bearing cap-to-cylinder head		12 ft. lbs.	16 Nm
Camshaft timing gear-to-camshaft		54 ft. lbs.	74 Nm
Chain tensioner-to-cylinder head		15 ft. lbs.	21 Nm
Exhaust manifold-to-cylinder head		29 ft. lbs.	39 Nm
Front exhaust pipe-to-rear TWC		34 ft. lbs.	46 Nm
Vibration damper-to-block		14 ft. lbs.	20 Nm
Timing chain cover-to-block		15 ft. lbs.	21 Nm
Crankshaft pulley-to-crankshaft		304 ft. lbs.	412 Nm
Water pump-to-block		15 ft. lbs.	21 Nm
Rear oil seal retainer-to-block		15 ft. lbs.	21 Nm
Driveplate-to-crankshaft		74 ft. lbs.	100 Nm
Transmission-to-block		53 ft. lbs.	72 Nm

86823C14

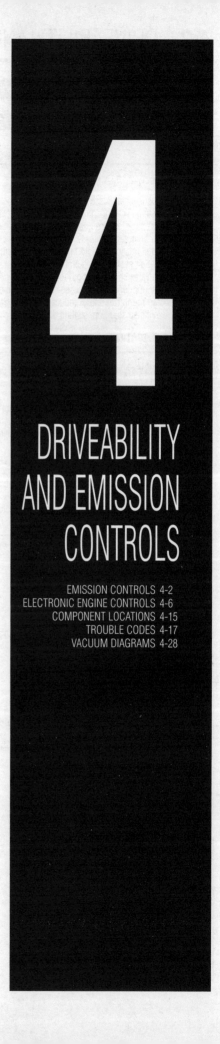

4

DRIVEABILITY AND EMISSION CONTROLS

EMISSION CONTROLS

Crankcase Ventilation System

OPERATION

▶ See Figures 1, 2, 3, 4 and 5

A closed, positive crankcase ventilation system is employed on all Toyota trucks. This system cycles incompletely burned fuel (which works its way past the piston rings) back into the intake manifold for reburning with the air/fuel mixture. The oil filler cap is sealed; and the air is drawn from the top of the crankcase into the intake manifold through a valve with a variable orifice.

The recirculation system relies on the integrity of the engine seals. Any air leak around the valve cover, head gasket, oil pan, dipstick, oil filler cap, air intake ducts, vacuum or breather hoses can introduce excess air into the fuel/air mixture, causing rough running or reduced efficiency. Likewise, a plugged hose or passage can cause sludging, stalling and oil leaks.

COMPONENT TESTING

Inspect the PCV system hoses and connections at each tune-up and replace any deteriorated hoses. Check the PCV valve at every tune-up and replace it at 30,000 mile (48,000 km) intervals.

The PCV valve is easily checked with the engine running at normal idle speed (warmed up).

1. Remove the PCV valve from the valve cover or intake manifold, but leave it connected to its hose.
2. Start the engine.
3. Place your thumb over the end of the valve to check for vacuum. If there is no vacuum, check for plugged hoses or ports. If these are open, the valve is faulty.
4. With the engine **OFF**, remove the valve completely. Shake it end-to-end,

listening for the rattle of the needle inside the valve. If no rattle is heard, the needle is jammed (probably due to oil sludge) and the valve should be replaced.

✲✲ CAUTION

Don't blow directly into the valve; petroleum deposits within the valve can be harmful.

An engine without crankcase ventilation is quickly damaged. It is important to check the PCV at regular intervals. When replacing a PCV valve you must use the correct one for the engine. Many valves look alike on the outside, but have different mechanical values. Putting the incorrect valve on a vehicle can cause a great deal of driveability problems. The engine computer assumes the valve is the correct one and may over adjust the ignition timing or fuel mixture.

REMOVAL & INSTALLATION

1. Pull the PCV valve from the valve cover or the intake manifold.
2. Remove the hose from the valve.
3. Check the valve for proper operation.
4. Inspect the rubber grommet the PCV valve fits into. If it is in any way deteriorated or oil soaked, replace it.
5. Push the valve into the rubber grommet. Make sure the valve is firmly into place.

Evaporative Emission Controls

OPERATION

The evaporative emission control system prevents the release of unburned hydrocarbons from the liquid gasoline vapor into the atmosphere. Evaporative

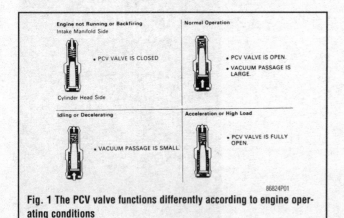
Fig. 1 The PCV valve functions differently according to engine operating conditions

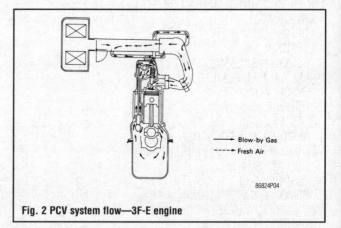
Fig. 2 PCV system flow—3F-E engine

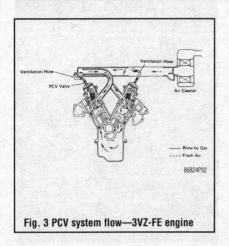
Fig. 3 PCV system flow—3VZ-FE engine

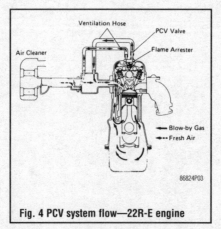
Fig. 4 PCV system flow—22R-E engine

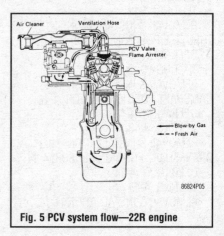
Fig. 5 PCV system flow—22R engine

fuel vapor from the tank is routed to the charcoal canister located in the engine compartment. The charcoal canister stores the vapor until the engine coolant temperature reaches 129°F (54°C). When the engine coolant temperature is above 129°F (54°C) and the throttle is open, a vacuum switching valve opens to allow vapors trapped in the canister to enter the intake manifold.

A fuel filler cap with a check valve allows air to enter the fuel tank as the fuel is used. This prevents a vacuum build-up in the fuel tank as the engine is running.

COMPONENT TESTING

Charcoal Canister

EXCLUDING 3RZ-FE

▶ **See Figures 6 and 7**

1. Label and disconnect the vacuum hose at the top of the canister. Remove the charcoal canister from the vehicle.
2. Remove the vacuum switching valve.
3. Look for cracks or damage to the outside of the canister.
4. Using a low pressure of compressed air 0.68 psi (5 kPa), blow into the tank pipe and check that air flows without any resistance from the other pipes.

5. Blow air into the purge pipe and check that air does not flow from the tank pipe and air flows without resistance from the other pipe. If there is any problems found, replace the canister.
6. Clean the filter in the canister by blowing 43 psi (296 kPa) of compressed air into the canister tank pipe while holding the other upper canister tank pipe closed.

→**Do not attempt to wash the canister. No activated carbon should come out of the canister while cleaning or inspecting the unit. If it does, replace the canister.**

7. Install the canister on the vehicle and reconnect the hoses.

3RZ-FE ENGINE

▶ **See Figures 8, 9, 10 and 11**

1. Label and disconnect the vacuum hose at the top of the canister. Remove the charcoal canister from the vehicle.

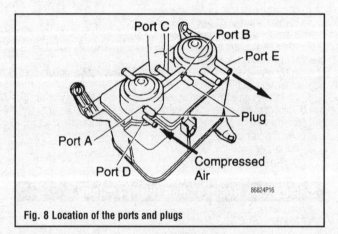

Fig. 8 Location of the ports and plugs

Fig. 6 Always label hoses and wires prior to removal

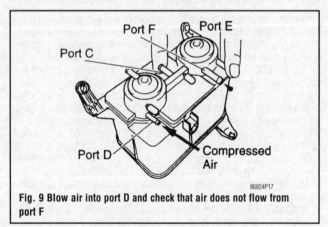

Fig. 9 Blow air into port D and check that air does not flow from port F

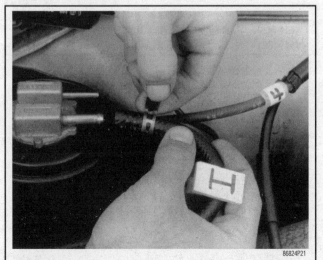

Fig. 7 The hoses are easily removed by sliding the clamp off the port end

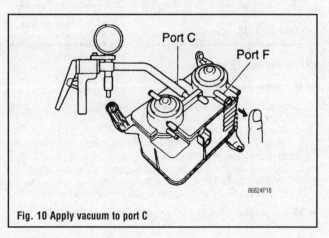

Fig. 10 Apply vacuum to port C

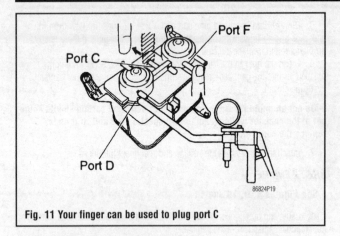

Fig. 11 Your finger can be used to plug port C

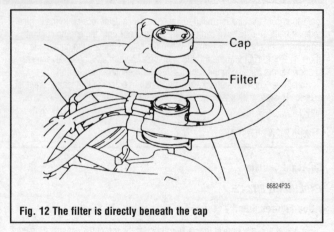

Fig. 12 The filter is directly beneath the cap

2. Remove the vacuum switching valve.
3. Look for cracks or damage to the outside of the canister.
4. Install the plugs to ports **A** and **B**.
5. While holding port **C** closed, blow air 0.21 psi (2 kPa) into port **D** and check that air does not flow from port **F**.
6. Apply vacuum 0.87 in. Hg (3 kPa) to port **C**, check that the vacuum does not decrease when port **F** is released.
7. While holding port **F** closed, apply 0.87 in. Hg (2.94 kPa) of vacuum to port **D**. Check that the vacuum does not decrease when port **C** is closed, and check that the vacuum decreases when port **C** is released. If a problem occurs, replace the charcoal canister.
8. Remove the plugs.
9. Reinstall the canister.

Exhaust Gas Recirculation (EGR) System

OPERATION

Oxides of nitrogen can be formed under conditions of high pressure and very high temperature. Reduction of one of these conditions reduces the production of NOx. A reduction of peak combustion temperature is accomplished by recirculating a small amount of exhaust gases into the combustion chamber.

This metering of exhaust gasses must be carefully monitored. Too much at the wrong time causes extremely poor driveability; too little and the emissions can increase dramatically. EGR function is overseen by an EGR modulator which controls the flow of vacuum (and therefore the flow of exhaust gasses), limiting it to times when the presence of exhaust gas will not adversely effect the running of the engine. The 22R and 22R-E use a Bimetal Vacuum Switching Valve (BVSV) to sense the coolant temperature; in this way, EGR function is limited to warm engine operation. All other engines use a Vacuum Solenoid Valve (VSV) connected to the Engine Control Module (ECM). The computer receives many inputs regarding engine status and temperature; when conditions are favorable, the VSV is activated, allowing EGR function.

Additionally, EGR function is controlled by manifold vacuum so that the valve only operates during periods of high-load or high rpm driving. A binding EGR valve (either open or closed) can cause stalling, missing or reduced fuel economy.

SYSTEM TESTING

2RZ-FE and 3RZ-FE Engines

♦ **See Figures 12, 13 and 14**

1. Remove the EGR vacuum modulator cap and filter.
2. Check the filter for damage or contamination.
3. With compressed air, clean the filter.
4. Reinstall the filter and cap. Install the filter with the coarser surface facing the atmospheric side (outwards).
5. Using a 3-way connector, attach a vacuum gauge to the hose between the EGR valve and the EGR vacuum modulator.
6. Start the engine and check that the engine starts and runs at idle.

➡ **The engine coolant temperature should be below 122°F (20°C).**

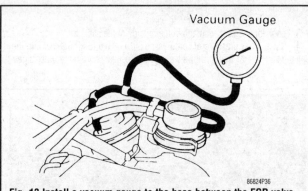

Fig. 13 Install a vacuum gauge to the hose between the EGR valve and vacuum modulator

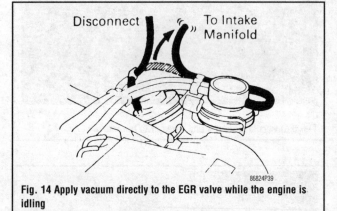

Fig. 14 Apply vacuum directly to the EGR valve while the engine is idling

7. Check that the vacuum gauge indicates zero at 3000 rpm.
8. Check that the EGR pipe is not hot.
9. Remove the 3-way connector with the vacuum hose.
10. Connect the vacuum hose from port **Q** of the EGR modulator to the EGR valve.
11. Plug the hose from the VSV to the EGR.
12. Remove the vacuum gauge, then reconnect the vacuum hoses to their proper locations.
13. Apply vacuum directly to the EGR valve with the engine at idle.
14. Check that the engine now idles rough or stalls.
15. Reconnect the vacuum hoses to their proper locations.

1FZ-FE Engines

1. Check the filter in the EGR modulator for contamination and clean using compressed air.

2. Using a 3-way connector, connect a vacuum gauge inline between the valve and vacuum pipe.
Check that the engine starts and idles smoothly.

➡**For accurate system testing, the coolant temperature should be below 117°F (47°C).**

3. Accelerate the engine to 2500 rpm and verify that the vacuum gauge reads 0 in. Hg.
4. Check that the EGR pipe is not hot.

➡**For accurate test results, the engine coolant temperature must now be above 127°F (53°C)**

5. Check that the vacuum gauge reads low vacuum at 2500 rpm.
6. Disconnect the vacuum hose from port **R** of the EGR vacuum modulator and connect port **R** directly to the intake manifold.
7. Check that the vacuum reading is high at 2500 rpm. As large amounts of vacuum enters the engine, the engine idle quality may be adversely affected.
8. Remove the vacuum gauge and reconnect the vacuum hoses in their original locations.
9. Apply vacuum directly to the EGR valve with the engine at idle.
10. If the EGR valve is operating properly, the engine should misfire, run rough and possibly stall. Reconnect the vacuum hoses in their proper locations.

22R-E and 3VZ-E Engines

1. Check the filter in the EGR modulator for contamination and clean using compressed air.
2. Using a 3-way connector, connect a vacuum gauge inline between the valve and vacuum pipe.
3. Check that the engine starts and idles smoothly.

➡**For accurate system testing, the coolant temperature should be below 118°F (48°C) on 3VZ-E engines or below 93°F (34°C) on 22R-E engines.**

4. Accelerate the engine to 3500 rpm and verify that the vacuum gauge reads 0.

➡**For accurate test results, the engine coolant temperature must be above 126°F (52°C) on 3VZ-E engines or above 104°F (40°C) on 22R-E engines.**

5. Check that the vacuum gauge reads low vacuum at 3500 rpm.
6. Disconnect the vacuum hose from port **R** of the EGR vacuum modulator and connect port **R** directly to the intake manifold.
7. Check that the vacuum reading is high at 3000 rpm. As large amounts of vacuum enters the engine, the engine idle quality may be adversely affected.
8. Remove the vacuum gauge and reconnect the vacuum hoses in their original locations.
9. Apply vacuum directly to the EGR valve with the engine at idle. If the EGR valve is operating properly, the engine should run rough or possibly stall.
10. Reconnect the vacuum hoses in their proper locations.

5VZ-FE Engine

▶ **See Figures 15 and 16**

1. Remove the EGR vacuum modulator cap and filter.
2. Check the filter for damage or contamination.
3. With compressed air, clean the filter.
4. Reinstall the filter and cap. Install the filter with the coarser surface facing the atmospheric side (outwards).
5. Using a 3-way connector, attach a vacuum gauge to the hose between the EGR valve and the VSV.
6. Start the engine and check that the engine starts and runs at idle.
7. Using the SST 09843–18020 or an equivalent jumper wire, connect terminals TE1 and E1 of the DLC1 under the hood.

➡**The engine coolant temperature should be below 113°F (45°C).**

8. Check that the vacuum gauge indicates zero at 2800 rpm.
9. Check that the EGR pipe is not hot.

➡**Allow the engine to run until the coolant temperature is above 176°F (80°C).**

10. Check that the vacuum hose gauge reads low vacuum at 2800 rpm.

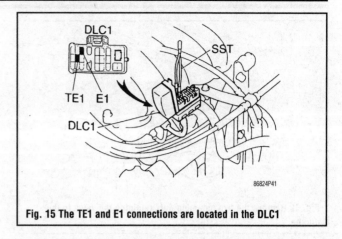

Fig. 15 The TE1 and E1 connections are located in the DLC1

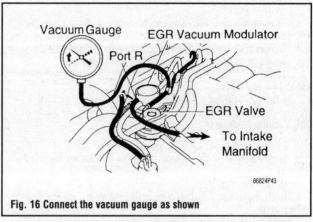

Fig. 16 Connect the vacuum gauge as shown

11. Disconnect the hose from port **R** of the EGR vacuum modulator and connect port **R** directly to the intake manifold with another hose.
12. Check the vacuum gauge, it should read high vacuum at 3500 rpm.
13. Disconnect terminals TE1 and E1 by removing the jumper wire.
14. Remove the vacuum gauge, and reconnect the vacuum hoses to their proper locations.
15. Apply vacuum directly to the EGR valve with the engine at idle.
16. Check that the engine idles rough or stalls.

REMOVAL & INSTALLATION

EGR Valve

▶ **See Figures 17 and 18**

1. Disconnect the EGR pipe, discard of the gasket.
2. Label and disconnect the vacuum hoses, EGR hose, water bypass hoses from the IAC valve and water bypass pipe.

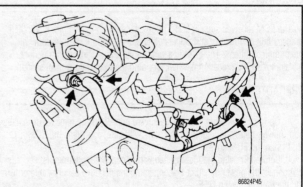

Fig. 17 Detach the EGR pipe from the valve and mounting points on the engine

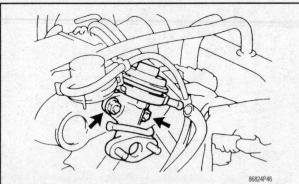

Fig. 18 Remove the two mounting nuts retaining the valve to the engine

3. Some models have an EGR sensor, remove the screws attaching this if applicable.
4. Remove the 2 nuts holding the valve to the engine. Discard the old gasket.
5. Inspect the valve for sticking and heavy carbon deposits.
6. With a new gasket, mount the valve to the engine. Tighten the mounting nuts to 14 ft. lbs. (19 Nm).
7. Install the remaining components in the reverse order of removal.

EGR Vacuum Modulator

1. Label and disconnect the hoses attached to the modulator.
2. Loosen and remove the nut holding the tube from the modulator to the EGR valve.
3. Remove the modulator.
4. Check the hoses for signs of deterioration and replace as necessary.
To install:
5. Install the modulator and tighten the nut securely.
6. Connect the vacuum hoses.

Vacuum Switching Valve (VSV)

EXCEPT 1FZ-FE ENGINE

1. Disconnect the negative battery cable.
2. Disconnect the VSV electrical connector.
3. Label and remove the vacuum hoses.
4. Remove the VSV. Installation is the reverse of the removal procedure. During installation, check the vacuum hoses for deterioration and replace as necessary.

1FZ-FE ENGINE

1. Disconnect the negative battery cable.
2. Remove the air intake chamber as follows:
   a. Drain the engine coolant. Disconnect the cruise control cable, accelerator cable and the throttle cable at the throttle body.
   b. Disconnect the vacuum hoses and the EGR gas temperature sensor connector at the EGR valve.
   c. Loosen the EGR pipe union nut. Remove the 2 nuts holding the EGR valve and air intake chamber.
   d. Remove the 2 stud bolts, EGR valve and vacuum modulator assembly.
   e. Remove the bolt holding the heater pipe and the air intake chamber. Disconnect the air cleaner hose, if still attached.
3. Label and disconnect the PCV hoses. Label and disconnect the remaining vacuum hoses. Disconnect the brake booster hose.
4. Disconnect the water bypass hose and the EVAP hose. Unsecure the connector from the emission control valve set assembly, throttle position sensor and the Idle Air Control (IAC) valve.
5. Remove the power steering reservoir tank mounting bolts and position tank aside. Disconnect the fluid lines should not be necessary.
6. Remove the engine oil dipstick and guide. Remove the O-ring from the end of the dipstick guide.
7. Disconnect the engine ground strap.
8. Remove the air intake chamber nuts and bolts and 2 gaskets.
9. Disconnect the vacuum hoses from the the the air intake chamber.
10. Remove the 4 mounting bolts and the emission control valve set assembly.
11. Disconnect the harness and 2 vacuum hoses. Remove the screws securing the VSV.
12. Installation is in the reverse order of the removal procedure.

EGR Gas Temperature Sensor

1. Disconnect the negative battery cable. Remove the vacuum modulator.
2. Disconnect the sensor electrical harness.
3. Unscrew the sensor to the EGR valve.
4. When installing the sensor, coat the threads with anti-seize compound and tighten the unit to 14 ft. lbs. (20 Nm). Engage the electrical connections.

Bimetal Vacuum Switching Valve (BVSV)

1. Drain the engine cooling system.
2. Unplug and label the vacuum hoses from the valve.
3. Remove the valve using an appropriate sized wrench.
4. Apply liquid sealer to the bottom threads and install the valve. Tighten the valve to 18 ft. lbs. (25 Nm).

ELECTRONIC ENGINE CONTROLS

Electronic Control Module (ECM)

OPERATION

The ECM receives signals from various sensors on the engine. It will then process this information and calculate the correct air/fuel mixture under all operating conditions. The ECM is a very fragile and expensive component. Always follow the precautions when servicing the electronic control system.

PRECAUTIONS

• Do not permit parts to receive a severe impact during removal or installation. Always handle all fuel injection parts with care, especially the ECM. DO NOT open the ECM cover!
• Before removing the fuel injected wiring connectors, terminals, ect., first disconnect the power by either disconnecting the negative battery cable or turning the ignition switch **OFF**.
• Do not be careless during troubleshooting as there are numerous amounts of transistor circuits; even a slight terminal contact can induce troubles.

• When inspecting during rainy days, take extra caution not to allow entry of water in or on the unit. When washing the engine compartment, prevent water from getting on the fuel injection parts and wiring connectors.

REMOVAL & INSTALLATION

▶ See Figures 19 and 20

1. Disconnect the negative battery cable.
2. Locate the ECM and release the lock, then pull out the connector. Pull on the connectors only!
3. Unbolt the ECM from its mounting area.
4. Mount the ECM on the vehicle in the proper location.
5. Fully insert the connector, then check that it is locked.

Oxygen Sensor

➡**During handling of the oxygen sensor, do not hit the end of the sensor. The sensor can be easily damaged by impact or rough handling. Additionally, do not allow the sensor to come in contact with water or petroleum products.**

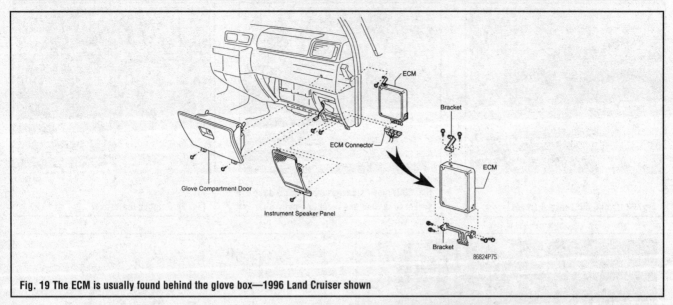

Fig. 19 The ECM is usually found behind the glove box—1996 Land Cruiser shown

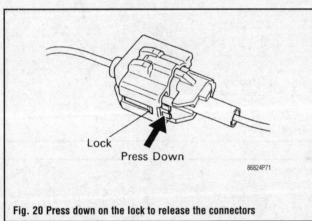

Fig. 20 Press down on the lock to release the connectors

OPERATION

The exhaust oxygen sensor, or O2S, is mounted in the exhaust stream where it monitors oxygen content in the exhaust gas. The oxygen content in the exhaust is a measure of the air/fuel mixture going into the engine. The oxygen in the exhaust reacts with the oxygen sensor to produce a voltage which is read by the ECM.

TESTING

1. Start the engine and bring it to normal operating temperature, then run the engine above 1200 rpm for two minutes.
2. Backprobe with a high impedance averaging voltmeter (set to the DC voltage scale) between the oxygen sensor (O2S) and battery ground.
3. Verify that the O2S voltage fluctuates rapidly between 0.40–0.60 volts.
4. If the O2S voltage is stabilized at the middle of the specified range (approximately 0.45–0.55 volts) or if the O2S voltage fluctuates very slowly between the specified range (O2S signal crosses 0.5 volts less than 5 times in ten seconds), the O2S may be faulty.
5. If the O2S voltage stabilizes at either end of the specified range, the ECM is probably not able to compensate for a mechanical problem such as a vacuum leak or a high float level (carbureted engines). These types of mechanical problems will cause the O2S to sense a constant lean or constant rich mixture. The mechanical problem will first have to be repaired, then the O2S test repeated.
6. Pull a vacuum hose located after the throttle plate. Voltage should drop to approximately 0.12 volts (while still fluctuating rapidly). This tests the ability of the O2S to detect a lean mixture condition. Reattach the vacuum hose.

7. Richen the mixture using a propane enrichment tool. Voltage should rise to approximately 0.90 volts (while still fluctuating rapidly). This tests the ability of the O2S to detect a rich mixture condition.
8. If the O2S voltage is above or below the specified range, the O2S and/or the O2S wiring may be faulty. Check the wiring for any breaks, repair as necessary and repeat the test.

REMOVAL & INSTALLATION

▶ **See Figures 21, 22, 23, 24 and 25**

1. Disconnect the negative battery cable.
2. Ensure that the engine and exhaust pipes are cold. Locate the oxygen sensor. Some sensors are on the front pipe others are on the catalyst, you may have more than one.
3. Spray a lubricant on the studs to ease removal.
4. Disconnect the negative battery cable and disconnect the oxygen sensor wiring.
5. Remove the oxygen sensor retaining nuts.
6. Remove the oxygen sensor and gasket.
7. Install the oxygen sensor with a new gasket. Apply a coating of anti-seize to the studs. Install the nuts and tighten to 14 ft. lbs. (20 Nm).
8. Connect the oxygen sensor wiring. Connect the negative battery cable.

Fig. 21 Use a penetrating lubricant to ease the removal of the mountng nuts

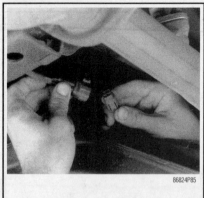

Fig. 22 Unplug the sensor harness

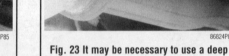

Fig. 23 It may be necessary to use a deep socket to remove the mounting nuts

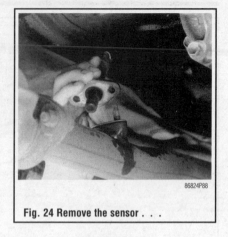

Fig. 24 Remove the sensor . . .

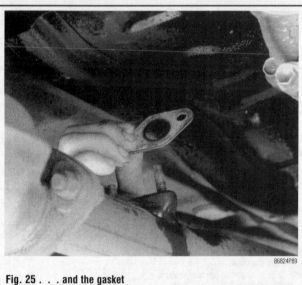

Fig. 25 . . . and the gasket

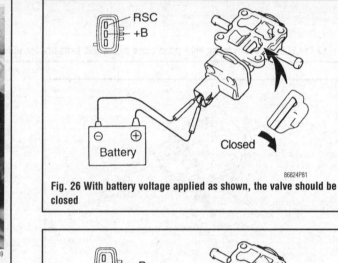

Fig. 26 With battery voltage applied as shown, the valve should be closed

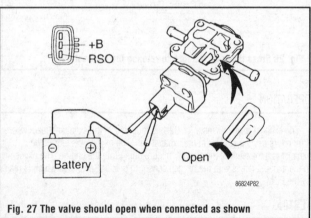

Fig. 27 The valve should open when connected as shown

Idle Air Control (IAC) Valve

OPERATION

The purpose of the Idle Air Control (IAC) system is to control engine idle speed while preventing stalls due to changes in engine load. The IAC assembly, mounted on the throttle body, controls bypass air around the throttle plate. By extending or retracting a conical valve, a controlled amount of air can move around the throttle plate. If rpm is too low, more air is diverted around the throttle plate to increase rpm.

During idle, the proper position of the IAC valve is calculated by the ECM based on battery voltage, coolant temperature, engine load, and engine rpm. If the rpm drops below a specified rate, the throttle plate is closed. The ECM will then calculate a new valve position.

The IAC only affects the engine's idle characteristics. If it is stuck fully open, idle speed is too high (too much air enters the throttle bore) If it is stuck closed, idle speed is too low (not enough air entering). If it is stuck somewhere in the middle, idle may be rough, and the engine won't respond to load changes.

TESTING

Excluding 1FZ-FE Engine

▶ **See Figures 26 and 27**

1. Using an ohmmeter, measure the resistance between terminal +B and other terminals (RSC and RSO). It should read between 17.0–24.5 ohms.
2. Connect the positive lead from the battery to terminal +B and negative lead to RSC, then check that the valve is closed.

3. Connect the positive lead from the battery to terminal +B and negative lead to RSO, then check that the valve is open.
4. If the valve did not perform as indicated, replace the IAC valve.

1FZ-FE Engine

▶ **See Figures 28 and 29**

1. Remove the IAC valve.
2. Apply battery voltage to terminals B1 and B2. Then, apply ground to terminals S1, S2, S3, S4 and S1 in sequence. Check that the valve moves towards the closed position.
3. Apply ground to terminals S4, S3, S2, S1 and S4 in sequence. Check that the valve moves towards the open position.
4. If the valve does not function as indicated, replace the IAC valve.
5. Install the IAC valve.

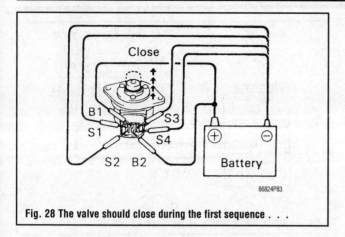

Fig. 28 The valve should close during the first sequence . . .

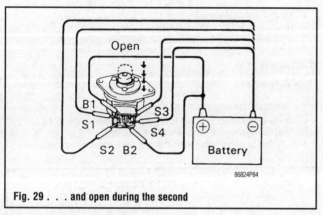

Fig. 29 . . . and open during the second

REMOVAL & INSTALLATION

▶ **See Figure 30**

1. Drain the coolant to a level below the throttle body.
2. Remove the throttle body.
3. Remove the screws and the IAC valve. Discard the old gasket or O-ring.
4. Using a new gasket or O-ring, attach the IAC valve and secure the mounting screws. Install the remaining components in the reverse order of removal.

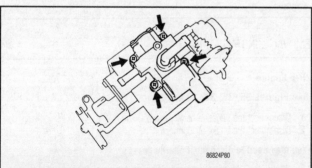

Fig. 30 Remove the screws mounting the valve to the throttle body

Engine Coolant Temperature (ECT) Sensor

OPERATION

The ECT is a thermistor (a resistor which changes value based on temperature) mounted in the engine coolant stream. As the temperature of the engine coolant changes, the resistance of the coolant sensor changes. Low coolant temperature produces a high resistance: 100,000 ohms at −40°F (−40°C). High temperature causes low resistance: 70 ohms at 284°F (140°C).

The ECM supplies a 5 volt signal to the coolant sensor and measures the voltage that returns. By measuring the voltage change, the ECM determines the engine coolant temperature. The voltage will be high when the engine is cold and low when the engine is hot. This information is used to control fuel management, IAC, spark timing, EGR, canister purge and other engine operating conditions.

TESTING

▶ **See Figures 31 and 32**

1. Remove the ECT sensor from the engine.
2. Fill a clean container with water and measure its temperature.
3. Place the sensor tip into the water. Measure the resistance of the ECT sensor with an ohmmeter. Compare the resistance with the temperature as per the graph.
4. If it is not within specifications, replace the sensor.
5. Install the ECT sensor.

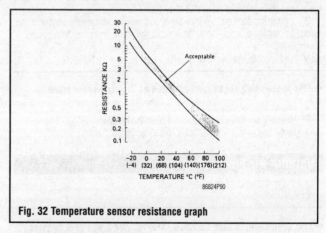

Fig. 31 Submerge the end of the temperature sensor in cold or hot water and check resistance

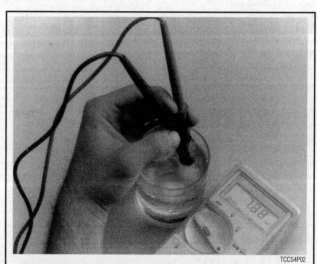

Fig. 32 Temperature sensor resistance graph

REMOVAL & INSTALLATION

Excluding 5VZ-FE Engine

1. Drain the engine cooling system.
2. Disconnect the engine wire protector from the brackets if used.
3. Detach the ECT sensor wiring.
4. Using a deep socket, remove the ECT sensor and gasket.
5. Using a new gasket and deep socket, install the ECT sensor. Tighten to 14 ft. lbs. (20 Nm).
6. Connect the ECT wiring to the sensor.

5VZ-FE Engine

1. Drain the engine cooling system.
2. Disconnect the upper radiator hose from the water outlet.
3. Disconnect the power steering air hose from the clamp.
4. Disconnect the three cord clamps of the spark plug wires from the belt cover.
5. Remove the mounting bolts for the timing belt cover and gasket.
6. Disconnect the camshaft position sensor harness.
7. Remove the fuel pipe.
8. Disconnect the ECT sensor harness.
9. Using a 19mm deep socket, remove the sensor and gasket.
10. Using a new gasket and deep socket, install the ECT sensor. Tighten to 14 ft. lbs. (20 Nm).
11. Reconnect the ECT wiring. Install the fuel pipe with four new gaskets, tighten to 25 ft. lbs. (34 Nm).
12. Install the remaining components in the reverse order of removal.

Intake Air Temperature (IAT) Sensor

OPERATION

The Intake Air Temperature (IAT) is a thermistor built into the air flow meter or is mounted on the air cleaner housing. A thermistor is a resistor which changes resistance value based intake air temperature. Low manifold air temperature produces a high resistance: 100,000 ohms at -40°F (-40°C). High temperature cause low resistance: 70 ohms at 284°F (140°C).

The ECM supplies a 5 volt signal to the IAT sensor through a resistor in the ECM and monitors the voltage. The voltage will be high when the manifold air is cold and low when the air is hot. By monitoring the voltage, the ECM calculates the air temperature and uses this data to help determine the fuel delivery and spark advance.

TESTING

▶ **See Figures 31 and 32**

1. With the IAT sensor exposed, measure the resistance between terminals E2 and THA of the air flow meter or the IAT sensor itself, with an ohmmeter. Check the ambient air temperature with a thermometer. Compare the resistance with the temperature as per the graph. Using a hair dryer, warm the air around the sensor and check the resistance once again.
2. If it sensor does not come within specifications, replace the sensor. If the sensor is contained inside the air flow meter, the entire unit must be replaced.

REMOVAL & INSTALLATION

➡ **This applies only for sensors mounted on the air cleaner housing.**

1. Detach the electrical harness.
2. Unscrew the sensor and remove it from the housing.
3. Install the sensor into the housing and attach the harness.

Volume Air Flow (VAF) Meter

OPERATION

➡ **This component is used on 22R-E, 1FZ-FE, 3VZ-E and 3F-E engines.**

This meter measures the amount of air flowing through the intake system. It is used by the ECM to calculate air/fuel mixture.

TESTING

22R-E, 1FZ-FE and 3VZ-E Engines

▶ **See Figure 33**

1. Disconnect the wiring from the VAF meter.
2. Using an ohmmeter, measure the resistance between terminals E2 and VS. Resistance should be as follows:

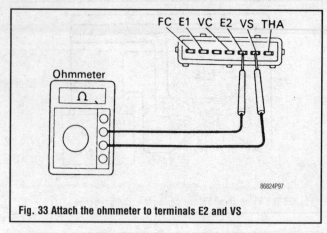

Fig. 33 Attach the ohmmeter to terminals E2 and VS

- 22R-E—200–400 ohms
- 1FZ-FE and 3VZ-E—200–600 ohms

3. If the test is not within specification, replace the unit.

3F-E Engines

▶ **See Figure 34**

1. Disconnect the wiring from the VAF meter.
2. Using an ohmmeter, measure the resistance between terminals E2 and VC.
3. The reading should be between 200–400 ohms.
4. If the test is not within specification, replace the unit.

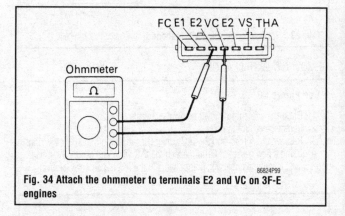

Fig. 34 Attach the ohmmeter to terminals E2 and VC on 3F-E engines

REMOVAL & INSTALLATION

22R-E Engine

▶ **See Figures 35, 36, 37, 38 and 39**

1. Disconnect the negative battery cable.
2. Disconnect the intake air connector.

➡ **You may need to disconnect nearby hoses.**

3. Unplug the VAF meter connector. Remove the air cleaner cap with the VAF meter still attached.
4. Remove the retainer bolts, 4 nuts, washers, VAF meter and gasket from the air cleaner cap.
   To install:
5. Attach the VAF meter with a new gasket to the air cleaner, then tighten until snug.
6. Connect the VAF meter wiring and install the remaining components in the reverse order of removal.

1FZ-FE Engine

1. Disconnect the negative battery cable.
2. Disconnect the VAF meter electrical connector and wire clamp.

Fig. 35 Loosen the clamp surrounding the intake air hose

Fig. 36 Slide the hose off of the sensor

Fig. 37 Detach the electrical harness from the sensor

Fig. 38 Remove the mounting nuts and washers

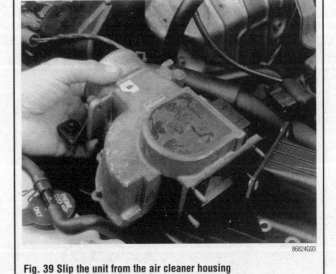

Fig. 39 Slip the unit from the air cleaner housing

3. Disconnect the cruise control actuator cable.
4. Remove the air cleaner cap, VAF meter and silencer as follows:
   a. Disconnect the air cleaner hose from the VAF meter.
   b. Disconnect the air hose from the air suction reed valve.
   c. Disconnect the 3 clamps, remove the wing nut, air cleaner cap, VAF meter and silencer.
5. Remove the 4 nuts, plate washers, VAF meter, gasket and 4 collars. Remove the 2 screws, bolt and bolt bracket. Disconnect the VAF meter from the air cleaner cap.

To install:

6. Install the VAF meter to the air cleaner cap. Install the bracket to the VAF meter and tighten the fasteners to 43 inch lbs. (5 Nm). Place a new gasket and 4 collars on the VAF meter. Install the VAF meter to the air cleaner cap and tighten retainer nuts to 7 ft. lbs. (10 Nm).
7. Install the remaining components in the reverse order of removal.

3VZ-E Engine

▶ See Figure 40

1. Disconnect the negative battery cable.
2. Disconnect the air hose and remove the 2 bolts from the resonator bracket. Disconnect the resonator bracket.
3. Disconnect the VAF meter connector. Disconnect the air cleaner hose connector.
4. Remove the air cleaner cap with the VAF meter assembly attached. Remove the retainers and the bracket from the meter assembly. Pry off the locking plate and and remove the 4 nuts, washers, lock plate, volume air flow meter and gasket.

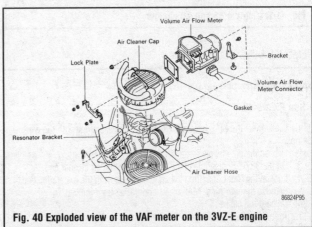

Fig. 40 Exploded view of the VAF meter on the 3VZ-E engine

Volume Air Flow Meter
Air Cleaner Cap
Bracket
Lock Plate
Volume Air Flow Meter Connector
Gasket
Resonator Bracket
Air Cleaner Hose

To install:

5. Install the VAF meter to the air cleaner cap. Install the meter, new gasket, lock plate, washer and 4 nuts. Pry the locking plate on the nut and install the bracket.
6. Install the remaining components in the reverse order of removal.

3F-E Engine

1. Disconnect the air flow meter wiring harness.
2. Remove the air cleaner hose.

3. Unbolt the air flow meter from the vehicle and discard the gasket.

4. Attach the air flow meter with a new gasket. Connect the air cleaner hose and the air flow meter harness.

Mass Air Flow (MAF) Meter

OPERATION

➡ **This component is found on 2RZ-FE, 3RZ-FE, 5VZ-FE and 1FZ-FE engines.**

This meter measures the amount of air flowing through the intake system. It is used by the ECM to calculate air/fuel mixture.

TESTING

▸ **See Figure 41**

1. Remove the air intake hose.
2. Turn the ignition switch to the **ON** position.
3. Using a voltmeter, attach the positive terminal probe to the terminal VG, and the negative probe to terminal E3.
4. Blow air into the MAF meter, then check that the voltage fluctuates.
5. Turn the ignition switch to the **OFF** position.
6. Install the air intake hose.

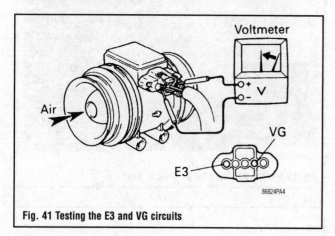

Fig. 41 Testing the E3 and VG circuits

REMOVAL & INSTALLATION

2RZ-FE and 3RZ-FE Engines

▸ **See Figure 42**

1. Disconnect the MAF harness and the IAT harness along with the wire clamps.
2. Loosen the are cleaner hose clamp.
3. On the 3RZ-FE, disconnect the air hose from the air cleaner cap.
4. Loosen the four clips, then the air cleaner cap with the meter attached.
5. Remove the four nuts, MAF meter and gasket from the cap.
6. Discard the gasket.
7. To install, place a new gasket on the air cleaner cap.
8. Install the MAF meter and tighten the nuts to 74 inch lbs. (8 Nm).

5VZ-FE Engine

▸ **See Figure 43**

1. Disconnect the air cleaner cap from the MAF meter.
2. Detach the harness from the meter.
3. Remove the mounting bolts for the meter. Lift off the meter and discard the gasket.
4. To install, place a new gasket in position, then tighten the meter to the cap assembly, 61 inch lbs. (7 Nm).

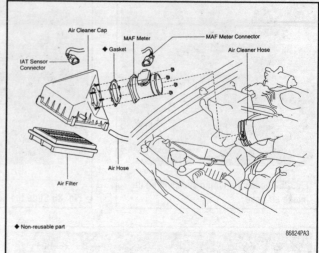

◆ Non-reusable part

86824PA3

Fig. 42 Exploded view of the MAF meter on the 2RZ-FE and 3RZ-FE engines

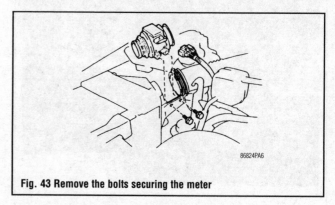

86824PA6

Fig. 43 Remove the bolts securing the meter

1FZ-FE Engine

1. Disconnect the MAF meter harness and wire clamps.
2. Separate the cruise control actuator cable from the unit.
3. Loosen the air cleaner hose.
4. Disconnect the 3 clips, then remove the wing nut, air cleaner cap and meter assembly.
5. Unbolt the meter from the cap. Discard the gasket.
6. Installation is the reverse of the removal procedure.

Throttle Position Sensor

OPERATION

Throttle position sensors are usually mounted on the side of the throttle body. The sensor detects the throttle valve opening angle. The voltage of terminal VTA increases in proportion to the opening angle of the throttle valve. The ECM judges the driving conditions from the input signals of the terminals, and uses them as one of the conditions for deciding the air/fuel ratio correction, power increases and fuel cut-off control.

TESTING

22R-E Engine

1. Confirm that the throttle valve is fully closed.
2. Connect an ohmmeter to terminals IDL and E2.

➡ **E2 is the bottom terminal, while terminal IDL is positioned directly above it.**

3. Insert a 0.0224 in. (0.57mm) thick feeler gauge between the throttle stop screw and lever. Check for continuity between terminals IDL and E2 of the sensor. Continuity should be present.

4. Insert a 0.0335 in. (0.85mm) thick feeler gauge between the throttle stop screw and lever. Check for continuity between terminals IDL and E2 of the sensor. No continuity should be present.

3VZ-E and 1FZ-FE Engines

1. Confirm that the throttle valve is fully closed.
2. Connect an ohmmeter to terminals IDL and E2.

➡**E2 is the bottom terminal, while terminal IDL is positioned directly above it.**

3. Insert a 0.020 in. (0.50mm) thick feeler gauge between the throttle stop screw and lever. Check for continuity between terminals IDL and E2 of the sensor. Continuity should be present.

4. Insert a 0.031 in. (0.80mm) thick feeler gauge between the throttle stop screw and lever. Check for continuity between terminals IDL and E2 of the sensor. Continuity should not be present.

2RZ-FE and 3RZ-FE Engines

1. Confirm that the throttle valve is fully closed.
2. Connect an ohmmeter to terminals IDL and E2.

➡**E2 is the bottom terminal, while terminal IDL is positioned directly above it.**

3. Insert a 0.022 in. (0.57mm) thick feeler gauge between the throttle stop screw and lever. Check for continuity between terminals IDL and E2 of the sensor. Continuity should be present.

4. Insert a 0.029 in. (0.74mm) thick feeler gauge between the throttle stop screw and lever. Check for continuity between terminals IDL and E2 of the sensor. No continuity should be present.

5VZ-FE Engine

1. Confirm that the throttle valve is fully closed.
2. Connect an ohmmeter to terminals IDL and E2.

➡**E2 is the bottom terminal, while terminal IDL is positioned directly above it.**

3. Insert a 0.013 in. (0.32mm) thick feeler gauge between the throttle stop screw and lever. Check for continuity between terminals IDL and E2 of the sensor. Continuity should be present.

4. Insert a 0.021 in. (0.54mm) thick feeler gauge between the throttle stop screw and lever. Check for continuity between terminals IDL and E2 of the sensor. No continuity should be present.

3F-E Engine

1. Confirm that the throttle valve is fully closed.
2. Connect an ohmmeter to terminals IDL and E2.

➡**E2 is the bottom terminal, while terminal IDL is positioned directly above it.**

3. Insert a 0.0303 in. (0.77mm) thick feeler gauge between the throttle stop screw and lever. Check for continuity between terminals IDL and E2 of the sensor. Continuity should be present.

4. Insert a 0.0429 in. (1.09mm) thick feeler gauge between the throttle stop screw and lever. Check for continuity between terminals IDL and E2 of the sensor. Continuity should not be present.

REMOVAL & INSTALLATION

▶ See Figures 44 and 45

1. Disconnect the negative battery cable.
2. Unplug the throttle position sensor electrical connector.
3. Loosen the sensor attaching screws and remove the sensor.
4. When installing, make sure the tangs to the sensor are in good shape, not bent. Place the sensor on the the throttle body and tighten the screws.

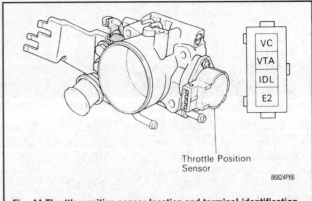

Throttle Position Sensor

86824P66

Fig. 44 Throttle position sensor location and terminal identification

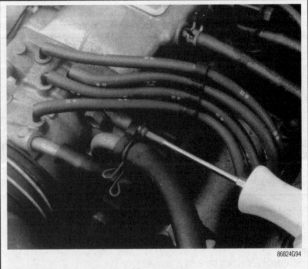

86824G94

Fig. 45 The sensor is secured by screws

Camshaft Position Sensor

OPERATION

The camshaft position sensor is installed on the front of the right cylinder head of the 5VZ-FE engine. The timing rotor has been integrated with the right bank camshaft timing pulley. While the camshaft rotates, the air gap between the protrusion on the timing rotor and the pick-up coil changes. This causes fluctuation in the magnetic field and generates a voltage signal in the pick-up coil.

TESTING

1. Disconnect the sensor harness.
2. Measure the resistance between the sensor terminals. It should be between 835–1400 ohms.
3. If not within specifications, replace the sensor.

REMOVAL & INSTALLATION

▶ See Figure 46

1. Drain the engine cooling system.
2. Disconnect the upper radiator hose from the water outlet.
3. Disconnect the power steering air hose from the clamp.
4. Disconnect the three cord clamps of the spark plug wires from the belt cover.
5. Remove the mounting bolts for the timing belt cover and gasket.
6. Disconnect the camshaft position sensor harness.

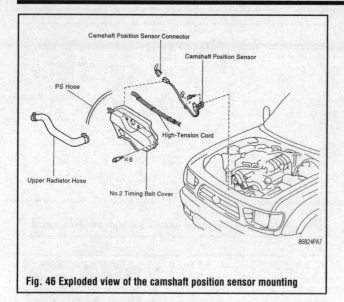

Fig. 46 Exploded view of the camshaft position sensor mounting

7. Remove the sensor.
To install:
8. Attach the camshaft position sensor to the engine, tighten to 69 inch lbs. (8 Nm).
9. Attach the wiring harness to the sensor and install the remaining components in the reverse order of removal.

Crankshaft Position Sensor

OPERATION

This sensor is essentially a pick-up coil. While the crankshaft rotates, the air gap between the protrusion on the timing rotor and the pick-up coil changes. This causes fluctuation in the magnetic field and generates a varying voltage signal in the pick-up coil. It is used by the ECM to determine crankshaft angle and speed based upon the voltage received.

TESTING

1FZ-FE Engine

1. Unplug the sensor connector.
2. Measure the resistance between the sensor terminals. It should be between 240–325 ohms.
3. If not as specified, replace the sensor.

5VZ-FE, 2RZ-FE and 3RZ-FE Engines

1. Unplug the sensor connector.
2. Measure the resistance between the sensor terminals. It should be between 2065–3225 ohms.
3. If not as specified, replace the sensor.

REMOVAL & INSTALLATION

5VZ-FE Engine

▶ **See Figure 47**

1. Disconnect the crankshaft position sensor harness.
2. Remove the bolt retaining the sensor to the vehicle.
3. To install, attach the sensor to the vehicle, tighten to 69 inch lbs. (8 Nm).

2VZ-FE and 3VZ-FE Engines

1. Remove the engine under cover.
2. Remove the alternator.
3. Unbolt the alternator bracket.

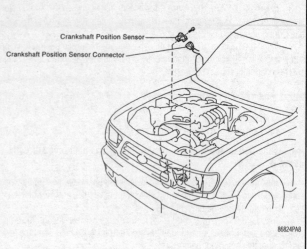

Fig. 47 Exploded view of the crankshaft position sensor mounting

4. Disconnect the crankshaft position sensor harness.
5. Remove the bolts securing the sensor and the sensor. Discard the O-ring.
6. When installing, apply a light coat of engine oil on the O-ring, install on the sensor tighten the bolts to 74 inch lbs. (9 Nm).

1FZ-FE Engine

▶ **See Figure 48**

1. Remove the engine under cover.
2. Unbolt the sensor protector.
3. Disconnect the crankshaft position sensor bracket and harness.
4. Remove the bolt, nuts and sensor.
5. When installing, attach the sensor and tighten the bolt to 14 ft. lbs. (20 Nm) and the nuts to 78 inch lbs. (9 Nm).

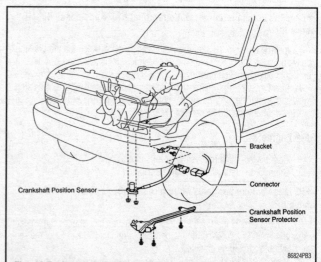

Fig. 48 Exploded view of the 1FZ-FE crankshaft position sensor mounting

Knock Sensor

OPERATION

Located in the engine block or cylinder head, the Knock Sensor (KS) retards ignition timing during a spark knock condition to allow the ECM to maintain maximum timing advance under most conditions.

TESTING

▶ **See Figure 49**

1. Remove the knock sensor(s).
2. Using an ohmmeter, check that there is no continuity between the terminal and the body. If there is continuity replace the sensor.

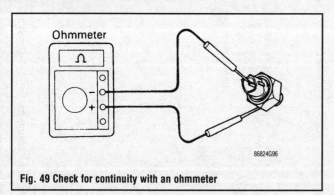

Fig. 49 Check for continuity with an ohmmeter

REMOVAL & INSTALLATION

▶ **See Figure 50**

➡A Toyota special tool or an equivalent socket, may be needed to remove and install the knock sensor. On the 2RZ-FE, 3RZ-FE and 1FZ-FE engines use SST 09816–30010. For the 5VZ-FE use SST 09817–16011.

Excluding 5VZ-FE Engines

1. Disconnect the sensor wire.
2. Remove the sensor with the aid of the appropriate SST tool or the equivalent.
3. Install the sensor and tighten to 33 ft. lbs. (44 Nm).

5VZ-FE Engines

4RUNNER

1. Drain the engine cooling system.
2. Remove the air cleaner hose.

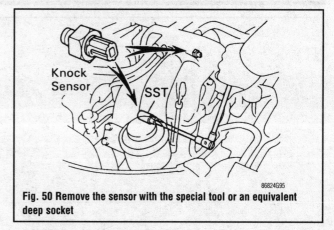

Fig. 50 Remove the sensor with the special tool or an equivalent deep socket

3. Remove the timing belt cover.
4. Remove the intake air connector.
5. Remove the fuel pressure regulator.
6. Remove the intake manifold.
7. Remove the water bypass pipe and knock sensor wire.
8. Using the appropriate SST or an equivalent, remove the two knock sensors.
9. Inspect the sensors.
10. Using the sensor tool, install the two sensors and tighten to 29 ft. lbs. (39 Nm).

EXCLUDING 4RUNNER

1. Drain the engine cooling system.
2. Remove the air cleaner hose.
3. Remove the intake air connector.
4. Remove the spark plug wires and the ignition coils.
5. Unbolt the timing belt and camshaft idler pulleys.
6. Unbolt the timing belt idler.
7. Remove the fuel pressure regulator.
8. Remove the intake manifold assembly.
9. Unbolt and remove the water bypass pipe. Disconnect the knock sensor wire.
10. With the aid of the SST 09816–30010 or equivalent socket, remove the sensor(s).
11. Inspect the sensor(s) for continuity.

COMPONENT LOCATIONS

▶ **See Figures 51, 52, 53 and 54**

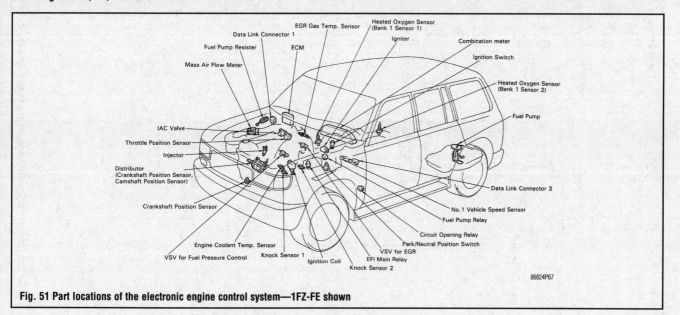

Fig. 51 Part locations of the electronic engine control system—1FZ-FE shown

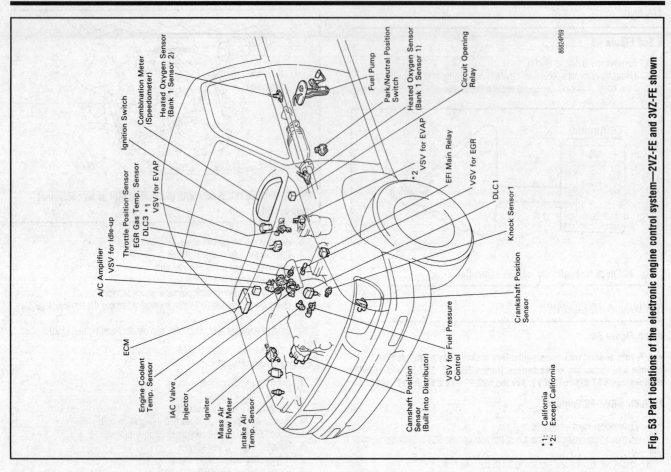

Fig. 53 Part locations of the electronic engine control system—2VZ-FE and 3VZ-FE shown

*1: California
*2: Except California

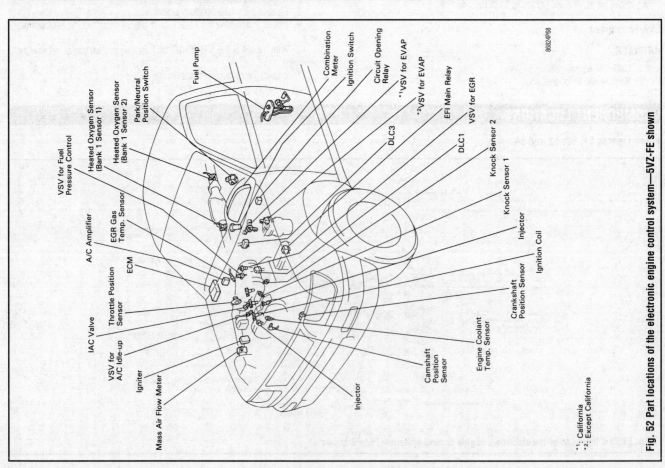

Fig. 52 Part locations of the electronic engine control system—5VZ-FE shown

*1: California
*2: Except California

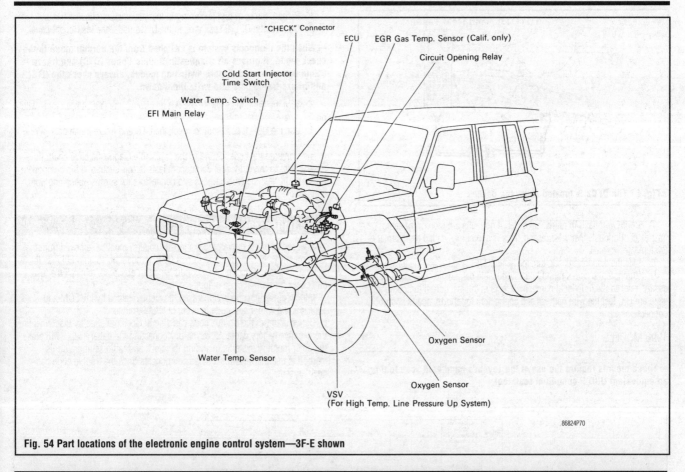

Fig. 54 Part locations of the electronic engine control system—3F-E shown

TROUBLE CODES

General Information

The ECM contains a built-in, self-diagnosis system which detects troubles within the engine signal network. Once a malfunction is detected, the Malfunction Indicator Lamp (MIL), located on the instrument panel, will light.

By analyzing various signals, the ECM detects system malfunctions related to the operating sensors. The ECM stores the failure code associated with the detected failure until the diagnosis system is cleared.

The MIL on the instrument panel informs the driver that a malfunction has been detected. The light will go out automatically once the malfunction has been cleared.

Data Link Connector (DLC)

▶ **See Figures 55, 56 and 57**

The DLC1 is located in the engine compartment. The DLC3 is located in the interior of the vehicle, under the driver's side dash.

Reading Codes

1989–95 MODELS

1. Make sure the battery voltage is at least 11 volts.
2. Make sure the throttle valve is fully closed.
3. Place the gear shift lever in Neutral. Turn all accessories off.
4. The engine should be at normal operating temperature.
5. Using a jumper wire, connect terminals TE1 and E1 of the Data Link Connector 1 (DLC1).
6. Turn the ignition switch **ON**, but do not start the engine. Read the diagnostic code by the counting the number of flashes of the malfunction indicator lamp.

Fig. 55 The DLC1 terminals are protected by a lid

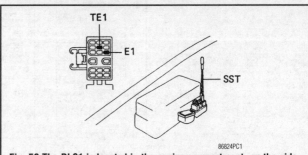

Fig. 56 The DLC1 is located in the engine compartment, on the side of the fuse block

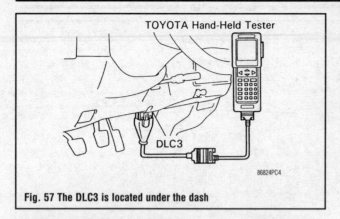

TOYOTA Hand-Held Tester

DLC3

86824PC4

Fig. 57 The DLC3 is located under the dash

7. Codes will flash in numerical order. If no faults are stored, the lamp flashes continuously every ½ second. This is sometimes called the Normal or System Clear signal.

8. After the diagnosis check, turn the ignition **OFF** and remove the jumper wire.

9. Compare the codes found to the applicable diagnostic code chart. If necessary, refer to the individual component tests in this section. If the component tests are OK, test the wire harness and connectors for shorts, opens and poor connections.

1996 MODELS

→**These models require the use of the Toyota's hand held scan tool or an equivalent OBD II compliant scan tool.**

1. Prepare the scan tool according to the manufacturers instructions.

2. Connect the OBD II scan tool, to the DLC3 under the instrument panel.

→**When the diagnosis system is switched from the normal mode to the check mode, it erases all Diagnostic Trouble Codes (DTC) and freeze frame data recorded. Before switching modes, always check the DTC and freeze frame data and write them down.**

3. Turn the ignition switch to the **ON** and switch the OBD II scan tool switch on.

4. Use the OBD II scan tool to check the DTC and freeze frame data. Write them down.

5. Compare the codes found to the applicable diagnostic code chart. If necessary, refer to the individual component tests in this section. If the component tests are OK, test the wire harness and connectors for shorts, opens and poor connections.

Clearing Trouble Codes

After repair of the circuit, the diagnostic code(s) must be removed from the ECM memory. With the ignition turned **OFF**, remove the 15 amp EFI fuse for 30 seconds or more. Once the time period has been observed, reinstall the fuse and check for normal code output.

If the diagnostic code is not erased, it will be retained by the ECM and appear along with a new code in event of future trouble.

Cancellation of the trouble code can also be accomplished by disconnecting the negative battery cable. However, disconnecting the battery cable will erase the other memory systems including the clock and radio settings. If this method is used, always reset these components once the trouble code has been erased.

Code No.	Number of blinks Malfunction Indicator Lamp	System	MIL	Diagnosis	Trouble Area
—	(continuous) FI1401	Normal	—	No trouble code is recorded.	————
12	FI1389	RPM Signal	ON	No NE signal is input to the ECM for 2 secs. or more after STA turns ON.	• Open or short in NE circuit • Distributor • Open or short in STA circuit • ECM
13	FI1390	RPM Signal	ON	NE signal is not input to ECM for 300 msec. or more when engine speed is 1,500 rpm or more.	• Open or short in NE circuit • Distributor • ECM
14	FI1391	Ignition Signal	ON	IGF signal from igniter is not input to ECM for 4 consecutive ignition.	• Open or short in IGF or IGT circuit from igniter to ECM • Igniter • ECM
21	FI1400	Main Oxygen Sensor Signal	ON	(1) Open or short in heater circuit of main oxygen sensor for 500 msec. or more. (HT) (2) At normal driving speed (below 60 mph and engine speed is above 1,700 rpm), amplitude of main oxygen sensor signal (OX1) is reduced to between 0.35—0.70 V continuously for 60 secs. or more.	• Open or short in heater circuit of main oxygen sensor • Main oxygen sensor heater • ECM • Open or short in main oxygen sensor circuit • Main oxygen sensor • ECM
22	FI1392	Engine Coolant Temp. Sensor Signal	ON	Open or short in engine coolant temp. sensor circuit for 500 msec. or more. (THW)	• Open or short in engine coolant temp. sensor circuit • Engine coolant temp. sensor • ECM
24	FI1611	Intake Air Temp. Sensor Signal	ON	Open or short in intake air temp. sensor circuit for 500 msec. or more. (THA)	• Open or short in intake air temp. circuit • Intake air temp. sensor • ECM
25	FI2562	Air-Fuel Ratio Lean Malfunction	ON	(1) Oxygen sensor output is less than 0.45 V for at least 90 secs. when oxygen sensor is warmed up (racing at 2,000 rpm). —Only for code 25. (2) When the air-fuel compensation value fluctuates more than 20% from the ECM set range within 60 secs. period while driving at 15 km/h (9 mph) or more at coolant temp. of 70°C (158°F) or above.	• Engine ground bolt loose • Open in E1 circuit • Open in injector circuit • Fuel line pressure (Injector blockage, etc.) • Open or short in oxygen sensor circuit • Oxygen sensor • Ignition system • Engine coolant temp. sensor • Volume air flow meter (Air intake) • ECM
26	FI2563	Air-Fuel Ratio Rich Malfunction	ON		• Engine ground bolt loose • Open in E1 circuit • Short in injector circuit • Fuel line pressure (Injector leakage, etc.) • Open or short in cold start injector circuit • Cold start injector • Open or short in oxygen sensor circuit • Oxygen sensor • Engine coolant temp. sensor • Volume air flow meter • Compression pressure • ECM

86824C55

Fig. 58 Diagnostic Trouble Codes—22R-E

DTC No.	Number of Malfunction Indicator Lamp Blinks	System	Malfunction Indicator Lamp Normal Mode	Malfunction Indicator Lamp Test Mode	Diagnosis	Trouble area
–		Normal	–	–	No malfunctions detected.	–
12		RPM Signal	ON	N.A.	No "NE" or "G1", "G2" signal to ECM within 2 seconds after cranking the engine.	• Distributor circuit • Distributor • Starter signal circuit • ECM
13		RPM Signal	ON	ON	No "NE" signal is to ECM for 0.1 sec. or more when engine speed is above 1,000 rpm.	• Distributor circuit • Distributor • ECM
14		Ignition Signal	ON	N.A.	No "IGF" signal to ECM 6 times in succession, and no signal input within 256 msec.	• Igniter and ignition coil circuit • Igniter and ignition coil • ECM
21		No.1 Heated Oxygen Sensor Signal	ON	ON	At normal driving speed (below 100 km/h and engine speed is above 1,700 rpm), amplitude of heated oxygen sensor signal (OX1) is reduced to between 0.35 – 0.70 V continuously for 60 seconds or more.	• Heated oxygen sensor circuit • Heated oxygen sensor • ECM
21		No.1 Heated Oxygen Sensor Heater Signal		ON	Open or short circuit in heated oxygen sensor heater. (HT1)	• Heated oxygen sensor circuit • Heated oxygen sensor • ECM
22		Engine Coolant Temp. Sensor Signal	ON	ON	Open or short circuit in engine coolant temp. sensor signal for 0.5 sec. or more. (THW)	• ECT sensor circuit • ECT sensor • ECM
24		Intake Air Temp. Sensor Signal	ON	ON	Open or short circuit in intake air temp. sensor signal for 0.5 sec. or more. (THA)	• IAT sensor circuit • IAT sensor • ECM
25		Air-Fuel Ratio Lean Malfunction	ON	ON	(1) Heated oxygen sensor output at 2,000 rpm is less than 0.45 V for at least 20 seconds when warmed up. Applies only to code 25 and for California models, excepting high-altitude areas. (2) When the engine speed varies by more than 15 rpm over the preceding crank angle period during a period of 20 seconds during idling with the intake air temp. 0°C (32°F) or above.	• Engine ground bolt loose • Open in E1 circuit • Injector circuit • Injector • Fuel line pressure • VAF meter • Air system • Heated oxygen sensor circuits • Heated oxygen sensors • Ignition system • ECM
26		Air-Fuel Ratio Rich Malfunction	ON	ON	(3) When the difference between the air-fuel ratio feedback compensation value of the front and rear exceeds 15 % of the two values in a period of 20 seconds while the engine speed is 2,000 rpm or more and the engine coolant temp. is between 70°C (158°F) and 95°C (203°F).	• Engine ground bolt loose • Open in E1 circuit • Injector circuit • Injector • Fuel line pressure • Compression pressure • ECT sensor • ECM

Fig. 60 Diagnostic Trouble Codes—3F-E

Code No.	Number of blinks Malfunction Indicator Lamp	System	MIL	Diagnosis	Trouble Area
27		Sub-Oxygen Sensor Signal	ON	(1) When sub-oxygen sensor is warmed up and full acceleration continued for 2 seconds, output of main oxygen sensor is 0.45 V or more (rich) and output of sub-oxygen sensor is 0.45 V or less (lean). (OX2) (2) Open or short detected continuously for 500 msec. or more in sub-oxygen sensor heater circuit.	• Short or open in sub-oxygen sensor circuit • Sub-oxygen sensor • Open or short in sub-oxygen sensor heater circuit • ECM
31		Volume Air Flow Meter Signal	ON	Open or short detected continuously for 500 msec. or more in volume air flow meter circuit • Open — VC or E2 • Short — VC —E2 or VS —VC	• Open or short in volume air flow meter circuit • Volume air flow meter • ECM
41		Throttle Position Sensor Signal	ON	Open or short detected in throttle position sensor signal (VTA) for 500 msec. or more.	• Open or short in throttle position sensor circuit • Throttle position sensor • ECM
42		Vehicle Speed Sensor Signal	OFF	SPD signal is not input to the ECM for at least 8 seconds during high load driving with engine speed between 2,200 and 5,000 rpm.	• Open or short in vehicle speed sensor circuit • Vehicle speed sensor • ECM
43		Starter Signal	OFF	Starter signal (STA) is not input to ECM even once until engine reaches 800 rpm or more when cranking.	• Open or short in starter signal circuit • Open or short in IG SW or main relay circuit • ECM
52		Knock Sensor Signal	ON	With engine speed 2,000 rpm or more signal from knock sensor is not input to ECM for 25 revolution. (KNK)	• Open or short in knock sensor circuit • Knock sensor (looseness, etc.) • ECM
53		Knock Control Signal	ON	The engine control computer (for knock control malfunction is detected.	• ECM
71		EGR System Malfunction	ON	With the coolant temp. at 65°C (149°F) or more, 50 seconds from start of EGR operation. The EGR gas temp. is less than 70°C (158°F) and the EGR gas temp. has risen less than 3°C during the 50 seconds.	• Open in EGR gas temp. sensor circuit • Open in VSV circuit for EGR • EGR vacuum hose disconnected, valve stuck • Cologged in EGR gas passage • ECM
51		Switch Condition Signal	OFF	Displayed when IDL contact OFF or shift position in "R", "D", "2", or "L" ranges with the check terminals E1 and TE1 connected.	• Throttle position sensor IDL circuit • PNP switch circuit • Accelerator pedal, cable • ECM

Fig. 59 Diagnostic Trouble Codes—22R-E continued

Fig. 62 (top table)

DTC No.	Number of Malfunction Indicator Lamp Blinks	System	Malfunction Indicator Lamp — Normal Mode	Malfunction Indicator Lamp — Test Mode	Diagnosis	Trouble area
53	(BE3935)	Knock Control Signal	ON	N.A.	No knock control signal to ECM for 6 crank revolutions with engine speed between 1,800 rpm and 5,200 rpm.	• ECM
55	(BE3935)	No. 2 Knock Sensor Signal (rear side)	ON	N.A.	No No. 2 knock sensor signal to ECM for 6 crank revolutions with engine speed between 1,800 rpm and 5,200 rpm.	• Open or short in No. 2 knock sensor circuit • No. 2 knock sensor (looseness) • ECM
71	(F12622)	EGR System Malfunction	ON	ON	60 seconds after the start of EGR operation, EGR gas temp. is less than 70°C (158°F) and the following (a)-(c) conditions also occurs: (a) Engine coolant temp.: 53°C (127°F) or more (b) Engine speed: 1,200 rpm or more (c) Intake air temp.: 0°C (32°F) or more	• EGR valve • EGR hose • EGR gas temp. sensor circuit • EGR gas temp. sensor • VSV for EGR • VSV circuit for EGR • ECM
81	(P09304)	TCM Communication	ON	N.A.	Open in ECT1 circuit for 2 sec. or more.	• ECT1 circuit
83	(P09304)	TCM Communication	ON	N.A.	Open in ESA1 circuit for 0.5 sec. after 0.5 sec. at idle.	• ESA1 circuit
84	(P09304)	TCM Communication	ON	N.A.	Open in ESA2 circuit for 0.5 sec. after 0.5 sec. at idle.	• ESA2 circuit
85	(P09304)	TCM Communication	ON	N.A.	Open in ESA3 circuit for 0.5 sec. after 0.5 sec. at idle.	• ESA3 circuit
51	(F11617)	Switch Condition Signal	N.A.	OFF	No "IDL" signal, "NSW" signal or "A/C" signal to ECM, with the DLC1 terminals E1 and TE1 connected.	• A/C switch circuit • A/C switch • A/C amplifier • TP sensor circuit • PNP switch circuit • PNP switch • Acceleration pedal and cable • ECM

86824C59

Fig. 62 Diagnostic Trouble Codes—3F-E continued

Fig. 61 (bottom table)

DTC No.	Number of Malfunction Indicator Lamp Blinks	System	Malfunction Indicator Lamp — Normal Mode	Malfunction Indicator Lamp — Test Mode	Diagnosis	Trouble area
28	(F12698)	No. 2 Heated Oxygen Sensor Signal	ON	ON	At normal driving speed (below 100 km/h and engine speed is above 1,700 rpm), amplitude of heated oxygen sensor signal (OX2) is reduced to between 0.35 – 0.70 V continuously for 60 seconds or more.	• Heated oxygen sensor circuit • Heated oxygen sensor • ECM
		No. 2 Heated Oxygen Sensor Heater Signal	ON		Open or short circuit in heated oxygen sensor heater. (HT2)	• Heated oxygen sensor circuit • Heated oxygen sensor • ECM
31	(F11612)	Volume Air Flow Meter Signal	ON		When idle contacts are closed and engine speed is 1,500 rpm or less, there is an open circuit in VC and VS signal or a short circuit between VS and E2.	• VAF meter circuit • VAF meter • ECM
32	(F11613)	Volume Air Flow Meter Signal	ON		Open circuit in E2 or short circuit between VC and VS.	• VAF meter circuit • VAF meter • ECM
35	(F12699)	BARO Sensor Signal	ON		Open or short circuit in BARO sensor signal for 60 seconds or more.	• ECM
41	(F11614)	Throttle Position Sensor Signal	ON		Open or short circuit in throttle position sensor signal.	• TP sensor circuit • TP sensor • ECM
42	(F11615)	Vehicle Speed Sensor Signal	OFF	OFF	No "SPD" signal for 8 seconds when engine speed 2,300 rpm or more and with vehicle not moving.	• Vehicle speed sensor circuit • Vehicle speed sensor • ECM
43	(F11616)	Starter Signal	N.A.	OFF	No "STA" signal to ECM until engine speed reaches 800 rpm with vehicle not moving.	• Ignition switch circuit • Ignition switch • ECM
52	(BE3935)	No. 1 Knock Sensor Signal (front side)	ON	N.A.	No No.1 knock sensor signal to ECM for 6 crank revolutions with engine speed between 1,600 rpm and 5,200 rpm.	• Open or short in No. 1 knock sensor circuit • No. 1 knock sensor (looseness) • ECM

86824C58

Fig. 61 Diagnostic Trouble Codes—3F-E continued

Fig. 64 Diagnostic Trouble Codes—3VZ-E and 3RZ-FE continued

DTC No.	Number of blinks Malfunction Indicator Lamp	System	Malfunction Indicator Lamp (Normal Mode / Test Mode)	Diagnosis	Trouble Area
27	F13294	Sub Heated Oxygen Sensor Signal	ON	(1) Open or short in heater circuit of sub heated oxygen sensor for 0.5 sec. or more. (HT2) (2) When sub heated oxygen sensor is warmed up and full acceleration continued for 2 seconds, output of main heated oxygen sensor is 0.45 V or more (rich) and output of sub heated oxygen sensor is 0.45 V or less (lean). (OX2)	• Open or short in heater circuit of sub heated oxygen sensor circuit • Short or open in sub heated oxygen sensor circuit • Sub heated oxygen sensor • ECM
31	F11612	Volume Air Flow Meter Signal	ON	At idling, open or short detected continuously for 0.5 sec. or more in volume air flow meter circuit. Open – VC / Short – VC – E2	• Open or short in volume air flow meter circuit • Volume air flow meter • ECM
32	F11613	Volume Air Flow Meter Signal	ON	Open or short detected continuously for 0.5 sec. or more in volume air flow meter circuit. Open – E2 / Short – VS – VC	
41	F11614	Throttle Position Sensor Signal	ON	Open or short detected in throttle position sensor signal (VTA) for 0.5 sec. or more. IDL contact is ON and VTA output exceeds 1.45 V	• Open or short in throttle position sensor circuit • Throttle position sensor • ECM
42	F11615	Vehicle Speed Sensor Signal	OFF	SPD signal is not input to the ECM for at least 8 seconds during high load driving with engine speed between 2,750 rpm and 4,000 rpm.	• Open or short in vehicle speed sensor circuit • Vehicle speed sensor • ECM
43	F11616	Starter Signal	N.A.	Starter signal (STA) is not input to ECM even once until engine reaches 800 rpm or more when cranking.	• Open or short in starter signal circuit • Open or short in IG SW or main relay circuit • ECM
52	F11618	Knock Sensor Signal	N.A.	With engine speed between 1,600 rpm – 5,200 rpm, signal from knock sensor is not input to ECM for 6 revolution. (KNK)	• Open or short in knock sensor circuit • Knock sensor (looseness, etc.) • ECM
53	F11619	Knock Control signal	N.A.	Engine speed is between 650 rpm and 5,200 rpm and engine control module (for knock control) malfunction is detected.	• ECM
71	F12622	EGR System Malfunction	ON	With the engine coolant temp. at 60°C (140°F) or more, 240 seconds from start of EGR operation. The EGR gas temp. is less than 55°C (131°F) and the EGR gas temp. has risen less than 20°C (36°C) during the 240 seconds.	• Open in EGR gas temp. sensor circuit • Open in VSV circuit for EGR • EGR vacuum hose disconnected, valve stuck • Clogged in EGR gas passage • ECM
51	F11617	Switch Condition Signal	OFF	Displayed when A/C is ON, IDL contact OFF or shift position in "R", "D", "2", or "L" positions with the DLC1 terminals E1 and TE1 connected.	• A/C switch circuit • Throttle position sensor IDL circuit • Park/Neutral position switch circuit • Accelerator pedal, cable • ECM

Fig. 63 Diagnostic Trouble Codes—3VZ-E and 3RZ-FE

DTC No.	Number of blinks Malfunction Indicator Lamp	System	Malfunction Indicator Lamp (Normal Mode / Test Mode)	Diagnosis	Trouble Area
–	F11401	Normal	–	Output when no other code is recorded.	
12	F11605	RPM Signal	N.A.	No G or NE signal is input to the ECM for 2 secs. or more after STA turns ON.	• Open or short in NE, G circuit • Distributor • Open or short in STA circuit • ECM
13	F11607	RPM Signal	ON	NE signal is not input to ECM for 0.1 sec. or more when engine speed is 1,000 rpm or more.	• Open or short in NE circuit • Distributor • ECM
14	F11608	Ignition Signal	ON	IGF signal from igniter is not input to ECM for 6 consecutive ignitions.	• Open or short in IGF or IGT circuit from igniter to ECM • Igniter • ECM
16	F13600	A/T Control Signal	N.A.	Normal signal is not output from ECU CPU.	• ECM
21	F11609	Heated Oxygen Sensor Signal	ON	(1) Open or short in heater circuit of heated oxygen sensor for 0.5 sec. or more. (HT) (2) At normal driving speed (below 60 mph and engine speed is above 1,500 rpm), amplitude of heated oxygen sensor signal (OX) is reduced to between 0.35 – 0.70 V continuously for 60 secs. or more.	• Open or short in heater circuit of heated oxygen sensor • Heated oxygen sensor heater • ECM • Open or short in heated oxygen sensor circuit • Heated oxygen sensor • ECM
22		Engine Coolant Temp. Sensor Signal	ON	Open or short in engine coolant temp. sensor circuit for 0.5 sec. or more. (THW)	• Open or short in engine coolant temp. sensor circuit • Engine coolant temp. sensor • ECM
24	F11611	Intake Air Temp. Sensor Signal	ON	Open or short in intake air temp. sensor circuit for 0.5 sec. or more. (THA)	• Open or short in intake air temp. circuit • Intake air temp. sensor • ECM
25	F12562	Air-Fuel Ratio Lean Malfunction	ON	Heated oxygen sensor output is less than 0.45 V for at least 90 secs. When heated oxygen sensor is warmed up (racing at 2,000 rpm) and drive at 50 – 100 km/h	• Engine ground bolt loose • Open in E1 circuit • Fuel line pressure (Injector blockage, etc.) • Open or short in heated oxygen sensor circuit • Heated oxygen sensor • Engine coolant temp. sensor • Volume air flow meter (Air intake) • Ignition system • ECM
26	F12563	Air-Fuel Ratio Rich Malfunction	ON	When the engine speed varies by more than 15 rpm over the preceding crankshaft position period during a period of 50 seconds during idling with the engine coolant temp. 75°C (167°F) or more.	• Engine ground bolt loose • Open in E1 circuit • Short in injector circuit • Fuel line pressure (Injector leakage, etc.) • Open or short in cold start injector circuit • Cold start injector • Open or short in heated oxygen sensor circuit • Heated oxygen sensor • Engine coolant temp. sensor • Volume air flow meter • Compression pressure • ECM

DTC No.	Detection Item	Diagnostic Trouble Code Detecting Condition
P0135	Heated Oxygen Sensor Heater Circuit Malfunction (Bank 1 Sensor 1)	When the heater operates, heater current exceeds 2 A or voltage drop for the heater circuit exceeds 5 V Heater current of 0.25 A or less when the heater operates
P0136	Heated Oxygen Sensor Circuit Malfunction (Bank 1 Sensor 2)	Voltage output of the heated oxygen sensor (bank 1 sensor 2) remains at 0.4 V or more or 0.5 V or less when the vehicle is driven at 50 km/h (31 mph) or more after the engine is warmed up
P0141	Heated Oxygen Sensor Heater Circuit Malfunction (Bank 1 Sensor 2)	Same as DTC No. P0135
P0170	Fuel Trim Malfunction	When the air fuel ratio feedback is stable after engine warming up, the fuel trim is considerably in error on the RICH side or the LEAN side (2 trip detection logic)
P0401	Exhaust Gas Recirculation Flow Insufficient Detected	After the engine is warmed up and run at 80 km/h (50 mph) for 3 to 5 minutes, small difference between value of EGR gas temp. sensor and ambient air temperature
P0402	Exhaust Gas Recirculation Flow Excessive Detected	EGR gas temp. sensor value is high during EGR cut-off when engine is cold and vacuum is applied to port E EGR valve is always open
P0420	Catalyst System Efficiency Below Threshold	After the engine is warmed up and the vehicle driven for 5 min. at 30 ~ 83 km/h (19 ~ 52 mph), the wave forms of the heated oxygen sensors, bank 1 sensor 1 and bank 1 sensor 2, have the same amplitude
P0441	Evaporative Emission Control System Incorrect Purge Flow	The proper response to the computer command does not occur

Fig. 66 Diagnostic Trouble Codes—2RZ-FE and 3RZ-FE continued

DTC No.	Detection Item	Diagnostic Trouble Code Detecting Condition
P0100	Mass Air Flow Circuit Malfunction	Open or short in mass air flow meter circuit with engine speed 4,000 rpm or less
P0101	Mass Air Flow Circuit Range/Performance Problem	Conditions (a) and (b) continue with engine speed 900 rpm or less: (2 trip detection logic) (a) Closed throttle position switch: ON (b) Mass air flow meter output > 2.2 V
P0110	Intake Air Temp. Circuit Malfunction	Open or short in intake air temp. sensor circuit
P0115	Engine Coolant Temp. Circuit Malfunction	Open or short in engine coolant temp. sensor circuit
P0116	Engine Coolant Temp. Circuit Range/Performance Problem	20 min. or more after starting engine, engine coolant temp. sensor value is 32°C (90°F) or less
P0120	Throttle/Pedal Position Sensor/ Switch "A" Circuit Malfunction	Condition (a) or (b) continues: (a) VTA< 0.1 V, and closed throttle position switch is OFF (b) VTA > 4.9 V
P0121	Throttle/Pedal Position Sensor/ Switch "A" Circuit Range/ Performance Problem	When closed throttle position switch is ON, condition (a) continues: (2 trip detection logic) (a) VTA > 2.0 V
P0125	Insufficient Coolant Temp. for Closed Loop Fuel Control	After the engine is warmed up, heated oxygen sensor output does not indicate RICH even once when conditions (a), (b) and (c) continue for at least 1.5 minutes: (a) Engine speed: 1,500 rpm or more (b) Vehicle speed: 40 ~ 100 km/h (25 ~ 62 mph) (c) Closed throttle position switch: OFF
P0130	Heated Oxygen Sensor Circuit Malfunction (Bank 1 Sensor 1)	Voltage output of heated oxygen sensor remains at 0.4 V or more, or 0.55 V or less, during idling after the engine is warmed up
P0133	Heated Oxygen Sensor Circuit Slow Response (Bank 1 Sensor 1)	Response time for the heated oxygen sensor's voltage output to change from rich to lean, or from lean to rich, is 1 sec. or more during idling after the engine is warmed up

Fig. 65 Diagnostic Trouble Codes—2RZ-FE and 3RZ-FE

DTC No.	Detection Item	Diagnostic Trouble Code Detecting Condition
P1300	Igniter Circuit Malfunction	No IGF signal to ECM for 4 consecutive IGT signals during engine running
P1335	Crankshaft Position Sensor Circuit Malfunction (during engine running)	No crankshaft position sensor signal to ECM with engine speed 1,000 rpm or more
P1500	Starter Signal Circuit Malfunction	No starter signal to ECM
P1600	ECM BATT Malfunction	Open in back up power source circuit
P1780	Park/Neutral Position Switch Malfunction	2 or more switches are ON simultaneously for "N", "2" and "L" position (2 trip detection logic) When driving under conditions (a) and (b) for 30 sec. or more the park/neutral position switch is ON (N position): (2 trip detection logic) (a) Vehicle speed: 70 km/h (44 mph) or more (b) Engine speed: 1,500 ~ 2,500 rpm

86824C63

Fig. 68 Diagnostic Trouble Codes—2RZ-FE and 3RZ-FE continued

DTC No.	Detection Item	Diagnostic Trouble Code Detecting Condition
P0500	Vehicle Speed Sensor Malfunction	No vehicle speed sensor signal to ECM under conditions (a): (a) Vehicle is being driven
P0505	Idle Control System Malfunction	Idle speed continues to vary greatly from the target speed
P0510	Closed Throttle Position Switch Malfunction	The closed throttle position switch does not turn ON even once when the vehicle is driven
P1300	Igniter Circuit Malfunction	No IGF signal to ECM for 4 consecutive IGT signals during engine running
P1335	Crankshaft Position Sensor Circuit Malfunction (during engine running)	No crankshaft position sensor signal to ECM with engine speed 1,000 rpm or more
P1500	Starter Signal Circuit Malfunction	No starter signal to ECM
P1600	ECM BATT Malfunction	Open in back up power source circuit
P1780	Park/Neutral Position Switch Malfunction	2 or more switches are ON simultaneously for "N", "2" and "L" position When driving under conditions (a) and (b) for 30 sec. or more the park/neutral position switch is ON (N position): (2 trip detection logic) (a) Vehicle speed: 70 km/h (44 mph) or more (b) Engine speed: 1,500 ~ 2,500 rpm

86824C62

Fig. 67 Diagnostic Trouble Codes—2RZ-FE and 3RZ-FE continued

DTC No.	Detection Item	Diagnostic Trouble Code Detecting Condition
P0130	Heated Oxygen Sensor Circuit Malfunction (Bank 1 Sensor 1)	Voltage output of heated oxygen sensor remains at 0.4 V or more, or 0.55V or less, during idling after the engine is warmed up
P0133	Heated Oxygen Sensor Circuit Slow Response (Bank 1 Sensor 1)	Response time for the heated oxygen sensor's voltage output to change from rich to lean, or from lean to rich, is 1 sec. or more during idling after the engine is warmed up
P0135	Heated Oxygen Sensor Heater Circuit Malfunction (Bank 1 Sensor 1)	When the heater operates, heater current exceeds 2 A or voltage drop for the heater circuit exceeds 5 V / Heater current of 0.25 A or less when the heater operates
P0136	Heated Oxygen Sensor Circuit Malfunction (Bank 1 Sensor 2)	Voltage output of the heated oxygen sensor (bank 1 sensor 2) remains at 0.4 V or more or 0.5 V or less when the vehicle is driven at 50 km/h (31 mph) or more after the engine is warmed up
P0141	Heated Oxygen Sensor Heater Circuit Malfunction (Bank 1 Sensor 2)	Same as DTC No. P0135
P0170	Fuel Trim Malfunction	When the air fuel ratio feedback is stable after engine warming up, the fuel trim is considerably in error on the RICH side or the LEAN side
P0300	Random/Multiple Cylinder Misfire Detected	Misfiring of random cylinders is detected during any particular 200 or 1,000 revolutions
P0301 P0302 P0303 P0304 P0305 P0306	Misfire Detected — Cylinder 1 — Cylinder 2 — Cylinder 3 — Cylinder 4 — Cylinder 5 — Cylinder 6	For any particular 200 revolutions of the engine, misfiring is detected which can cause catalyst overheating (This causes MIL to blink) / For any particular 1,000 revolutions of the engine, misfiring is detected which cause a deterioration in emissions

Fig. 70 Diagnostic Trouble Codes—5VZ-FE continued

DTC No.	Detection Item	Diagnostic Trouble Code Detecting Condition
P0100	Mass Air Flow Circuit Malfunction	Open or short in mass air flow meter circuit with engine speed 4,000 rpm or less
P0101	Mass Air Flow Circuit Range/Performance Problem	Conditions (a) and (b) continue with engine speed 900 rpm or less: (a) Closed throttle position switch: ON (b) Mass air flow meter output > 2.2 V
P0110	Intake Air Temp. Circuit Malfunction	Open or short in intake air temp. sensor circuit
P0115	Engine Coolant Temp. Circuit Malfunction	Open or short in engine coolant temp. sensor circuit
P0116	Engine Coolant Temp.Circuit Range/Performance Problem	20 min. or more after starting engine, engine coolant temp. sensor value is 35°C (95°F) or warmed up
P0120	Throttle/Pedal Position Sensor/Switch "A" Circuit Malfunction	Condition (a) or (b) continues: (a) VTA < 0.1 V, and closed throttle position switch is OFF (b) VTA > 4.9 V
P0121	Throttle/Pedal Position Sensor/Switch "A" Circuit Range/Performance Problem	When closed throttle position switch is ON, condition (a) continues: (a) VTA > 2.0 V
P0125	Insufficient Coolant Temp. for Closed Loop Fuel Control	After the engine is warmed up, heated oxygen sensor output does not indicate RICH even once when conditions (a), (b) and (c) continue for at least 1.5 minutes: (a) Engine speed: 1,500 rpm or more (b) Vehicle speed: 40 km/h (25 mph) or more (c) Closed throttle position switch: OFF

Fig. 69 Diagnostic Trouble Codes—5VZ-FE

DTC No.	Detection Item	Diagnostic Trouble Code Detecting Condition
P0420	Catalyst system Efficiency Below Threshold	After the engine is warmed up and the vehicle driven for 17 min. at 30 ~ 83 km/h (19 ~ 52 mph), the waveforms of the heated oxygen sensors, bank 1 sensor 1 and bank 1 sensor 2, have the same amplitude
P0441	Evaporative Emission Control System Incorrect Purge Flow	The proper response to the computer command does not occur
P0500	Vehicle Speed Sensor Malfunction	No vehicle speed sensor signal to ECM under conditions (a): (a) Vehicle is being driven
P0505	Idle Control System Malfunction	Idle speed continues to vary greatly from the target speed
P0510	Closed Throttle Position Switch Malfunction	The closed throttle position switch does not turn ON even once when the vehicle is driven

Fig. 72 Diagnostic Trouble Codes—5VZ-FE continued

DTC No.	Detection Item	Diagnostic Trouble Code Detecting Condition
P0325	Knock Sensor 1 Circuit Malfunction	No knock sensor 1 signal to ECM with engine speed 2,000 rpm or more
P0330	Knock Sensor 2 Circuit Malfunction	No knock sensor 2 signal to ECM with engine speed 2,000 rpm or more
P0335	Crankshaft Position Sensor "A" Circuit Malfunction	No crankshaft position sensor signal to ECM during cranking
		No crankshaft position sensor signal to ECM with engine speed 600 rpm or more
P0340	Camshaft Position Sensor Circuit Malfunction	No camshaft position sensor signal to ECM during cranking
		No camshaft position sensor signal to ECM during engine running
P0401	Exhaust Gas Recirculation Flow Insufficient Detected	After the engine is warmed up and run at 80 km/h (50 mph) for 3 to 5 minutes, the EGR gas temperature sensor value does not exceed 60°C (140°F) above the ambient air temperature
P0402	Exhaust Gas Recirculaton Flow Excessive Detected	EGR gas temp. sensor value is high during EGR cut-off when engine is cold (Race engine at about 4,000 rpm without load so that vacuum is applied to port E)
		EGR valve is always open

Fig. 71 Diagnostic Trouble Codes—5VZ-FE continued

DTC No.	Detection Item	Trouble Area
P0100	Mass Air Flow Circuit Malfunction	• Open or short in mass air flow meter circuit • Mass air flow meter • ECM
P0101	Mass Air Flow Circuit Range/Performance Problem	• Mass air flow meter
P0110	Intake Air Temp. Circuit Malfunction	• Open or short in intake air temp. sensor circuit • Intake air temp. sensor • ECM
P0115	Engine Coolant Temp Circuit Malfunction	• Open or short in engine coolant temp. sensor circuit • Engine coolant temp. sensor • ECM
P0116	Engine Coolant Temp. Circuit Range/Performance Problem	• Engine coolant temp. sensor • Cooling system
P0120	Throttle/Pedal Position Sensor/Switch "A" Circuit Malfunction	• Open or short in throttle position sensor circuit • Throttle position sensor • ECM
P0121	Throttle/Pedal Position Sensor/Switch "A" Circuit Range/Performance Problem	• Throttle position sensor
P0125	Insufficient Coolant Temp. for Closed Loop Fuel Control	• Open or short in heated oxygen sensor circuit • Heated oxygen sensor
P0130	Heated Oxygen Sensor Circuit Malfunction (Bank 1 Sensor 1)	• Heated oxygen sensor • Fuel trim malfunction
P0133	Heated Oxygen Sensor Circuit Slow Response (Bank 1 Sensor 1)	• Heated oxygen sensor
P0135	Heated Oxygen Sensor Heater Circuit Malfunction (Bank 1 Sensor 1) (Bank 1 Sensor 2)	• Open or short in heater circuit of heated oxygen sensor • Heated oxygen sensor heater • ECM
P0136	Heated Oxygen Sensor Circuit Malfunction (Bank 1 Sensor 2)	• Heated oxygen sensor

86824CA1

Fig. 74 Diagnostic Trouble Codes—1FZ-FE

DTC No.	Detection Item	Diagnostic Trouble Code Detecting Condition
P1300	Igniter Circuit Malfunction	No IGF signal to ECM for 6 consecutive IGT signals during engine running
P1335	Crankshaft Position Sensor Circuit Malfunction (during engine running)	No crankshaft position sensor signal to ECM with engine speed 1,000 rpm or more
P1500	Starter Signal Circuit Malfunction	No starter signal to ECM
P1600	ECM BATT Malfunction	Open in back up power source circuit
P1605	Knock Control CPU Malfunction	Engine control computer malfunction (for knock control)
		Two or more switches are ON simultaneously for "N", "2" and "L" position
P1780	Park/Neutral Position Switch Malfunction	When driving under conditions (a) and (b) for 30 sec. or more, the park/neutral position switch is ON (N position): (a) Vehicle speed: 70 km/h (44 mph) or more (b) Engine speed: 1,500 ~ 2,500 rpm

86824C58

Fig. 73 Diagnostic Trouble Codes—5VZ-FE continued

DTC No.	Detection Item	Trouble Area
P0401	Exhaust Gas Recirculation Flow Insufficient Detected	• EGR valve stuck closed • Short in VSV circuit for EGR • Open in EGR gas temp. sensor circuit • EGR hose disconnected • ECM
P0402	Exhaust Gas Recirculation Flow Excessive Detected	• EGR valve stuck open • EGR VSV open malfunction • Open in VSV circuit for EGR • Short in EGR gas temp. sensor circuit • ECM
P0420	Catalyst System Efficiency Below Threshold	• Three-way catalytic converter • Open or short in heated oxygen sensor circuit • Heated oxygen sensor
P0500	Vehicle Speed Sensor Malfunction	• Open or short in vehicle speed sensor circuit • Vehicle speed sensor • Combination meter • ECM
P0505	Idle Control System Malfunction	• IAC valve is stuck or closed • Open or short in IAC valve circuit • Open or short A/C signal circuit • Air intake (hose loose)
P0510	Closed Throttle Position Switch Malfunction	• Open in closed throttle position switch circuit • Closed throttle position switch • ECM

86824CA3

Fig. 76 Diagnostic Trouble Codes—1FZ-FE continued

DTC No.	Detection Item	Trouble Area
P0141	Heated Oxygen Sensor Heater Circuit Malfunction (Bank 1 Sensor 2)	• Same as DTC No. P0135
P0171	System too Lean (Fuel Trim)	• Air intake (hose loose) • Fuel line pressure, leak • Injector blockage • Heated oxygen sensor malfunction • Mass air flow meter • Engine coolant temp. sensor
P0172	System too Rich (Fuel Trim)	• Fuel line pressure • Injector blockage, leak • Heated oxygen sensor malfunction • Mass air flow meter • Engine coolant temp. sensor
P0300	Random/Multiple Cylinder Misfire Detected	• Ignition system • Injector • Fuel line pressure
P0301 P0302 P0303 P0304 P0305 P0306	Misfire Detected — Cylinder 1 — Cylinder 2 — Cylinder 3 — Cylinder 4 — Cylinder 5 — Cylinder 6	• EGR • Compression pressure • Valve clearance not to specification • Valve timing • Mass air flow meter • Engine coolant temp. sensor
P0325	Knock Sensor 1 Circuit Malfunction	• Open or short in knock sensor 1 circuit • Knock sensor 1 (looseness) • ECM
P0330	Knock Sensor 2 Circuit Malfunction	• Open or short in knock sensor 2 circuit • Knock sensor 2 (looseness) • ECM
P0335	Crankshaft Position Sensor "A" Circuit Malfunction	• Open or short in crankshaft position sensor circuit for NE signal • Crankshaft position sensor for NE signal • Starter • ECM
P0340	Camshaft Position Sensor Circuit Malfunction	• Open or short in camshaft position sensor circuit • Camshaft position sensor • ECM
P0385	Crankshaft Position Sensor "B" Circuit Malfunction	• Open or short in crankshaft position sensor circuit for NE2 signal • Crankshaft position sensor for NE2 signal • ECM

86824CA2

Fig. 75 Diagnostic Trouble Codes—1FZ-FE continued

VACUUM DIAGRAMS

Following is a listing of vacuum diagrams for most of the engine and emissions package combinations covered by this manual. Because vacuum circuits will vary based on various engine and vehicle options, always refer first to the vehicle emission control information label, if present. Should the label be missing, or should vehicle be equipped with a different engine from the vehicle's original equipment, refer to the diagrams below for the same or similar configuration.

If you wish to obtain a replacement emissions label, most manufacturers make the labels available for purchase. The labels can usually be ordered from a local dealer.

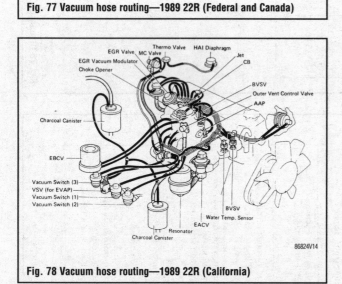

Fig. 77 Vacuum hose routing—1989 22R (Federal and Canada)

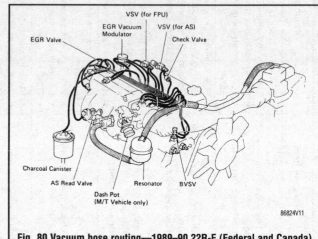

Fig. 80 Vacuum hose routing—1989–90 22R-E (Federal and Canada)

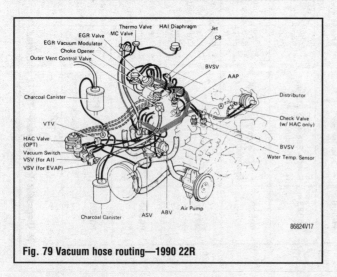

Fig. 78 Vacuum hose routing—1989 22R (California)

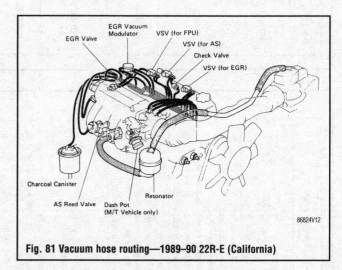

Fig. 81 Vacuum hose routing—1989–90 22R-E (California)

Fig. 79 Vacuum hose routing—1990 22R

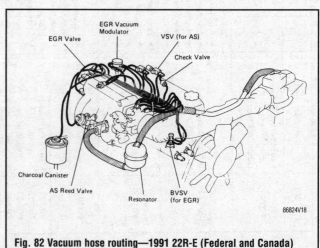

Fig. 82 Vacuum hose routing—1991 22R-E (Federal and Canada)

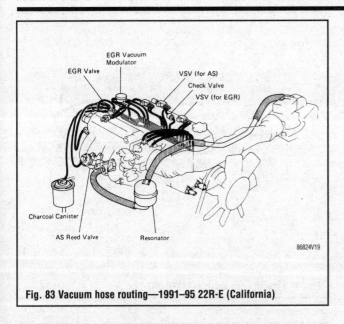

Fig. 83 Vacuum hose routing—1991–95 22R-E (California)

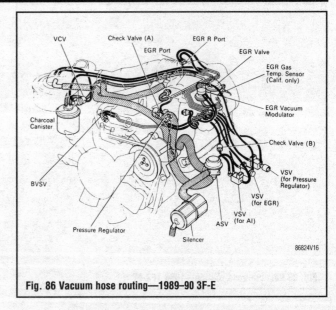

Fig. 86 Vacuum hose routing—1989–90 3F-E

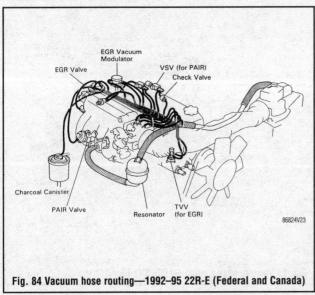

Fig. 84 Vacuum hose routing—1992–95 22R-E (Federal and Canada)

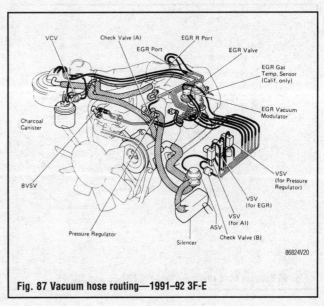

Fig. 87 Vacuum hose routing—1991–92 3F-E

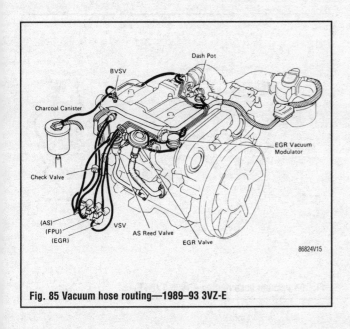

Fig. 85 Vacuum hose routing—1989–93 3VZ-E

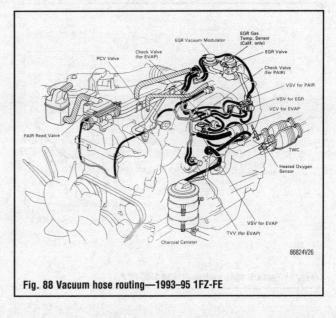

Fig. 88 Vacuum hose routing—1993–95 1FZ-FE

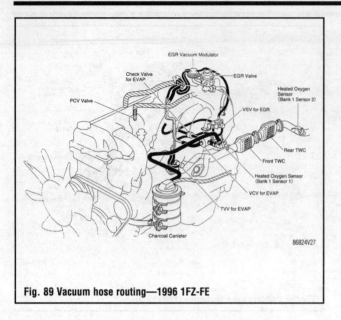

Fig. 89 Vacuum hose routing—1996 1FZ-FE

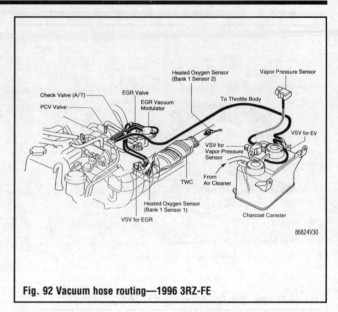

Fig. 92 Vacuum hose routing—1996 3RZ-FE

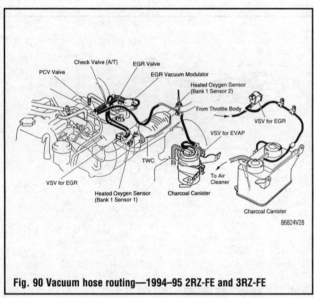

Fig. 90 Vacuum hose routing—1994–95 2RZ-FE and 3RZ-FE

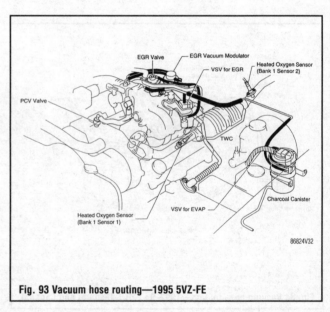

Fig. 93 Vacuum hose routing—1995 5VZ-FE

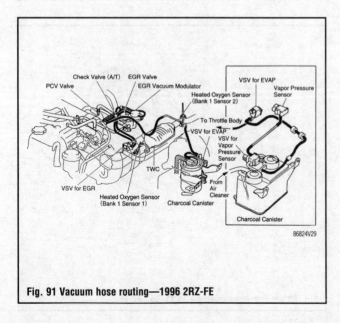

Fig. 91 Vacuum hose routing—1996 2RZ-FE

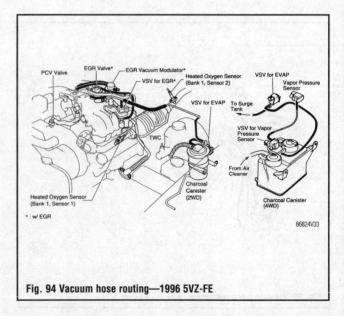

Fig. 94 Vacuum hose routing—1996 5VZ-FE

5

FUEL SYSTEM

BASIC FUEL SYSTEM DIAGNOSIS

When there is a problem starting or driving a vehicle, two of the most important checks involve the ignition and the fuel systems. The questions most mechanics attempt to answer first, "is there spark?" and "is there fuel?" will often lead to solving most basic problems. For ignition system diagnosis and testing, please refer to the information on engine electrical components and ignition systems found earlier in this manual. If the ignition system checks out (there is spark), then you must determine if the fuel system is operating properly (is there fuel?).

CARBURETED FUEL SYSTEM

Understanding the Fuel System

An automotive fuel system consists of everything between the fuel tank and the carburetor. This includes the tank itself, the fuel lines, one fuel filter, a mechanical fuel pump, and the carburetor.

With the exception of the carburetor, the fuel system is quite simple in operation. Fuel is drawn from the tank through the fuel line by the fuel pump, which forces it through the fuel filter to the carburetor, where it is distributed to the cylinders.

Precautions

- Before working on the fuel system, disconnect the negative battery cable.
- Keep gasoline off rubber or leather parts.
- When working around fuel, keep away from possible fire hazards; DO NOT SMOKE.
- Keep your work area clean to avoid contamination of the carburetor and its components.
- Work on only one component group at a time to avoid any confusion between similar looking parts.
- Be careful not to mix-up or loose any clips or springs.

Mechanical Fuel Pump

All 22R carbureted engines are equipped with a mechanically operated fuel pump of diaphragm construction (some models use two different types of pump). A separate fuel filter is incorporated into the fuel line (see Section 1 for its required service). The fuel pump is located on the right side of the cylinder head.

REMOVAL & INSTALLATION

▶ **See Figure 1**

1. Disconnect the negative battery cable.
2. Drain the radiator coolant.

✳✳ CAUTION

When draining coolant, keep in mind that cats and dogs are attracted by ethylene glycol antifreeze, and are quite likely to drink any that is left in an uncovered container or in puddles on the ground. This will prove fatal in sufficient quantity. Always drain coolant into a sealable container. Coolant may be reused unless it is contaminated or several years old.

3. Disconnect the upper radiator hose and wire it out of the way.
4. Label, disconnect and plug the three fuel lines from the fuel pump.
5. Unscrew the two fuel pump retaining bolts. Remove the fuel pump, gaskets and insulator.
6. Inspect the insulator for any cracks or damage. Replace if necessary.
7. Place new gaskets on either side of the insulator. Insert the pump arm through the hole of the insulator/gaskets and mount on the engine. Tighten the two mounting screws.

➡**Always use new gaskets when installing the fuel pump.**

TESTING

Fuel pumps should always be tested on the vehicle. The larger line between the pump and tank is the suction side of the system and the smaller line, between the pump and carburetor is the pressure side. A leak in the pressure side would be apparent because of dripping fuel. A leak in the suction side is usually only apparent because of a reduced volume of fuel delivered to the pressure side.

✳✳ CAUTION

Gasoline, both liquid and vapor, is extremely explosive. Extinguish all open flames in the area including smoking materials, heaters, welders, etc. Contain all spillage and keep away from sources of high temperature. Have a dry powder (Type B-C) fire extinguisher within arm's reach and know how to use it.

1. Tighten any loose line connections and look for any kinks or restrictions.
2. Disconnect the fuel line at the carburetor or fuel pump. Disconnect the distributor-to-coil primary wire. Place a container at the end of the fuel line and crank the engine a few revolutions. If little or no fuel flows from the line, either the fuel pump is inoperative or the line is plugged. Blow through the lines with compressed air and try the test again. Reconnect the line.
3. If fuel flows in good volume, check the fuel pump pressure to be sure.
4. Attach a pressure gauge to the pressure side of the fuel line. On trucks equipped with a vapor return system, squeeze off the return hose.
5. Run the engine at idle and note the reading on the gauge. Stop the engine and compare the reading with the specifications listed in the "Tune-Up Specifications" chart. If the pump is operating properly, the pressure will be as specified and will be constant at idle speed. If pressure varies sporadically or is too high or low, the pump should be replaced.
6. Remove the pressure gauge.

Carburetor

The carburetor is the most complex part of the fuel system. Carburetors vary greatly in construction, but they all operate basically the same way; their job is to supply the correct mixture of fuel and air to the engine in response to varying conditions.

Despite their complexity in operation, carburetors function because of a simple physical principle; the venturi principle. Air is drawn into the engine by the pumping action of the pistons. As the air enters the top of the carburetor, it passes through a venturi, which is nothing more than a restriction in the throttle bore. The air speeds up as it passes through the venturi, causing a slight drop in pressure. This pressure drop pulls fuel from the float bowl through a nozzle into the throttle bore, where it mixes with the air and forms a fine mist, which is distributed to the cylinders through the intake manifold.

There are multiple systems (air/fuel circuits) in a carburetor that make it work; the

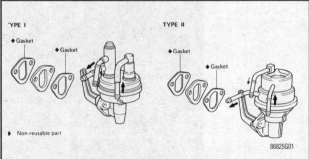

TYPE I   TYPE II

◆Gasket   ◆Gasket   ◆Gasket   ◆Gasket

▶ Non-reusable part

86825G01

Fig. 1 Mechanical fuel pump types with their insulators and gaskets

- Float
- Main Metering
- Idle
- Low-Speed
- Accelerator Pump
- Power
- Choke system

The way these systems are arranged in the carburetor determines the carburetor's size and shape.

Carburetors all function in the same fashion; larger engines have larger carburetors to move more air and fuel, but the principle is still the same. Older units don't have as many external linkages and controls to manage emissions and driveability, but the principle is still the same.

It's important to remember that carburetors seldom give trouble during normal operation. Other than changing the fuel and air filters and making sure the idle speed is OK at every tune-up, there's not much maintenance you can perform on the average carburetor. Quality of fuel and presence of water in the system will affect the carburetor; dirt particles in the fuel can clog the jets and water causes rust and corrosion. If the vehicle is to be parked or stored for a long period of time, drain the carburetor to prevent the evaporating fuel from gumming up the system.

The carburetors used on Toyota trucks are conventional 2 bbl, downdraft types similar to domestic carburetors. The main circuits are:

- Primary—for normal operational requirements
- Secondary—to supply high speed/high load fuel needs
- Float—to supply fuel to the primary and secondary circuits
- Accelerator—to supply fuel for quick and safe acceleration
- Choke—for reliable starting in cold weather
- Power valve—for fuel economy

ADJUSTMENTS

Fast Idle

OFF THE TRUCK

▶ See Figures 2 and 3

➡An angle gauge tool is used to properly make this adjustment.

1. Remove the carburetor.
2. Set the throttle shaft lever to the first step of the fast idle cam. Make certain the choke plate (blade) is completely closed.
3. Attach the blade angle tool to the primary throttle blade. Adjust the primary throttle blade angle to 23° (Calif.) or 24.5° (Federal and Canada) from horizontal by turning the fast idle screw.
4. Remove the angle tool from the carburetor and install the carburetor as previously outlined.

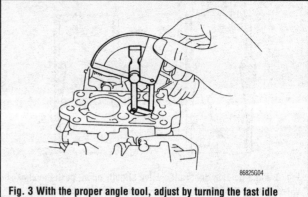

Fig. 3 With the proper angle tool, adjust by turning the fast idle screw

ON THE TRUCK

▶ See Figures 4, 5, 6 and 7

1. Start the engine and allow it to reach normal operating temperature.
2. Stop the engine and connect a tachometer to the engine.
3. Remove the air cleaner assembly.
4. Disconnect and plug the vacuum hoses for the Hot Air Intake (HAI) and Mixture Control (MC) systems if so equipped. Disconnect and plug the vacuum hose from the choke opener diaphragm and Exhaust Gas Recirculation (EGR) valve; plug the ends.

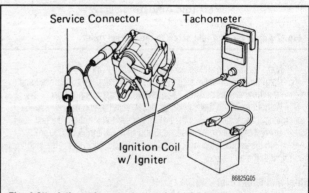

Fig. 4 Attach the tachometer test probe to the ignition coil negative terminal

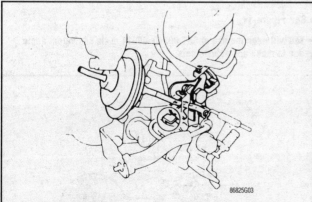

Fig. 2 Set the throttle shaft lever to the first step of the fast idle cam

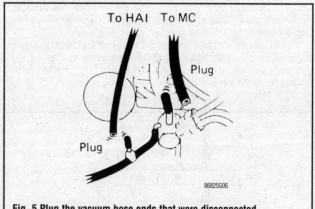

Fig. 5 Plug the vacuum hose ends that were disconnected

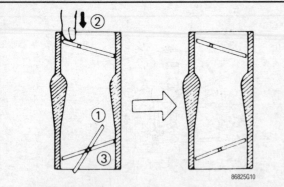

Fig. 6 While holding the throttle valve slightly open, push the choke valve closed. Hold it closed as you release the throttle valve

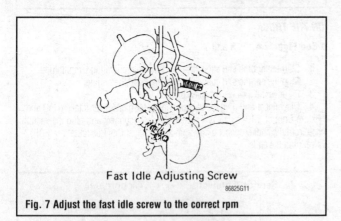

Fast Idle Adjusting Screw

Fig. 7 Adjust the fast idle screw to the correct rpm

5. Make certain that all accessories are switched OFF.

6. Open the throttle valve slightly, push the choke plate fully closed and release the throttle. This sets the carburetor in the fast idle position.

7. Without touching the accelerator pedal, start the engine and read the tachometer. If necessary, adjust the fast idle speed to by turning the fast idle screw. Correct fast idle speed: Federal, 3000 rpm; California, 2600 rpm.

8. Stop the engine. Reconnect the vacuum hoses, disconnect the tachometer and reinstall the air cleaner.

Float and Fuel Level

▶ **See Figures 8, 9 and 10**

Float level adjustments are unnecessary if the fuel level falls within the lines on the sight glass when the engine is running. The sight glass is located on the side of the carburetor and is literally a window to the float bowl. Removing the air cleaner is usually required for access, although it can be done with an extension or dental mirror.

With the carburetor off the engine, there are two float level adjustments which may be made. One is done with the air horn inverted, so that the float is in a fully raised position; the other is with the air horn in an upright position, so that the float falls to the bottom of its travel.

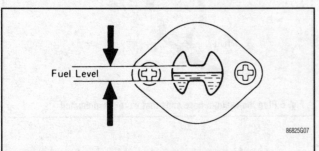

Fuel Level

Fig. 8 The sight glass is located on the side of the carburetor

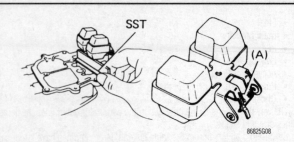

SST (A)

Fig. 9 Bend portion (A) of the float to achieve a 0.386 in. (9.8mm) gap

The float level is measured either with a special carburetor float level gauge, which comes with a rebuilding kit, or with a standard wire gauge.

1. Turn the air horn upside down and let the float hang down by its own weight.

2. Using a special float gauge SST 09240–00014 or equivalent, check the clearance between the tip of the float and the flat surface of the air horn. The clearance should be 0.386 in. (9.8mm).

➡**This measurement should be made without the gasket on the air horn.**

3. If the float clearance is not within specifications, adjust it by bending the upper (A) float tab.

4. Lift up the float and check the clearance between the air horn and the float bottom. A Vernier caliper works well for this measurement. The clearance should be 48mm (1.89 inches)

5. If the clearance is not within specifications, adjust it by bending the lower float tab (B).

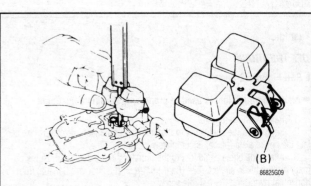

(B)

Fig. 10 Bend portion (B) of the float to set 1.89 in. (48mm) of clearance between the air horn and the float bottom

Choke Unloader

▶ **See Figure 11**

➡**You will need the use of SST 09240–00014 or its equivalent angle gauge to make any adjustments.**

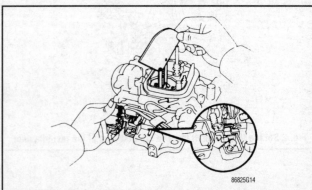

Fig. 11 Bend the primary throttle arm 45° from the horizontal plane

The unloader adjustment is made with the primary throttle valve fully open. With the valve open, check the choke valve angle with a special gauge supplied in the rebuilding kit or with a gauge of the proper angle fabricated out of cardboard. The angle of the choke valve opening should be 45°.

To adjust the angle, bend the fast idle lever until the proper measurement is achieved.

Choke Opener

▶ See Figures 12 and 13

1. Apply vacuum to the choke opener diaphragm.
2. Check that the fast idle cam is releasing to the fourth step. Adjust it by bending the choke opener lever A.
3. Disconnect the vacuum hose.
4. Close the choke valve, then set the fast idle lever to the first step.
5. Check that there is clearance between the choke opener lever and the fast idle cam.

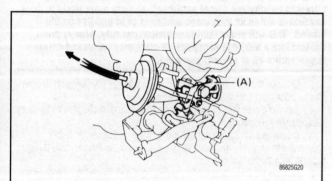

Fig. 12 With vacuum attached to the diaphragm, check that the fast idle cam is released to the fourth step. Adjust by bending the choke opener lever A

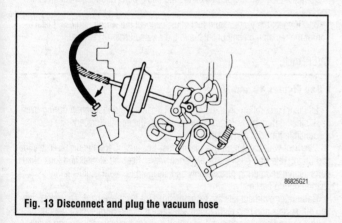

Fig. 13 Disconnect and plug the vacuum hose

Choke Breaker

1. With the engine **OFF** and the air cleaner removed, apply vacuum to the choke breaker diaphragm.
2. Close the choke by hand, and remove your hand. With an angle gauge, check that the angle of the choke plate is 42° from horizontal.
3. If the angle is incorrect, adjust it by bending the choke breaker link.

Idle Mixture Screw

▶ See Figures 14, 15, 16, 17 and 18

As stated in section 2, the mixture adjusting screw (MAS) should be the very last item you try to adjust during tune-up or troubleshooting. The MAS is concealed behind a plug; the plug cannot be removed with the carburetor on the truck. If adjustment is to be done, great care must be taken during removal of the plug; clearances are very tight and damage to the carburetor can occur.

1. Tag and disconnect all hoses and linkages attached to the carburetor.
2. Remove the carburetor.
3. Plug each carburetor vacuum port to prevent entry of metal particles.
4. Mark the center of the MAS plug with a punch. Drill a 0.256 in. (6.5mm) hole in the center of the plug.

✳✳ WARNING

The head of the screw is only 0.04 in. (1mm) below the plug—drill carefully and slowly to avoid damage.

5. The plug may come out with the drill at this time. If not, use a small screwdriver to reach through the hole and gently turn the adjusting screw all the way in. Do NOT overtighten the screw; just tighten it until it touches bottom.
6. Use a 0.295 in. (7.5mm) drill to force the plug off.
7. Remove the adjusting screw. Inspect the tip for any damage; remove any steel particles. If the drill has damaged the top of the screw, it must be replaced.
8. Reinstall the adjusting screw. Turn it all the way in, just touching bottom.
9. Once the screw has bottomed, turn it out 3½ turns.

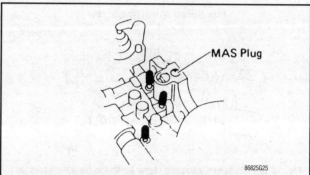

Fig. 14 After plugging all of the carburetor lines, punch a hole in the center of the MAS plug

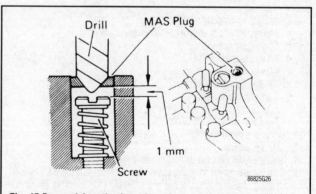

Fig. 15 Be careful, notice how close the screw is when you drill 0.256 in. (6.5mm) into the plug

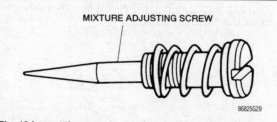

Fig. 16 Inspect the screw for any damage during removal; if any is found, replace the screw

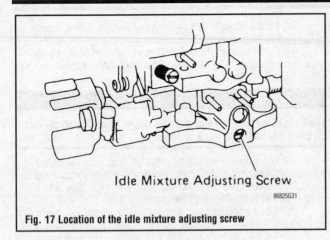

Idle Mixture Adjusting Screw

86825G31

Fig. 17 Location of the idle mixture adjusting screw

Idle Speed Adjusting Screw

86825G32

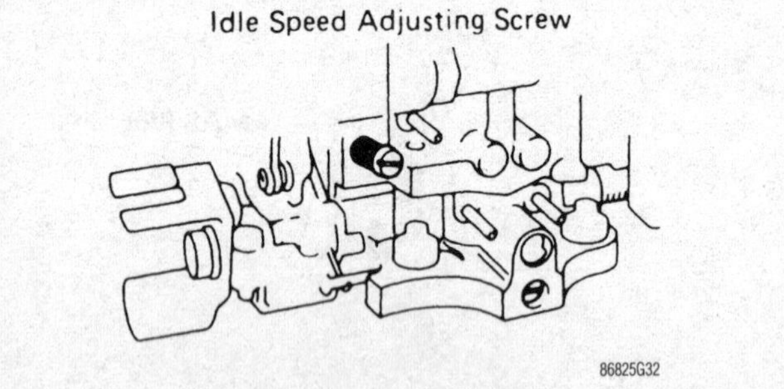

Fig. 18 The idle speed adjusting screw protrudes from the carburetor

10. Reinstall the carburetor and air cleaner.

11. Before adjusting idle speed and mixture, ALL the following conditions must be met:
   a. Air cleaner properly installed.
   b. Engine running at normal operating temperature.
   c. Choke fully opened.
   d. All accessories OFF.
   e. All vacuum lines connected.
   f. Ignition timing correctly set.
   g. Transmission in neutral.
   h. Fuel level should be approximately centered in float glass.
   i. For Calif. vehicles, EBCV off.

12. Connect a tachometer.

13. Start the engine.

14. Turn the MAS until the highest possible rpm is achieved.

15. Turn the idle speed adjusting screw until 740 rpm is achieved. The idle speed adjusting screw is located above and to the left (10 o'clock) of the mixture adjusting screw.

16. Repeat Steps 13 and 14 several times. When the idle does not rise no matter how much the MAS is turned, proceed to the next Step.

17. Turn the MAS screw IN to set the idle to 700 rpm.

18. Turn the engine OFF. Remove the air cleaner.

19. Gently tap a new plug into place over the MAS plug. Check and adjust the fast idle speed as described in this section.

➡The method used to set the idle mixture and speed is also known as the Lean Drop Method.

Automatic Choke Inspection

The automatic choke system sets and opens completely automatically, with no input from the accelerator pedal. The driver no longer sets the choke; it is set by the expansion or contraction of the bimetal coil. Refer to section 4, Emission Controls, for complete inspection procedures.

REMOVAL & INSTALLATION

1. Disconnect the negative battery cable.
2. Loosen the radiator drain plug and drain the coolant into a suitable container.

✳✳ CAUTION

When draining coolant, keep in mind that cats and dogs are attracted by ethylene glycol antifreeze, and are quite likely to drink any that is left in an uncovered container or in puddles on the ground. This will prove fatal in sufficient quantity. Always drain coolant into a sealable container. Coolant may be reused unless it is contaminated or several years old.

3. Unscrew the mounting screws and remove the air filter housing. Disconnect all hoses and lines leading from the air cleaner.
4. Tag and disconnect all fuel, vacuum, coolant and electrical lines or hoses leading from the carburetor.
5. Disconnect the accelerator linkage from the carburetor. On trucks equipped with an automatic transmission, disconnect the throttle cable linkage running from the transmission.
6. Remove the four carburetor mounting bolts and lift off the carburetor and its gasket.

➡Cover the manifold opening with a clean rag to prevent objects from falling into the engine.

7. Install the carburetor, tighten the mounting bolts and reconnect all linkages.
8. Connect the vacuum and fuel lines; connect the wiring harness. Installation of the remaining components is in the reverse order of removal.

OVERHAUL

▶ **See Figures 19 and 20**

Efficient carburetion depends on careful cleaning and inspection during overhaul, since dirt, gum, water, or varnish in or on the carburetor parts are often responsible for poor performance.

Overhaul your carburetor in a clean, dust-free area. Carefully disassemble the carburetor, referring often to the exploded views. Keep all similar and look-alike parts segregated during disassembly and cleaning to avoid accidental interchange during assembly.

Carburetor overhaul kits are recommended for each overhaul. These kits contain all gaskets and new parts to replace those that deteriorate most rapidly. Failure to replace all parts supplied with the kit (especially gaskets) can result in poor performance and a leaks.

➡The following procedure is organized so that only one group of components is worked on at a time. This will help eliminate confusion of parts or sub-assemblies on the bench. Always keep parts in order; take great care not to lose small parts or clips.

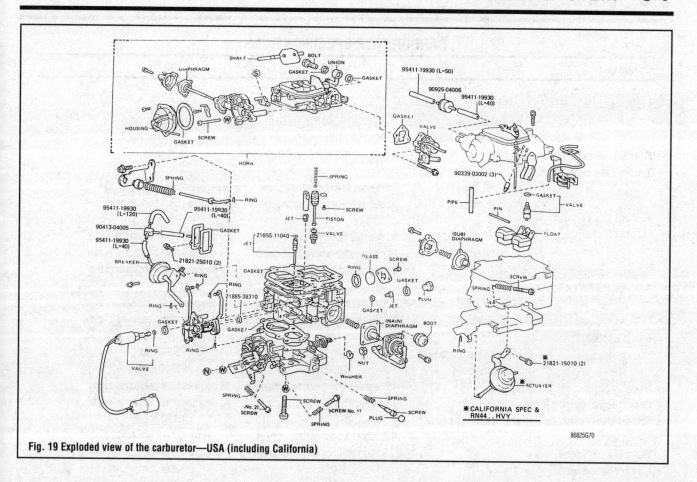

Fig. 19 Exploded view of the carburetor—USA (including California)

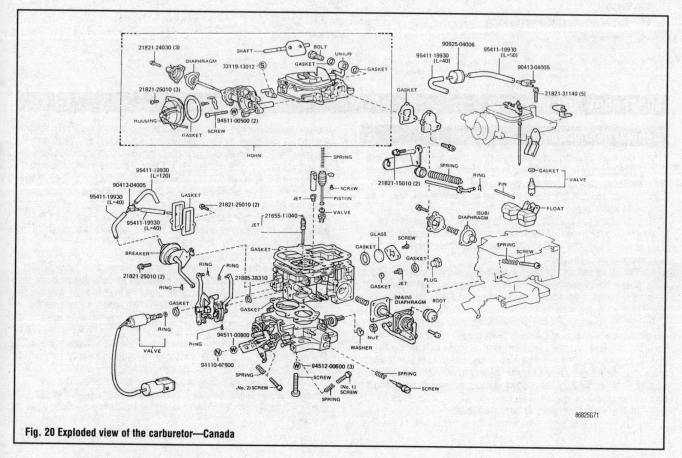

Fig. 20 Exploded view of the carburetor—Canada

CARBURETOR SPECIFICATIONS

Component		U.S.	Metric
Float level			
	Raised position	0.386 in.	9.8mm
	Lowered position	1.89 in.	48mm
Float lip clearance		0.04 in.	1mm
Throttle valve closed angle			
	Primary	9 degrees from horizontal plane	9 degrees from horizontal plane
	Secondary	20 degrees from horizontal plane	20 degrees from horizontal plane
Throttle valve full open angle			
	Primary	90 degrees from horizontal plane	90 degrees from horizontal plane
	Secondary	90 degrees from horizontal plane	90 degrees from horizontal plane
Secondary touch angle		59 degrees from horizontal plane	59 degrees from horizontal plane
Fast idle angle			
	Federal and Canada	24.5 degrees from horizontal plane	24.5 degrees from horizontal plane
	California	23 degrees from horizontal plane	23 degrees from horizontal plane
Fast idle speed			
	Federal and Canada	3000 rpm	3000 rpm
	California	2600 rpm	2600 rpm
Unloader angle		45 degrees from horizontal plane	45 degrees from horizontal plane
Choke breaker opening angle		42 degrees from horizontal plane	42 degrees from horizontal plane
Choke heater resistance	at 68 deg F (20 deg C)	20-22 ohms	20-22 ohms
Idle-up angle		16.5 degrees from horizontal plane	16.5 degrees from horizontal plane
Dashpot touch angle		24.5 degrees from horizontal plane	24.5 degrees from horizontal plane
Dashpot setting speed		3000 rpm	3000 rpm
Idle speed		700 rpm	700 rpm
Idle mixture adjusting screw presetting		screw-out 3 1/2 turns	screw-out 3 1/2 turns
Idle mixture speed		740 rpm	740 rpm

86825C55

MULTI-PORT FUEL INJECTION

General Information

Most of the Toyota trucks that are covered by this manual are equipped with Multi–Port Electronic Fuel Injection system. The Electronic Fuel Injection System is particularly good because it requires no venturi restriction. That means that it has a greater capacity for total power than an equivalent carburetor since air flows through it more freely.

Also, carburetors are limited in their ability to accurately measure airflow because they do not compensate for changes in barometric pressure or temperature, and cannot accurately and quickly respond to throttle changes. This system overcomes those limitations.

Additionally, by injecting the fuel with nearly 40 psi (275 kPa) pressure right at the intake ports, the engine is fed a more consistently atomized mixture with the injection system. This improves performance and fuel economy, especially during cold engine operation.

The heart of the system is a fuel supply loop which consists of a fuel pump and damper which supply fuel under a relatively smooth and constant pressure through a filter. Fuel is supplied to a loop which passes by each fuel injector (one for each cylinder). The loop returns to the fuel tank via a pressure regulator which measures the pressure or vacuum (depending upon throttle opening) that exists in the intake manifold. The pressure regulator, by precisely controlling the return of fuel to the tank, maintains pressure in the fuel line at 38–44 psi (265–304 kPa) above the pressure in the intake manifold.

Each fuel injector consists of a nozzle and a tiny electric solenoid valve. The system employs electronics to determine how long the solenoid should be open each time it inputs fuel. By varying the electric pulse sent to the injector and, therefore, the time each injector stays open, the system can precisely supply the amount of fuel the engine requires. For example, if the driver should open the throttle and double the amount of airflow to the engine, the injectors would stay open twice as long as before the driver accelerated. Air entering into the system flows through the air cleaner, and then enters the air flow meter. This device measures the flow of air through a precisely designed flap, which opens or closes with airflow against the tension of a spring. A temperature measuring device compensates for the increased density of cold air.

From here, the air passes into the throttle body. This part contains the throttle which controls airflow according to the driver's demands, much the same as in an ordinary carburetion system. There are switches which tell the electronic microprocessor which controls the system when the truck is at full throttle or at idle, for appropriate changes in fuel/air mixture. When the driver releases the throttle pedal, the throttle body closes off entirely, and air for engine idle is routed through the auxiliary air valve to the engine.

Whenever there is a change in airflow, it is reflected as a change in the air flow meter's signal to the microprocessor and a change in the width (or time) of the pulse the microprocessor sends to the injectors.

The system is extremely complex and requires specialized training and equipment to fully service and repair. You can, however, make a few simple checks and adjustments, provided you adhere to the safety precautions in this section.

FUEL DELIVERY

The MFI system supplies the optimum air/fuel mixture to the engine under all various operating conditions. System fuel, is injected into the intake air passage of the cylinder head. The amount of fuel injected is controlled by the intermittent injection system where the electromagnetic injection valve (fuel injector) opens only for a short period of time, depending on the amount of air required for 1 cycle of operation. During system operation, the amount injected is determined by the duration of an electric pulse sent to the fuel injector, which permits precise metering of the fuel.

All the operating conditions of the engine are converted into electric signals, resulting in additional features of the system, such as improved adaptability and easier addition of compensating element. The MFI system also incorporates the following features:

• Reduced emission of exhaust gases

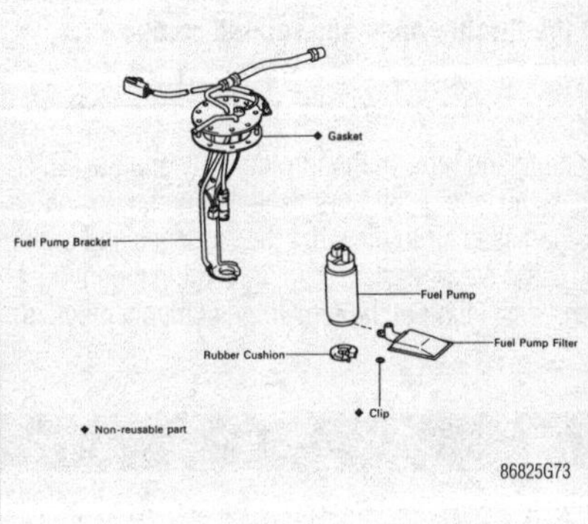

Fig. 21 The electric fuel pump has a filter; check it for any clogging or tears

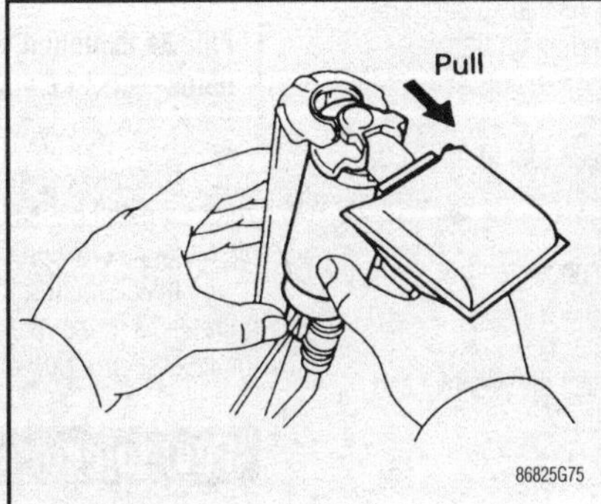

• Reduction in fuel consumption
• Increased engine output
• Superior acceleration and deceleration
• Superior starting and warm-up performance in cold weather since compensation is made for coolant and intake air temperature

Safety Precautions

• Do not connect the battery, or jumper cables, with reverse polarity.
• Never disconnect the battery while the engine is running.
• Turn **OFF** the ignition switch before disengaging any connector in the system.
• Keep all connectors dry.
• If handling electronic parts, do not drop them or permit them to be banged against another part while removing or installing them.

Relieving Fuel System Pressure

To relieve the fuel pressure, wrap a clean towel or rag around either the cold start injector fitting or the union bolt (banjo bolt) at the fuel delivery rail. Slowly loosen the bolt holding the fuel line and allow pressurized fuel to escape into the rag. Do NOT allow fuel to spray from the fitting; always wrap it in a rag.

➡Pressure may also be relieved at the fuel filter if desired. Tighten the union bolt.

Electric Fuel Pump

All Toyota fuel-injected vehicles are equipped with an electric fuel pump. The pump is located inside the fuel tank. For a fuel injection system to work properly, the pump must develop pressures well above those of a mechanical fuel pump. This high pressure is maintained within the lines even when the engine is not running. Extreme caution must be used to safely release the pressurized fuel before any work is begun.

※※ CAUTION

Always relieve the fuel pressure within the system before any work is begun on any fuel component. Failure to safely relive the pressure may result in fire and/or serious injury.

REMOVAL & INSTALLATION

◗ **See Figures 21, 22 and 23**

1. Disconnect the negative battery cable.
2. Safely relieve the fuel pressure within the system. Drain the fuel from the tank, then remove the fuel tank.
3. Remove the pump retaining bolts, then pull the fuel pump bracket up and out of the fuel tank.

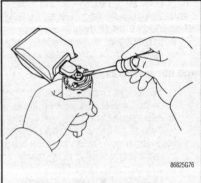

Fig. 22 The pump should can be pulled from the lower side of the bracket in most cases

Fig. 23 The filter has a retaining clip; pry it off to separate the filter and pump

4. Remove the gasket from the bracket.
5. Remove the two nuts and then disconnect the wires at the fuel pump.
6. Pull the fuel pump out of the lower side of the bracket. Disconnect the pump from the fuel hose.
7. Remove the rubber cushion and the clip. Disconnect the fuel pump filter from the pump.
8. Install the fuel pump and filter into the bracket, install the assembly and tighten the mounting bolts to 30–35 inch lbs. (3–4 Nm). Make sure to use a new gasket. Install the fuel tank.

TESTING

1989–95 Models

1. Turn the ignition switch to the **ON** position, but don't start the engine.
2. With a jumper wire, connect terminals B+ and FP of the Data Link Connector (DLC1). This should cause the fuel pump to run; check that there is pressure in the fuel return line from the pressure regulator. You should also hear fuel return noises.
3. Turn the ignition switch **OFF** and remove the jumper wire. If the fuel pump did not function or if there was no pressure in the line, check the fuses, links, EFI main relay, ECM and wiring connections.

1996 Models

2RZ-FE AND 3RZ-FE ENGINE

1. To check the fuel pump operation:
    a. Remove the fuse cover on the instrument panel.
    b. Connect the Toyota hand-held tester or its equivalent scan tool to the Data Link Connector (DLC3).

➡**Do not start the engine.**

    c. Turn the ignition switch ON and push the tester main switch on.
    d. On the Toyota tester select the active test mode.

e. Refer to the tester instructions for further details.

f. If you do not have a Toyota hand-held tester, connect the positive and negative leads from the battery to the fuel pump connector.

g. Check that there is pressure in the fuel inlet hose from the fuel filter. If there is pressure, you will hear the sound of fuel flowing.

h. If there is no pressure, check the following:
- M-fuse (AM2 30A)
- Fuses (EFI 15A, IGN 7.5A)
- EFI main relay
- Fuel pump
- ECM
- Wiring connections
    i. Turn the ignition switch to **LOCK**.
    j. Disconnect the Toyota hand-held tester from the DLC3.
2. Check the fuel pressure.
    a. Disconnect the negative cable from the battery.
    b. Wrap a shop towel around the delivery pipe, then slowly loosen the union bolt holding the fuel pipe to the delivery pipe and gasket.
    c. Install the SST 09268–45012 (pressure gauge) or its equivalent to the delivery pipe with the two gaskets and the SST (union and union bolt). Tighten to 22 ft. lbs. (29 Nm).
    d. Be sure to wipe off any spilled fuel.
    e. Attach the Toyota hand-held tester or equivalent scan tool to the DLC3.

➡**DO NOT start the engine!**

f. Turn the ignition switch ON and push the tester main switch on.
g. On the Toyota tester select the active test mode.
h. Refer to the tester instructions for further details.
i. If you do not have a Toyota hand-held tester, connect the positive and negative leads from the battery to the fuel pump connector.
j. Reconnect the negative battery cable.
k. Turn the ignition switch ON.
l. Measure the fuel pressure, it should read 38–44 psi (265–304 kPa).
m. If the pressure is too high, replace the pressure regulator.
n. If the pressure is too low, check the following:
- Fuel connections and hoses
- Fuel pump
- Fuel filter
- Pressure regulator
- Injectors
    o. Disconnect the tester from the DLC3.
    p. Start the engine.
    q. Measure the fuel pressure at idle. It should read 31–37 psi (206–255 kPa). If the pressure is not as specified, check the sensing hose and pressure regulator.
    r. Stop the engine.
    s. Check that the fuel pressure remains at 21 psi (147 kPa) for 5 minutes after the engine has stopped. If the pressure is not as specified, check the pump, pressure regulator and/or injectors.
    t. After checking the fuel pressure, disconnect the negative cable from the battery, then carefully remove the SST 09268–45012 or equivalent, to prevent gas from splashing.
    u. Reconnect the fuel inlet pipe to the delivery pipe with new gaskets and the union bolt. Tighten to 22 ft. lbs. (29 Nm).
    v. Reconnect the negative cable to the battery.
    w. Check for fuel leaks.

Throttle Body

REMOVAL & INSTALLATION

♦ **See Figure 24**

1. Disconnect the negative battery cable.
2. Drain the engine coolant..
3. Tag and disconnect all lines, hoses or wires that lead from the throttle body. On models equipped with an automatic transmission, disconnect the throttle control cable.
4. Remove the air intake connector and the air cleaner hose with resonator.

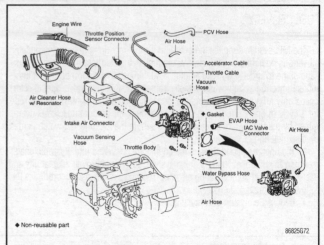

♦ Non-reusable part

86825G72

Fig. 24 Exploded view of the throttle body and related components—3RZ-FE shown

5. Unscrew the mounting bolts and remove the throttle body and gasket. The cast metal parts may be cleaned with a soft brush and carburetor solvent. If possible, use low pressure compressed air to blow out the passages and tubing.

6. Install the throttle body using a new gasket, then tighten the mounting bolts 9–13 ft. lbs. (12–18 Nm). Installation of the remaining components is in the reverse order of removal.

Fuel Injectors

OPERATION

There is one fuel injector for each cylinder and a cold start injector which provides additional fuel during start-up. The injectors spray fuel into the intake port, in front of the intake valve. When the injector is energized, the coil pulls the plunger up, opening the needle valve and allowing the pressurized fuel to pass through the injector. Injectors operation is controlled by the EFI computer. The injectors operate at low pressure and are open for only a fraction of a second at a time.

REMOVAL & INSTALLATION

22R-E, 3VZ-E and 3F-E Engines

1. Disconnect the negative battery cable.
2. Safely discharge the fuel pressure in the system. Place a suitable container under the intake manifold to catch any dripping fuel.
3. On 3VZ-E and 22R-E engines, drain the coolant.
4. Disconnect the accelerator and throttle cable.
5. Remove the air intake chamber. Removal of the throttle body is required on 22R-E engine. On the 3F-E engine, make particularly certain all vacuum hoses and wires are labeled; they are numerous.
6. Tag and disconnect all hoses and wires which interfere with injector removal.
7. On 3VZ-E and 3F-E engines, the main engine wiring harness must be moved out of the way. This will require unplugging various connectors at components on the engine until the harness can be repositioned. Label everything before disconnecting it.
8. Unplug the wiring connectors from the tops of the fuel injectors and remove the two plastic clamps that hold the wiring harness to the fuel delivery pipe.
9. For the 22R-E and 3F-E engines, disconnect the fuel line to the fuel delivery rail. Carefully remove the fuel rail (with the injectors attached) from the engine. Do not allow the injectors to fall out during removal.
10. For the 3VZ-E engine, disconnect the vacuum hose from the Bimetal Vacuum Switching Valve (BVSV). Remove the four union bolts and Nos. 2 and 3 fuel pipes. Pull the injectors from the delivery pipes.

To install:

11. Insert new insulators into the injector holes on the intake manifold.

12. Install the grommet and a new O-ring to the delivery pipe end of each injector.

13. Apply a thin coat of gasoline to the O-ring on each injector and then press them into the delivery pipe. Make absolutely certain the injector is squarely and firmly mounted in the pipe; it must not be crooked.

14. Install the injectors together with the delivery pipe in the intake manifold. Tighten the mounting bolts to 11–15 ft. lbs. (15–20 Nm). Once installed, make sure the injectors turn freely in their mounts. If one doesn't turn, remove the injector and check the O-ring.

15. Installation of the remaining components is in the reverse order of removal.

2RZ-FE and 3RZ-FE Engines

1. Disconnect the negative battery cable.

2. Safely discharge the fuel pressure in the system. Place a suitable container under the intake manifold to catch any dripping fuel.

3. Remove the throttle body.

4. Unplug the four injector connectors, crankshaft position sensor and knock sensor connectors.

5. Disconnect the Data Link Connector (DLC1) and wire clamp from the brackets.

6. Remove the delivery pipe and injectors from the engine.

7. Disconnect the vacuum sensing hose from the fuel pressure regulator.

8. Place a suitable container and a shop rag under and around the delivery pipe to keep the fuel from spurting out. Then slowly loosen the union bolt and 2 gaskets, disconnect the fuel inlet pipe from the delivery pipe.

9. Remove the 2 bolts and the delivery pipe together with the 4 injectors.

➡Be careful not to drop the injectors when you are removing the injector pipe.

10. Remove the 4 insulators from the spacers.

11. Pull out the injectors from the delivery pipe.

12. Remove the O-ring and grommet from each injector.

To install:

13. Install a new grommet to the injector.

14. Apply a light coat of gasoline to the new O-ring, then install it to the injector.

15. While turning the injector to the left and right, install it to the delivery pipe. Install the injectors.

16. Position the injector connector upward.

17. Place the new insulators in position on the spacers. Place the injectors together with the delivery pipe into position on the cylinder head.

18. Temporarily install the bolts holding the delivery pipe to the head.

19. Check the the injectors rotate smoothly. If they do not, recheck the O-ring installation. Replace the O-rings if necessary.

20. Position the injector upward, then tighten the bolts holding the delivery pipe to the head to 15 ft. lbs. (21 Nm).

21. Connect the fuel inlet pipe to the delivery pipe with new gaskets and the union bolt. Tighten to 22 ft. lbs. (29 Nm).

22. Installation of the remaining components is in the reverse order of removal.

TESTING

▸ **See Figures 25, 26 and 27**

Proper testing of the fuel injectors requires specific equipment not usually available outside a dealership. It is recommended that any checking or testing of the injectors, other than that included in this section, be left to a properly equipped service facility.

Injector operation can be checked with the injectors installed in the engine. A sound scope, which is a stethoscope-like device available from most auto tool and parts jobbers, is recommended, although a piece of good rubber hose will suffice.

With the engine running or cranking, check each injector for normal operating noise (a buzzing or humming), which changes in proportion to engine rpm. If a sound scope is not available to you, check injector operation by touching

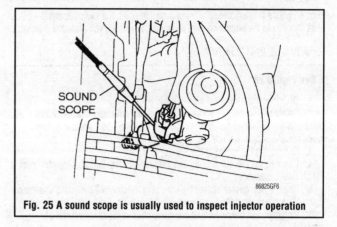

Fig. 25 A sound scope is usually used to inspect injector operation

Fig. 26 If a sound scope is not available, check the operation with your finger

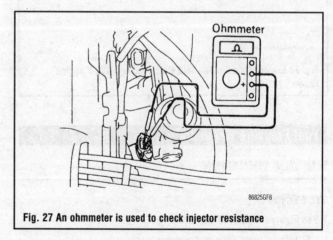

Fig. 27 An ohmmeter is used to check injector resistance

each injector with your finger. It should be buzzing. If no sound or an unusual sound is heard, check the wiring connector, or have the injector checked professionally.

With the engine **OFF**, measure injector resistance by unplugging the wiring connector from the injector, and connecting an ohmmeter across the injector terminals. Check the continuity at both terminals; resistance should be: cold 12–16 ohms. Resistance will vary with injector temperature.

Cold Start Injector

OPERATION

During cold engine starting, the cold start injector is used to supply additional fuel to the intake manifold to aid in initial start-up. The opening and closing of the injector is determined by the start injector time switch. When the engine coolant temperature falls below a certain point, the switch opens the cold start injector. As the engine coolant warms up, the switch will close the injector.

REMOVAL & INSTALLATION

▶ **See Figure 28**

1. Disconnect the negative battery cable.
2. Safely relieve the fuel pressure. Remove the cold start injector union bolt(s) on the delivery pipe. Before removing the union bolt, place a suitable container and shop towel under it to catch any escaping fuel.
3. Disconnect the wiring at the injector.
4. Unscrew the mounting bolts and then remove the cold start injector from the air intake chamber.
5. Using a new gasket, install the cold start injector into the intake chamber. Tighten the mounting bolts to between 48–69 inch lbs. (5–8 Nm).
6. Installation of the remaining components is in the reverse order of removal.

➡**Always use a new gasket when reinstalling the injector.**

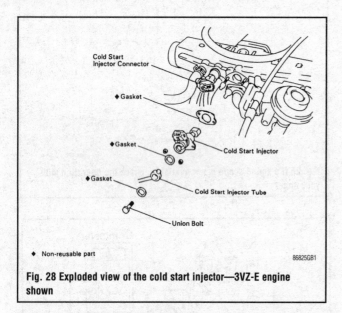

Fig. 28 Exploded view of the cold start injector—3VZ-E engine shown

Fuel Pressure Regulator

REMOVAL & INSTALLATION

3VZ-E Engine

▶ **See Figures 29, 30, 31 and 32**

1. Label all hoses and wiring being removed.
2. Drain the engine cooling system.
3. Disconnect these cables:
   - If equipped, cruise control and actuator cable with bracket
   - Accelerator cable
   - Throttle cables on automatic transmissions
4. Disconnect the air cleaner hose.
5. Unhook the vacuum sensing hose.
6. Place a suitable container or shop towel under the fuel pressure regulator. Disconnect the fuel return hose from the regulator.
7. Remove the air intake chamber:
   a. Disconnect the throttle position sensor wiring.
   b. Disconnect the canister vacuum hose from the throttle body.
   c. Disconnect the PCV hose from the union bolt.
   d. Remove the No. 4 water bypass hose from the union of the intake manifold.
   e. Detach the No. 5 water bypass hose from the water bypass pipe.

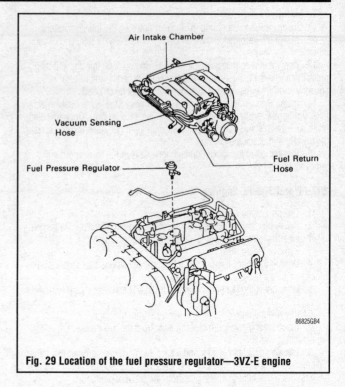

Fig. 29 Location of the fuel pressure regulator—3VZ-E engine

Fig. 30 With the aid of gasoline as a lubricant, install the new O-ring

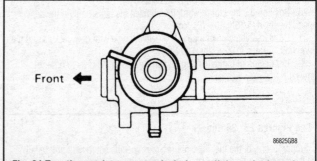

Fig. 31 Turn the regulator counterclockwise until the outlet faces in the direction indicated

f. Unhook the cold start injector connection.
g. Remove the union bolt, the 2 gaskets and the injector tube.
h. Disconnect the vacuum hose from the gas filter.
i. Disconnect the EGR gas temperature sensor wiring.
j. Remove the intake chamber stay and throttle cable bracket.
k. Remove EGR valve with pipes and discard of the two gaskets.

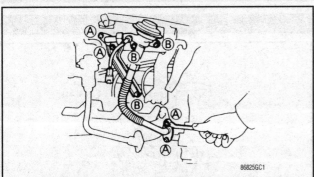

Fig. 32 Position new gaskets on the valve and intake, then tighten the mounting bolts as specified: "A" to 22 ft. lbs. (29 Nm), and "B" to 13 ft. lbs. (18 Nm)

l. Disconnect the No. 1 air hose from the PAIR reed valve.

m. Detach the 4 vacuum hoses from the air pipes.

n. Remove the bolts securing the accelerator cable, then remove the cable.

o. Unbolt the intake chamber and discard of the gasket.

8. Loosen the locknut, then remove the fuel pressure regulator.

To install:

9. Fully loosen the locknut of the pressure regulator, then apply a light coat of gasoline to the new O-ring. Install; the O-ring onto the regulator.

10. Thrust the fuel pressure regulator completely into the the delivery pipe by hand.

11. Turn the regulator counterclockwise until the outlet faces in the direction indicated.

12. Tighten the locknut to 22 ft. lbs. (29 Nm).

13. Installation of the remaining components is in the reverse order of removal. When installing the air intake chamber stay and throttle cable bracket. Tighten the nuts and bolts A: 22 ft. lbs. (29 Nm) and B:13 ft. lbs. (18 Nm).

14. To check for fuel leakage do the following:

a. With the ignition switch in the **ON** position, use a special service tool 09843-18020 or its equivalent to connect the terminals of the FP and +B of the DLC1.

b. Check for a fuel leakage.

22R-E Engines

▶ **See Figures 33 and 34**

1. Disconnect the vacuum sensing hose.

2. Remove the No. 1 EGR pipe.

3. Using a suitable container and shop towel, place them under the regulator. Slowly remove the fuel hose to the regulator.

4. Loosen the locknut, then remove the regulator.

5. Installation is in the reverse order of removal, tighten the regulator locknut to 22 ft. lbs. (29 Nm).

3F-E Engine

▶ **See Figures 34 and 35**

1. Disconnect the vacuum sensing hose.

2. Place a suitable container and shop towel under the regulator. Slowly remove the union bolt and discard the gaskets, then disconnect the return pipe.

3. Remove the bolts retaining the regulator, then pull out the unit.

4. When installing, apply a light coat of gasoline to a new O-ring, then install it onto the regulator.

5. Attach the pressure regulator to the engine, and tighten the bolts to 48 inch lbs. (5 Nm).

6. Using new gaskets, install the return pipe then the union bolt, tighten to 14 ft. lbs. (19 Nm).

2RZ-FE and 3RZ-FE Engines

▶ **See Figure 36**

1. Disconnect the vacuum sensing hose from the pressure regulator.

2. Place a suitable container and shop towel under the regulator, then disconnect the return pipe.

3. Remove the two mounting bolts retaining the regulator to the engine. Discard of the O-ring.

4. On installation, apply a light coat of gasoline to a new O-ring, then install it onto the regulator.

5. Attach the pressure regulator to the engine, tighten the bolts to 78 inch lbs. (9 Nm).

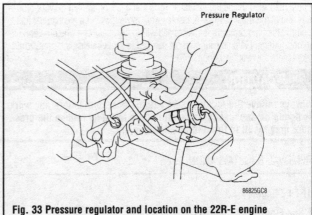

Fig. 33 Pressure regulator and location on the 22R-E engine

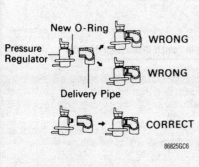

Fig. 34 Apply a light coating of gasoline to a new O-ring, then carefully install it onto the regulator

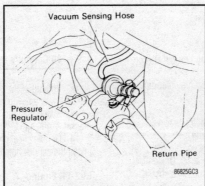

Fig. 35 Pressure regulator and location on the 3F-E engine

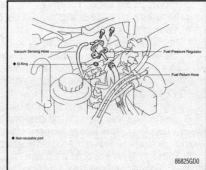

Fig. 36 Exploded view of the pressure regulator mounting on the 2VZ-FE and 3VZ-FE engines

SEQUENTIAL FUEL INJECTION

General Information

Sequential firing of the fuel injectors according to the engine firing order is the most accurate and desirable method of regulating multi-port injection. However, it is also the most complex and expensive to design and manufacture. In this system, the injectors are controlled individually. Each cylinder receives one charge every two revolutions just before the intake valve opens. This means that the mixture is never static in the intake manifold. Along with the mixture adjustments that can be made almost simultaneously between the firing of one injector and the next. A camshaft signal sensor or a special distributor reference pulse informs the ECM when the No. 1 cylinder is on the compression stroke. If the sensor fails or the distributor reference pulse is interrupted in any way, some injector systems shut down, while others revert to pulsing the injectors simultaneously.

Relieving Fuel System Pressure

To relieve the fuel pressure, wrap a clean towel or rag around either the cold start injector fitting or the union bolt (banjo bolt) at the fuel delivery rail. Slowly loosen the bolt holding the fuel line and allow pressurized fuel to escape into the rag. Do NOT allow fuel to spray from the fitting; always wrap it in a rag.

➡Pressure may also be relieved at the fuel filter if desired. Tighten the union bolt.

Electric Fuel Pump

All Toyota fuel-injected vehicles are equipped with an electric fuel pump. The pump is located inside the fuel tank. For a fuel injection system to work properly, the pump must develop pressures well above those of a mechanical fuel pump. This high pressure is maintained within the lines even when the engine is not running. Extreme caution must be used to safely release the pressurized fuel before any work is begun.

✳✳ CAUTION

Always relieve the fuel pressure within the system before any work is begun on any fuel component. Failure to safely relieve the pressure may result in fire and/or serious injury.

REMOVAL & INSTALLATION

1FZ-FE Engine

▶ **See Figures 37, 38 and 39**

1. Disconnect the negative battery cable.
2. Remove the rear seats of the vehicle.
3. Remove the scuff plate, side garnish and step plate.
4. Remove the floor mats.
5. Remove the floor service hole cover.
6. Disconnect the fuel pipe and hose from the fuel pump bracket. Disconnect the pump and sensor gauge connection as follows:
    a. Remove the union bolt and gaskets, then disconnect the outer pipe from the pump bracket.
    b. Detach the return hose from the pump bracket.
7. Remove the 8 mounting bolts and then remove the pump bracket assembly from the gas tank. Discard the gasket from the pump bracket. Be careful not to bend the arm of the sender gauge.
8. Pull off the lower side of the pump from the pump bracket. Disconnect the pump wiring, then disconnect the fuel hose from the pump. This should release the pump from the unit.
9. With the aid of a flat-bladed tool, remove the clip retaining the filter to the pump. This will allow the filter to slide off the pump.
To install:
10. Attach filter with the C-clip onto the pump.
11. Push the lower half of the pump into the bracket and attach the wiring. Connect the fuel hose to the pump.

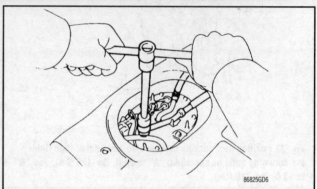
Fig. 37 Remove the union bolt and gaskets, then disconnect the outlet pipe

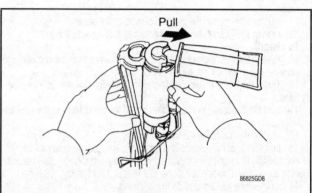
Fig. 38 Pull the pump in an outward motion from the lower side of the bracket

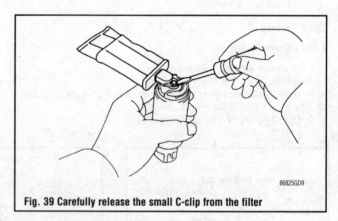
Fig. 39 Carefully release the small C-clip from the filter

12. With all of the pump parts in place on the bracket, slide the unit into the gas tank with a new gasket. Tighten to 35 inch lbs. (4 Nm).
13. Attach the union bolt with new gaskets along with the outlet pipe to the bracket. Tighten to 22 ft. lbs. (29 Nm).
14. Double check that all hoses are secure and that all wiring is intact, then install the service hole cover and the remaining components.

5VZ-FE Engine

1. Remove the fuel tank.
2. Remove the fuel pump bracket or assembly from the tank. Disconnect the pump wiring from the clamp as follows:
    a. Remove the 8 mounting bolts and pull the bracket assembly out of the tank.
    b. Remove and discard of the gasket on the pump bracket.

3. Pull off the lower side of the pump from the pump bracket. Disconnect the pump wiring, then disconnect the fuel hose from the pump. This should release the pump from the unit.

4. With the aid of a flat-bladed tool, remove the clip retaining the filter to the pump. This will allow the filter to slide off the pump.

To install:

5. Attach filter with the C-clip onto the pump and push the lower half of the pump into the bracket and attach the wiring. Connect the fuel hose to the pump.

6. With all of the pump parts in place on the bracket, slide the unit into the gas tank with a new gasket. Tighten to 35 inch lbs. (4 Nm). Install the fuel tank.

TESTING

1FZ-FE Engine

1993–95 MODELS

▶ See Figures 40 and 41

1. Disconnect the battery cables, negative first. Remove the battery from the vehicle.
2. Remove the rear seats.
3. Remove the scuff plate, side garnish, step plate and carpet mats.
4. Remove the floor service hole cover.
5. Disconnect the fuel pump and sender gauge connector.
6. Check the resistance with the aside of an ohmmeter on terminals 5 and 6. The reading should be 0.2–3.0 ohms at 68°F (20°C). If the reading is not within specifications, replace the pump.
7. To inspect the pump operation, connect the positive lead from the battery to terminal 6 of the connector. Attach the negative lead to terminal 5, then check how the pump operates.

➡These tests must be performed quickly (within 10 seconds) to prevent the coil from burning out. Keep the pump away from the battery and always perform switching at the battery side.

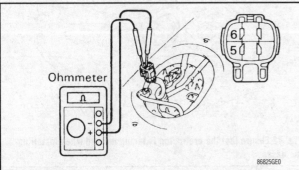

86825GE0

Fig. 40 Measure the resistance across terminals 5 and 6 of the fuel tank sender

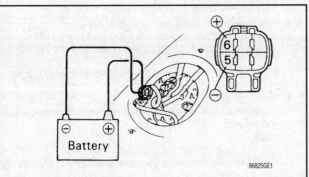

86825GE1

Fig. 41 Use a 12V battery to energize the fuel tank sender by placing the positive lead to terminal 6 and the negative lead to terminal 5

8. Reconnect the pump and sender gauge wiring.
9. Reinstall the floor service hole cover, floor mats, step plate, side garnish and scuff plates.
10. Install the rear seats, tighten to 29 ft. lbs. (39 Nm).
11. Reconnect the negative battery cable.

1996 MODELS

1. Check the fuel pump operation as follows:
   a. Remove the fuse cover on the instrument panel.
   b. Connect the Toyota hand-held tester or its equivalent scan tool to the DLC3.

➡Do not start the engine.

   c. Turn the ignition switch ON and push the tester main switch on.
   d. On the Toyota tester select the active test mode.
   e. Refer to the tester instructions for further details.
   f. If you do not have a Toyota hand-held tester, connect the positive and negative leads from the battery to the fuel pump connector.
   g. Check that there is pressure in the fuel inlet hose from the fuel filter. If there is pressure, you will hear the sound of fuel flowing.
   h. If there is no pressure, check the following:
- Fusible links (MAIN 2.0L, AM2 0.3P)
- Fuses (EFI 15A, IGN 7.5A)
- EFI main relay
- Fuel pump
- ECM
- Wiring connections
   i. Turn the ignition switch to LOCK.
   j. Disconnect the Toyota hand-held tester from the DLC3.
2. Check the fuel pressure as follows:
   a. Check the battery positive voltage is above 12 volts.
   b. Disconnect the negative cable from the battery.
   c. Wrap a shop towel around the delivery pipe, then slowly loosen the union bolt holding the fuel pipe to the delivery pipe and gasket.
   d. Install the SST 09268–45012 (pressure gauge) or its equivalent to the delivery pipe with the two gaskets and the SST (union and union bolt). Tighten to 22 ft. lbs. (29 Nm).
   e. Be sure to wipe off any spilled fuel.
   f. Attach the Toyota hand-held tester or equivalent scan tool to the DLC3. Reconnect the negative cable to the battery.
   g. Turn the ignition switch **ON**.
   h. Measure the fuel pressure, it should read 38–44 psi (265–304 kPa).
   i. If the pressure is too high, replace the pressure regulator.
   j. If the pressure is too low, check the following:
- Fuel connections and hoses
- Fuel pump
- Fuel filter
- Pressure regulator
   k. Disconnect the tester from the DLC3.
   l. Reinstall the fuse box cover on the instrument panel.
   m. Start the engine.
   n. Disconnect the vacuum sensing hose from the fuel pressure regulator, then plug the hose end.
   o. Measure the pressure at idle, 38–44 psi (265–304 kPa).
   p. Reconnect the vacuum sensing hose to the fuel pressure regulator.
   q. Measure the pressure at idle, 31–37 psi (226–255 kPa). If the pressure is not as specified, check the fuel; pump, pressure regulator and/or injectors.
   r. Stop the engine.
   s. Check that the fuel pressure remains as specified for 5 minuets after the engine has stopped. 21 psi (147 kPa). If the pressure is not as specified, check the pump, pressure regulator and/or injectors.
   t. After checking the fuel pressure, disconnect the negative cable from the battery, then carefully remove the SST 09268–45012 or equivalent, to prevent gas from splashing.
   u. Reinstall the fuel pipe with new gaskets and the union bolts, tighten the bolts to 22–25 ft. lbs. (29–34 Nm).
   v. Reconnect the negative cable to the battery.
   w. Check for fuel leaks.

5VZ-FE Engine

1. Check the fuel pump operation as follows:
   a. Remove the fuse cover on the instrument panel.
   b. Connect the Toyota hand-held tester or its equivalent scan tool to the DLC3.

➡ **Do not start the engine.**

   c. Turn the ignition switch ON and push the tester's main switch on.
   d. On the Toyota tester select the active test mode.
   e. Refer to the tester instructions for further details.
   f. If you do not have a Toyota hand-held tester, connect the positive and negative leads from the battery to the fuel pump connector.
   g. Check that there is pressure in the fuel inlet hose from the fuel filter. If there is pressure, you will hear the sound of fuel flowing.
   h. If there is no pressure, check the following:
- H-fuse (ALT 80)
- M-fuse (AM1 40A, AM2 30A)
- Fuses (EFI 15A, IGN 7.5A)
- EFI main relay
- Fuel pump
- ECM
- Wiring connections
   i. Turn the ignition switch to **LOCK**.
   j. Disconnect the Toyota hand-held tester from the DLC3.
2. Check the fuel pressure as follows:
   a. Be sure that the battery positive voltage is above 12V.
   b. Disconnect the negative cable from the battery.
   c. Remove the bolts, then disconnect the No. 2 timing belt cover.
   d. Wrap a shop towel around the delivery pipe, then slowly loosen the union bolt holding the fuel pipe to the delivery pipe and gasket.
   e. Install the SST 09268–45012 (pressure gauge) or its equivalent to the delivery pipe with the two gaskets and the SST (union and union bolt). Tighten to 22 ft. lbs. (29 Nm).
   f. Be sure to wipe off any spilled fuel.
   g. Attach the Toyota hand-held tester or equivalent scan tool to the DLC3. Reconnect the negative cable to the battery.
   h. Start the engine.
   i. Measure the fuel pressure, it should read 38–44 psi (265–304 kPa).
   j. If the pressure is too high, replace the pressure regulator.
   k. If the pressure is too low, check the following:
- Fuel connections and hoses
- Fuel pump
- Fuel filter
- Pressure regulator
- Vacuum Switching Valve (VSV) for fuel pressure control
   l. Disconnect the tester from the DLC3.
   m. If removed, install the No. 2 timing belt cover, tighten to 80 inch lbs. (9 Nm).
   n. Start the engine.
   o. Disconnect the vacuum sensing hose from the fuel pressure regulator, then plug the hose end.
   p. Measure the pressure at idle, 38–44 psi (265–304 kPa).
   q. Reconnect the vacuum, sensing hose to the fuel pressure regulator.
   r. Measure the pressure at idle, 31–38 psi (226–265 kPa). If the pressure is not as specified, check the fuel; pump, pressure regulator and/or injectors.
   s. Stop the engine.
   t. Check that the fuel pressure remains at 21 psi (147 kPa) for 5 minutes after the engine has stopped. If the pressure is not as specified, check the pump, pressure regulator and/or injectors.
   u. After checking the fuel pressure, disconnect the negative cable from the battery, then carefully remove the SST 09268–45012 or equivalent, to prevent gas from splashing.
   v. Remove the other union bolt, 3 gaskets and fuel pipe from the delivery pipes.
   w. Reinstall the fuel pipe with new gaskets and the union bolts, tighten the bolts to 22–25 ft. lbs. (29–34 Nm).
   x. Reconnect the negative cable to the battery.
   y. Check for leaks.

Throttle Body

REMOVAL & INSTALLATION

5VZ-FE Engine

▶ **See Figure 42**

1. Depressurize the fuel system.
2. Drain the cooling system.
3. Disconnect the cables:
- Accelerator—1
- Throttle cable on A/T—2
- Cruise control actuator—3
4. Remove the air cleaner hose.
5. Disconnect the throttle position sensor harness.
6. Separate the idle air control (IAC) connector.
7. Tag and disconnect the following vacuum hoses from the throttle body:
- Throttle opener
- EGR vacuum modulator
8. Tag and disconnect the following hoses from the idle air control valve:
- Water bypass; on EGR vehicles, from the EGR valve
- Water bypass; without EGR, from the bypass pipe
- Water hose from the intake manifold
- Air assist
9. Disconnect the ignition coil wiring.
10. Unbolt the throttle body, remove and discard of the gasket.
11. To install, place a new gasket on the air intake chamber facing the protrusion upwards, tighten to 13 ft. lbs. (18 Nm). Install the remaining components in the reverse order of removal.

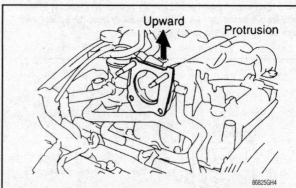

Fig. 42 Ensure that the protrusion is facing upward when installing the gasket

1FZ-FE Engine

1. Depressurize the fuel system.
2. Drain the engine cooling system.
3. Disconnect the PCV and air cleaner hoses.
4. Remove the control cables from the throttle body.
5. Disconnect the throttle position sensor and idle air control wiring.
6. Tag and disconnect all vacuum and water hoses leading to the throttle body.
7. Remove the bolts, and disconnect the throttle body from the air intake chamber.
8. Remove the throttle body gasket and discard.
9. Disconnect the both of the water bypass hoses from the throttle body, then remove the unit.

To install:
10. Attach the water bypass hoses to the throttle body.
11. Install the throttle body with a new gasket, then tighten to 15 ft. lbs. (21 Nm).
12. Install all of the previously tagged vacuum and water hoses.

INSPECTION

Always clean the throttle body once removed from the vehicle. Wash and clean the cast parts with a soft brush in carburetor cleaner. Using compressed air, clean all of the passages and apertures in the throttle body.

➡ **To prevent any damage, do not clean the throttle position sensor.**

Fuel Injectors

REMOVAL & INSTALLATION

1FZ-FE Engine

▸ **See Figures 43 thru 51**

1. Depressurize the fuel system.
2. Drain the engine cooling system.
3. Disconnect the cruise control accelerator cable and throttle cable.
4. Detach the No. 2 PCV hose.
5. Remove the air cleaner hose.
6. Remove the EGR vacuum modulator and valve.
7. Unbolt the heater inlet pipe and air intake chamber.
8. Disconnect the PCV hose.
9. Remove the vacuum sensing hose.
10. Remove the water bypass and the EVAP hoses.
11. Disconnect the brake booster hose.
12. Disconnect the throttle position sensor, idle air control valve, emission control valve set assembly and EGR gas temperature sensor wiring.
13. Remove the bolts retaining the power steering reservoir tank.
14. Remove the engine oil dipstick.
15. Unbolt the ground strap.
16. Remove the intake manifold stay.
17. Disconnect the vacuum hoses to the Thermal Vacuum Valve (TVV).
18. Remove the water bypass hose from the cylinder head.
19. Remove the air intake chamber and discard of the gaskets.
20. Remove the 2 union bolts, gaskets and fuel inlet pipe.

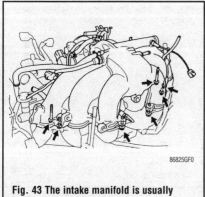

Fig. 43 The intake manifold is usually retained by 6 bolts and 2 nuts

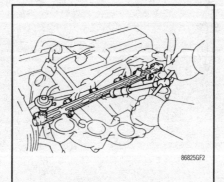

Fig. 44 The injectors are removed with the fuel delivery pipe

Fig. 45 Install a new grommet and lubricated O-ring on the injector

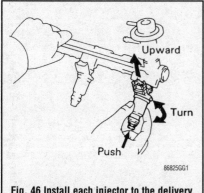

Fig. 46 Install each injector to the delivery pipe by pushing while twisting the injector

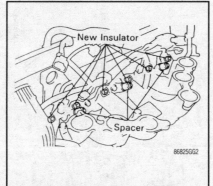

Fig. 47 Place new insulators and spacers in the indicated positions

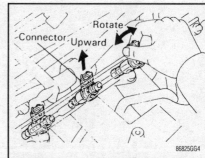

Fig. 48 Rotate the injectors to ensure their smooth movement. An injector that binds has been improperly installed and may leak

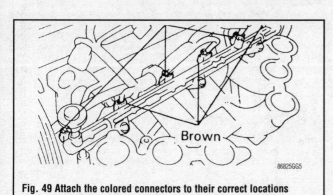

Fig. 49 Attach the colored connectors to their correct locations

21. Remove the delivery pipe and injectors as follows:
    a. Unsecure the 6 injector connectors.
    b. Remove the bolts and delivery pipe together with the 6 injectors. Be careful not to drop the injectors when removing the pipe.
    c. Remove the insulators and spacers on the intake manifold.
    d. Pull the injectors out of the delivery pipe. Remove the O-rings and grommets from each injector.
22. Inspect the injectors.
To install:
23. Install new grommets on each injector, then apply a light coat of gasoline to the new O-rings and install them.
24. While turning the injector left and right, install it to the delivery pipe. Install all 6 injectors.
25. Position the injector connector upwards. Place new insulators and spacers in position on the intake manifold.

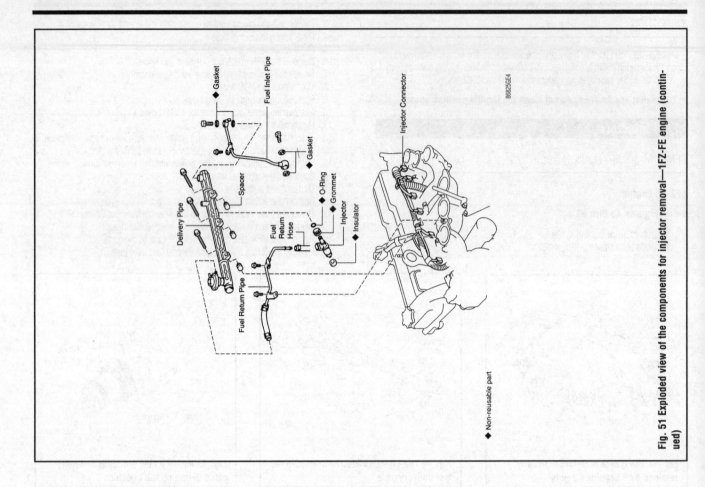

Fig. 51 Exploded view of the components for injector removal—1FZ-FE engine (continued)

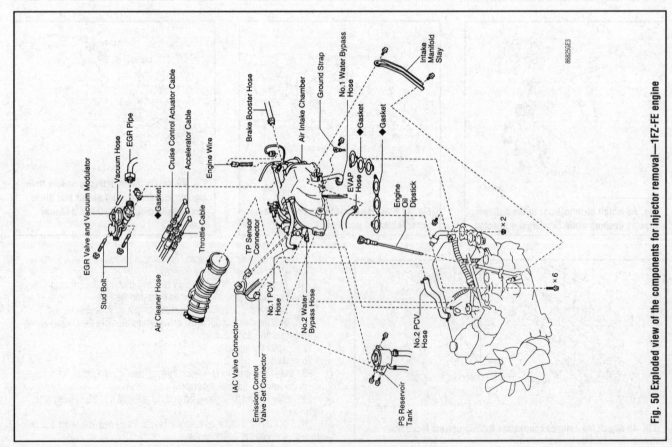

Fig. 50 Exploded view of the components for injector removal—1FZ-FE engine

26. Temporarily install the bolts holding the delivery pipe to the intake manifold.

27. Check the the injectors rotate smoothly. If they do not, recheck the position of the O-rings, or replace them.

28. Position the injector connector upward, then tighten the bolts holding the delivery pipe to the intake manifold to 15 ft. lbs. (21 Nm).

29. Attach the injector connectors.

30. Install the fuel inlet pipe with new gaskets, tighten the union bolt to 22 ft. lbs. (29 Nm) and the bolt to 14 ft. lbs. (20 Nm).

31. Install the fuel return pipe with the bolts and tighten to 14 ft. lbs. (20 Nm).

32. Attach the fuel hose to the pressure regulator.

33. Install the air intake chamber with new gaskets, tighten to 15 ft. lbs. (21 Nm).

34. Installation of the remaining components is in the reverse order of removal.

5VZ-FE Engine

▶ **See Figures 52, 53, 54, 55 and 56**

1. Depressurize the fuel system.
2. Remove the air cleaner hose.
3. Remove the air intake chamber.
4. Remove the fuel pressure regulator.
5. Disconnect the fuel inlet pipe.
6. Unsecure the injector connections.
7. Unbolt and remove the delivery pipes together with the injectors.
8. Remove the spacers from the intake manifold.
9. Pull the injectors out of the delivery pipes, then remove the O-rings and grommets from each injector.

To install:

10. Install new grommets and O-rings on each injector. Apply a light coat of gasoline on the O-rings.

   a. While turning the injector clockwise and counterclockwise, push it into the delivery pipe. Install all of the injectors.

   b. Position the injector connector outward.

   c. Place the spacers into position on the intake manifold. Temporarily install the bolts to hold the delivery pipes to the intake manifold.

   d. Check that the injectors rotate smoothly. If they do not, the O-rings have probably been installed incorrectly. If this has occurred, replace the O-rings with new ones.

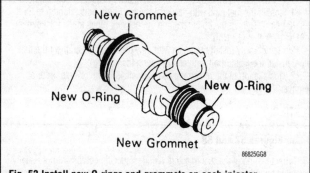

Fig. 53 Install new O-rings and grommets on each injector

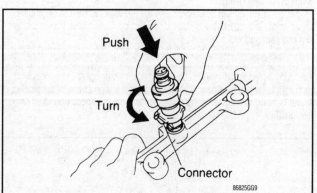

Fig. 54 While twisting the injector clockwise and counterclockwise, push it into its delivery pipe

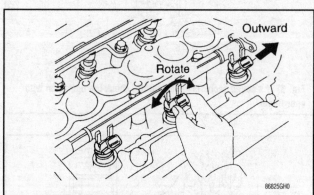

Fig. 55 Check that the injectors rotate smoothly. If they do not, the O-rings have probably been installed incorrectly

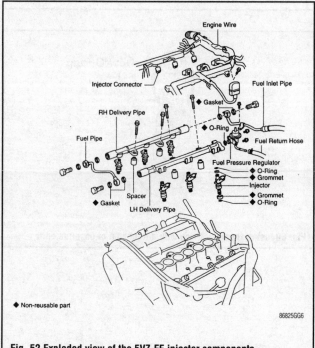

Fig. 52 Exploded view of the 5VZ-FE injector components

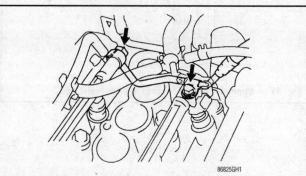

Fig. 56 Temporarily install the union with new gaskets, then connect the fuel pipe. Install the clamp bolt and tighten to 71 inch lbs. (8 Nm)

e. Position the injector outward, then attach the injector connectors.

11. Install the fuel pipe with new gaskets and union bolts, tighten to 25 ft. lbs. (34 Nm). Tighten the bolts retaining the delivery pipes to the intake manifold to 10 ft. lbs. (13 Nm).

12. Temporarily install the union with new gaskets, then connect the fuel pipe. Install the clamp bolt, tighten to 71 inch lbs. (8 Nm).

13. Installation of the remaining components is in the reverse order of removal.

TESTING

▶ **See Figures 57 and 58**

Correct testing of the fuel injectors requires specific equipment not usually available outside a dealership or a fuel lab. It is recommended that any checking or testing of the injectors, other than that included below, be left to a properly equipped service facility.

Injector operation can be checked with the injectors installed in the engine. A sound scope is needed here, a stethoscope-like device available from most auto tool and parts jobbers.

With the engine running or cranking, check each injector for normal operating noise (a buzzing or humming), which changes in proportion to engine rpm. If a sound scope is not available to you, check injector operation by touching each injector with your finger. It should be buzzing. If no sound or an unusual sound is heard, check the wiring connector, or have the injector checked professionally.

With the engine **OFF**, measure injector resistance by unplugging the wiring connector from the injector, and connecting an ohmmeter across the injector terminals. Check the continuity at both terminals; resistance should be: cold 12–16 ohms. Resistance will vary with injector temperature.

Fuel Pressure Regulator

REMOVAL & INSTALLATION

1FZ-FE Engine

▶ **See Figures 59, 60 and 61**

1. Depressurize the fuel system.
2. Disconnect the vacuum sensing hose from the pressure regulator.
3. Wrap a towel around the regulator then disconnect the fuel return hose.
4. Loosen the locknut, then remove the regulator. Discard of the O-ring.

To install:

5. Fully loosen the locknut on the regulator.
6. Apply a light coat of gasoline to a new O-ring, then install it onto the regulator.
7. Insert the regulator into the delivery pipe by hand. Turn the regulator counterclockwise until the fuel outlet port faces in the direction indicated.
8. Tighten the locknut to 18 ft. lbs. (25 Nm) and install the remaining components in the reverse order of removal. Use new gaskets on the return pipe.

Fig. 57 If a sound scope is not available, check the operation with your finger

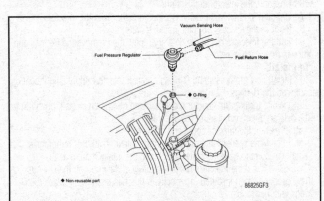

Fig. 59 The regulator is in plain view, on top of the fuel rail—1FZ-FE engine

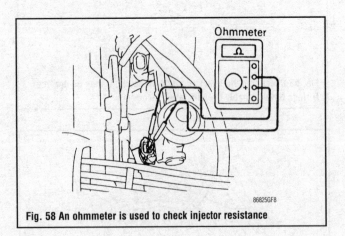

Fig. 58 An ohmmeter is used to check injector resistance

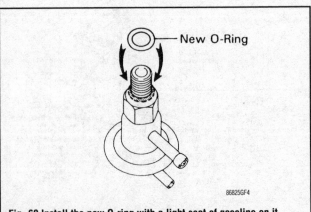

Fig. 60 Install the new O-ring with a light coat of gasoline on it

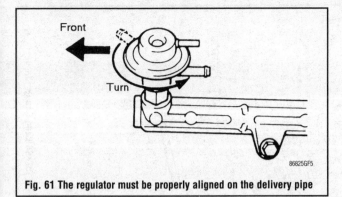

Fig. 61 The regulator must be properly aligned on the delivery pipe

5VZ-FE Engine

▸ See Figure 62

1. Depressurize the fuel system.
2. Remove the air cleaner hose.
3. Remove the intake air connector.
4. Disconnect the fuel return hose from the pressure regulator. Place a shop rag under the regulator to catch any leakage.
5. Remove the bolt holding the engine wire to the left-hand valve cover.
6. Disconnect the protector from the bracket on the right side valve cover, then lift up the engine wire.
7. Remove the 2 bolts, then pull out the regulator. Remove and discard the

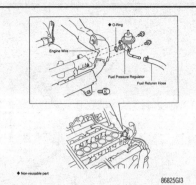

Fig. 62 The O-ring must be placed correctly onto the regulator or the unit will not fit properly

O-ring from the unit.

To install:

8. Apply a light coat of gasoline to the new O-ring, then install it on the regulator.
9. Attach the regulator to the left side delivery pipe.
10. Check that the pressure regulator rotates smoothly. If it does not, the O-ring may be the cause. Remove the regulator and install a new O-ring.
11. Attach the regulator, then tighten to 71 inch lbs. (8 Nm) and install the remaining components.

FUEL TANK

Tank Assembly

REMOVAL & INSTALLATION

▸ See Figures 63 and 64

The gas tanks on all Toyota trucks are mounted basically the same. Each one has a shield, straps, inlet pipe and so on. The procedure below is basic, and should help ease your way to the tank removal.

1. Remove the negative battery cable.
2. Raise the vehicle and support it with jackstands.
3. Remove the drain plug and drain any remaining fuel into a suitable container. Remember that it is possible to have several gallons in the tank; be prepared with sufficient containers. It is best to run the tank as low on fuel as possible before removing it, but Empty on the gauge may still leave 2 or 3 gallons in the tank.
4. Disconnect the plug from the pump and sending unit. Remove the gravel shield from the tank.
5. Disconnect the fuel lines. Plug all the lines to prevent fuel from leaking.
6. Disconnect the filler neck and vent line.

7. Remove the fuel tank protector.
8. Remove the bolts holding the tank to the vehicle and carefully lower the tank. It is recommended that a jack with a broad piece of lumber be placed to support the tank.
9. Remove the fuel pump assembly.

✳✳ CAUTION

Even though the tank has been drained, it still contains highly explosive fuel vapor. Immediately place the tank outside; never store it in the house or garage. Unless the tank is to be immediately reinstalled, use a hose to fill the tank with water as full as possible. This will flush remaining fuel and vapor from the tank and also carry off any dirt which has accumulated.

To install:

10. Make certain the fuel tank is completely dry if it has been flushed. Install the pump assembly.
11. Install the fuel tank. Tighten the tank bolts.
12. Connect the lines and hoses. Make certain the hoses are not crimped or pinched. Make sure the clamps are correctly seated.
13. Connect the negative battery cable.

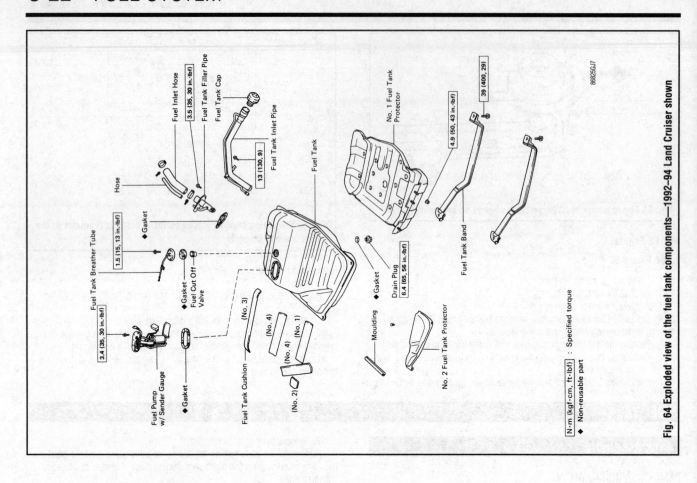

Fig. 64 Exploded view of the fuel tank components—1992–94 Land Cruiser shown

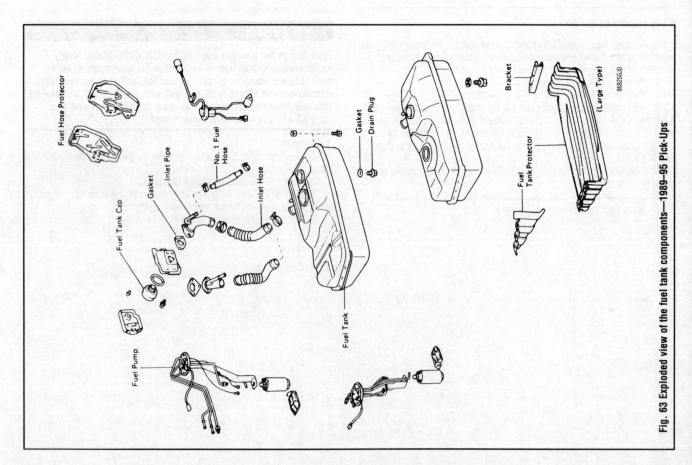

Fig. 63 Exploded view of the fuel tank components—1989–95 Pick-Ups

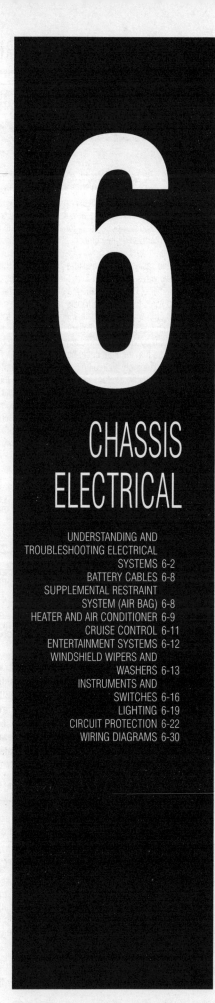

6

CHASSIS
ELECTRICAL

UNDERSTANDING AND TROUBLESHOOTING ELECTRICAL SYSTEMS

Basic Electrical Theory

♦ **See Figure 1**

For any 12 volt, negative ground, electrical system to operate, the electricity must travel in a complete circuit. This simply means that current (power) from the positive (+) terminal of the battery must eventually return to the negative (-) terminal of the battery. Along the way, this current will travel through wires, fuses, switches and components. If, for any reason, the flow of current through the circuit is interrupted, the component fed by that circuit will cease to function properly.

Perhaps the easiest way to visualize a circuit is to think of connecting a light bulb (with two wires attached to it) to the battery—one wire attached to the negative (-) terminal of the battery and the other wire to the positive (+) terminal. With the two wires touching the battery terminals, the circuit would be complete and the light bulb would illuminate. Electricity would follow a path from the battery to the bulb and back to the battery. It's easy to see that with longer wires on our light bulb, it could be mounted anywhere. Further, one wire could be fitted with a switch so that the light could be turned on and off.

The normal automotive circuit differs from this simple example in two ways. First, instead of having a return wire from the bulb to the battery, the current travels through the frame of the vehicle. Since the negative (-) battery cable is attached to the frame (made of electrically conductive metal), the frame of the vehicle can serve as a ground wire to complete the circuit. Secondly, most automotive circuits contain multiple components which receive power from a single circuit. This lessens the amount of wire needed to power components on the vehicle.

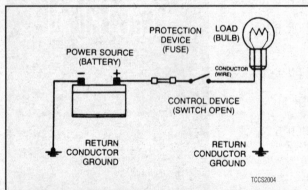

TCCS2004

Fig. 1 This example illustrates a simple circuit. When the switch is closed, power from the positive (+) battery terminal flows through the fuse and the switch, and then to the light bulb. The light illuminates and the circuit is completed through the ground wire back to the negative (-) battery terminal. In reality, the two ground points shown in the illustration are attached to the metal frame of the vehicle, which completes the circuit back to the battery

HOW DOES ELECTRICITY WORK: THE WATER ANALOGY

Electricity is the flow of electrons—the subatomic particles that constitute the outer shell of an atom. Electrons spin in an orbit around the center core of an atom. The center core is comprised of protons (positive charge) and neutrons (neutral charge). Electrons have a negative charge and balance out the positive charge of the protons. When an outside force causes the number of electrons to unbalance the charge of the protons, the electrons will split off the atom and look for another atom to balance out. If this imbalance is kept up, electrons will continue to move and an electrical flow will exist.

Many people have been taught electrical theory using an analogy with water. In a comparison with water flowing through a pipe, the electrons would be the water and the wire is the pipe.

The flow of electricity can be measured much like the flow of water through a pipe. The unit of measurement used is amperes, frequently abbreviated as amps (a). You can compare amperage to the volume of water flowing through a pipe.

When connected to a circuit, an ammeter will measure the actual amount of current flowing through the circuit. When relatively few electrons flow through a circuit, the amperage is low. When many electrons flow, the amperage is high.

Water pressure is measured in units such as pounds per square inch (psi); The electrical pressure is measured in units called volts (v). When a voltmeter is connected to a circuit, it is measuring the electrical pressure.

The actual flow of electricity depends not only on voltage and amperage, but also on the resistance of the circuit. The higher the resistance, the higher the force necessary to push the current through the circuit. The standard unit for measuring resistance is an ohm. Resistance in a circuit varies depending on the amount and type of components used in the circuit. The main factors which determine resistance are:

• Material—some materials have more resistance than others. Those with high resistance are said to be insulators. Rubber materials (or rubber-like plastics) are some of the most common insulators used in vehicles as they have a very high resistance to electricity. Very low resistance materials are said to be conductors. Copper wire is among the best conductors. Silver is actually a superior conductor to copper and is used in some relay contacts, but its high cost prohibits its use as common wiring. Most automotive wiring is made of copper.

• Size—the larger the wire size being used, the less resistance the wire will have. This is why components which use large amounts of electricity usually have large wires supplying current to them.

• Length—for a given thickness of wire, the longer the wire, the greater the resistance. The shorter the wire, the less the resistance. When determining the proper wire for a circuit, both size and length must be considered to design a circuit that can handle the current needs of the component.

• Temperature—with many materials, the higher the temperature, the greater the resistance (positive temperature coefficient). Some materials exhibit the opposite trait of lower resistance with higher temperatures (negative temperature coefficient). These principles are used in many of the sensors on the engine.

OHM'S LAW

There is a direct relationship between current, voltage and resistance. The relationship between current, voltage and resistance can be summed up by a statement known as Ohm's law.

Voltage (E) is equal to amperage (I) times resistance (R): $E = I \times R$
Other forms of the formula are $R = E/I$ and $I = E/R$

In each of these formulas, E is the voltage in volts, I is the current in amps and R is the resistance in ohms. The basic point to remember is that as the resistance of a circuit goes up, the amount of current that flows in the circuit will go down, if voltage remains the same.

The amount of work that the electricity can perform is expressed as power. The unit of power is the watt (w). The relationship between power, voltage and current is expressed as:

Power (w) is equal to amperage (I) times voltage (E): $W = I \times E$

This is only true for direct current (DC) circuits; The alternating current formula is a tad different, but since the electrical circuits in most vehicles are DC type, we need not get into AC circuit theory.

Electrical Components

POWER SOURCE

Power is supplied to the vehicle by two devices: The battery and the alternator. The battery supplies electrical power during starting or during periods when the current demand of the vehicle's electrical system exceeds the output capacity of the alternator. The alternator supplies electrical current when the engine is running. Just not does the alternator supply the current needs of the vehicle, but it recharges the battery.

The Battery

In most modern vehicles, the battery is a lead/acid electrochemical device consisting of six 2 volt subsections (cells) connected in series, so that the unit is capable of producing approximately 12 volts of electrical pressure. Each sub-section consists of a series of positive and negative plates held a short distance apart in a solution of sulfuric acid and water.

The two types of plates are of dissimilar metals. This sets up a chemical reaction, and it is this reaction which produces current flow from the battery when its positive and negative terminals are connected to an electrical load . The power removed from the battery is replaced by the alternator, restoring the battery to its original chemical state.

The Alternator

On some vehicles there isn't an alternator, but a generator. The difference is that an alternator supplies alternating current which is then changed to direct current for use on the vehicle, while a generator produces direct current. Alternators tend to be more efficient and that is why they are used.

Alternators and generators are devices that consist of coils of wires wound together making big electromagnets. One group of coils spins within another set and the interaction of the magnetic fields causes a current to flow. This current is then drawn off the coils and fed into the vehicles electrical system.

GROUND

Two types of grounds are used in automotive electric circuits. Direct ground components are grounded to the frame through their mounting points. All other components use some sort of ground wire which is attached to the frame or chassis of the vehicle. The electrical current runs through the chassis of the vehicle and returns to the battery through the ground (-) cable; if you look, you'll see that the battery ground cable connects between the battery and the frame or chassis of the vehicle.

➡It should be noted that a good percentage of electrical problems can be traced to bad grounds.

PROTECTIVE DEVICES

▶ See Figure 2

It is possible for large surges of current to pass through the electrical system of your vehicle. If this surge of current were to reach the load in the circuit, the surge could burn it out or severely damage it. It can also overload the wiring, causing the harness to get hot and melt the insulation. To prevent this, fuses, circuit breakers and/or fusible links are connected into the supply wires of the electrical system. These items are nothing more than a built-in weak spot in the system. When an abnormal amount of current flows through the system, these protective devices work as follows to protect the circuit:

• Fuse—when an excessive electrical current passes through a fuse, the fuse "blows" (the conductor melts) and opens the circuit, preventing the passage of current.

• Circuit Breaker—a circuit breaker is basically a self-repairing fuse. It will open the circuit in the same fashion as a fuse, but when the surge subsides, the circuit breaker can be reset and does not need replacement.

• Fusible Link—a fusible link (fuse link or main link) is a short length of special, high temperature insulated wire that acts as a fuse. When an excessive electrical current passes through a fusible link, the thin gauge wire inside the link melts, creating an intentional open to protect the circuit. To repair the circuit, the link must be replaced. Some newer type fusible links are housed in plug-in modules, which are simply replaced like a fuse, while older type fusible links must be cut and spliced if they melt. Since this link is very early in the electrical path, it's the first place to look if nothing on the vehicle works, yet the battery seems to be charged and is properly connected.

✳✳ CAUTION

Always replace fuses, circuit breakers and fusible links with identically rated components. Under no circumstances should a component of higher or lower amperage rating be substituted.

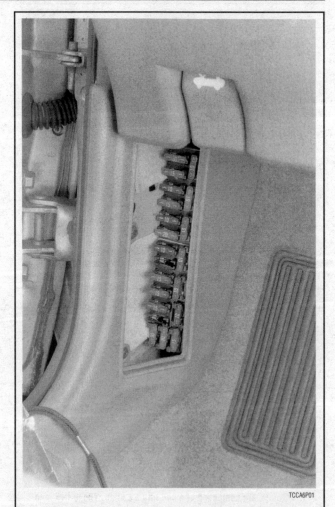

TCCA6P01

Fig. 2 Most vehicles use one or more fuse panels. This one is located on the driver's side kick panel

SWITCHES & RELAYS

▶ See Figures 3 and 4

Switches are used in electrical circuits to control the passage of current. The most common use is to open and close circuits between the battery and the various electric devices in the system. Switches are rated according to the amount of amperage they can handle. If a sufficient amperage rated switch is not used in a circuit, the switch could overload and cause damage.

Some electrical components which require a large amount of current to operate use a special switch called a relay. Since these circuits carry a large amount of current, the thickness of the wire in the circuit is also greater. If this large wire were connected from the load to the control switch, the switch would have to carry the high amperage load and the fairing or dash would be twice as large to accommodate the increased size of the wiring harness. To prevent these problems, a relay is used.

Relays are composed of a coil and a set of contacts. When the coil has a current passed though it, a magnetic field is formed and this field causes the contacts to move together, completing the circuit. Most relays are normally open, preventing current from passing through the circuit, but they can take any electrical form depending on the job they are intended to do. Relays can be considered "remote control switches." They allow a smaller current to operate devices that require higher amperages. When a small current operates the coil, a larger current is allowed to pass by the contacts. Some common circuits which may use relays are the horn, headlights, starter, electric fuel pump and other high draw ciruits.

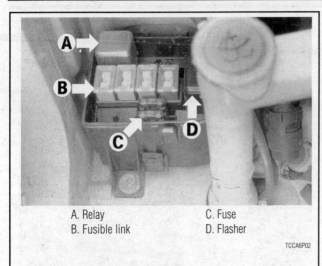

A. Relay   C. Fuse
B. Fusible link   D. Flasher

TCCA6P02

Fig. 3 The underhood fuse and relay panel usually contains fuses, relays, flashers and fusible links

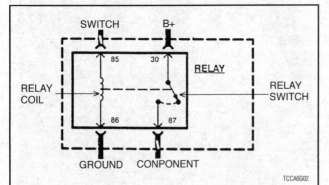

TCCA6G02

Fig. 4 Relays are composed of a coil and a switch. These two components are linked together so that when one operates, the other operates at the same time. The large wires in the circuit are connected from the battery to one side of the relay switch (B+) and from the opposite side of the relay switch to the load (component). Smaller wires are connected from the relay coil to the control switch for the circuit and from the opposite side of the relay coil to ground

LOAD

Every electrical circuit must include a "load" (something to use the electricity coming from the source). Without this load, the battery would attempt to deliver its entire power supply from one pole to another. This is called a "short circuit." All this electricity would take a short cut to ground and cause a great amount of damage to other components in the circuit by developing a tremendous amount of heat. This condition could develop sufficient heat to melt the insulation on all the surrounding wires and reduce a multiple wire cable to a lump of plastic and copper.

WIRING & HARNESSES

The average vehicle contains meters and meters of wiring, with hundreds of individual connections. To protect the many wires from damage and to keep them from becoming a confusing tangle, they are organized into bundles, enclosed in plastic or taped together and called wiring harnesses. Different harnesses serve different parts of the vehicle. Individual wires are color coded to help trace them through a harness where sections are hidden from view.

Automotive wiring or circuit conductors can be either single strand wire, multi-strand wire or printed circuitry. Single strand wire has a solid metal core

and is usually used inside such components as alternators, motors, relays and other devices. Multi-strand wire has a core made of many small strands of wire twisted together into a single conductor. Most of the wiring in an automotive electrical system is made up of multi-strand wire, either as a single conductor or grouped together in a harness. All wiring is color coded on the insulator, either as a solid color or as a colored wire with an identification stripe. A printed circuit is a thin film of copper or other conductor that is printed on an insulator backing. Occasionally, a printed circuit is sandwiched between two sheets of plastic for more protection and flexibility. A complete printed circuit, consisting of conductors, insulating material and connectors for lamps or other components is called a printed circuit board. Printed circuitry is used in place of individual wires or harnesses in places where space is limited, such as behind instrument panels.

Since automotive electrical systems are very sensitive to changes in resistance, the selection of properly sized wires is critical when systems are repaired. A loose or corroded connection or a replacement wire that is too small for the circuit will add extra resistance and an additional voltage drop to the circuit.

The wire gauge number is an expression of the cross-section area of the conductor. Vehicles from countries that use the metric system will typically describe the wire size as its cross-sectional area in square millimeters. In this method, the larger the wire, the greater the number. Another common system for expressing wire size is the American Wire Gauge (AWG) system. As gauge number increases, area decreases and the wire becomes smaller. An 18 gauge wire is smaller than a 4 gauge wire. A wire with a higher gauge number will carry less current than a wire with a lower gauge number. Gauge wire size refers to the size of the strands of the conductor, not the size of the complete wire with insulator. It is possible, therefore, to have two wires of the same gauge with different diameters because one may have thicker insulation than the other.

It is essential to understand how a circuit works before trying to figure out why it doesn't. An electrical schematic shows the electrical current paths when a circuit is operating properly. Schematics break the entire electrical system down into individual circuits. In a schematic, usually no attempt is made to represent wiring and components as they physically appear on the vehicle; switches and other components are shown as simply as possible. Face views of harness connectors show the cavity or terminal locations in all multi-pin connectors to help locate test points.

CONNECTORS

♦ See Figures 5 and 6

Three types of connectors are commonly used in automotive applications—weatherproof, molded and hard shell.

• Weatherproof—these connectors are most commonly used where the

TCCA6P03

Fig. 5 Hard shell (left) and weatherproof (right) connectors have replaceable terminals

Fig. 6 Weatherproof connectors are most commonly used in the engine compartment or where the connector is exposed to the elements

connector is exposed to the elements. Terminals are protected against moisture and dirt by sealing rings which provide a weathertight seal. All repairs require the use of a special terminal and the tool required to service it. Unlike standard blade type terminals, these weatherproof terminals cannot be straightened once they are bent. Make certain that the connectors are properly seated and all of the sealing rings are in place when connecting leads.

• Molded—these connectors require complete replacement of the connector if found to be defective. This means splicing a new connector assembly into the harness. All splices should be soldered to insure proper contact. Use care when probing the connections or replacing terminals in them, as it is possible to create a short circuit between opposite terminals. If this happens to the wrong terminal pair, it is possible to damage certain components. Always use jumper wires between connectors for circuit checking and NEVER probe through weatherproof seals.

• Hard Shell—unlike molded connectors, the terminal contacts in hardshell connectors can be replaced. Replacement usually involves the use of a special terminal removal tool that depresses the locking tangs (barbs) on the connector terminal and allows the connector to be removed from the rear of the shell. The connector shell should be replaced if it shows any evidence of burning, melting, cracks, or breaks. Replace individual terminals that are burnt, corroded, distorted or loose.

Test Equipment

Pinpointing the exact cause of trouble in an electrical circuit is most times accomplished by the use of special test equipment. The following describes different types of commonly used test equipment and briefly explains how to use them in diagnosis. In addition to the information covered below, the tool manufacturer's instructions booklet (provided with the tester) should be read and clearly understood before attempting any test procedures.

JUMPER WIRES

❋❋ CAUTION

Never use jumper wires made from a thinner gauge wire than the circuit being tested. If the jumper wire is of too small a gauge, it may overheat and possibly melt. Never use jumpers to bypass high resistance loads in a circuit. Bypassing resistances, in effect, creates a short circuit. This may, in turn, cause damage and fire. Jumper wires should only be used to bypass lengths of wire or to simulate switches.

Jumper wires are simple, yet extremely valuable, pieces of test equipment. They are basically test wires which are used to bypass sections of a circuit. Although jumper wires can be purchased, they are usually fabricated from

lengths of standard automotive wire and whatever type of connector (alligator clip, spade connector or pin connector) that is required for the particular application being tested. In cramped, hard-to-reach areas, it is advisable to have insulated boots over the jumper wire terminals in order to prevent accidental grounding. It is also advisable to include a standard automotive fuse in any jumper wire. This is commonly referred to as a "fused jumper". By inserting an in-line fuse holder between a set of test leads, a fused jumper wire can be used for bypassing open circuits. Use a 5 amp fuse to provide protection against voltage spikes.

Jumper wires are used primarily to locate open electrical circuits, on either the ground (-) side of the circuit or on the power (+) side. If an electrical component fails to operate, connect the jumper wire between the component and a good ground. If the component operates only with the jumper installed, the ground circuit is open. If the ground circuit is good, but the component does not operate, the circuit between the power feed and component may be open. By moving the jumper wire successively back from the component toward the power source, you can isolate the area of the circuit where the open is located. When the component stops functioning, or the power is cut off, the open is in the segment of wire between the jumper and the point previously tested.

You can sometimes connect the jumper wire directly from the battery to the "hot" terminal of the component, but first make sure the component uses 12 volts in operation. Some electrical components, such as fuel injectors or sensors, are designed to operate on about 4 to 5 volts, and running 12 volts directly to these components will cause damage.

TEST LIGHTS

▶ **See Figure 7**

The test light is used to check circuits and components while electrical current is flowing through them. It is used for voltage and ground tests. To use a 12 volt test light, connect the ground clip to a good ground and probe wherever necessary with the pick. The test light will illuminate when voltage is detected. This does not necessarily mean that 12 volts (or any particular amount of voltage) is present; it only means that some voltage is present. It is advisable before using the test light to touch its ground clip and probe across the battery posts or terminals to make sure the light is operating properly.

❋❋ WARNING

Do not use a test light to probe electronic ignition, spark plug or coil wires. Never use a pick-type test light to probe wiring on computer controlled systems unless specifically instructed to do so. Any wire insulation that is pierced by the test light probe should be taped and sealed with silicone after testing.

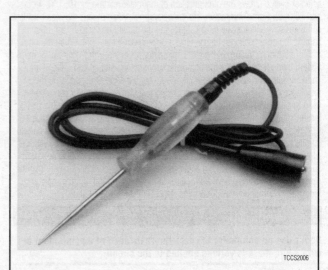

Fig. 7 A 12 volt test light is used to detect the presence of voltage in a circuit

Like the jumper wire, the 12 volt test light is used to isolate opens in circuits. But, whereas the jumper wire is used to bypass the open to operate the load, the 12 volt test light is used to locate the presence of voltage in a circuit. If the test light illuminates, there is power up to that point in the circuit; if the test light does not illuminate, there is an open circuit (no power). Move the test light in successive steps back toward the power source until the light in the handle illuminates. The open is between the probe and a point which was previously probed.

The self-powered test light is similar in design to the 12 volt test light, but contains a 1.5 volt penlight battery in the handle. It is most often used in place of a multimeter to check for open or short circuits when power is isolated from the circuit (continuity test).

The battery in a self-powered test light does not provide much current. A weak battery may not provide enough power to illuminate the test light even when a complete circuit is made (especially if there is high resistance in the circuit). Always make sure that the test battery is strong. To check the battery, briefly touch the ground clip to the probe; if the light glows brightly, the battery is strong enough for testing.

➡A self-powered test light should not be used on any computer controlled system or component. The small amount of electricity transmitted by the test light is enough to damage many electronic automotive components.

MULTIMETERS

Multimeters are an extremely useful tool for troubleshooting electrical problems. They can be purchased in either analog or digital form and have a price range to suit any budget. A multimeter is a voltmeter, ammeter and ohmmeter (along with other features) combined into one instrument. It is often used when testing solid state circuits because of its high input impedance (usually 10 megaohms or more). A brief description of the multimeter main test functions follows:

• Voltmeter—the voltmeter is used to measure voltage at any point in a circuit, or to measure the voltage drop across any part of a circuit. Voltmeters usually have various scales and a selector switch to allow the reading of different voltage ranges. The voltmeter has a positive and a negative lead. To avoid damage to the meter, always connect the negative lead to the negative (-) side of the circuit (to ground or nearest the ground side of the circuit) and connect the positive lead to the positive (+) side of the circuit (to the power source or the nearest power source). Note that the negative voltmeter lead will always be black and that the positive voltmeter will always be some color other than black (usually red).

• Ohmmeter—the ohmmeter is designed to read resistance (measured in ohms) in a circuit or component. Most ohmmeters will have a selector switch which permits the measurement of different ranges of resistance (usually the selector switch allows the multiplication of the meter reading by 10, 100, 1,000 and 10,000). Some ohmmeters are "auto-ranging" which means the meter itself will determine which scale to use. Since the meters are powered by an internal battery, the ohmmeter can be used like a self-powered test light. When the ohmmeter is connected, current from the ohmmeter flows through the circuit or component being tested. Since the ohmmeter's internal resistance and voltage are known values, the amount of current flow through the meter depends on the resistance of the circuit or component being tested. The ohmmeter can also be used to perform a continuity test for suspected open circuits. In using the meter for making continuity checks, do not be concerned with the actual resistance readings. Zero resistance, or any ohm reading, indicates continuity in the circuit. Infinite resistance indicates an opening in the circuit. A high resistance reading where there should be none indicates a problem in the circuit. Checks for short circuits are made in the same manner as checks for open circuits, except that the circuit must be isolated from both power and normal ground. Infinite resistance indicates no continuity, while zero resistance indicates a dead short.

✳✳ WARNING

Never use an ohmmeter to check the resistance of a component or wire while there is voltage applied to the circuit.

• Ammeter—an ammeter measures the amount of current flowing through a circuit in units called amperes or amps. At normal operating voltage, most circuits have a characteristic amount of amperes, called "current draw" which can be measured using an ammeter. By referring to a specified current draw rating, then measuring the amperes and comparing the two values, one can determine what is happening within the circuit to aid in diagnosis. An open circuit, for example, will not allow any current to flow, so the ammeter reading will be zero. A damaged component or circuit will have an increased current draw, so the reading will be high. The ammeter is always connected in series with the circuit being tested. All of the current that normally flows through the circuit must also flow through the ammeter; if there is any other path for the current to follow, the ammeter reading will not be accurate. The ammeter itself has very little resistance to current flow and, therefore, will not affect the circuit, but it will measure current draw only when the circuit is closed and electricity is flowing. Excessive current draw can blow fuses and drain the battery, while a reduced current draw can cause motors to run slowly, lights to dim and other components to not operate properly.

Troubleshooting Electrical Systems

When diagnosing a specific problem, organized troubleshooting is a must. The complexity of a modern automotive vehicle demands that you approach any problem in a logical, organized manner. There are certain troubleshooting techniques, however, which are standard:

• Establish when the problem occurs. Does the problem appear only under certain conditions? Were there any noises, odors or other unusual symptoms? Isolate the problem area. To do this, make some simple tests and observations, then eliminate the systems that are working properly. Check for obvious problems, such as broken wires and loose or dirty connections. Always check the obvious before assuming something complicated is the cause.

• Test for problems systematically to determine the cause once the problem area is isolated. Are all the components functioning properly? Is there power going to electrical switches and motors. Performing careful, systematic checks will often turn up most causes on the first inspection, without wasting time checking components that have little or no relationship to the problem.

• Test all repairs after the work is done to make sure that the problem is fixed. Some causes can be traced to more than one component, so a careful verification of repair work is important in order to pick up additional malfunctions that may cause a problem to reappear or a different problem to arise. A blown fuse, for example, is a simple problem that may require more than another fuse to repair. If you don't look for a problem that caused a fuse to blow, a shorted wire (for example) may go undetected.

Experience has shown that most problems tend to be the result of a fairly simple and obvious cause, such as loose or corroded connectors, bad grounds or damaged wire insulation which causes a short. This makes careful visual inspection of components during testing essential to quick and accurate troubleshooting.

Testing

OPEN CIRCUITS

▶ See Figure 8

This test already assumes the existence of an open in the circuit and it is used to help locate the open portion.

1. Isolate the circuit from power and ground.
2. Connect the self-powered test light or ohmmeter ground clip to the ground side of the circuit and probe sections of the circuit sequentially.
3. If the light is out or there is infinite resistance, the open is between the probe and the circuit ground.
4. If the light is on or the meter shows continuity, the open is between the probe and the end of the circuit toward the power source.

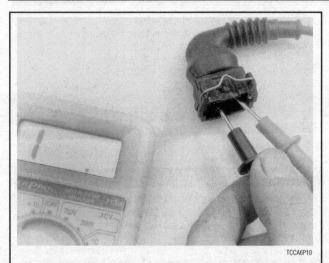

Fig. 8 The infinite reading on this multimeter indicates that the circuit is open

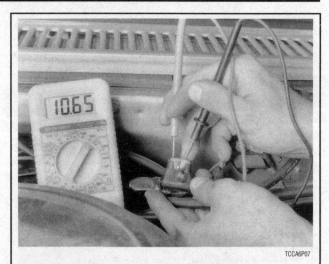

Fig. 9 This voltage drop test revealed high resistance (low voltage) in the circuit

SHORT CIRCUITS

➡Never use a self-powered test light to perform checks for opens or shorts when power is applied to the circuit under test. The test light can be damaged by outside power.

1. Isolate the circuit from power and ground.
2. Connect the self-powered test light or ohmmeter ground clip to a good ground and probe any easy-to-reach point in the circuit.
3. If the light comes on or there is continuity, there is a short somewhere in the circuit.
4. To isolate the short, probe a test point at either end of the isolated circuit (the light should be on or the meter should indicate continuity).
5. Leave the test light probe engaged and sequentially open connectors or switches, remove parts, etc. until the light goes out or continuity is broken.
6. When the light goes out, the short is between the last two circuit components which were opened.

VOLTAGE

This test determines voltage available from the battery and should be the first step in any electrical troubleshooting procedure after visual inspection. Many electrical problems, especially on computer controlled systems, can be caused by a low state of charge in the battery. Excessive corrosion at the battery cable terminals can cause poor contact that will prevent proper charging and full battery current flow.

1. Set the voltmeter selector switch to the 20V position.
2. Connect the multimeter negative lead to the battery's negative (-) post or terminal and the positive lead to the battery's positive (+) post or terminal.
3. Turn the ignition switch **ON** to provide a load.
4. A well charged battery should register over 12 volts. If the meter reads below 11.5 volts, the battery power may be insufficient to operate the electrical system properly.

VOLTAGE DROP

▶ See Figure 9

When current flows through a load, the voltage beyond the load drops. This voltage drop is due to the resistance created by the load and also by small resistances created by corrosion at the connectors and damaged insulation on the wires. The maximum allowable voltage drop under load is critical, especially if there is more than one load in the circuit, since all voltage drops are cumulative.

1. Set the voltmeter selector switch to the 20 volt position.
2. Connect the multimeter negative lead to a good ground.

3. Operate the circuit and check the voltage prior to the first component (load).
4. There should be little or no voltage drop in the circuit prior to the first component. If a voltage drop exists, the wire or connectors in the circuit are suspect.
5. While operating the first component in the circuit, probe the ground side of the component with the positive meter lead and observe the voltage readings. A small voltage drop should be noticed. This voltage drop is caused by the resistance of the component.
6. Repeat the test for each component (load) down the circuit.
7. If a large voltage drop is noticed, the preceding component, wire or connector is suspect.

RESISTANCE

▶ See Figures 10 and 11

✳✳ WARNING

Never use an ohmmeter with power applied to the circuit. The ohmmeter is designed to operate on its own power supply. The normal 12 volt electrical system voltage could damage the meter!

Fig. 10 Checking the resistance of a coolant temperature sensor with an ohmmeter. Reading is 1.04 kilohms

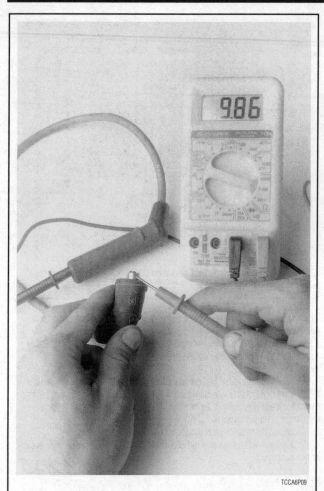

Fig. 11 Spark plug wires can be checked for excessive resistance using an ohmmeter

1. Isolate the circuit from the vehicle's power source.
2. Ensure that the ignition key is **OFF** when disconnecting any components or the battery.
3. Where necessary, also isolate at least one side of the circuit to be checked, in order to avoid reading parallel resistances. Parallel circuit resistances will always give a lower reading than the actual resistance of either of the branches.
4. Connect the meter leads to both sides of the circuit (wire or component) and read the actual measured ohms on the meter scale. Make sure the selector switch is set to the proper ohm scale for the circuit being tested, to avoid misreading the ohmmeter test value.

Wire and Connector Repair

Almost anyone can replace damaged wires, as long as the proper tools and parts are available. Wire and terminals are available to fit almost any need. Even the specialized weatherproof, molded and hard shell connectors are now available from aftermarket suppliers.

Be sure the ends of all the wires are fitted with the proper terminal hardware and connectors. Wrapping a wire around a stud is never a permanent solution and will only cause trouble later. Replace wires one at a time to avoid confusion. Always route wires exactly the same as the factory.

➡**If connector repair is necessary, only attempt it if you have the proper tools. Weatherproof and hard shell connectors require special tools to release the pins inside the connector. Attempting to repair these connectors with conventional hand tools will damage them.**

BATTERY CABLES

Disconnecting the Cables

✳✳ WARNING

Make sure the battery cables are connected to the correct battery terminals. Reverse polarity may damage electrical components, particularly the (expensive) engine control unit.

Before performing work on the vehicle, it is usually required to disconnect the battery from the electrical system. Whether this is to prevent a mistaken startup or electrical power from being supplied to the system components, the certain measure is to shut off the power at the source.

To disconnect the battery cable(s), perform the following:

1. Make sure the ignition switch is **OFF**.

➡**Disconnecting the negative battery cable FIRST with the ignition turned OFF is required to prevent unwanted surges through the electrical system. If a voltage spike is fed through the system, delicate and expensive componentry (like the engine control unit) could be destroyed in fractions of a second.**

2. Use the appropriate size wrench(es) to loosen the hardware on the cable clamps. ALWAYS begin with the negative battery cable.
3. If necessary, use a puller tool designed for battery cable clamps and remove the clamp, then isolate it from the battery terminal.
4. Remove the positive battery cable, if required, once the negative battery cable has been removed.

SUPPLEMENTAL RESTRAINT SYSTEM (AIR BAG)

General Information

SYSTEM OPERATION

✳✳ CAUTION

The 1994–96 Toyota trucks are equipped with a Supplemental Restraint System (SRS), which is comprised of a driver's and pas- senger's (on some models) air bag. Failure to carry out the service operations in the correct sequence could cause the SRS to unexpectedly deploy during servicing the SRS may fail to operate when needed. Before servicing, be sure to read the following items carefully.

When a vehicle is involved in a frontal collision in the hatched area and the shock is larger than a predetermined level, the SRS is activated automatically. A safing sensor is designed to tip at a smaller deceleration rate than the air bag sensor. The ignition is caused when a current flows to the squib, which occurs when a safing sensor and the air bag sensor trip simultaneously. When a deceleration force acts on the sensors, two sqibs in the driver's air bag and passenger's air bag (if applicable) ignite and generate gas. The gas that discharges into the driver's and passenger's air bags rapidly increases the pressure inside the bags, breaking open the steering wheel pad and (if applicable) instrument panel.

Bag inflation then ends, and the bags deflate as the gas is discharged through the holes at the bag's rear or side.

SYSTEM COMPONENTS

Steering Wheel Air Bag

The inflator and bag of the SRS are stored in the steering wheel pad and cannot be disassembled. The inflator contains a igniter charge, sqib, and gas generate. This inflates the bag when instructed by the center air bag sensor.

Front Passenger Air Bag

The inflator and bag of the SRS are stored in the instrument panel and cannot be disassembled. The inflator contains a igniter charge, sqib, and gas generate. This inflates the bag when instructed by the center air bag sensor.

Spiral Cable

The spiral cable is located in the combination switch. This cable is used as an electrical joint from the vehicle body side to the steering wheel.

SRS Warning Light

The SRS warning light is located on the combination meter. It turns to alert the driver of trouble in the system when a malfunction is detected in the air bag sensor assembly. In normal operating conditions when the ignition switch is turned to the **ON** or **ACC** position, the light turns on for about 6 seconds and then turns off.

Air Bag Sensor Assembly

▶ **See Figure 12**

The air bag sensor assembly is mounted on the floor inside the console box on Land Cruisers, on the air bag sensor cover on Tacoma and inside each fender on the T100. The air bag sensor assembly consists of an air bag sensor,

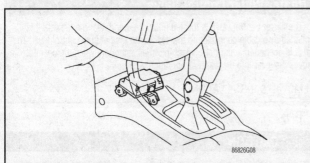

86826G08

Fig. 12 On Land Cruisers, the air bag sensor is mounted on the floor in the center console

safing sensor, diagnosis circuit and ignition control. It receives signals from the air bag sensors and judges whether the SRS must be activated or not.

SRS Connectors

All connectors in the SRS are colored yellow to distinguish them from the other connectors. These connectors have special functions are specifically designed for the SRS. These connectors use durable gold-platted terminals.

SYSTEM PRECAUTIONS

1. Work must be started after 90 seconds from the time the ignition switch is turned to the **LOCK** position and the negative battery cable has been disconnected. The SRS is equipped with a back-up power source so that if work is started within 90 seconds of disconnecting the negative battery cable, the SRS may deploy. When the negative terminal cable is disconnected from the battery, memory of the clock and radio will be canceled. Before you start working, make a note of the contents memorized by the audio memory system. When you have finished working, reset the audio systems and adjust the clock. Never use a back-up power supply from outside the vehicle.

2. In the event that of a minor frontal collision where the air bag does not deploy, the steering wheel pad, front air bag sensors and center air bag sensor assembly should be inspected.

3. Before repairs, remove the air bag sensors if shocks are likely to be applied to the sensors during repairs.

4. Never disassemble and repair the steering wheel pad, front air bag sensors or center air bag sensors.

5. Do not expose the steering wheel pad, front air bag sensors or center air bag sensor assembly directly to flames or hot air.

6. If the steering wheel pad, front air bag sensors or center air bag sensor assembly have been dropped, or there are cracks, dents or other defects in the case, bracket or connectors, have them replaced with new ones.

7. Information labels are attached to the periphery of the SRS components. Follow the instructions of the notices.

8. After arming the system, check for proper operation of the SRS warning light.

9. If the wiring harness in the SRS system is damaged, have the entire harness assembly replaced.

DISARMING THE SYSTEM

Work must be started only after 90 seconds from the time the ignition switch is turned to the **LOCK** position and the negative battery cable has been disconnected. The SRS is equipped with a back-up power source so that if work is started within 90 seconds of disconnecting the negative battery cable, the SRS may deploy. When the negative terminal cable is disconnected from the battery, memory of the clock and radio will be canceled. Before you start working, make a note of the contents memorized by the audio memory system. When you have finished work, reset the audio systems as before and adjust the clock. To avoid erasing the memory of each system, never use a back-up power supply from outside the vehicle.

ARMING THE SYSTEM

Once the negative battery cable is reconnected, the system is armed. Turn the ignition **ON**. Check that the SRS lamp turns off after about 6 seconds.

HEATER AND AIR CONDITIONER

The heater core is a small heat exchanger located inside the truck, similar to the radiator at the front of the truck. Coolant is circulated from the engine through the heater core and back to the engine. The heater fan blows fresh, outside air through the heater core; the air is heated and sent on to the interior of the truck.

About the only time the heater core will need removal is for replacement due to clogging or leaking. Thankfully, this doesn't happen too often; removing the heater core can be a major task. Some are easier than others, but all require working in unusual positions inside the truck and fitting tools into very cramped quarters behind the dashboard. In some cases, the dashboard must be removed during the procedure, another major project.

✳✳ CAUTION

On models equipped with air conditioning, the heater and air conditioner are adjacent but completely separate units. Be certain when working under the dashboard that only the heater hoses are disconnected. The air conditioning hoses are under pressure. If disconnected, the escaping refrigerant will freeze any surface with which it comes in contact, including your skin and eyes.

Blower Motor

REMOVAL & INSTALLATION

▶ **See Figure 13**

The blower motor and fan is mounted in the bottom of the heater case. Depending on model and equipment, it may be possible to remove the blower without removing the heater case. If there is insufficient clearance below the case to get the motor and fan free, the case must be removed.

1. On Land Cruisers with the 1FZ-FE engine you may need to remove the right side scuff plate to gain access to the motor. Remove the linkage on the lower cover, then remove the cover while pushing the locking protrusion.
2. Disconnect the electrical harness from the blower motor.
3. Disconnect the ductwork from the casing if it will interfere with motor removal. The motor may have a smaller, flexible air exchange hose on it; remove it.
4. Remove the blower motor fasteners and lower the blower motor out of the heater case.
5. Installation is in the reverse of removal. On the 1FZ-FE Land Cruisers, attach the lower cover, linkage and scuff plate.

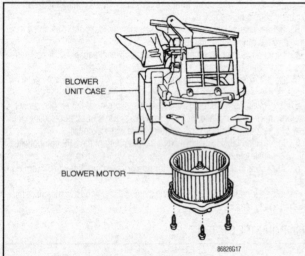

BLOWER UNIT CASE

BLOWER MOTOR

86826G17

Fig. 13 The blower motor is mounted in the blower unit case. You should be able to see it from under the dash

Heater Unit

REMOVAL & INSTALLATION

➡**Have alot of towels handy for the leakage of coolant that may spill into your interior.**

1. Discharge the A/C system (if applicable), using an approved recovery/recycling machine.
2. Remove the cooling unit (if applicable).
3. Drain the cooling system.

✴✴ CAUTION

When draining engine coolant, keep in mind that cats and dogs are attracted to ethylene glycol antifreeze and could drink any that is left in an uncovered container or in puddles on the ground. This will prove fatal in sufficient quantity. Always drain coolant into a sealable container. Coolant should be reused unless it is contaminated or is several years old.

4. Disconnect the hoses to the heater core. Tag each one so that it may be correctly reinstalled.
5. On some models you will need to remove the glove box assembly. On others, the instrument panel and reinforcement.

6. At the heater case under the dash, disconnect the ductwork from the case.
7. Disconnect the heater control cables from the heater case. Lift the spring clip holding the control cable to allow the cable to be manipulated. Don't deform or crimp the cables.
8. Disconnect the wiring harness to the blower fan.
9. Remove the three or four bolts holding the heater case to the dash. Because the inlet tubes project through the firewall, the unit will not fall straight down. It will need to be pulled into the passenger compartment and then brought downward.
10. Installation of the heater case is in the reverse order of removal..
11. Connect the control cables to their linkages. Each must be adjusted so that the motion of the dashboard lever causes the correct corresponding motion in the air door. Both the control lever and the door must reach maximum travel at the same time if full function is to be achieved.
   a. Set the dashboard control to FRESH. Lift the adjustment clip and adjust the air inlet cable towards the heater unit until the linkage is at its maximum travel on Pick-Up and 4Runner. For Land Cruiser, adjust the cable away from the case. Release the adjustment clip.
   b. Set the dashboard control to VENT (Pick-Up) or DEF (4Runner and Land Cruiser). Lift the adjustment clip and adjust the air flow control cable towards the heater unit until the linkage is at its maximum travel. Release the adjustment clip.
   c. Set the temperature control to COOL. Lift the adjustment clip and adjust the air mix control damper control cable away from the heater case until the linkage is at its maximum travel on . For Land Cruiser, adjust the cable towards the case.

Heater Core

REMOVAL & INSTALLATION

1. Drain the cooling system.
2. Remove the heater unit assembly. Have alot of towels handy for the leakage of coolant that may spill into your interior.
3. Remove the fasteners securing the core to the unit, remove the plates and clamps. Pull upwards to remove the core from the unit.
4. Installation of the core is in the reverse order of removal. Check the cooling system.

Heater Water Control Valve

The heater control valve is located under the hood near the firewall. It is operated by a cable connected to the dashboard temperature control lever. The valve is opened to admit hot coolant into the heater core when the operator moves the lever into the warm or hot range.

REMOVAL & INSTALLATION

1. Drain the engine coolant.

✴✴ CAUTION

When draining engine coolant, keep in mind that cats and dogs are attracted to ethylene glycol antifreeze and could drink any that is left in an uncovered container or in puddles on the ground. This will prove fatal in sufficient quantity. Always drain coolant into a sealable container. Coolant should be reused unless it is contaminated or is several years old.

2. Disconnect the control cable from the valve.
3. Disconnect the inlet and outlet hoses from the valve.
4. If equipped, remove the harness connector.
5. If the valve is held by a retaining bolt, remove it and remove the valve.
To install:
6. Reinstall in reverse order. Use new hose clamps.
7. Move the temperature selector lever to COOL. On T100 and Tacoma move it to WARM. Adjust the valve lever to the corresponding position and connect the cable. Lift the spring clip and adjust the cable so that the valve achieves full travel to the closed position.

Control Cables

REMOVAL & INSTALLATION

1. Remove the control assembly from the dash.
2. Tag and remove the cables from the back of the control assembly.
3. Disconnect the cable from the component it is attached to.
4. When installing attach the cable to its appropriate location in the back of the control assembly. Then route it to the appropriate component. Adjust the cable.
5. Check for proper operation.

ADJUSTMENTS

First move the control levers to the left then right. Check for any stiffness and binding through the full range of the levers.

Air Inlet Damper Control

Set the air inlet damper to the FRESH position, then install the control cable and lock the clamp.

Mode Damper Control

Set the mode damper and the control cable to the FACE position on T100, Tacoma, and 1996 4Runner. Set it to DEF on all other models. Clamp the section of the control cable and install the cable to the damper control lever.

Air Mix Damper Control

Set the air mix damper and the control lever to the COOL position or WARM on the Tacoma and T100. Install the control cable and lock the clamp.

Water Valve Control

Set the water valve in the COOL position or WARM on the Tacoma and T100. While pushing the outer cable in the direction, clamp the outer cable to the water valve bracket.

Air Conditioning Components

REMOVAL & INSTALLATION

Repair or service of air conditioning components is not covered by this manual, because of the risk of personal injury or death, and because of the legal ramifications of servicing these components without the proper EPA certification and experience. Cost, personal injury or death, environmental damage, and legal considerations (such as the fact that it is a federal crime to vent refrigerant into the atmosphere), dictate that the A/C components on your vehicle should be serviced only by a Motor Vehicle Air Conditioning (MVAC) trained, and EPA certified automotive technician.

➡If your vehicle's A/C system uses R-12 refrigerant and is in need of recharging, the A/C system can be converted over to R-134a refrigerant (less environmentally harmful and expensive). Refer to Section 1 for additional information on R-12 to R-134a conversions, and for additional considerations dealing with your vehicle's A/C system.

CRUISE CONTROL

System Diagnosis

INDICATOR CHECK

Turn the ignition switch **ON**. Check that the CRUISE indicator light comes on when the cruise control main switch is turned on, and that the indicator light turns off when the main switch is turned off.

DIAGNOSTIC TROUBLE CODES (DTC)

If a malfunction occurs in the system during cruise control driving, the ECM cancels the cruise control, and blinks the CRUISE indicator light 5 times to inform the driver of a malfunction. At the same time, the malfunction is stored in the memory as a diagnostic trouble code.

Reading Codes

▶ **See Figures 14, 15 and 16**

1. Turn the ignition switch **ON**.
2. Using SST 09843-18020 or an equivalent jumper wire, connect terminals

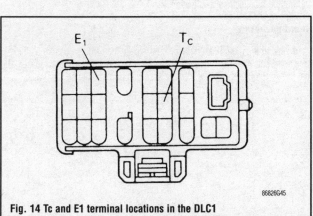

Fig. 14 Tc and E1 terminal locations in the DLC1

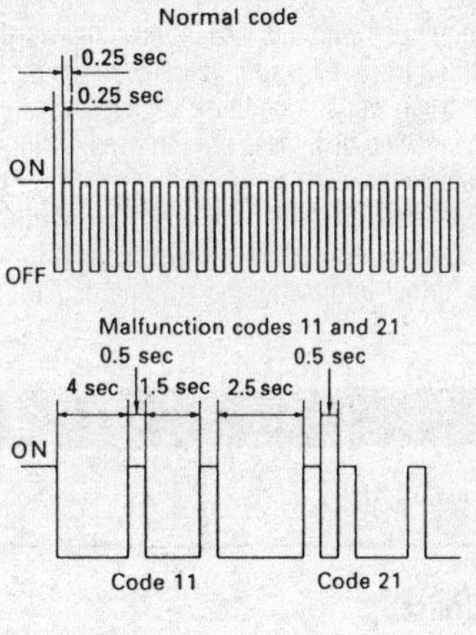

Fig. 15 Blink pattern examples for normal system and codes 11 and 21

Code No.	CRUISE MAIN Indicator Light Blinking Pattern	Diagnosis
—	ON OFF	Normal
11	ON OFF	• Duty ratio of 100% output to motor acceleration side. • Overcurrent (short) in motor circuit.
12	ON OFF	• Overcurrent (short) in magnet clutch circuit. • Open in magnet clutch circuit.
13	ON OFF	• Open in actuator motor circuit. • Position sensor detects abnormal voltage. • Position sensor signal value does not change when the motor operates.
21	ON OFF	• Speed signal is not input to the Cruise Control ECU.
*23	ON OFF	• Actual vehicle speed has dropped by 16 km/h (10 mph) or more below the set speed during cruising.
32	ON OFF	• Short in control switch circuit.
34	ON OFF	• Voltage abnormality in control switch circuit.
41	ON OFF	• When 41 code is indicated, replace the cruise control ECU.

Fig. 16 Cruise control diagnostic codes

TC and E1 of the DLC1. The DLC1 is located under the hood next to the fuse-box.

3. Read the diagnostic trouble code on the CRUISE indicator light. Write the codes down and compare them to the chart.
4. Turn the ignition **OFF**.
5. Disconnect the jumper wire.

ENTERTAINMENT SYSTEMS

✳✳ WARNING

Never operate the radio without a speaker; severe damage to the output transistors will result. If the speaker must be replaced, use a speaker of the correct impedance (ohms) or the output transistors will be damaged and require replacement.

Some audio systems have an anti-theft system built into them. The system requires the operator to select and program a 3-digit code into the unit. If no code is installed, the anti-theft system is inoperative; the audio unit functions as a normal unit, even after power is restored after disconnection. Complete instructions for installing the code are found in the owners manual for the vehicle. As long as power remains connected to the unit, it may be used in the normal fashion. Once power is disconnected from the unit, the audio unit will not function until the correct code is re-entered. This will render the unit useless if stolen; there is no way to retrieve the stored code.

The anti-theft code may be canceled or changed following an exact procedure explained in the vehicles owners manual. If an error is made during the procedure, the designation ERR appears in the digital window. Once 10 errors have occurred, the word HELP appears and the system will not work regardless of codes or power. If the HELP message appears, the unit must be taken to a Toyota dealer to be reset.

All of the units equipped with this system have the words ANTI THEFT SYSTEM visible on the front of the unit, usually on the tape player door. Care must be taken when working on these vehicles anytime the battery cable must be disconnected or if fuses are removed during other test procedures. If its your vehicle, chances are good that you know the code (if one was installed); if its a friends vehicle, you could be in trouble. Always check with the owner before beginning any work which could interrupt power to the audio unit. For any further information on your particular unit, see your owners manual.

Radio

REMOVAL & INSTALLATION

Excluding Land Cruiser

1. If audio unit carries the designation ANTI THEFT SYSTEM, make certain you or the vehicle owner knows the security code. If the code is not re-entered after installation, the unit will not operate.
2. On some models you may need to remove the heater control assembly, Following the procedures given earlier in this section.
3. Remove the bolts holding the audio unit to the dash. Pull the unit out of the lower dash enough to allow access to the rear and sides.
4. Unplug the antenna cable and wiring connectors. Remove the radio with its bracket.
5. Installation of the radio is in the reverse order of removal.

Clearing Codes

After completing repairs, the diagnostic code can be cleared by removing the ECM-B fuse for 10 seconds or more. Make sure the ignition switch is **OFF**. Check that the normal code is displayed after connecting the fuse.

Land Cruiser

1. Disconnect the negative battery cable. Remove the ashtray. Remove the ashtray holder.
2. Remove the 3 screws holding the lower instrument panel trim. The trim panel is also held by six clips; each must be gently pried loose.

✳✳ WARNING

The trim panel is long and thin; it is easily broken by rough handling.

3. Once the panel is loose, unplug the wiring connectors to the switches and remove the panel from the area.
4. Remove the screws holding the audio unit to the dash.
5. Pull the unit outward enough to allow access to the sides and back. Unplug the antenna cable and the electrical connectors. Remove the radio unit.
6. Installation is in the reverse order of removal.

Speakers

REMOVAL & INSTALLATION

➡**Always disconnect the negative battery cable before attempting to remove the speakers.**

Dash Mounted

Dash mounted speakers can be accessed after removing the appropriate trim panel. These panels are usually retained by screws and clips. Be sure you have removed all of the attaching screws before prying the panel from the dash. Do not use excessive force on the panel as this will only lead to damage. Once the panel has been removed, loosen the speaker attaching bolts/screws, then pull the speaker from the dash and unplug the electrical connection.

Door Mounted

Door mounted speakers can be accessed after removing the door panel. These panels are usually retained by screws and clips. Be sure you have removed all of the attaching screws before prying the panel from the door. A special tool can be purchased for this purpose. Do not use excessive force on the panel as this will only lead to damage. Once the panel has been removed, loosen the speaker attaching bolts/screws, then pull the speaker from its mount and unplug the electrical connection.

Rear Speakers

Removing the rear speakers involves basically the same procedure as the front speakers. Remove the appropriate trim panel, then remove the speaker. The rear speakers on some models can be accessed from inside the rear hatch.

WINDSHIELD WIPERS AND WASHERS

Windshield Wiper Blade

REMOVAL & INSTALLATION

➡ **Wiper blade element replacement is covered in Section 1.**

Toyota has two types of wiper blades. The screw-on type and the clip-on type.

To remove the clip-on type, lift up the wiper arm from the windshield. Lift up on the spring release tab on the wiper blade-to-wiper arm connector, then pull the blade assembly off the wiper arm.

To remove the screw-on type, lift up the wiper arm from the windshield. Loosen and remove the two screws retaining the blade to the arm, then lift the blade assembly off the wiper arm.

Windshield Wiper Arm

REMOVAL & INSTALLATION

▶ **See Figures 18 and 19**

1. There may be a cover over the nut, remove this to access the nut. With the arm in the down position, unscrew the nut which secures it to the pivot. Carefully pull the arm upward and off the pivot.

2. Install the arm by placing the arm onto the linkage shaft. Make sure it is seated correctly; if not correctly aligned, the blade will slap the bodywork at the top or bottom of its stroke. Tighten the nut to approximately 15 ft. lbs. (20 Nm).

➡ **If one wiper arm does not move when turned on or only moves a little bit, check the retaining nut at the bottom of the arm. The extra effort of moving snow or wet leaves off the glass can cause the nut to come loose—the pivot will move without moving the arm.**

Fig. 19 Loosen and remove the retaining nut

Front Wiper Motor

REMOVAL & INSTALLATION

▶ **See Figures 20, 21, 22 and 23**

The wiper motor is located in the engine compartment and is secured to the firewall.

1. Disconnect the wiring from the wiper motor and unbolt it from the firewall.

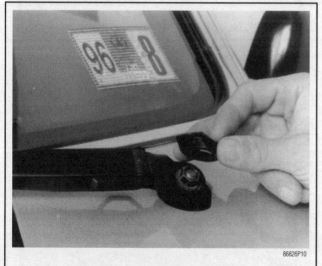

Fig. 18 Remove the cover concealing the nut

Fig. 20 Unplug the wiring harness from the motor assembly

Fig. 21 Remove the bolts securing the wiper motor from the firewall

Fig. 22 Pull the motor from the firewall, then using a prybar . . .

Fig. 23 . . . separate the linkage from the motor

2. Gently pry the wiper link from the crank arm. Its a ball and socket arrangement, but it may be tight.

3. Remove the motor.

4. Install the motor, tighten the mounting nuts to 47 inch lbs. (6 Nm). Attach the electrical lead.

Rear Wiper Motor

REMOVAL & INSTALLATION

4Runner

TOP MOUNTED

1. Make certain the ignition switch is **OFF**.
2. Pop up the acorn nut cover, remove the nut and pull off the rear wiper.
3. Remove the pivot nut.
4. Inside, above the tailgate, pop out the clips and remove the wiper motor cover.
5. Disconnect the electrical lead and the washer hose, remove the mounting bolts and lift out the wiper motor.
6. Install the motor and then tighten the pivot nut to 8 ft. lbs. (11 Nm). Tighten the wiper arm acorn nut to 48 inch lbs. Snap the cover down.

BOTTOM MOUNTED

♦ **See Figure 24**

1. Remove the back door trim, plate and door glass run.
2. Separate the weatherstrip from the door, then remove the rear glass.
3. Remove the rear wiper arm.
4. Disconnect the wiring harness from the motor.
5. Remove the nut securing the packing holder.
6. Loosen the wiper motor seat bolts, then the motor assembly.
7. Remove the washer nozzle.

To install:

8. Install the washer nozzle and motor. Tighten the motor bolts to 48 inch lbs. (5 Nm). Install the packing holder and nut, tighten to 48 inch lbs. (5 Nm).

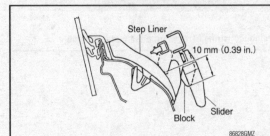

Fig. 24 Push the slider 0.39 in. (10mm) towards the turning slide, then tighten the wiper arm set nut

9. When tightening the rear arm, push the slider 0.39 in. (10mm) towards the turning slide, then tighten the wiper arm set nut with the step liner on the block as shown in the illustration. Tighten to 48 inch lbs. (5 Nm).

Land Cruiser

1. Insure the ignition switch is **OFF**.
2. On some models you will need to remove the pull handle and pull handle bezel.
3. If present, remove the backdoor trim.
4. Lift the cover at the base of the rear wiper arm, remove the retaining nut and remove the wiper arm.
5. Remove the large nut on the wiper motor axle (the part that the arm bolts to).
6. Disconnect the wiper motor wire harness.
7. Remove the retaining bolts holding the motor and remove the motor. Don't bend or damage nearby linkages or components.
8. Unbolt the wiper control relay, then disconnect the harness.
9. Reassemble in reverse order.

Wiper Linkage

REMOVAL & INSTALLATION

♦ **See Figures 25, 26 and 27**

1. Remove the wiper motor.
2. Separate the wiper arms by removing their retaining nuts and working them off their shafts.

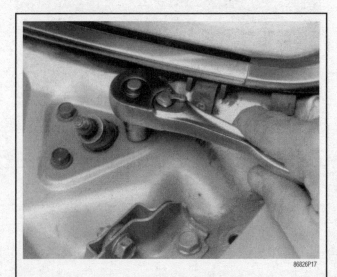

Fig. 25 Remove the wiper linkage bolts retaining the ends

Fig. 26 Unclamp the grilles to lift the linkage up and out

Fig. 28 Unbolt the reservoir bracket from the upper radiator support

Fig. 27 Slide the linkage through the hole in the cowl

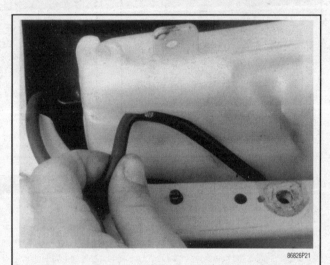

Fig. 29 Lift the unit from the engine compartment, then slide the hose from the unit

3. Remove the nuts/bolts and spacers holding the wiper shafts.

4. Unclamp the grilles that cover the linkage, then push the shafts down into the body cavity. Pull the linkage out of the cavity through the wiper motor hole.

5. Install the wiper motor as it was removed.

Washer Fluid Reservoir

REMOVAL & INSTALLATION

▶ See Figures 28 and 29

1. Unbolt the washer reservoir from the upper radiator support.

2. Lift the unit from the engine compartment, then slide the hose from the unit.

3. Unhook the wiring harness for the pump from the side of the reservoir.

4. Disconnect the wiring harness from the pump, then remove the pump from the reservoir. Check the condition of the grommet that the pump sits into.

5. Install the reservoir in the reverse order of removal.

Washer Pump

REMOVAL & INSTALLATION

▶ See Figures 30 thru 35

1. Remove the washer reservoir from the vehicle.

2. Unhook the wiring harness for the pump from the side of the reservoir.

3. Disconnect the wiring harness form the pump, then remove the pump from the reservoir. Check the condition of the grommet that the pump sits into.

4. Install the pump into the reservoir, ensure the grommet is in good condition.

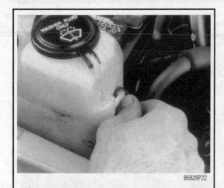

Fig. 30 Unhook the wiring harness for the pump from the side of the reservoir

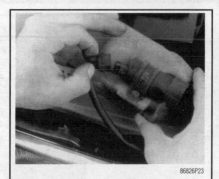

Fig. 31 Disconnect the wiring harness form the pump, then remove the pump from the reservoir

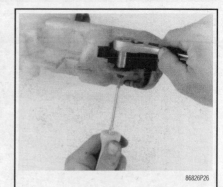

Fig. 32 Remove the bolts retaining the pump to the reservoir

Fig. 33 Pull the pump from the grommet

Fig. 34 Remove the hose from the pump

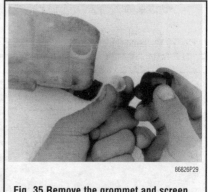

Fig. 35 Remove the grommet and screen, then inspect for deterioration

INSTRUMENTS AND SWITCHES

To keep our terms straight, we'll use Toyotas names for various dash components. The instrument panel is what you may call the dashboard; it runs completely across the front of the vehicle. The instrument panel is covered with a soft vinyl surface called the safety pad. The safety pad is the part you see; the instrument panel is the framework under it. All the instruments and warning lamps in front of the driver are contained in the combination meter, sometimes called the gauge set or instrument cluster. The combination meter is removed as a unit.

Generally, numbered components use a numbering system with item 1 on the left side of the vehicle. Knowing how many numbered components there are can be helpful, too. For example, if heater ducts Nos. 1, 2 and 3, must be removed, its a fair bet that No. 2 is in the center. If the procedure only refers to Nos. 1 and 2, one may be on the left side and 2 on the right of the passenger compartment.

When disassembling components, always suspect the hidden screw or clip. Much of the fit and finish in the interior is accomplished by using concealed retainers to keep panels in place. Don't force anything during removal; if any resistance is felt, search out the hidden connector. Some of the panels assemble only in the correct order; pay attention. Take note of which bolts and screws go into each retainer; a too-long bolt can damage wiring or components behind the assembly being held.

Finally, understand that this is a lengthy project. Work slowly and carefully so as not to damage anything. Label or mark each electrical connector as it is disconnected; many of the plastic connector shells can be marked with an indelible laundry pen or similar marker. As a panel or component is removed, disconnect the wiring running to switches or components held by the panel. Be careful working around wires and harnesses; most are held by retainers and do not allow a lot of slack.

Instrument Cluster

REMOVAL & INSTALLATION

Pick-Up and 4Runner

1989–95 MODELS

1. Disconnect the negative battery cable.
2. Remove the steering wheel.
3. Remove the upper and lower steering column covers.
4. Remove the hood release lever from the panel. Its held by two screws.
5. Separate the cowl side trim on each side. You might refer to it as the kick panel.
6. Remove the lower center instrument panel cover.
7. Remove the ignition key cylinder cover.
8. Detach the No. 1 lower finish panel; this is the panel holding the left speaker and running under the steering column.
9. Remove the heater duct to No. 2 outlet.
10. Pull off the heater control knobs, remove the A/C switch if equipped and gently pry up on the heater control unit to remove it. The prying tool should have a taped or protected edge.
11. For all models except Pick-Ups with 4 speed manual transmissions, remove the 3 screws holding the meter trim. Pull the trim or surround panel outward and disconnect the connectors. On 4Runner, remove the two screws and the cup holder from the panel.
12. For Pick-Ups with 4 speed manual transmissions, the trim is

held only by two screws. Disconnect the connectors and remove the trim panel.

13. Remove the No.1 air outlet. Its held by 2 screws.

14. Remove the four screws holding the combination meter. Work it out of the panel until access is gained to the rear. Disconnect the speedometer cable and remove the electrical connectors. Remove the meter assembly, placing it in a safe location out of the work area.

15. Install the instrument panel in the reverse order of removal.

1996 MODELS

✷✷ CAUTION

Your vehicle may contain an air bag system. Read the procedure in this section of how to disarm the system before proceeding.

1. Disable the SRS system.
2. Disconnect the negative battery cable.
3. Remove the steering wheel.
4. Remove the cowl side trim and the front door scuff plate.
5. Remove the retainers securing the lower finish panel, then remove the hood and fuel tank release lever screws.
6. Remove the ignition switch bezel.
7. Detach the No. 2 and No. 1 heater-to-register duct.
8. Remove the cluster finish panel, then remove.
9. Disconnect the harness for the combination meter. Remove the screws securing the meter, then remove it from the vehicle.
10. Installation is in the reverse order or removal.

Tacoma

✷✷ CAUTION

Your vehicle may contain an air bag system. Read the procedure in this section of how to disarm the system before proceeding.

1. Disable the SRS system.
2. Disconnect the negative battery cable.
3. Remove the steering wheel.
4. Remove the following:
 a. Steering column cover
 b. Hood lock release lever
 c. Combination switch
 d. Fuse box cover
5. Remove the lower left hand finish panel.
6. Remove the following parts:
 a. Ignition switch bezel
 b. No. 2 heater-to-register
 c. Steering column
7. Remove the cup holder and the heater control knobs.
8. With the aid of a flat-bladded tool, remove the heater control panel.
9. Disconnect the hazard harness.
10. Remove the mounting screws for the center cluster finish panel.
11. Detach the heater control assembly.
12. Remove the radio.
13. Remove the screws mounting the cluster finish panel, then lower the panel.
14. Remove the combination meter screws and the speedometer cable.
15. Disconnect the harness.
16. Installation is in the reverse order of removal.

T100

✷✷ CAUTION

Your vehicle may contain an air bag system. Read the procedure in this section of how to disarm the system before proceeding.

1. Disable the SRS system.
2. Disconnect the negative battery cable.
3. Remove the front pillar garnish.
4. Unscrew the front door scuff plate and the cowl side trims.

5. Remove the steering wheel.
6. Remove the following:
 a. Steering column cover
 b. Disconnect the hood lock release lever
 c. Detach the No. 1 lower finish panel
 d. Remove the combination switch
 e. Remove the glove box door
 f. Detach the No. 2 lower finish panel
 g. Remove the lower center panel
7. Remove the screws retaining the center cluster finish panel, then detach the harness from the unit.
8. Remove the stereo opening cover.
9. Remove the screws securing the meter, then remove the combination meter from the dash.
10. Installation is in the reverse order of removal.

Land Cruiser

1989–90 MODELS

1. Disconnect the negative battery cable.
2. Remove the steering wheel.
3. Remove the Nos. 2 and 3 air ducts. These are above the drivers knees under the dash.
4. Disconnect the throttle cable from the accelerator pedal and retainer. Remove the cable set nut and remove the cable.
5. Remove the hood release lever and the fuel lid opener lever.
6. Remove the heater control lever knobs. Remove the ash tray.
7. Remove the five screws and two clips holding the center instrument finish panel. Remove the panel and disconnect the wiring harnesses.
8. Remove the seven screws holding the instrument cluster trim panel. Loosen the panel, disconnect the wiring and unhook the speedometer cable.
9. Remove the radio.
10. Remove the glove box door. Remove the courtesy switch and light.
11. Remove the four screws holding the glove box; remove the latch striker and remove the glove box.
12. Remove the mounting bolt holding the EFI computer; pull it out gently and disconnect the wire harness. Place the computer in a very safe location out of the work area.
13. Remove air ducts Nos. 1, 3 and 4; each is held by a single screw.
14. At the side defrosters, remove only the lower screw and disconnect the defroster hose.
15. Remove the four screws and four nuts holding the safety pad. Loosen two clips and remove the pad.
16. Remove the heater control panel.
17. Disconnect the electrical connectors at the mirror control switch, the dimmer and 4WD control.
18. Remove the fuse box; its held by two screws.
19. Remove the two upper side mounting bolts holding the steering column.
20. The instrument panel is held by 15 bolts. Once the bolts are removed, move the panel towards the rear of the vehicle to remove it. Once removed, the various components still on the panel may be removed if desired.
21. Installation is in the reverse order of removal.

1991–95 MODELS

✷✷ CAUTION

Your vehicle may contain an air bag system. Read the procedure in this section of how to disarm the system before proceeding.

1. Disable the SRS system.
2. Disconnect the negative battery cable.
3. Remove the steering wheel.
4. Apply strips of protective tape on the inside of each windshield pillar. This will protect the trim during removal.
5. Remove the upper and lower steering column covers.
6. Remove the hood release and fuel door release levers.
7. Unscrew the lower trim panel below the steering column.
8. Remove the heater duct running to the left dash outlet (No. 2). The duct is the one under the dash, above the drivers knees.
9. Remove the combination switch from the steering column. The combination switch is the lighting control/turn signal/wiper control assembly.

10. Remove the turn signal bracket from the steering column; its the piece just below the combination switch.

11. Remove the ashtray and remove the ashtray holder.

12. Remove the three screws holding the lower instrument panel trim. The trim panel is held by six clips which must be released to remove the panel.

13. Remove the instrument cluster trim; it is held by six screws and two clips.

14. Remove the four screws holding the combination meter and gently loosen it; disconnect the electrical harnesses and unhook the speedometer cable. Place the instrument cluster in a safe location out of the work area.

15. Installation is in the reverse order of removal.

1996 MODELS

✸✸ CAUTION

Your vehicle may contain an air bag system. Read the procedure in this section of how to disarm the system before proceeding.

---

1. Disable the SRS system.
2. Disconnect the negative battery cable.
3. Remove the steering wheel.
4. Apply strips of protective tape on the inside of each windshield pillar. This will protect the trim during removal.
5. Remove the upper and lower steering column covers.
6. Remove the hood release and fuel door release levers.
7. Remove the fuse box opening cover.
8. Remove the retaining screws for the lower trim panel below the steering column.
9. Remove the lower instrument panel.
10. Disconnect the No. 2 heater-to-register duct.
11. Loosen the screws and remove the fuse block.
12. Detach the No. 2 center cluster finish panel.
13. Remove the steering column.

14. Detach the cluster finish panel, then the combination meter.

15. Remove the center cluster finish panel assembly with the clock attached.

16. With the aid of a taped prytool, take off the 2 claws, then remove the cup holder hole cover.

17. Remove the ashtray.

18. Remove the center cluster finish panel with the heater control assembly, then disconnect the harness.

19. Remove the screws retaining the heater control assembly from the center cluster finish panel.

20. Remove the following:
   a. Radio
   b. Glove compartment door
   c. Speaker panel
   d. Speaker
   e. Front console box
   f. Rear console box

21. Loosen and remove the 5 screws and 9 bolts holding the instrument panel.

22. Remove the lower instrument panel reinforcement, then the No. 1 brace and the instrument panel.

23. Installation is in the reverse order of removal.

Speedometer, Tachometer and Gauges

REMOVAL & INSTALLATION

▶ **See Figure 36**

1. Remove the gauge cluster from the dashboard.
2. Remove the lens covering the gauges.
3. Remove the printed circuit board from the back of the cluster. This can be done by removing any attaching screws and by removing the bulbs from the unit.

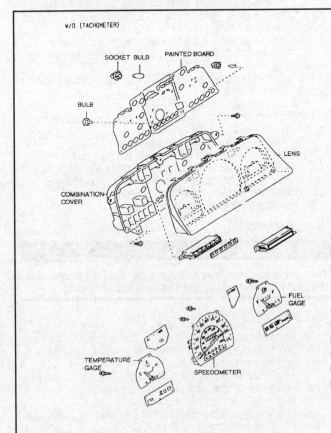
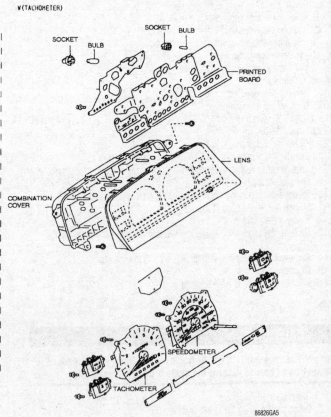

Fig. 36 Exploded view of the instrument cluster assembly

86826GA5

4. Remove the gauge attaching screws.
5. Installation is the reverse of removal.

Combination Switch

REMOVAL & INSTALLATION

▶ **See Figures 37, 38 and 39**

The individual switches within the combination switch are individually replaceable after the combination assembly is removed and disassembled. It is not difficult but is exacting, requiring careful removal of wire terminals from the multi-pin connector.

1. Insure the ignition is switched **OFF**.
2. Remove the steering wheel pad and remove the steering wheel. Use of a steering wheel puller is strongly recommended. Always scribe matchmarks on the wheel hub and the steering column so the wheel can be reinstalled straight.
3. Remove the upper and lower steering covers. Remove the retaining screws holding the combination switch to the column.
4. Trace the harness down the column, then unplug the connector. Release the harness from the harness clamp(s).
5. Remove the combination switch from the column and place it on the workbench.
6. Disconnect the air bag harness (if equipped).
7. On the newer models, you will be able to remove the two mounting screws retaining the arm (wiper ect.) to replace.
8. Otherwise, identify the component to be replaced. Identify each wire color running from the switch (to be replaced) to the connector. These wires (only) will need to be removed so the new ones can be installed.
9. On all but 1989–90 Land Cruiser, the cover on the back of the connector must be raised to allow the wire terminal to be removed.
10. Diagram each wire and terminal to be removed. The new wiring must be installed exactly as original. From the front or plug end, insert a small pick or probe between the locking lug and the wire terminal. Pry down the locking lug with the tool and pull the terminal out from the rear. Repeat the procedure for each wire from the switch to be replaced.
11. With the wires free, remove the retaining screws holding the switch to be changed and remove it. The lighting control stalk has a spring and small ball with it; don't lose them.

To install:
12. Assemble the new component(s) onto the switch. Retaining screws should be tight enough to hold but not over tight. Components with cracked plastic won't stay in place.
13. Track the new wires into the harness and place each wire loosely in its correct terminal socket.
14. Use the removal tool to push on the back of the terminal (NOT on the wire) and force the terminal into the connector. Each terminal must be pushed in until it overrides the locking lug. Test each one by gently pulling the wire backwards; it should be positively locked in place.
15. Fit the combination switch assembly onto the steering column and install the retaining screws. Install the upper and lower steering column covers.
16. Install the wire harness into the clips and retainers; connect the connectors.
17. Align the matchmarks and reinstall the steering wheel and pad.
18. Turn the ignition switch **ON** and check the function of each control item.

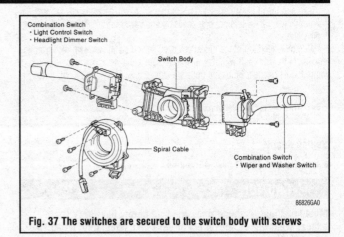

Fig. 37 The switches are secured to the switch body with screws

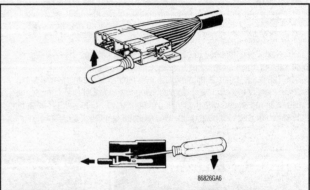

Fig. 38 From the front or plug end, insert a small pick or probe between the locking lug and the wire terminal

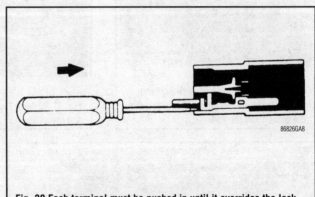

Fig. 39 Each terminal must be pushed in until it overrides the locking lug

LIGHTING

Headlights

Headlights, like any other lighting device, can fail due to broken filaments. The front of any truck is the worst possible location for a lighting device since it is subject to impact, extensive temperature change and severe vibration—all of which shorten the life of the light. The front of the truck is also where good lighting is needed the most so its not uncommon to have to replace a headlight during the life of the vehicle.

There are two general styles of headlamps, the sealed beam and replaceable bulb type. The sealed beam is by far the most common and includes almost all of the circular and rectangular lamps found on vehicles built through the 1980s. The sealed beam is so named because it includes the lamp (filament), the reflector and the lens in one sealed unit. Sealed beams are available in several sizes and shapes.

The replaceable bulb is the newer technology. Using a small halogen bulb, only the lamp is replaced, while the lens and reflector are part of the body of the car. This is generally the style found on wrap-around or "European" lighting systems. While the replaceable bulbs are more expensive than sealed beams, they generally produce more and better light. The fixed lenses and reflectors can be engineered to allow better frontal styling and better light distribution for a particular vehicle.

It is quite possible to replace a headlight of either type without affecting the alignment (aim) of the light. Sealed beams mount into a bracket (bucket) to which springs are attached. The adjusting screws control the position of the bucket which in turn aims the light. Replaceable bulbs simply fit into the back of the reflector. The lens and reflector unit are aimed by separate adjusting screws.

Take a moment before disassembly to identify the large adjusting screws (generally two for each lamp, one above and one at the side) and don't change their settings.

With the exception of the oldest vehicles, sealed beams are removed from the outside of the car. Start with the outer trim pieces and work your way in to the lamp and its retainer. Bulb type units are almost always replaced from under the hood.

REMOVAL & INSTALLATION

Sealed Beams

See Figures 40, 41, 42, 43 and 44

CAUTION

Most headlight retaining rings and trim bezels have very sharp edges. Wear gloves. Never pry or push on a headlamp; it can shatter suddenly.

1. Make sure the headlight switch is in the off position.
2. Loosen the parking and side turn signal unit retaining screws. Then remove the units unplugging the connectors.
3. Next, open the hood to gain access to the grille clips. Using a flat-bladded tool, push on the clips to release them and the grille.
4. The sealed beam is held in place by a retainer and either 2 or 4 small screws. Identify these screws before applying any tools.DO NOT confuse the small retaining screws with the larger aiming screws. There will be two aiming screws or adjustors for each lamp. (One adjustor controls the up/down motion and the other controls the left/right motion.) Identify the adjustors and avoid them during removal. If they are not disturbed, the new headlamp will be in identical aim to the old one.
5. Using a small screwdriver (preferably magnetic) and a pair of taper-nose pliers if necessary, remove the small screws in the headlamp retainer. DON'T drop the screws; they vanish into unknown locations.

A good kitchen or household magnet placed on the shank of the screwdriver will provide enough grip to hold the screw during removal.

6. Remove the retainer and the headlamp may be gently pulled free from its mounts. Detach the connector from the back of the sealed beam unit and remove the unit from the car.

CAUTION

The retainers can have very sharp edges. Wear gloves.

7. Installation is in the reverse order of removal.

European Headlamps

RETAINING RING TYPE

1. With the ignition switch **OFF** and the headlamp switch OFF, raise and prop the hood.
2. On Pick-Ups and 4Runners, remove the grille and headlamp assembly.
3. On Land Cruisers:
   a. If the right headlamp is to be changed, lift out the coolant reserve tank. Place the tank in an out of the way location but NOT on the battery or engine. Don't spill any coolant.

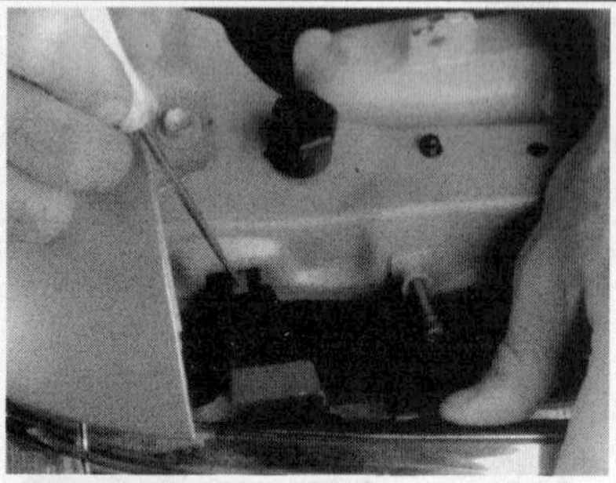

86826P88

Fig. 40 After removing the parking lamp screws, push on the grille clips to release the grille from the truck

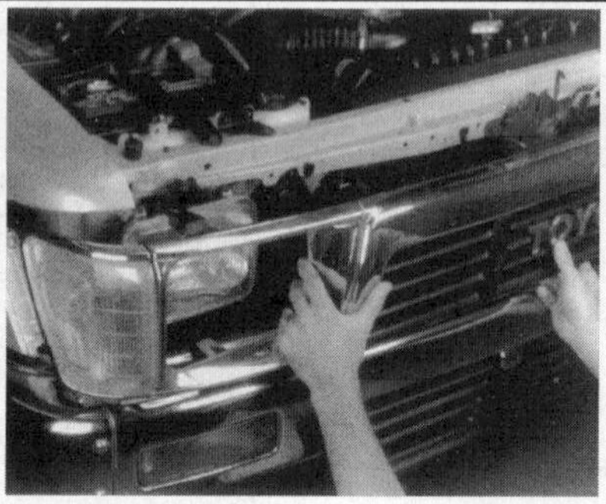

86826P30

Fig. 41 Remove the grille from the truck and set aside

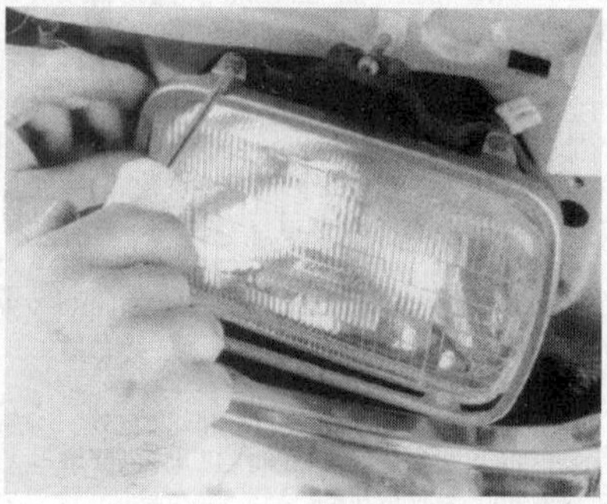

86826P31

Fig. 42 Remove the headlight bezel which is secured by screws

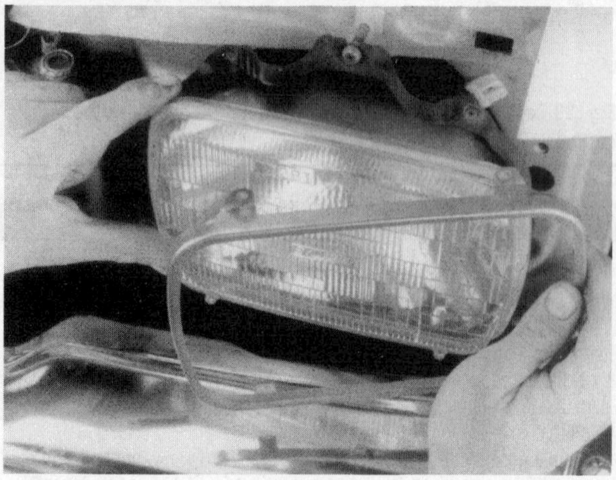

86826P32

Fig. 43 Be careful when removing the headlight bezels, they can be sharp

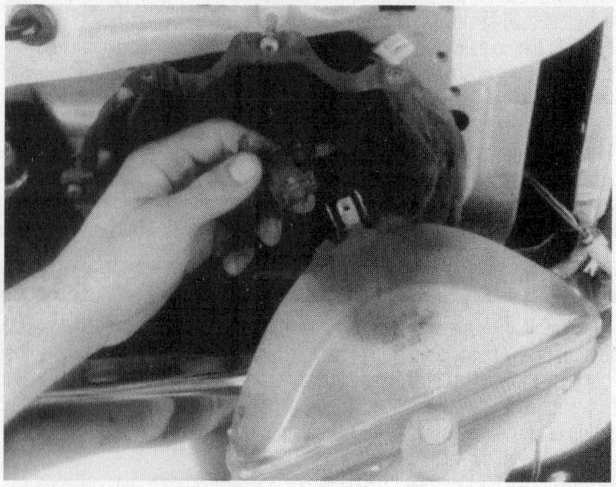

86826P33

Fig. 44 Pull the light out, then unplug the wiring harness from behind it

b. If the left headlamp is to be changed:

c. Remove the air duct (3 bolts).

d. Remove the air suction silencer and the air switching valve. Label the hoses so they go back in the correct places.

e. Remove the underhood cooling fan. Its held by 3 bolts; don't forget to disconnect the electrical harness.

4. Unplug the connector from the back of the lamp. The connector may be very tight and require wiggling to loosen; support the lamp while with your other hand while wiggling the connector loose.

5. Turn the bulb socket about 45° counterclockwise to loosen it. Remove the bulb.

6. Install the new bulb with the socket pointing up. Make certain all three guide tabs fit into their correct slots. Turn the socket clockwise to lock it in place.

✳✳ WARNING

Hold the new bulb by the base; do NOT touch the glass part with fingers or gloves. The grease left on the glass will form hot spots and shorten the life of the bulb. If the glass is accidentally touched, clean the glass with alcohol.

RUBBER COVER TYPE

▶ See Figure 45

1. With the ignition switch **OFF** and the headlamp switch OFF, raise and prop the hood.
2. Unplug the connector for the bulb.
3. Turn the plastic cover counterclockwise, then remove it.
4. Remove the rubber cover.
5. Release the bulb retaining spring, then remove the bulb.
6. When installing, align the tabs of the bulb with the cutout of the mounting hole and install the bulb.
7. Install the retaining spring, then attach the rubber cover. Make sure the rubber cover is snug on the connector and the headlight body.
8. Install the plastic cover with the ON mark facing upwards. Turn it clockwise, then insert the connector.
9. Aiming is not necessary after replacing these types of bulbs.

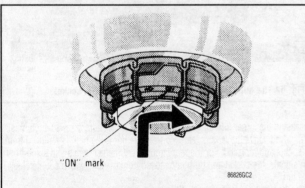

Fig. 45 Install the plastic cover with the ON mark facing upwards. Turn it clockwise, then insert the connector

AIMING

Excluding Land Cruiser

▶ See Figure 46

The head lamps should be aimed using a special alignment tool, however this procedure may be used for temporary adjustment. Local regulations may vary regarding head lamp aiming, consult with your local authorities.

1. Verify the tires are at their proper inflation pressure. Clean the head lamp lenses and make sure there are no heavy loads in the trunk. The gas tank should be filled.

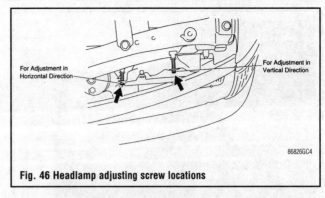

Fig. 46 Headlamp adjusting screw locations

2. Position the vehicle on a level surface facing a flat wall 25 ft. (7.7 m) away.

3. Measure and record the distance from the floor to the center of the head lamp. Place a strip of tape across the wall at this same height.

4. Place strips of tape on the wall, perpendicular to the first measurement, indicating the vehicle centerline and the centerline of both head lamps.

5. Rock the vehicle side-to-side a few times to allow the suspension to stabilize.

6. Turn the lights on, adjust the head lamps to achieve a high intensity pattern.

Land Cruiser

▶ See Figures 46 and 47

1. Before checking the headlamp aim:
   a. Be sure that the body around the headlamp is not deformed.
   b. Park the vehicle on a level spot.
   c. Bounce in the vehicle a few times to get the truck at a level point.
2. Look at the beam angle gauge (vertical movement). The bubble of the gauge should not move out of the center of the gauge by more than two marks on either the upward or downward scale of the gauge.
3. Look at the beam angle gauge (horizontal movement). The red mark should not move by more than one mark on either side of the gauge.
4. If the error is over the specified marks, take your vehicle to a specialized shop to have the headlamps adjusted professionally.

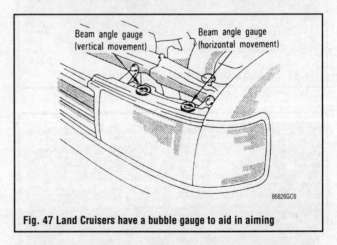

Fig. 47 Land Cruisers have a bubble gauge to aid in aiming

Signal, Marker and Interior Lamps

REMOVAL & INSTALLATION

▶ See Figures 48, 49, 50, 51 and 52

The lens is removed to allow access to the bulb. External lenses usually have a rubber gasket around them to keep dust and water out of the housing; the gasket must be present and in good condition at reinstallation. Exterior

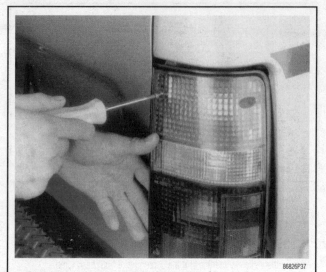

Fig. 48 Remove the screws securing the lens . . .

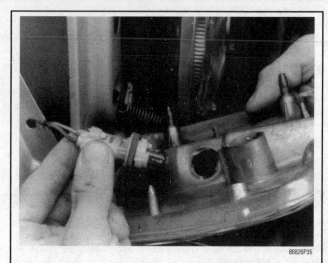

Fig. 51 On these types of lenses, twist then pull the socket and bulb from the housing

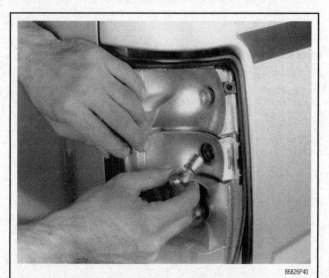

Fig. 49 . . . then twist and pull the bulb from the socket

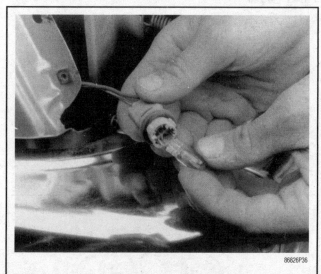

Fig. 52 The bulb can now simply be pulled from the socket

lenses and the larger interior ones are held by one or more screws which must be removed. Once the lens is removed from the body, the bulb is removed from the socket and replaced. For the rear lamps, front marker lamps and some front turn signals, the socket and bulb is removed from the lens with a counterclockwise turn.

Smaller interior lenses usually fit in place with plastic clips and must be pried or popped out of place. A small, flat, plastic tool is ideal for this job; if other tools are used, care must be taken not to break the lens or the clip.

The bulbs used on Toyota trucks are all US standard and may be purchased at any auto store or dealer. Because of the variety of lamps used on any vehicle, take the old one with you when shopping for the replacement.

On some models the lens can be replaced separately. On others, you have to replace the lens with the plastic backing attached.

Fig. 50 The front parking light assemblies are secured by screws

CIRCUIT PROTECTION

Fuses

REPLACEMENT

▶ See Figure 53

The fuse block is located below the left side of the instrument panel on all vehicles. Additional fuses are found on the underhood relay board. The radio or audio unit is protected by an additional fuse in the body of the unit. In the event that anything electrical isn't working, the fuse should be the first item checked.

The underdash or underhood fusebox contains a fuse puller which can be used to grip and remove the fuse. The fuse cannot be checked while in the fuseblock; it must be removed. View the fuse from the side, looking for a broken element in the center. Sometimes the break is hard to see; if you can't check the fuse with an ohmmeter for continuity, replace the fuse.

If a fuse should blow, turn **OFF** the ignition switch and also the circuit involved. Replace the fuse with one of the same amperage rating, and turn on the switches. If the new fuse immediately blows out, the circuit should be tested for shorts, broken insulation, or loose connections.

➡ Do not use fuses of a higher amperage than recommended.

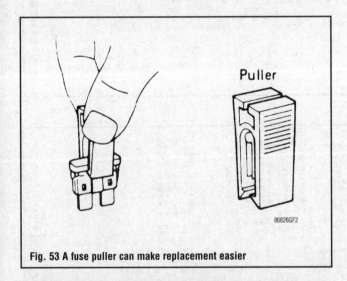

Fig. 53 A fuse puller can make replacement easier

Circuit Breakers

REPLACEMENT

▶ See Figure 54

Many circuits, particularly high amperage ones, are protected by resetable circuit breakers. Much like the circuit breakers in a household system, these units will trip when too much current attempts to pass through. On Toyota vehicles, the breaker does not automatically reset, but may be reset manually.

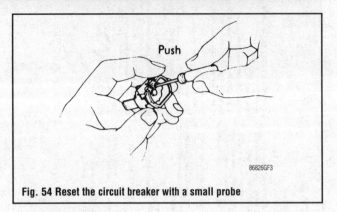

Fig. 54 Reset the circuit breaker with a small probe

To reset the breaker:
1. Disconnect the negative battery cable.
2. Remove the circuit breaker.
3. Use a needle probe or other thin tool to reach into the reset hole and push the reset button.
4. Once reset, use an ohmmeter to check for continuity at the pins of breaker. If no continuity is present, the breaker did not reset.
5. Reinstall the circuit breaker and connect the negative battery cable.

Relays

REPLACEMENT

As vehicles rely more and more on electronic systems and electrically operated options, the number of relays grows steadily. Many relays are located in logical positions on the relay and fuse board under the dash or in the engine compartment. However, many relays are located throughout the vehicle, often near the component they control. During diagnosis of a circuit, always suspect a failed relay until proven otherwise.

Flashers

REPLACEMENT

The turn signal flasher is located in the convenience center, under the dash, on the left side kick panel. In all cases, replacement is made by unplugging the old unit and plugging in a new one.

Fuse and Circuit Breaker Applications

▶ See Figures 55 thru 70

The following diagrams are to be used as a reference for locations. To obtain the exact fuse type, look in your owner's manual or on the label under the lid of each fuse box.

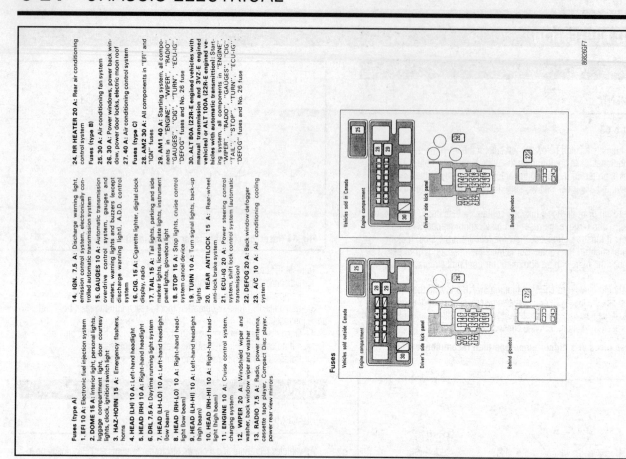

Fuses (type A)

1. **EFI** 10 A: Electronic fuel injection system
2. **DOME** 15 A: Interior light, personal lights, luggage compartment light, door courtesy lights, clock, ignition switch light
3. **HAZ-HORN** 15 A: Emergency flashers, horns
4. **HEAD (LH)** 10 A: Left-hand headlight
5. **HEAD (RH)** 10 A: Right-hand headlight
6. **DRL** 7.5 A: Daytime running light system
7. **HEAD (LH-LO)** 10 A: Left-hand headlight (low beam)
8. **HEAD (RH-LO)** 10 A: Right-hand headlight (low beam)
9. **HEAD (LH-HI)** 10 A: Left-hand headlight (high beam)
10. **HEAD (RH-HI)** 10 A: Right-hand headlight (high beam)
11. **ENGINE** 10 A: Cruise control system, charging system
12. **WIPER** 20 A: Windshield wiper and washer, back window wiper and washer
13. **RADIO** 7.5 A: Radio, power antenna, cassette tape player, Compact Disc player, power rear view mirrors

14. **IGN.** 7.5 A: Discharge warning light, emission control system, electronically controlled automatic transmission system
15. **GAUGES** 10 A: Automatic transmission overdrive control system, gauges and meters, warning lights and buzzers (except discharge warning light), A.D.D. control system
16. **CIG.** 15 A: Cigarette lighter, digital clock display, radio
17. **TAIL** 15 A: Tail lights, parking and side marker lights, license plate lights, instrument panel lights, glovebox light
18. **STOP** 15 A: Stop lights, cruise control system cancel device
19. **TURN** 10 A: Turn signal lights, back-up lights
20. **REAR ANTILOCK** 15 A: Rear-wheel anti-lock brake system
21. **ECU-IG** 20 A: Power steering control system, shift lock control system (automatic transmission)
22. **DEFOG** 20 A: Back window defogger
23. **A/C** 10 A: Air conditioning cooling system

24. **RR HEATER** 20 A: Rear air conditioning control system

Fuses (type B)

25. **30 A**: Air conditioning fan system
26. **30 A**: Power windows, power back window, power door locks, electric moon roof
27. **40 A**: Air conditioning control system

Fuses (type C)

28. **AM2** 30 A: All components in "EFI" and "IGN" fuses
29. **AM1** 40 A: Starting system, all components in "ENGINE", "WIPER", "RADIO", "GAUGES", "CIG", "TURN", "ECU-IG", "DEFOG" fuses and No. 26 fuse
30. **ALT** 80A (22R-E engined vehicles with manual transmission and 3VZ-E engined vehicles) or ALT 100A (22R-E engined vehicles with automatic transmission): Starting system, all components in "ENGINE", "WIPER", "RADIO", "GAUGES", "CIG", "TAIL", "STOP", "TURN", "ECU-IG", "DEFOG" fuses and No. 26 fuse

Fig. 56 Fuse and circuit breaker locations—1992–94 4Runner

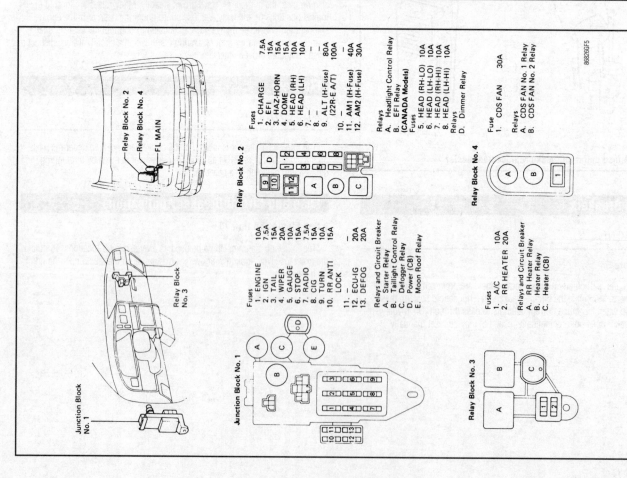

Junction Block No. 1

Relay Block No. 3

Junction Block No. 1

Fuses
1. ENGINE 10A
2. IGN 7.5A
3. TAIL 15A
4. WIPER 20A
5. GAUGE 15A
6. STOP 7.5A
7. RADIO 15A
8. CIG 15A
9. TURN 10A
10. RR ANTILOCK 15A
11. — —
12. ECU-IG 20A
13. DEFOG 20A

Relays and Circuit Breaker
A. Starter Relay
B. Taillight Control Relay
C. Defogger Relay
D. Power (CB)
E. Moon Roof Relay

Relay Block No. 2

Fuses
1. CHARGE 7.5A
2. EFI 15A
3. HAZ-HORN 15A
4. DOME 15A
5. HEAD (RH) 10A
6. HEAD (LH) 10A
7. — —
8. — —
9. ALT (H-Fuse) 80A
 ALT (H-Fuse A/T) 100A
10. — —
11. AM1 (H-Fuse) 40A
12. AM2 (H-Fuse) 30A

Relays
A. Headlight Control Relay
B. EFI Relay
(CANADA Models)
Fuses
5. HEAD (RH-LO) 10A
6. HEAD (LH-LO) 10A
7. HEAD (RH-HI) 10A
8. HEAD (LH-HI) 10A
Relays
D. Dimmer Relay

Relay Block No. 4

Fuse
1. CDS FAN 30A

Relays
A. CDS FAN No. 1 Relay
B. CDS FAN No. 2 Relay

Relay Block No. 3

Fuses
1. A/C 10A
2. RR HEATER 20A

Relays and Circuit Breaker
A. RR Heater Relay
B. Heater Relay
C. Heater (CB)

Fig. 55 Fuse and circuit breaker locations—1989–91 4Runner

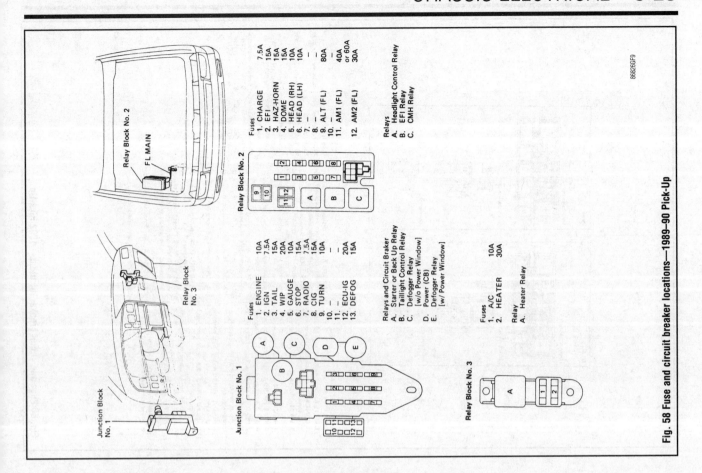

Relay Block No. 2

FL MAIN

Relay Block No. 3

Junction Block No. 1

Fuses
1. CHARGE — 7.5A
2. EFI — 15A
3. HAZ-HORN — 15A
4. DOME — 10A
5. HEAD (RH) — 10A
6. HEAD (LH) — 10A
7. — — —
8. — — —
9. ALT (FL) — 80A
10. ALT (FL) — —
11. AM1 (FL) — 40A or 60A
12. AM2 (FL) — 30A

Relays
A. Headlight Control Relay
B. EFI Relay
C. CMH Relay

Relay Block No. 2

Junction Block No. 1

Fuses
1. ENGINE — 10A
2. IGN — 7.5A
3. TAIL — 15A
4. WIP — 20A
5. GAUGE — 10A
6. STOP — 15A
7. RADIO — 7.5A
8. CIG — 15A
9. TURN — 10A
10. — — —
11. — — —
12. ECU-IG — 20A
13. DEFOG — 15A

Relays and Circuit Braker
A. Starter or Back Up Relay
B. Taillight Control Relay
C. Defogger Relay [w/o Power Window]
D. Power (CB)
E. Defogger Relay [w/ Power Window]

Relay Block No. 3

Fuses
1. A/C — 10A
2. HEATER — 30A

Relay
A. Heater Relay

Fig. 58 Fuse and circuit breaker locations—1989–90 Pick-Up

86826GF9

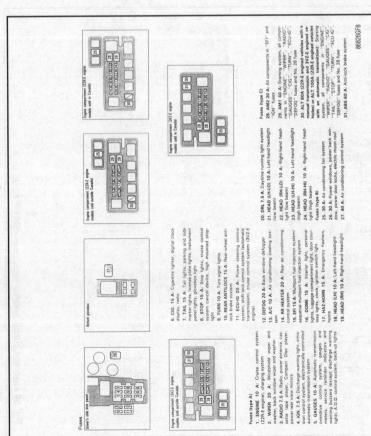

Fuses

Driver's side kick panel

Behind glovebox

Engine compartment (22R-E engine models sold outside Canada)

Engine compartment (22R-E engine models sold in Canada)

Engine compartment (3VZ-E engine models sold outside Canada)

Engine compartment (3VZ-E engine models sold in Canada)

Fuses (type A)
1. ENGINE 10 A: Cruise control system (22R-E engine), charging system
2. WIPER 20 A: Windshield wiper and washer, back window wiper and washer
3. RADIO 7.5 A: Radio, power antenna, cassette tape player, Compact Disc player, power rear view mirrors
4. IGN 7.5 A: Discharge warning light, emission control system, electronically controlled automatic transmission system
5. GAUGES 10 A: Automatic transmission, overdrive control system, gauges and meters, service reminder indicators and warning buzzers (except discharge warning light), back-up control system, back-up lights

6. CIG. 15 A: Cigarette lighter, digital clock display, radio
7. TAIL 15 A: Tail lights, parking and side marker lights, license plate lights, instrument panel lights, glovebox light
8. STOP 15 A: Stop lights, cruise control system cancel device, high mounted stop-light
9. TURN 10 A: Turn signal lights
10. REAR ANTILOCK 15 A: Rear-wheel anti-lock brake system
11. ECU-IG 20 A: Power steering control system, shift lock control system (automatic transmission), cruise control system (3VZ-E engine)
12. DEFOG 20 A: Back window defogger
13. A/C 10 A: Air conditioning cooling system
14. RR HEATER 20 A: Rear air conditioning control system
15. EFI 15 A: Multiport fuel injection system/ sequential multiport fuel injection system
16. DOME 15 A: Interior light, personal lights, luggage compartment light, door courtesy lights, clock, ignition switch light
17. HAZ-HORN 15 A: Emergency flashers, horn
18. HEAD (LH) 10 A: Left-hand headlight
19. HEAD (RH) 10 A: Right-hand headlight

20. DRL 7.5 A: Daytime running light system
21. HEAD (LH-LO) 10 A: Left-hand headlight (low beam)
22. HEAD (RH-LO) 10 A: Right-hand headlight (low beam)
23. HEAD (LH-HI) 10 A: Left-hand headlight (high beam)
24. HEAD (RH-HI) 10 A: Right-hand headlight (high beam)

Fuses (type B)
26. 30 A: Air conditioning fan system
26. 30 A: Power windows, power back window, power door locks, electric sun roof control system
27. 40 A: Air conditioning control system

Fuses (type C)
28. AM2 30 A: All components in "EFI" and "IGN" fuse
29. AM1 40 A: Starting system, all components in "ENGINE", "WIPER", "RADIO", "GAUGES", "CIG", "TURN", "ECU-IG", "DEFOG" fuses and No. 26 fuse
30. ALT 80A (22R-E engine) engined vehicles with manual transmission and 3VZ-E engined vehicles or ALT 100A (22R-E engined vehicles with an automatic transmission): Starting system, all components in "ENGINE", "WIPER", "RADIO", "GAUGES", "CIG", "TURN", "ECU-IG", "DEFOG" fuses and No. 26 fuse
31. ABS 60 A: Anti-lock brake system

86826GF8

Fig. 57 Fuse and circuit breaker locations—1995 4Runner

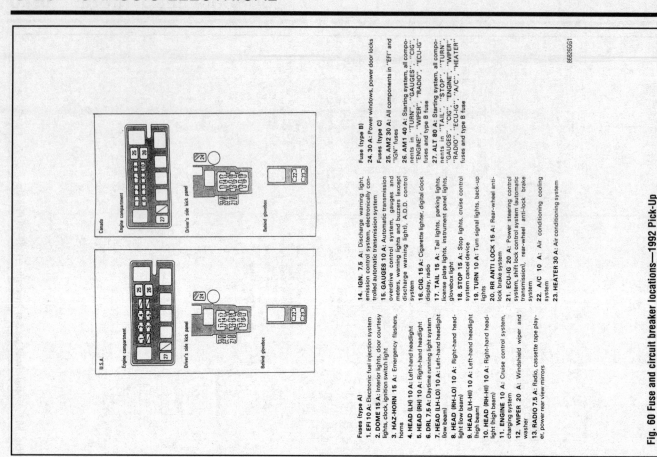

Fuses (type A)

1. **EFI** 10 A: Electronic fuel injection system
2. **DOME** 15 A: Interior lights, door courtesy lights, clock, ignition switch light
3. **HAZ-HORN** 15 A: Emergency flashers, horns
4. **HEAD (LH)** 10 A: Left-hand headlight
5. **HEAD (RH)** 10 A: Right-hand headlight
6. **DRL** 7.5 A: Daytime running light system
7. **HEAD (LH-LO)** 10 A: Left-hand headlight (low beam)
8. **HEAD (RH-LO)** 10 A: Right-hand headlight (low beam)
9. **HEAD (LH-HI)** 10 A: Left-hand headlight (high beam)
10. **HEAD (RH-HI)** 10 A: Right-hand headlight (high beam)
11. **ENGINE** 10 A: Cruise control system, charging system
12. **WIPER** 20 A: Windshield wiper and washer
13. **RADIO** 7.5 A: Radio, cassette tape player, power rear view mirrors

14. **IGN.** 7.5 A: Discharge warning light, emission control system, electronically controlled automatic transmission system
15. **GAUGES** 10 A: Automatic transmission overdrive control system, gauges and meters, warning lights and buzzers (except discharge warning light), A.D.D. control system
16. **CIG.** 15 A: Cigarette lighter, digital clock display, radio
17. **TAIL** 15 A: Tail lights, parking lights, license plate lights, instrument panel lights, glovebox light
18. **STOP** 15 A: Stop lights, cruise control system cancel device
19. **TURN** 10 A: Turn signal lights, back-up lights
20. **RR ANTI LOCK** 15 A: Rear-wheel anti-lock brake system
21. **ECU-IG** 20 A: Power steering control system, shift lock control system (automatic transmission), rear-wheel anti-lock brake system
22. **A/C** 10 A: Air conditioning cooling system
23. **HEATER** 30 A: Air conditioning cooling system

Fuse (type B)

24. **30 A:** Power windows, power door locks

Fuses (type C)

25. **AM2** 30 A: All components in "IGN" and "IGN" fuses
26. **AM1** 40 A: Starting system, all components in "TURN" "GAUGES" "CIG" "ENGINE" "WIPER" "RADIO" "ECU-IG" fuses and type B fuse
27. **ALT** 80 A: Starting system, all components in "TAIL" "STOP" "TURN" "GAUGES" "CIG" "ENGINE" "WIPER" "RADIO" "ECU-IG" "A/C" "HEATER" fuses and type B fuse

Fig. 60 Fuse and circuit breaker locations—1992 Pick-Up

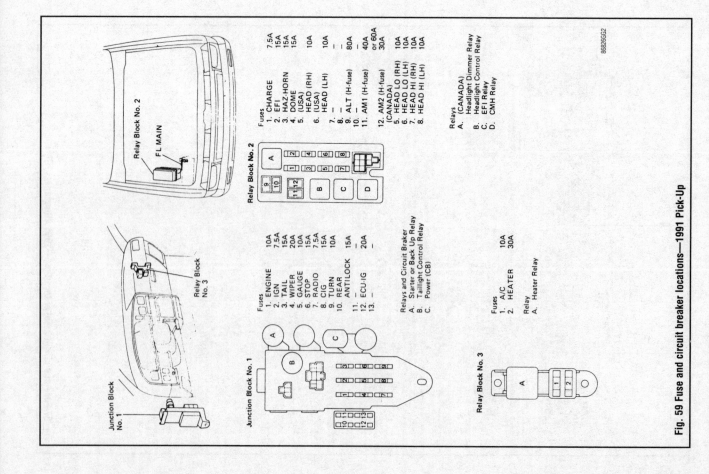

Fig. 59 Fuse and circuit breaker locations—1991 Pick-Up

Fuses

1.	CHARGE	7.5A
2.	EFI	15A
3.	HAZ-HORN	15A
4.	DOME	10A
	(USA)	
5.	HEAD (RH)	10A
	(USA)	
6.	HEAD (LH)	
7.	–	–
8.	–	–
9.	ALT (H-fuse)	80A
10.	–	–
11.	AM1 (H-fuse)	40A or 60A
12.	AM2 (H-fuse)	30A
	(CANADA)	
5.	HEAD LO (RH)	10A
6.	HEAD LO (LH)	10A
7.	HEAD HI (RH)	10A
8.	HEAD HI (LH)	10A

Relays

A. (CANADA)
   Headlight Dimmer Relay
B. Headlight Control Relay
C. EFI Relay
D. CMH Relay

Fuses

1.	ENGINE	10A
2.	IGN	7.5A
3.	TAIL	15A
4.	WIPER	20A
5.	GAUGE	10A
6.	STOP	15A
7.	RADIO	7.5A
8.	CIG	15A
9.	TURN	10A
10.	REAR ANTILOCK	15A
11.	–	–
12.	ECU-IG	20A
13.	–	–

Relays and Circuit Braker

A. Starter or Back Up Relay
B. Taillight Control Relay
C. Power (CB)

Fuses

1.	A/C	10A
2.	HEATER	30A

Relay

A. Heater Relay

Fuses

Engine compartment (U.S.A.)

Engine compartment (Canada)

Driver's side kick panel

86826GG5

Fuses (type A)

1. **HEAD (LH) 10 A:** Left-hand headlight
2. **HEAD (RH) 10 A:** Right-hand headlight
3. **A/C 10 A:** Air conditioning cooling system
4. **EFI 15 A:** Multiport fuel injection system/ sequential multiport fuel injection system, electronically controlled automatic transmission system
5. **HAZ-HORN 15 A:** Emergency flashers, horns
6. **DOME 15 A:** Interior light, personal lights, step lights, ignition switch light, radio, cassette tape player, Compact Disc player, power antenna, clock
7. **HEAD (LH-HI) 10 A:** Left-hand headlight (high beam)
8. **HEAD (RH-HI) 10 A:** Right-hand headlight (high beam)
9. **HEAD (LH-LO) 10 A (3VZ-E engine):** Left-hand headlight (low beam)
10. **HEAD (RH-LO) 10 A (3VZ-E engine):** Right-hand headlight (low beam)
11. **ENG 10 A:** Charging system
12. **IGN 7.5 A:** Charging system, discharge warning light, multiport fuel injection system/sequential multiport fuel injection system
13. **TAIL 15 A:** Tail lights, parking lights, license plate lights, instrument panel lights, glovebox light
14. **WIP 20 A:** Windshield wiper and washer
15. **GAUGE 10 A:** Automatic transmission overdrive control system, gauges and meters, service reminder indicators and warning buzzers (except discharge warning light), A.D.D. control system, cruise control system, back-up lights, power door lock system
16. **STOP 15 A:** Stop lights, High-mounted stoplight, cruise control system, electronically controlled automatic transmission system
17. **RADIO 7.5 A:** Radio, cassette tape player, Compact Disc player, power antenna, power rear view mirrors

Fuse (type C)

28. **ALT 80 A:** All components in "A/C", "TAIL", "STOP", "ECU-B", "AM1" and "HEATER" fuses

18. **CIG 15 A:** Cigarette lighter, digital clock display, shift lock control system (automatic transmission)
19. **TURN 10 A:** Turn signal lights, emergency flashers
20. **ECU-B 15 A:** Rear-wheel anti-lock brake system, SRS airbag system, cruise control system, daytime running light system
21. **DRL 7.5 A (Canada):** Daytime running light system

ABS 15 A (U.S.A): Rear-wheel anti-lock brake system
22. **ECU-IG 20 A:** Rear-wheel anti-lock brake system
23. **OBD II 7.5 A (3RZ-FE engine):** Check connector

Fuses (type B)

24. **AM1 40 A:** Starting system, all components in "ENG", "IGN", "WIP", "GAUGE", "RADIO", "CIG", "TURN" and "PWR" fuses
25. **AM2 30 A:** Starting system, all components in "ENG", "IGN", "WIP", "GAUGE".

Fig. 62 Fuse and circuit breaker locations—T100

Fuses

U.S.A.

Engine compartment

Driver's side kick panel

Behind glovebox

Canada

Engine compartment

Driver's side kick panel

Behind glovebox

86826GG4

Fuses (type A)

1. **EFI 15 A:** Multiport fuel injection system/ sequential multiport fuel injection system
2. **DOME 15 A:** Interior lights, door courtesy lights, clock, ignition switch light
3. **HAZ-HORN 15 A:** Emergency flashers, horns
4. **HEAD (LH) 10 A:** Left-hand headlight
5. **HEAD (RH) 10 A:** Right-hand headlight
6. **DRL 7.5 A:** Daytime running light system
7. **HEAD (LH-LO) 10 A:** Left-hand headlight (low beam)
8. **HEAD (RH-LO) 10 A:** Right-hand headlight (low beam)
9. **HEAD (LH-HI) 10 A:** Left-hand headlight (high beam)
10. **HEAD (RH-HI) 10 A:** Right-hand headlight (high beam)
11. **ENGINE 10 A:** Cruise control system, charging system
12. **WIPER 20 A:** Windshield wiper and washer
13. **RADIO 7.5 A:** Radio, cassette tape player, power rear view mirrors

Fig. 61 Fuse and circuit breaker locations—1993-95 Pick-Up

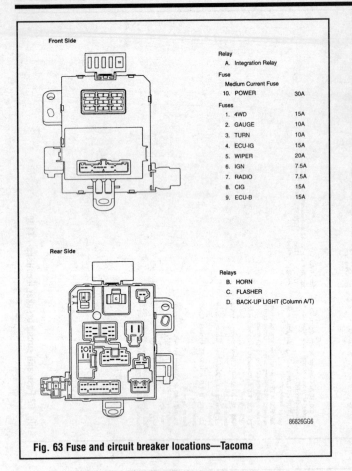

Front Side

Relay
A. Integration Relay

Fuse
Medium Current Fuse
10. POWER   30A

Fuses
1. 4WD   15A
2. GAUGE   10A
3. TURN   10A
4. ECU-IG   15A
5. WIPER   20A
6. IGN   7.5A
7. RADIO   7.5A
8. CIG   15A
9. ECU-B   15A

Rear Side

Relays
B. HORN
C. FLASHER
D. BACK-UP LIGHT (Column A/T)

86826GG6

Fig. 63 Fuse and circuit breaker locations—Tacoma

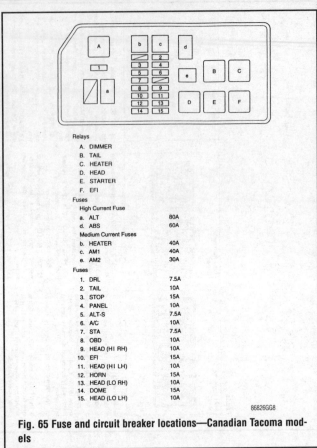

Relays
A. DIMMER
B. TAIL
C. HEATER
D. HEAD
E. STARTER
F. EFI

Fuses
High Current Fuse
a. ALT   80A
d. ABS   60A
Medium Current Fuses
b. HEATER   40A
c. AM1   40A
e. AM2   30A

Fuses
1. DRL   7.5A
2. TAIL   10A
3. STOP   15A
4. PANEL   10A
5. ALT-S   7.5A
6. A/C   10A
7. STA   7.5A
8. OBD   10A
9. HEAD (HI RH)   10A
10. EFI   15A
11. HEAD (HI LH) 10A
12. HORN   15A
13. HEAD (LO RH) 10A
14. DOME   15A
15. HEAD (LO LH) 10A

86826GG8

Fig. 65 Fuse and circuit breaker locations—Canadian Tacoma models

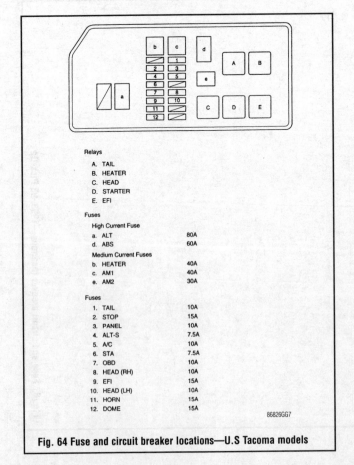

Relays
A. TAIL
B. HEATER
C. HEAD
D. STARTER
E. EFI

Fuses
High Current Fuse
a. ALT   80A
d. ABS   60A
Medium Current Fuses
b. HEATER   40A
c. AM1   40A
e. AM2   30A

Fuses
1. TAIL   10A
2. STOP   15A
3. PANEL   10A
4. ALT-S   7.5A
5. A/C   10A
6. STA   7.5A
7. OBD   10A
8. HEAD (RH)   10A
9. EFI   15A
10. HEAD (LH)   10A
11. HORN   15A
12. DOME   15A

86826GG7

Fig. 64 Fuse and circuit breaker locations—U.S Tacoma models

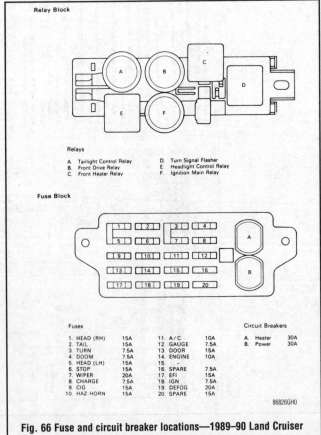

Relay Block

Relays
A. Taillight Control Relay
B. Front Drive Relay
C. Front Heater Relay
D. Turn Signal Flasher
E. Headlight Control Relay
F. Ignition Main Relay

Fuse Block

Fuses			Circuit Breakers	
1. HEAD (RH)	15A	11. A/C	10A	A. Heater	30A
2. TAIL	15A	12. GAUGE	7.5A	B. Power	30A
3. TURN	7.5A	13. DOOR	15A		
4. DOOM	7.5A	14. ENGINE	10A		
5. HEAD (LH)	15A	15. --			
6. STOP	15A	16. SPARE	7.5A		
7. WIPER	20A	17. EFI	15A		
8. CHARGE	7.5A	18. IGN	7.5A		
9. CIG	15A	19. DEFOG	20A		
10. HAZ-HORN	15A	20. SPARE	15A		

86826GH0

Fig. 66 Fuse and circuit breaker locations—1989–90 Land Cruiser

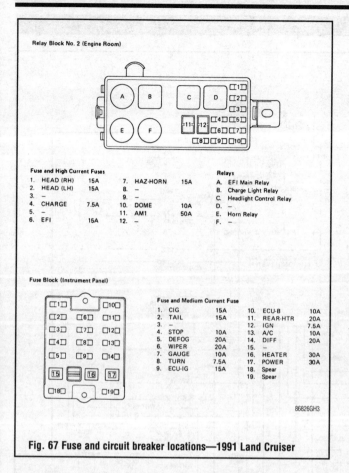

Relay Block No. 2 (Engine Room)

Fuse and High Current Fuses

1.	HEAD (RH)	15A	7.	HAZ-HORN	15A
2.	HEAD (LH)	15A	8.	–	
3.	–		9.	–	
4.	CHARGE	7.5A	10.	DOME	10A
5.	–		11.	AM1	50A
6.	EFI	15A	12.	–	

Relays

A. EFI Main Relay
B. Charge Light Relay
C. Headlight Control Relay
D. –
E. Horn Relay
F. –

Fuse Block (Instrument Panel)

Fuse and Medium Current Fuse

1.	CIG	15A	10.	ECU-B	10A
2.	TAIL	15A	11.	REAR-HTR	20A
3.	–		12.	IGN	7.5A
4.	STOP	10A	13.	A/C	10A
5.	DEFOG	20A	14.	DIFF	20A
6.	WIPER	20A	15.	–	
7.	GAUGE	10A	16.	HEATER	30A
8.	TURN	7.5A	17.	POWER	30A
9.	ECU-IG	15A	18.	Spear	
			19.	Spear	

86826GH3

Fig. 67 Fuse and circuit breaker locations—1991 Land Cruiser

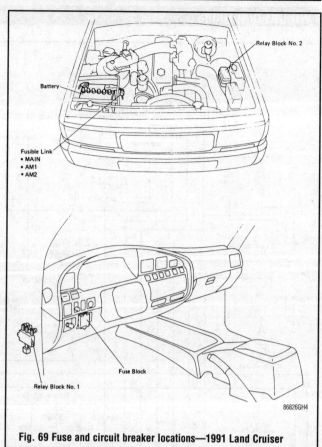

86826GH4

Fig. 69 Fuse and circuit breaker locations—1991 Land Cruiser

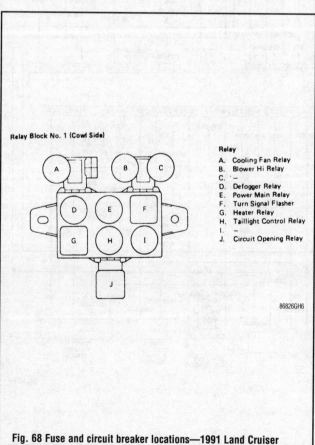

Relay Block No. 1 (Cowl Side)

Relay

A. Cooling Fan Relay
B. Blower Hi Relay
C. –
D. Defogger Relay
E. Power Main Relay
F. Turn Signal Flasher
G. Heater Relay
H. Taillight Control Relay
I. –
J. Circuit Opening Relay

86826GH6

Fig. 68 Fuse and circuit breaker locations—1991 Land Cruiser

Fuses

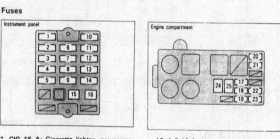

1. **CIG 15 A:** Cigarette lighter, power rear view mirrors, digital clock display, radio, cassette tape player, power antenna, automatic transmission shift lock system

2. **TAIL 15 A:** Tail lights, license plate lights, parking and front side marker lights, instrument panel lights, clock, glovebox light

3. **FOG 15 A:** No circuit

4. **STOP 10 A:** Stop lights, multiport fuel injection system/sequential multiport fuel injection system, cruise control cancel device, automatic transmission shift lock system

5. **DEFOG 20 A:** Rear window deffoger

6. **WIPER 20 A:** Windshield wipers and washer, rear window wiper and washer

7. **GAUGE 10 A:** Gauges and meters, service reminder indicators and warning buzzers (except discharge and open door warning lights), back-up lights

8. **TURN 7.5 A:** Turn signal lights

9. **ECU-IG 15 A:** Cruise control system

10. **ECU-B 10 A:** No circuit

11. **REAR-HTR 20 A:** Air conditioning system

12. **IGN 7.5 A:** Multiport fuel injection system/sequential multiport fuel injection system, emission control system

13. **A.C 10 A:** Air conditioning system

14. **DIFF 30 A:** Differential lock system

15. **FL HEATER 30 A:** Air conditioning system

16. **FL POWER 30 A:** Power windows, power door lock system, electric moon roof

17. **CHARGE 7.5 A:** Charging system, discharge warning light

18. **EFI 15 A:** Multiport fuel injection system/sequential multiport fuel injection system

19. **CDS-FAN 20 A:** Electric cooling fan

20. **HEAD (RH) 15 A:** Right-hand headlights

21. **HEAD (LH) 15 A:** Left-hand headlights

22. **HAZ-HORN 15 A:** Emergency flashers, horns

23. **DOME 10 A:** Interior lights, personal light, luggage compartment light, ignition switch light, open door warning light, clock, radio, cassette tape player, power antenna, vanity lights

24. **AM 1 50 A:** All components in "CIG", "WIPER", "GAUGE", "TURN", "ECU-IG", "REAR-HTR", "IGN", "DIFF", and "FL POWER" circuits

25. **ABS 60 A:** Anti-lock brake system

86826GH7

Fig. 70 Fuse and circuit breaker locations—1992–95 Land Cruiser

WIRING DIAGRAMS

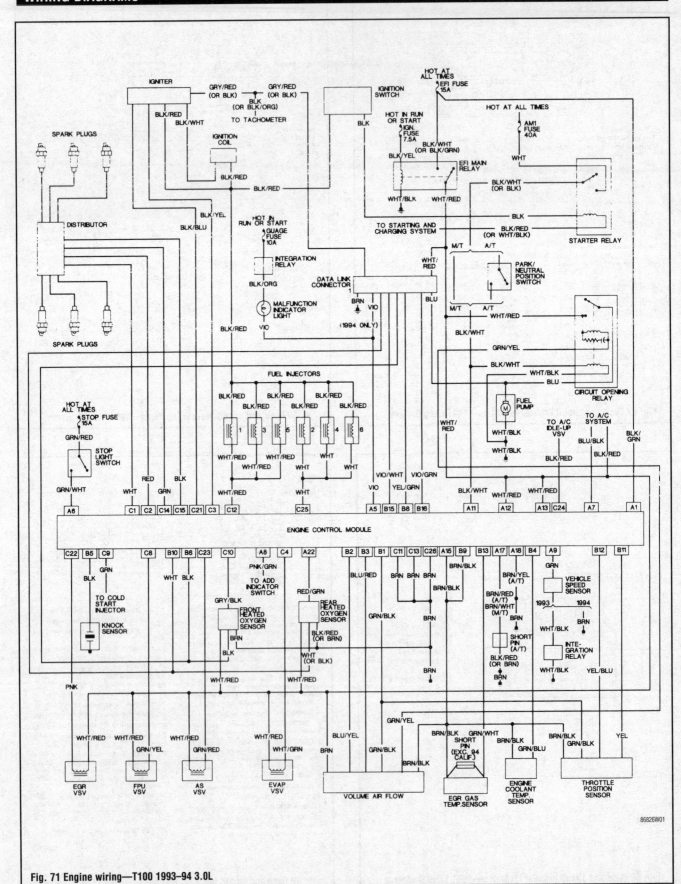

Fig. 71 Engine wiring—T100 1993–94 3.0L

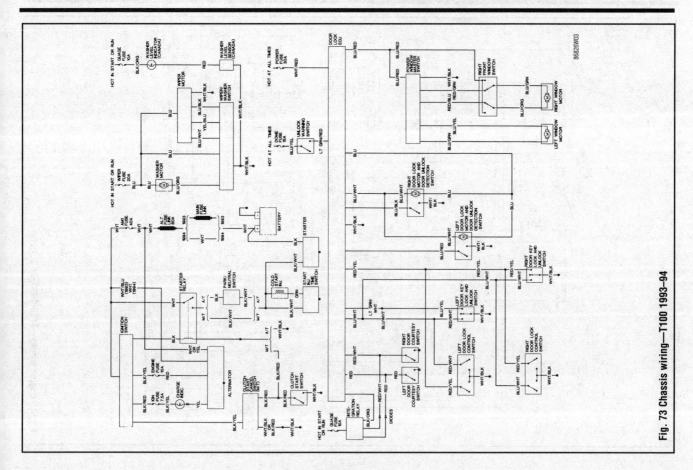

Fig. 73 Chassis wiring—T100 1993-94

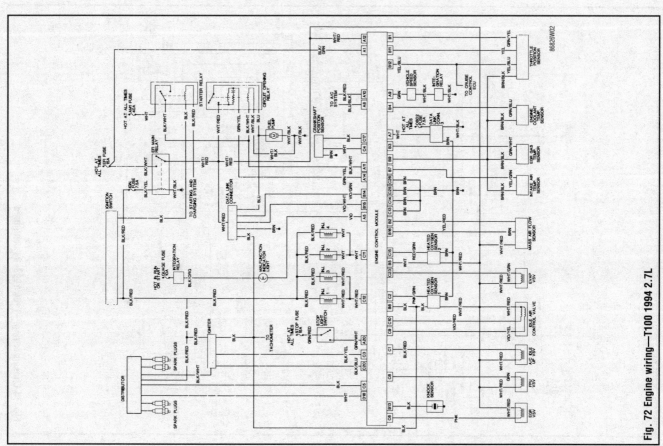

Fig. 72 Engine wiring—T100 1994 2.7L

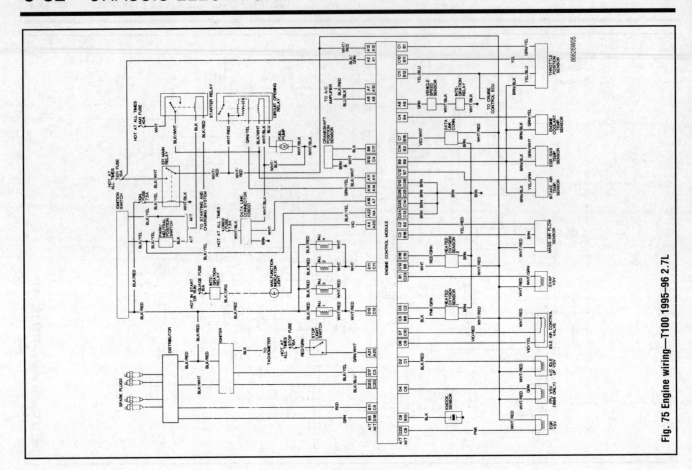

Fig. 75 Engine wiring—T100 1995–96 2.7L

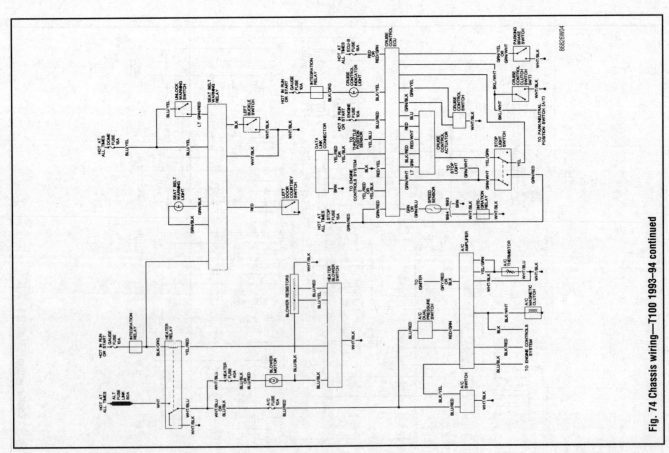

Fig. 74 Chassis wiring—T100 1993–94 continued

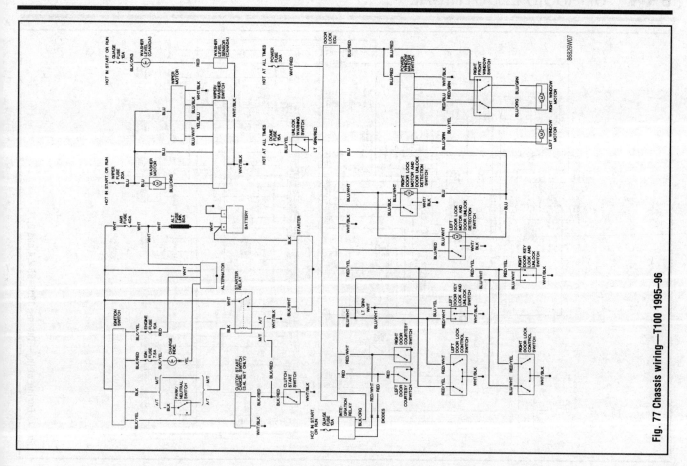

Fig. 77 Chassis wiring—T100 1995-96

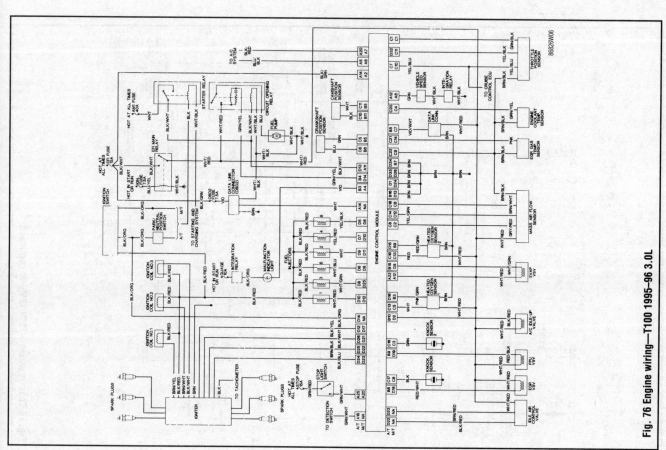

Fig. 76 Engine wiring—T100 1995-96 3.0L

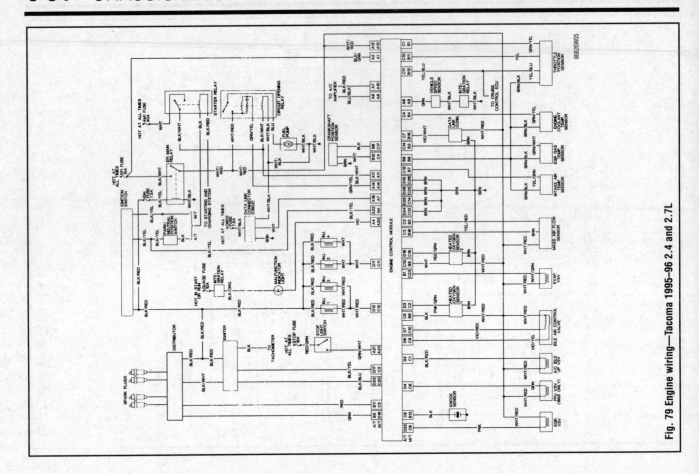

Fig. 79 Engine wiring—Tacoma 1995–96 2.4 and 2.7L

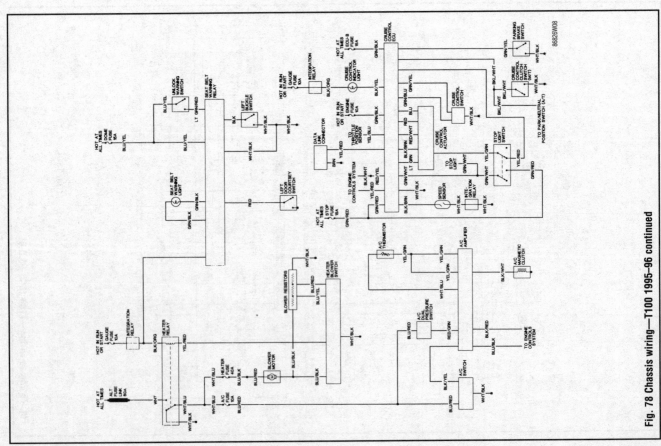

Fig. 78 Chassis wiring—T100 1995–96 continued

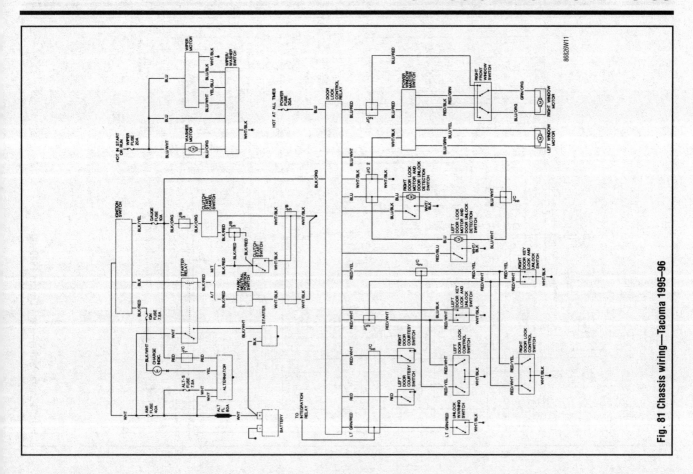

Fig. 81 Chassis wiring—Tacoma 1995–96

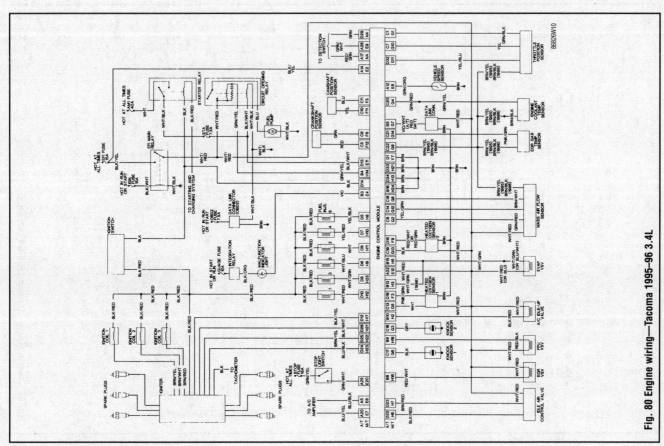

Fig. 80 Engine wiring—Tacoma 1995–96 3.4L

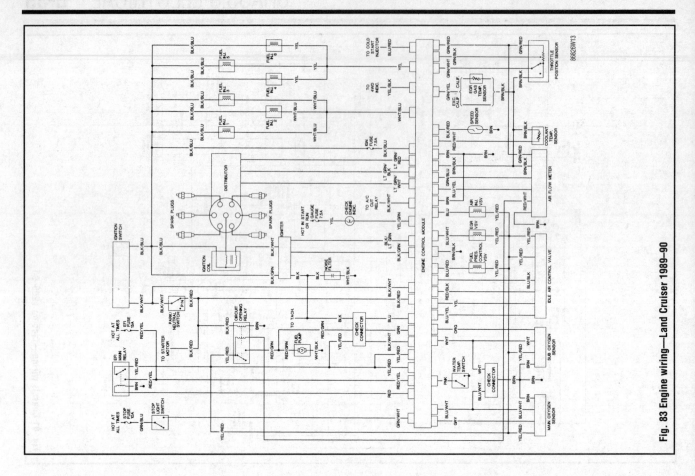

Fig. 83 Engine wiring—Land Cruiser 1989-90

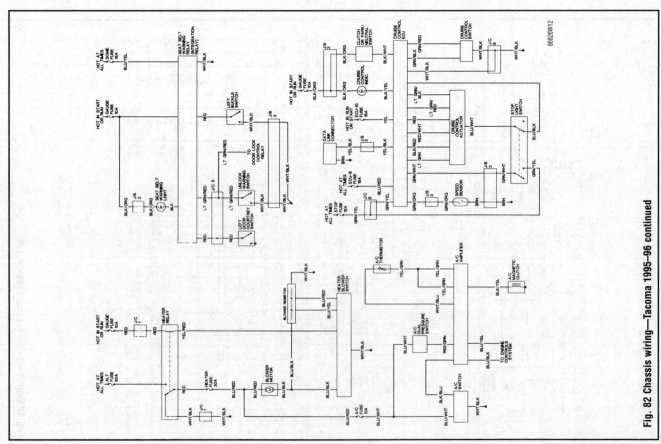

Fig. 82 Chassis wiring—Tacoma 1995-96 continued

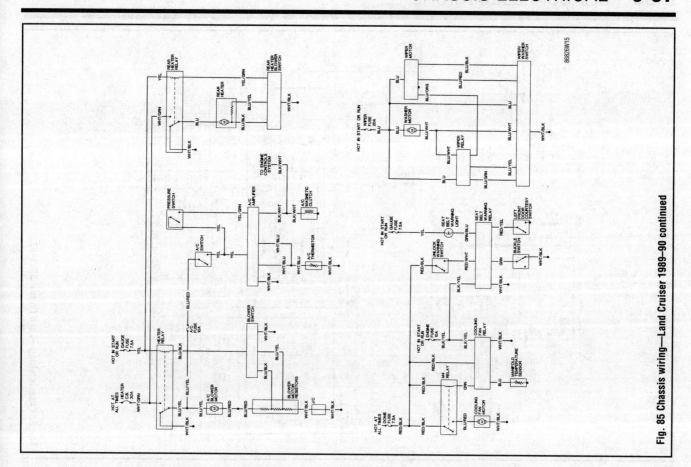

Fig. 85 Chassis wiring—Land Cruiser 1989–90 continued

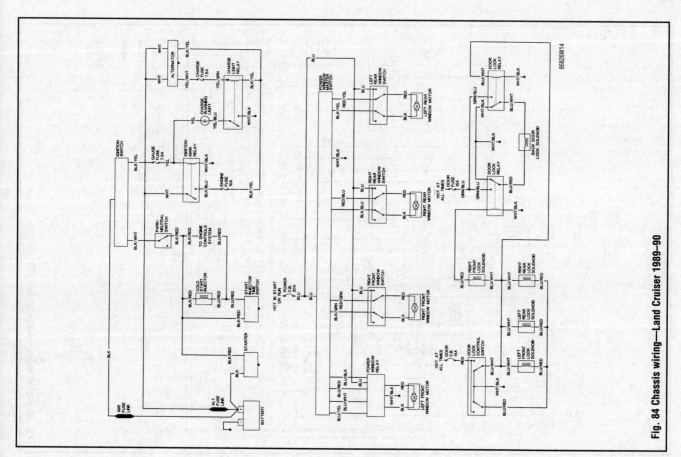

Fig. 84 Chassis wiring—Land Cruiser 1989–90

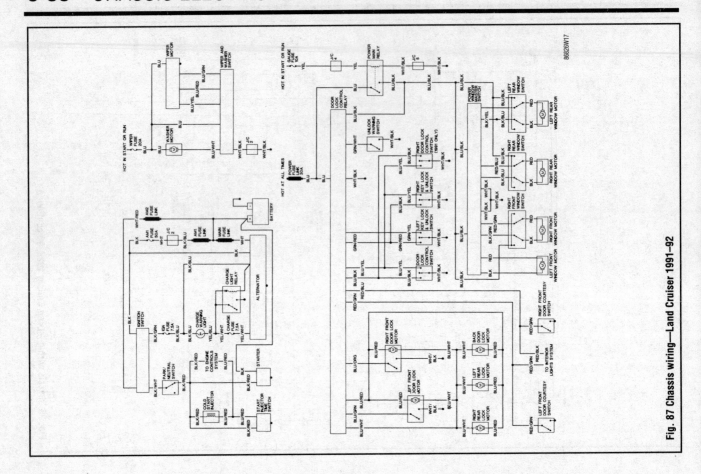

Fig. 87 Chassis wiring—Land Cruiser 1991–92

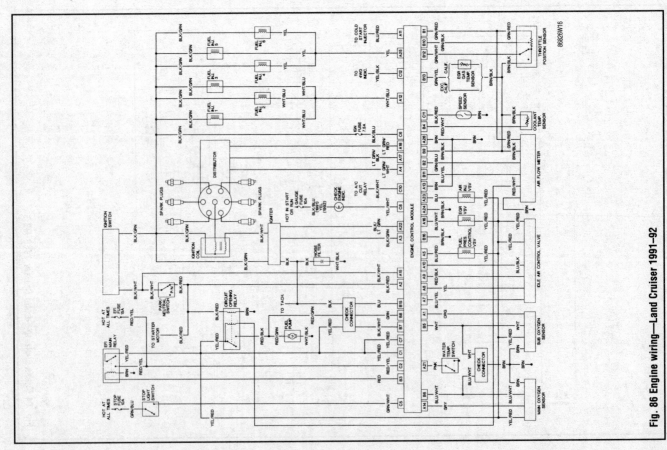

Fig. 86 Engine wiring—Land Cruiser 1991–92

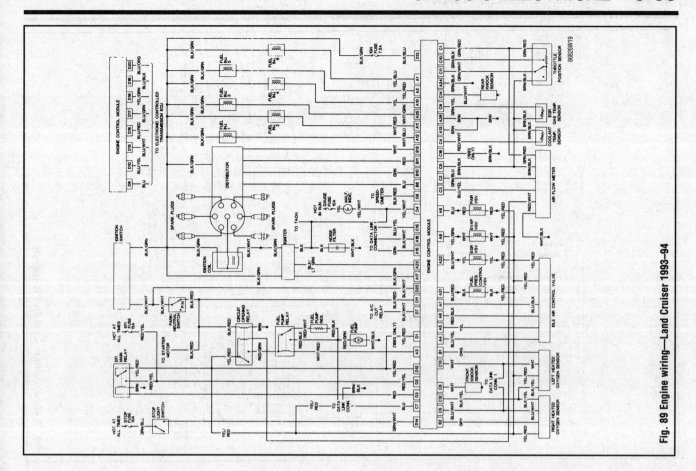

Fig. 89 Engine wiring—Land Cruiser 1993–94

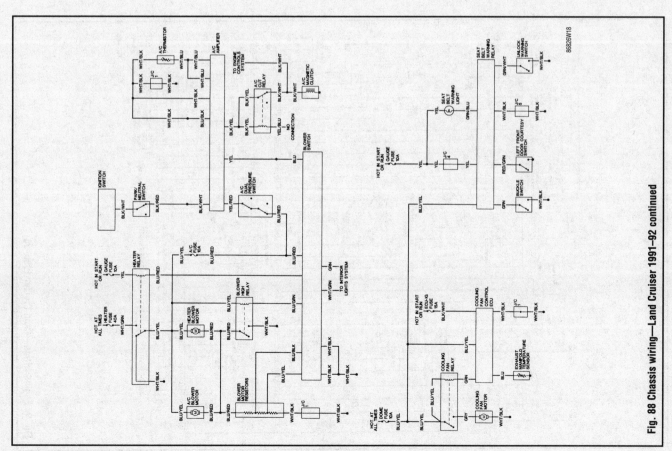

Fig. 88 Chassis wiring—Land Cruiser 1991–92 continued

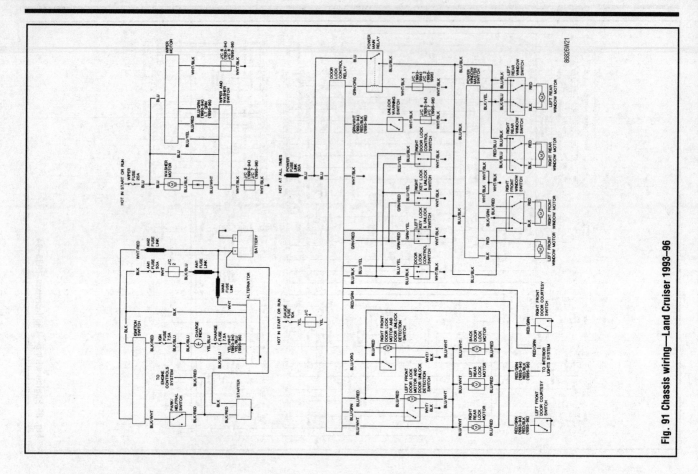

Fig. 91 Chassis wiring—Land Cruiser 1993–96

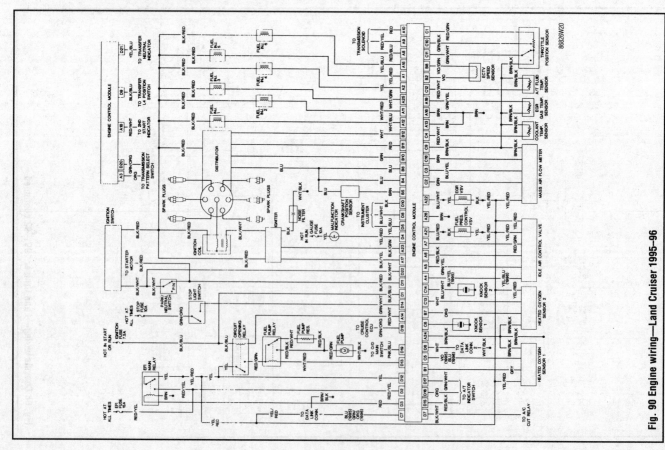

Fig. 90 Engine wiring—Land Cruiser 1995–96

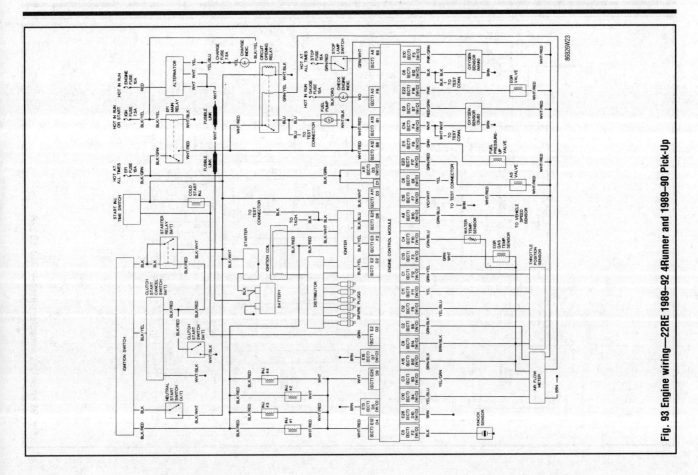

Fig. 93 Engine wiring—22RE 1989–92 4Runner and 1989–90 Pick-Up

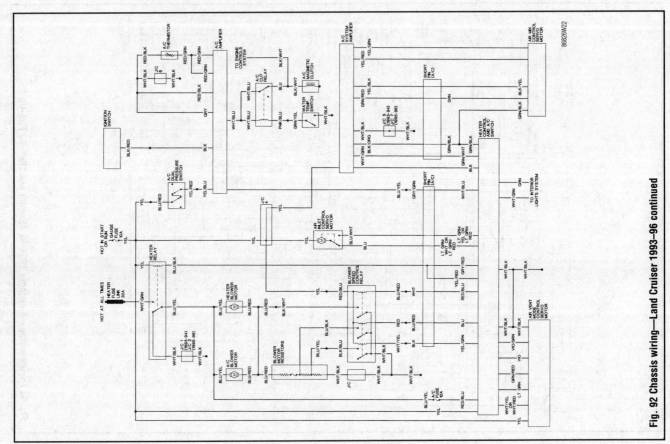

Fig. 92 Chassis wiring—Land Cruiser 1993–96 continued

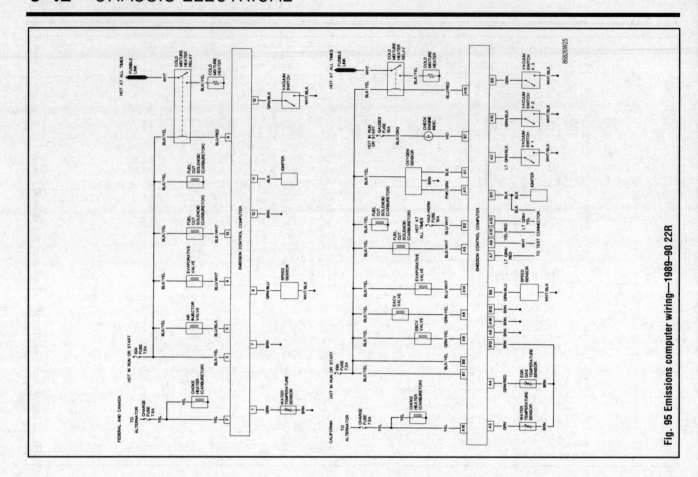

Fig. 95 Emissions computer wiring—1989-90 22R

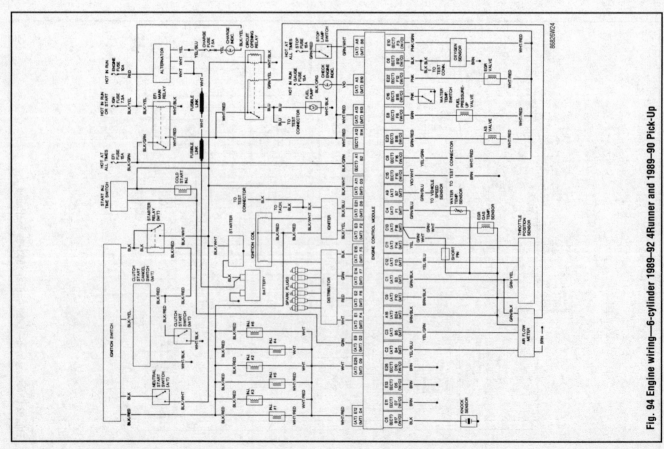

Fig. 94 Engine wiring—6-cylinder 1989-92 4Runner and 1989-90 Pick-Up

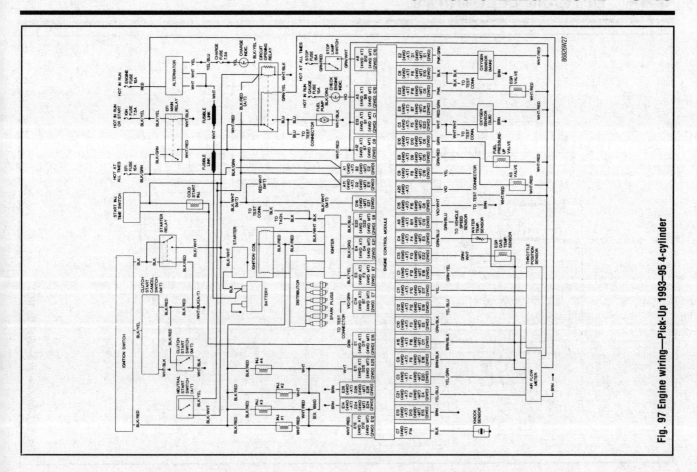

Fig. 97 Engine wiring—Pick-Up 1993-95 4-cylinder

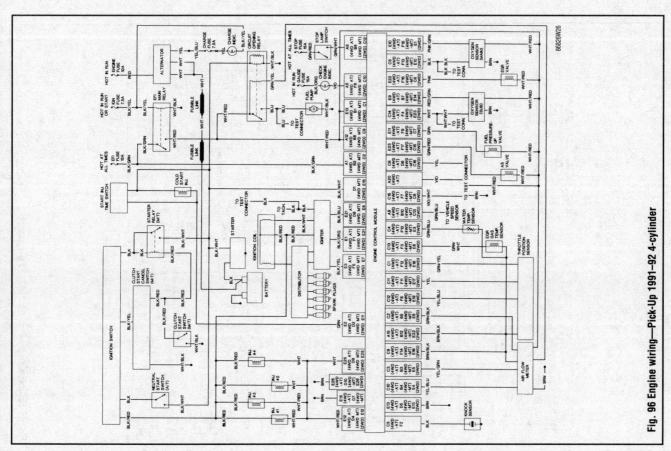

Fig. 96 Engine wiring—Pick-Up 1991-92 4-cylinder

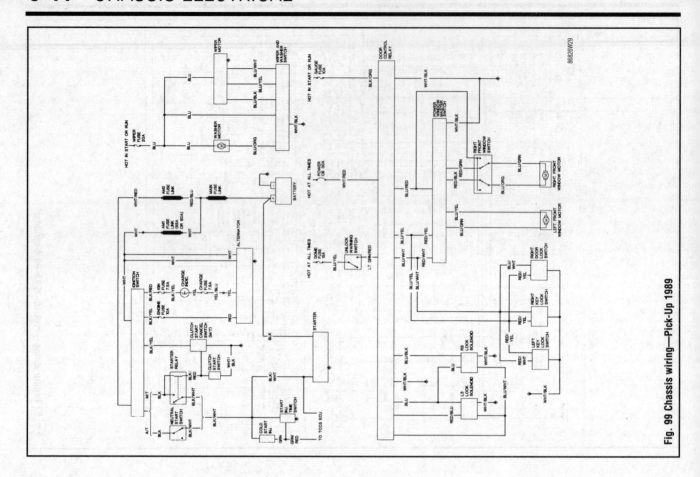

Fig. 99 Chassis wiring—Pick-Up 1989

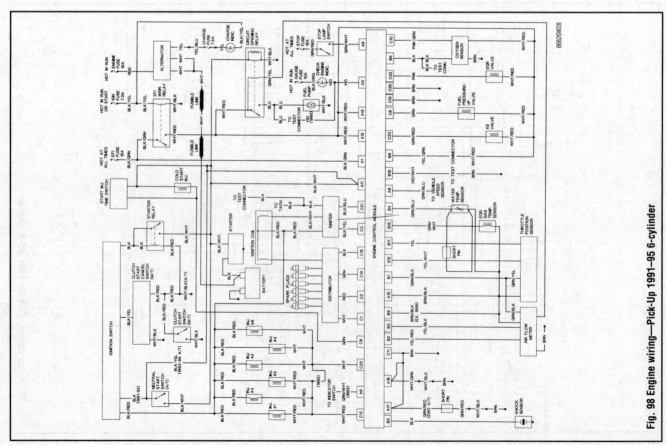

Fig. 98 Engine wiring—Pick-Up 1991–95 6-cylinder

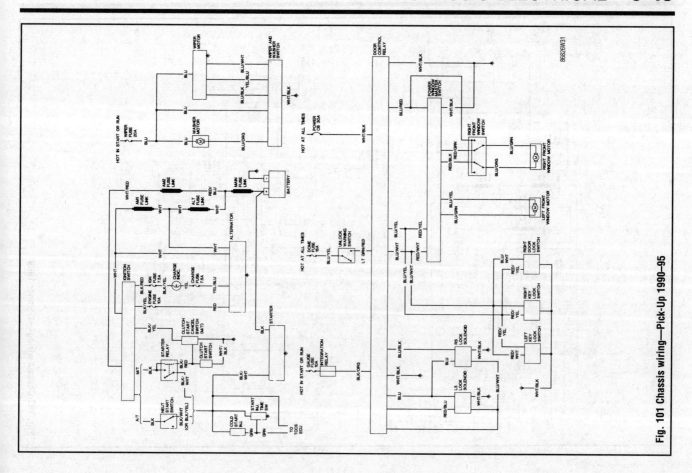

Fig. 101 Chassis wiring—Pick-Up 1990-95

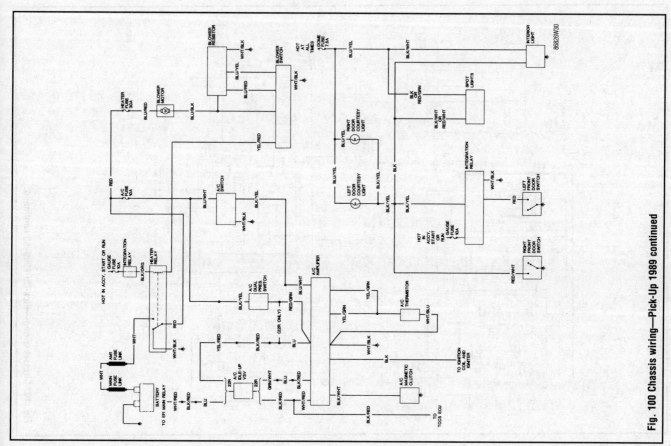

Fig. 100 Chassis wiring—Pick-Up 1989 continued

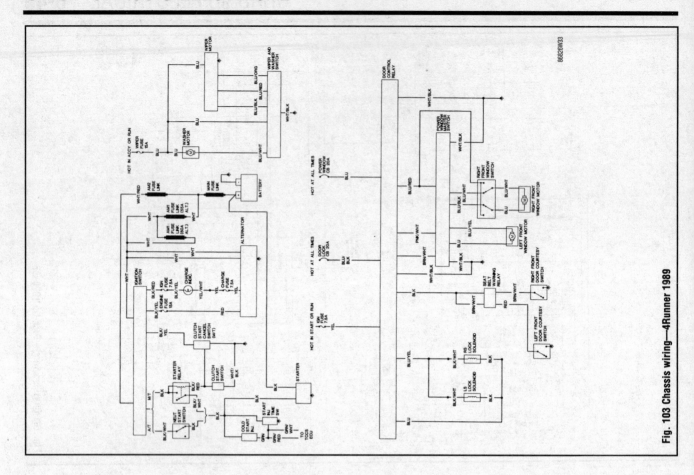

Fig. 103 Chassis wiring—4Runner 1989

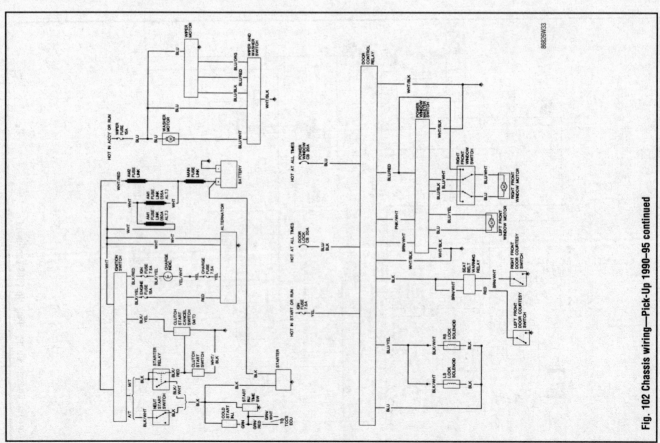

Fig. 102 Chassis wiring—Pick-Up 1990–95 continued

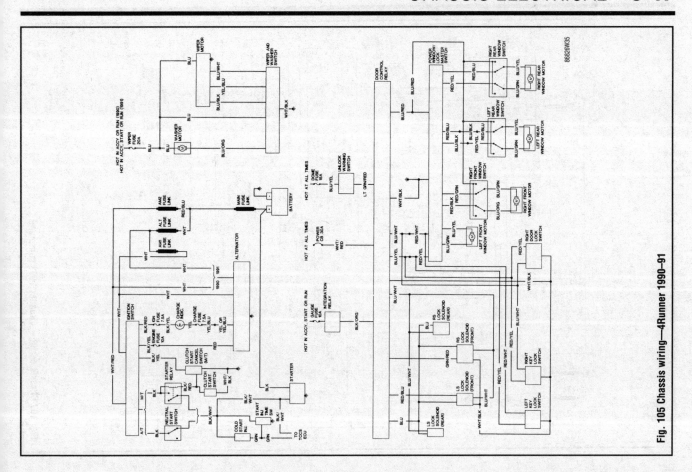

Fig. 105 Chassis wiring—4Runner 1990-91

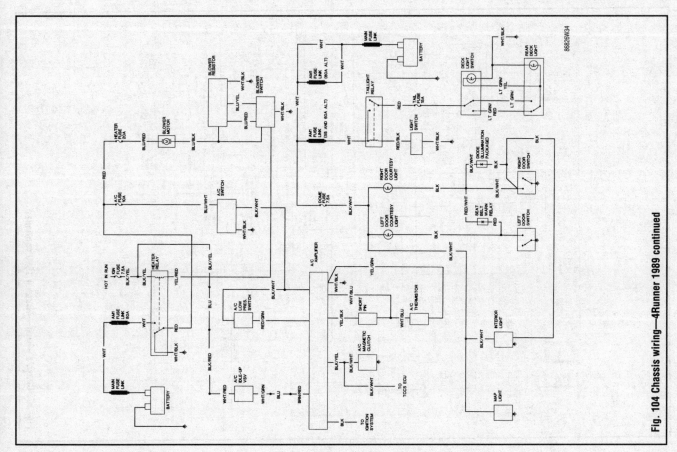

Fig. 104 Chassis wiring—4Runner 1989 continued

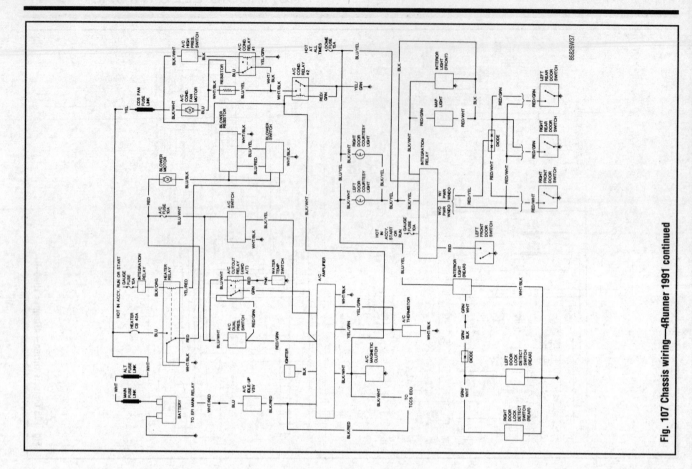

Fig. 107 Chassis wiring—4Runner 1991 continued

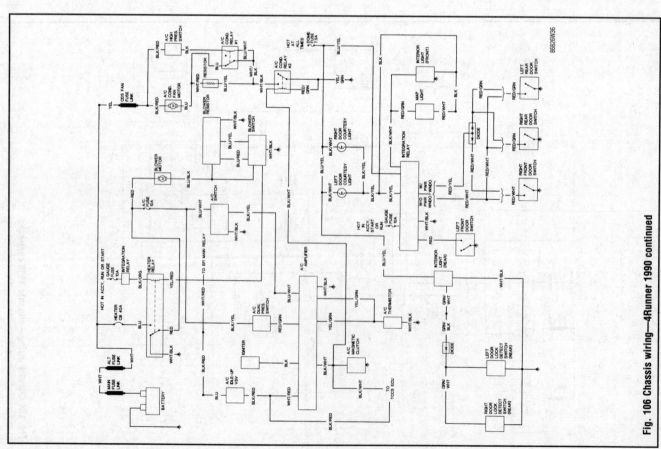

Fig. 106 Chassis wiring—4Runner 1990 continued

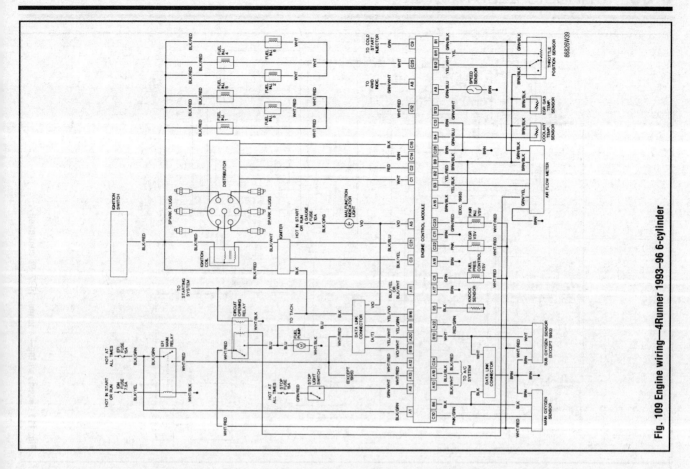

Fig. 109 Engine wiring—4Runner 1993-96 6-cylinder

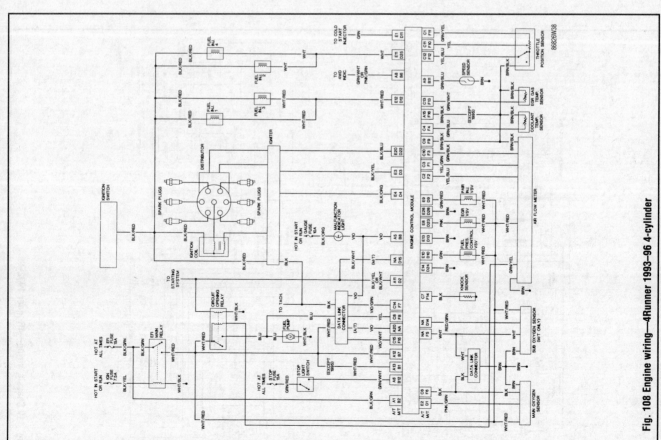

Fig. 108 Engine wiring—4Runner 1993-96 4-cylinder

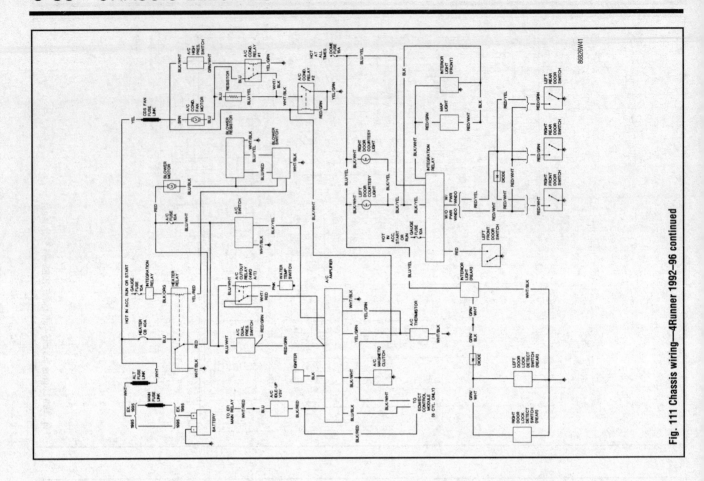

Fig. 111 Chassis wiring—4Runner 1992–96 continued

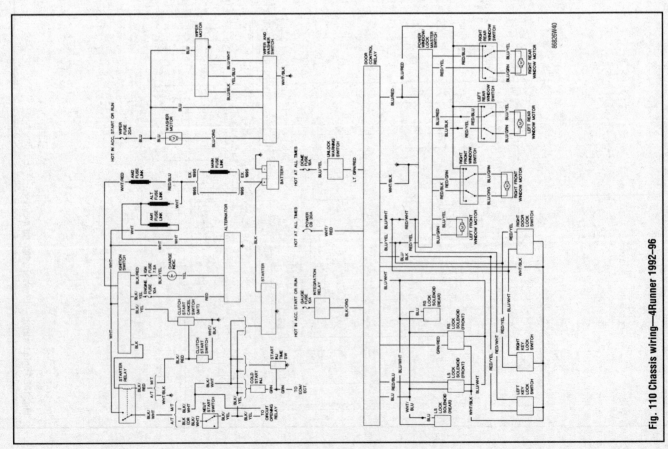

Fig. 110 Chassis wiring—4Runner 1992–96

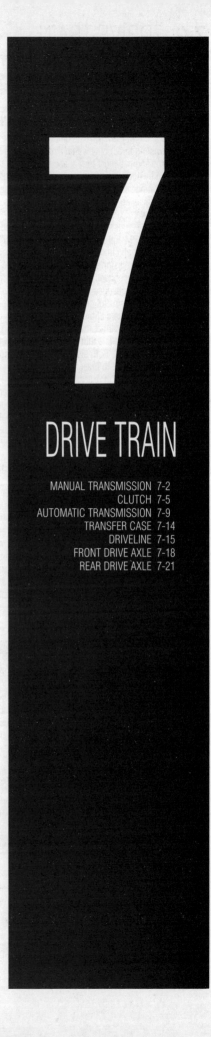

7

DRIVE TRAIN

MANUAL TRANSMISSION

Understanding the Manual Transmission

Because of the way an internal combustion engine breathes, it can produce torque (or twisting force) only within a narrow speed range. Most overhead valve pushrod engines must turn at about 2500 rpm to produce their peak torque. Often by 4500 rpm, they are producing so little torque that continued increases in engine speed produce no power increases.

The torque peak on overhead camshaft engines is, generally, much higher, but much narrower.

The manual transmission and clutch are employed to vary the relationship between engine RPM and the speed of the wheels so that adequate power can be produced under all circumstances. The clutch allows engine torque to be applied to the transmission input shaft gradually, due to mechanical slippage. The vehicle can, consequently, be started smoothly from a full stop.

The transmission changes the ratio between the rotating speeds of the engine and the wheels by the use of gears. 4-speed or 5-speed transmissions are most common. The lower gears allow full engine power to be applied to the rear wheels during acceleration at low speeds.

The clutch driveplate is a thin disc, the center of which is splined to the transmission input shaft. Both sides of the disc are covered with a layer of material which is similar to brake lining and which is capable of allowing slippage without roughness or excessive noise.

The clutch cover is bolted to the engine flywheel and incorporates a diaphragm spring which provides the pressure to engage the clutch. The cover also houses the pressure plate. When the clutch pedal is released, the driven disc is sandwiched between the pressure plate and the smooth surface of the flywheel, thus forcing the disc to turn at the same speed as the engine crankshaft.

The transmission contains a mainshaft which passes all the way through the transmission, from the clutch to the driveshaft. This shaft is separated at one point, so that front and rear portions can turn at different speeds.

Power is transmitted by a countershaft in the lower gears and reverse. The gears of the countershaft mesh with gears on the mainshaft, allowing power to be carried from one to the other. Countershaft gears are often integral with that shaft, while several of the mainshaft gears can either rotate independently of the shaft or be locked to it. Shifting from one gear to the next causes one of the gears to be freed from rotating with the shaft and locks another to it. Gears are locked and unlocked by internal dog clutches which slide between the center of the gear and the shaft. The forward gears usually employ synchronizers; friction members which smoothly bring gear and shaft to the same speed before the toothed dog clutches are engaged.

Identification

The manual transmissions found in Pick-Ups, Tacomas, T100s and 4Runners are denoted by their Toyota model designations: G40, G57, G58, W46, W55, W56, W59, R150 and R150F. The G40 and W46 units are 4-speed transmissions; the others are 5-speeds. The G58, W56, W59 and R150F are used in 4WD applications and are connected to the transfer case. The R150F is coupled to the 3VZ-E engine in 4WD applications. See Section 1 for more information on models.

Adjustments

All models utilize a floor-mounted shifter and an internally-mounted shift linkage. No external adjustments are either necessary or possible.

Back-Up Light Switch

REMOVAL & INSTALLATION

1. Position the shift lever in neutral.
2. Raise the truck and support it on safety stands.
3. Unplug the electrical connector at the switch.
4. Unscrew the back-up light switch from the extension housing.
5. Installation is the reverse of removal. Tighten the switch to 60 inch lbs. (7 Nm).

Shift Lever

REMOVAL & INSTALLATION

▶ See Figures 1 thru 7

1. Unscrew the shifter knob.
2. Remove the console box if necessary.
3. Remove the boot retainer mounting screws.
4. Lift the bezel up and over the shifter shaft.
5. Pull the boot up the shaft.
6. Press down on the shift lever cap and rotate it counterclockwise to remove.
7. Put out the shift lever.
8. Installation is in the reverse order of removal. When installing the lever, push down onto the retainer and with a clockwise motion, turn the shaft to lock into position.

Transmission Assembly

REMOVAL & INSTALLATION

Pick-Up and 4Runner

2WD MODELS

▶ See Figure 8

1. Disconnect the negative battery cable.
2. Remove the four bolts holding the fan shroud.
3. Remove the shift lever boot retainer. Pull up the boot and cover the shift lever cap with a cloth. Press down on the shift lever cap and turn it counterclockwise to remove it. Remove the shift lever.
4. Raise the truck and support it on safety stands. Double check the supports for secure placement.
5. Drain the fluid from the transmission.
6. Remove the driveshaft and plug the extension housing to prevent oil seepage.
7. Disconnect the speedometer cable.
8. Disconnect the back-up light switch wire.
9. Remove the exhaust pipe clamp from the bracket; disconnect the exhaust pipe at the manifold and remove the clamp bracket from the clutch housing. For vehicles with 3VZ-E engines, remove the pipe bracket from the clutch housing first, then disconnect the pipe at the manifold and from the front of the catalytic converter.
10. Remove the clutch slave cylinder mounting bolts and pipe bracket; position the cylinder out of the way. Do not disconnect the fluid line.
11. Remove the four bolts holding the stabilizer (sway) bar.
12. Remove the four bolts holding the frame auxiliary crossmember.
13. Remove the 4 rear engine mount bolts at the extension housing. Position a block of wood on a floor jack and raise the engine slightly. Remove the 4 rear mount-to-support member bolts and remove the rear engine mount. Remove the rear engine mount from the transmission.
14. Tape a piece of wood about 8/10 in. (20mm) thick onto the front crossmember and then lower the transmission.
15. Remove the starter. Remove the stiffener plate bolts.
16. Remove the remaining transmission housing bolts.
17. With a floor jack under the transmission case, pull it toward the rear and slowly lower the front end until it can be removed from the truck. On models with the 3VZ-E engine and R150 transmission, turn the transmission about 45° clockwise. Slide it toward the rear, lower the front and then remove it from the truck.
18. Installation is in the reverse order of removal. Please note the following steps:
19. Tighten the transmission bolts to 53 ft. lbs. (72 Nm) and the stiffener plate bolts to 27 ft. lbs. (37 Nm). Tighten the starter bolt to 29 ft. lbs. (39 Nm).
20. Install the rear engine mount and bracket. Tighten the bolts to 19 ft. lbs. (25 Nm).

Fig. 1 Unscrew the shifter knob

Fig. 2 While pulling back the carpeting, loosen and remove the bezel mounting screws

Fig. 3 Lift the bezel while pulling the carpet aside

Fig. 4 Lift the bezel up and over the shifter shafts

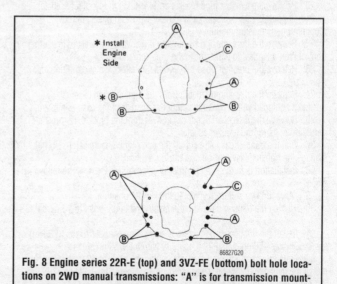

Fig. 5 Lift the boot up the shifter shaft to gain access to the lever cap

Fig. 6 Press down on the shift lever cap and rotate it counterclockwise

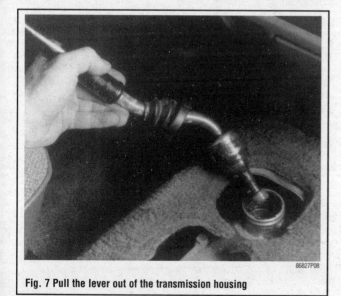

Fig. 7 Pull the lever out of the transmission housing

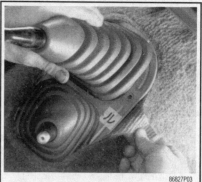

Fig. 8 Engine series 22R-E (top) and 3VZ-FE (bottom) bolt hole locations on 2WD manual transmissions: "A" is for transmission mounting, "B" is for the stiffener, and "C" is for the starter

21. Raise the engine slightly and position the rear engine mount in the support member and tighten the bolts to 43 ft. lbs. (59 Nm).

22. Lower the transmission until it rests on the rear mount. Install the 4 bolts and tighten them to 22 ft. lbs. (29 Nm).

23. Install the frame auxiliary crossmember, tightening the bolts to 70 ft. lbs. (95 Nm).

4WD MODELS

➡The manual transmission is removed with the transfer case attached.

1. Disconnect the battery cables at the battery, negative cable first.

2. Remove the retaining bolts holding the fan shroud.

3. On 3VZ-E engines, loosen the clamp on the heater hose at the firewall and move the clamp up the heater hose out of the way.

4. Remove the four screws holding the shift lever boot retainer. Pull up the boot and cover the shift lever cap with a cloth. Press down on the shift lever cap, rotate it counterclockwise and remove the shift lever.

5. Using needle nose pliers, remove the transfer case shift lever retainer snapring, then remove the shift lever.

6. Raise the vehicle and support is securely with jackstands. Double check the placement and security of the stand.

7. Drain the lubricant from both the transmission and the transfer case.

8. Matchmark the driveshaft flanges and the differential pinion flanges to indicate their relationships. These marks must be aligned during installation. Remove the four bolts from each end of the front driveshaft and remove the driveshaft assembly. On vehicles with G58 and R150F transmissions, it is necessary to remove the propeller shaft dust cover before the shaft is removed.

9. Matchmark the rear driveshaft and the slip yoke to indicate their relationships. These marks must be aligned during installation.

10. Remove the four bolts from the rearward flange of the rear driveshaft. Lower the driveshaft out of the vehicle. Remove the four bolts from the slip yoke flange then remove the flange and yoke assembly.

11. Disconnect the speedometer cable, reverse lamp switch connector and transfer switch connector.

12. On 3VZ-E engines, remove the exhaust pipe clamp, the exhaust bracket from the clutch housing and disconnect the exhaust pipe at the exhaust manifold. Disconnect the exhaust pipe from the front of the catalyst and remove the pipe. On 22R-E engines, remove the bolts holding the exhaust pipe clamp to the transmission and disconnect the exhaust pipe at the manifold. Support the pipe out of the way with wire or rope.

13. Disconnect the clutch slave cylinder retaining bolts and move the cylinder out of the way with the line attached. On 22R-E engines, the piping bracket must also be removed.

14. On 3VZ-E engines, support the front differential with a floor jack; remove the 3 bolts holding the front differential.

15. Remove the 4 bolts holding the stabilizer (sway) bar brackets.

16. Remove the 4 bolts from the rear engine mount. Raise the transmission slightly with a jack. Remove the eight bolts from the side frame rails and remove the No. 2 crossmember.

17. For 22R-E engines, place a piece of wood between the engine oil pan and the front axle.

18. Lower the transmission and transfer case.

19. Remove the two bolts holding the starter; place or hang the starter alongside the motor.

20. Remove the exhaust pipe bracket and the bolts in the stiffener plate.

21. Support the transmission and transfer case with a jack. Remove the remaining transmission mounting bolts.

22. Remove the transmission by moving it rearward; once clear, lower the front of the transmission and remove the unit from the vehicle.

23. Remove the breather hose between the transfer case and the transmission case.

24. Remove the rear engine mount from the transmission case.

25. If equipped with a dynamic damper at the rear of the case, remove it.

26. Remove the driveshaft upper dust cover from the bracket. Remove the mounting bolts for the transfer adapter.

27. Pull the transfer case straight off the end of the transmission. Do not damage the transfer case rear oil seal during removal.

28. Installation is in the reverse order of removal. Please note the following steps:

29. Move the two shift fork shafts to the High-4 position.

30. Tighten the transfer case to transmission retaining bolts to 28 ft. lbs. (38 Nm).

31. Install the dust cover bolt to the bracket. Tighten to 17 ft. lbs. (23 Nm) for R150F and G58 units and 29 ft. lbs. (39 Nm) for W56 units.

32. Install the rear engine mount, tightening the bolts to 19 ft. lbs. (25 Nm). Install the dynamic damper if it was removed and tighten the bolts to 27 ft. lbs. (37 Nm). Install the breather hose.

33. Install the transmission and stiffener bolts. On 3VZ-E engines, install the starter. Tighten the transmission bolts to 53 ft. lbs. (72 Nm), the stiffener bolts to 27 ft. lbs. (37 Nm). Tighten the starter bolts to 29 ft. lbs. (39 Nm).

34. Raise the transmission slightly with a jack. Install the No. 2 frame crossmember and tighten the bolts to 70 ft. lbs. (95 Nm).

35. Lower the transmission assembly. Install the 4 bolts to the engine rear mount; tighten them to 9 ft. lbs. (13 Nm).

36. For 3VZ-E engines, install the front differential bolts. Tighten the bolt holding the carrier cover to the frame to 108 ft. lbs. (147 Nm); tighten the others to 123 ft. lbs. (167 Nm).

T100 and Tacoma

RF150

On vehicles with the RF150 and RF150F transmissions, the engine and transmission are removed as an assembly. Refer to Section 3 for removal procedures. Separate the transmission from the engine as follows:

1. Unbolt the rear end plate.

2. Remove the starter.

3. Loosen and remove the transmission mounting bolts from the engine. Pull out the transmission toward the rear.

4. Unbolt the rear engine mounting.

5. Installation is in the reverse order of removal. Tighten the rear engine mounting bolts to 18 ft. lbs. (25 Nm). Tighten the engine to transmission mounting bolts to 53 ft. lbs. (73 Nm).

6. Install the rear end and stiffener plates with the mounting bolts, tighten to 27 ft. lbs. (37 Nm).

RF150F

1. Remove the transmission shift lever.

2. Raise the vehicle, then drain the transmission fluid.

3. Remove the front propeller shaft.

4. Remove the rear propeller shaft.

5. Disconnect the speedometer cable.

6. Unplug the back-up light switch and the 4WD position switch connector.

7. Remove the 2 bolts, then slave cylinder.

8. Remove the connection and wire from the starter. Remove the bolt for the starter lower side with the clutch line bracket.

9. Remove the front exhaust pipe, separate the oxygen sensor from the pipe.

10. Remove the starter.

11. Remove the bolts and nuts retaining the rear end and right and left stiffener plates.

12. Unbolt the stabilizer bar.

13. Jack up the transmission slightly.

14. Remove the bolts from the engine rear mounting, then remove the bolts and nuts for the crossmember.

15. Remove the 6 transmission mounting bolts from the engine. Disconnect the wire clamps from the transmission. Pull the transmission down and towards the rear.

16. Remove the 4 bolts and the engine rear mounting.

17. Remove the transfer adapter rear mounting bolts. Pull the transfer case straight up, then remove it from the transmission.

➡**Take care not to damage the adapter rear oil seal with the transfer input gear slide.**

18. Installation is in the reverse order of removal. Please note the following steps.

19. Attach the transfer case to the transmission. Tighten the adapter rear mounting bolts 27 ft. lbs. (37 Nm).

20. Tighten the rear engine mounting bolts to 18 ft. lbs. (25 Nm).

21. Place the transmission/transfer case into position. Attach the wire clamps to the transmission, then tighten the 6 rear engine mounting bolts to 53 ft. lbs. (72 Nm).

22. Install the nuts and bolts for the crossmember, tighten to 70 ft. lbs. (95 Nm). Install the 4 bolts to the rear engine mounting, tighten to 9 ft. lbs. (13 Nm).

23. Attach the rear end plate and stiffener plates. Tighten to 27 ft. lbs. (37 Nm).

W59—2-WHEEL DRIVE

On vehicles with the W59 transmission, the engine and transmission are removed as an assembly. Refer to Section 3 for removal procedures. Separate the transmission from the engine as follows:

1. Remove the transmission with the engine.

2. Unbolt the rear end and stiffener plates.

3. Remove the starter.

4. Remove the transmission mounting bolts from the engine. Pull out the transmission toward the rear.

5. Remove the rear engine mount.

6. Installation is in the reverse order of removal. Tighten the rear engine mount bolts to 48 ft. lbs. (65 Nm). Tighten the transmission mounting bolts to 53 ft. lbs. (73 Nm).

7. Install the rear end and stiffener plates with the mounting bolts, tighten to 27 ft. lbs. (37 Nm).

W59—4-WHEEL DRIVE

1. Remove the transmission shift lever.
2. Raise the vehicle, then drain the transmission fluid.
3. Remove the front propeller shaft.
4. Remove the rear propeller shaft.
5. Disconnect the speedometer cable.
6. Disconnect the back-up light switch and the 4WD position switch connector.
7. Remove the slave cylinder.
8. Remove the connection and wire from the starter. Remove the bolt for the starter lower side with the clutch line bracket.
9. Remove the front exhaust pipe.
10. Remove the bolts and nuts retaining the rear end and stiffener plates.
11. If applicable, remove the stabilizer bar.

12. Support the transmission rear side. Remove the bolts from the engine rear mounting. Disconnect the O-ring, then remove the bolts and nuts for the crossmember.

13. Remove the rear mounting.

14. Jack up the transmission slightly, then remove the starter.

15. Remove the 4 transmission mounting bolts from the engine. Pull the transmission down and towards the rear.

16. Remove the transfer adapter rear mounting bolts. Pull the transfer case straight up, then remove it from the transmission.

➡**Take care not to damage the adapter rear oil seal with the transfer case input gear slide.**

17. Installation is in the reverse order of removal. Please note the following steps.

18. Tighten the transfer case adapter rear mounting bolts 17 ft. lbs. (24 Nm). Tighten the 4 rear engine mounting bolts to 53 ft. lbs. (72 Nm). Tighten the rear mounting boltsto 48 ft. lbs. (65 Nm).

19. Install a new O-ring, then install the nuts and bolts for the crossmember. Tighten to 48 ft. lbs. (65 Nm). Install the rear engine mounting bolts, tighten to 14 ft. lbs. (19 Nm).

20. Attach the rear end plate, tighten to 27 ft. lbs. (37 Nm).

CLUTCH

✳✳ CAUTION

The clutch driven disc may contain asbestos, which has been determined to be a cancer causing agent. Never clean clutch surfaces with compressed air. Avoid inhaling any dust from any clutch surface! When cleaning clutch surfaces, use a commercially available brake cleaning fluid.

Adjustments

PEDAL HEIGHT

◆ **See Figure 9**

The pedal height measurement is gauged from the angled section of the floorboard to the center of the clutch pedal pad. If necessary, adjust the pedal height by loosening the locknut and turning the pedal stop bolt which is located above the pedal toward the driver's seat. Tighten the locknut after the adjustment.

Correct pedal height from the floor pan sheet (not the carpet) is:
- 1989–95 2WD Pick-Up—6.0827 in. (154.5mm)

- 1989–95 4WD Pick-Up—5.9646 in. (151.5mm)
- 1989–95 4Runner—6.201 in. (157.5mm)
- 1995–96 Tacoma—6.89–7.28 in. (175.0–185.0mm)
- 1995–96 T100—6.087–6.480 in. (154.6–164.6mm)

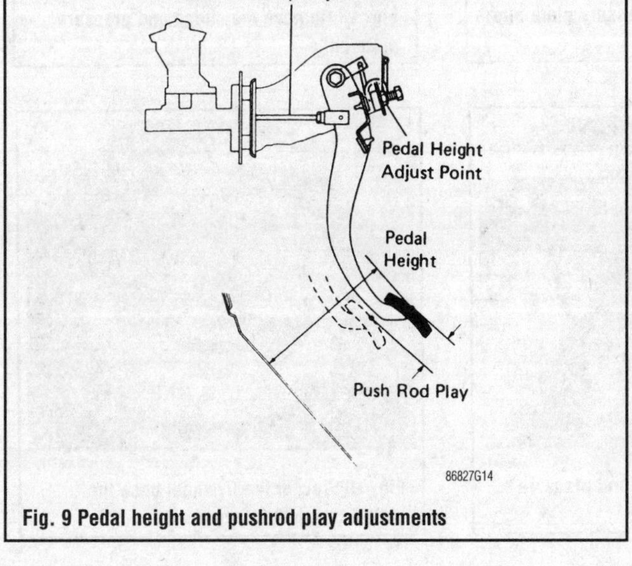

FREE-PLAY

◆ **See Figure 10**

Check the pedal free-play to see if it is correct, push in on until the beginning of the clutch resistance is felt. To adjust, loosen the locknut and turn the pushrod until the free-play is correct. Tighten the locknut. After adjusting the free-play, check the pedal height.
- 1989–95 Pick-Up—0.20–0.59 in. (5.0–15.0mm)
- 1989–95 4Runner—0.20–0.59 in. (5.0–15.0mm)
- 1995–96 Tacoma—0.197–0.591 in. (5.0–15.0mm)
- 1993–96 T100—0.197–0.591 in. (5.0–15.0mm)

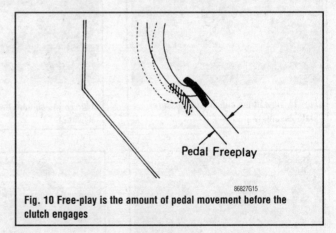

Pedal Freeplay

86827G15

Fig. 10 Free-play is the amount of pedal movement before the clutch engages

PEDAL PUSHROD PLAY

The pedal pushrod play is the distance between the clutch master cylinder piston and the pedal pushrod located above the pedal towards the firewall. Since it is nearly impossible to measure this distance at the source, it must be measured at the pedal pad.

If necessary, adjust the pedal play by loosening the pedal pushrod locknut and turning the pushrod. Tighten the locknut after the adjustment to 0.039–0.197 in. (1.0–5.0mm).

Push Rod Play Adjust Point

Pedal Height
Adjust Point

Pedal
Height

Push Rod Play

86827G14

Fig. 9 Pedal height and pushrod play adjustments

Driven Disc and Pressure Plate

REMOVAL & INSTALLATION

▶ See Figures 11 thru 36

❋❋ CAUTION

The clutch driven disc may contain asbestos, which has been determined to be a cancer causing agent. Avoid inhaling any dust from any clutch surface! When cleaning clutch surfaces, use a commercially available brake cleaning fluid. Never clean clutch surfaces with compressed air.

1. Remove the transmission.
2. Matchmark the clutch cover (pressure plate) and flywheel, indicating their relationship.
3. Loosen the clutch cover-to-flywheel retaining bolts one turn at a time in a crisscross pattern. The pressure on the clutch disc must be released GRADUALLY.
4. Remove the clutch cover-to-flywheel bolts. Remove the clutch cover and the clutch disc.
5. If the clutch throwout bearing is to be replaced, do so at this time as follows:

    a. Remove the bearing retaining clip(s) and remove the bearing and hub.

    b. Remove the release fork and the boot.

    c. The bearing is press fit to the hub. Turn the bearing by hand while placing it under some pressure; check for freedom of motion and lack of grinding or resistance. The bearing is permanently lubricated and cannot be disassembled or greased.

    d. Clean all parts; lightly grease the input shaft splines and all of the contact points.

    e. Install the bearing/hub assembly, fork, boot, and retaining clip(s) in their original locations.

6. Inspect the flywheel surface for cracks, heat scoring (blue marks), and warpage. If oil is present on the flywheel surface, this indicates that either the engine rear oil seal or the transmission front oil seal is leaking. If necessary, replace the seal(s). If in doubt concerning the condition of the flywheel, consult an automotive machine shop.
7. Before installing any new parts, make sure that they are clean. During installation, do not get grease or oil on any of the components, as this will shorten clutch life considerably. Grease or fingerprints may be cleaned with an evaporative cleaner such as the type used on brake linings.

To install:

8. Position the clutch disc against the flywheel. The long side of the splined section faces the flywheel.
9. Install the clutch cover over the disc and install the bolts loosely. Align the matchmarks made. If a new or rebuilt clutch cover assembly is installed, use the matchmark on the old cover assembly as a reference.

➡ Whenever the clutch disc is replaced, replacement of the pressure plate (clutch cover) and release bearing is highly recommended.

10. Align the clutch disc with the flywheel using a clutch aligning tool. These handy tools are available in many auto parts stores at a reasonable price. Do NOT attempt to align the clutch disc by eye; use an alignment tool.
11. With the clutch aligning tool installed, tighten the clutch cover bolts gradually in a star pattern, as is done with lug nuts. Final tighten the bolts to 14 ft. lbs. (19 Nm).
12. Apply molybdenum disulfide grease or multi-purpose grease to the release fork contact points, the pivot and the clutch disc splines. Install the boot, fork, and bearing on the transmission input shaft.
13. Install the transmission.

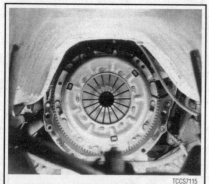

Fig. 11 View of the clutch and pressure plate assembly

Fig. 12 Matchmark the pressure plate and flywheel

Fig. 13 Remove the clutch and pressure plate bolts

Fig. 14 Carefully pry the clutch and pressure plate assembly away from the flywheel

Fig. 15 Remove the clutch and pressure plate

Fig. 16 View of the flywheel once the clutch assembly is removed

Fig. 17 Lock the flywheel in place, then remove the flywheel bolts

Fig. 18 Removing the flywheel from the crankshaft

Fig. 19 Add a threadlocking agent to the flywheel bolts upon installation

Fig. 20 Be sure that the flywheel surface is clean, before installing the clutch

Fig. 21 Place a straightedge across the flywheel surface, then use a feeler gauge to check for warpage

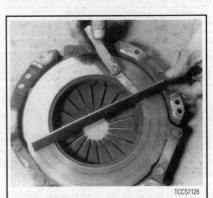

Fig. 22 Checking the pressure plate for excessive wear

Fig. 23 Install a clutch alignment arbor, to align the clutch assembly during installation

Fig. 24 Clutch plate installed with the arbor in place

Fig. 25 Clutch plate and pressure plate installed with the alignment arbor in place

Fig. 26 Pressure plate-to-flywheel bolt holes should align

Fig. 27 Apply locking agent to the clutch assembly bolts

Fig. 28 Be sure to use a torque wrench to tighten all bolts in a star pattern

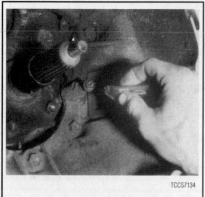

Fig. 29 Grease the clutch release fork ball

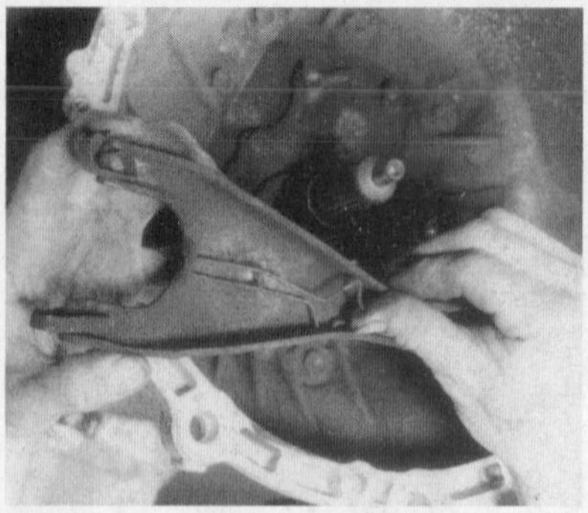

Fig. 30 View of the clutch release fork; check it for signs of damage

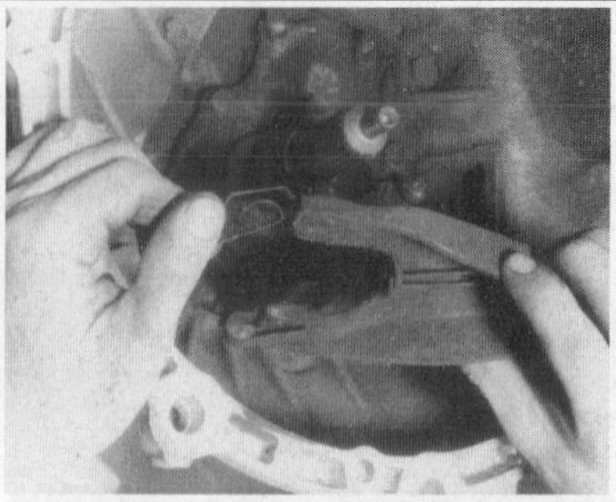

Fig. 31 View of the clutch release fork bearing clips; make sure these are not bent or broken

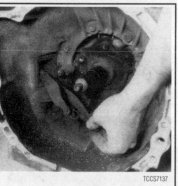

Fig. 32 Removing the clutch release fork bearing clips

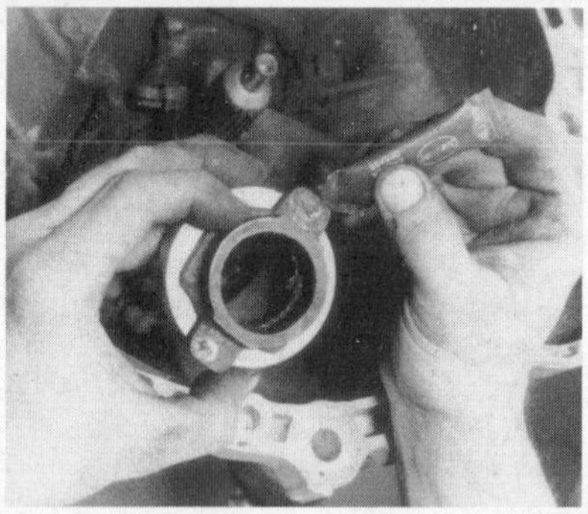

Fig. 33 Grease the throwout bearing assembly at the outer contact points

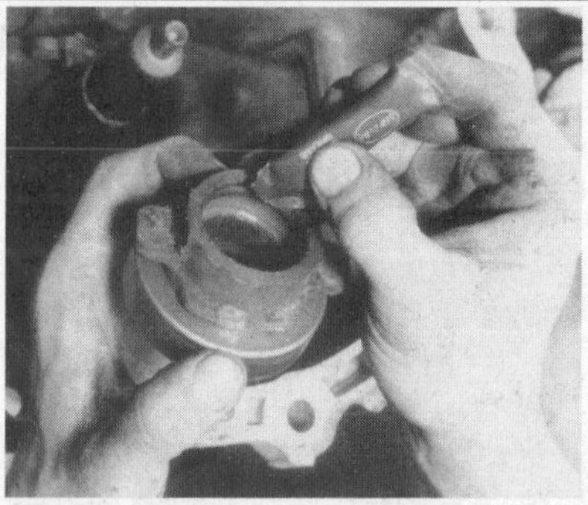

Fig. 34 Grease the throwout bearing assembly at the inner contact points

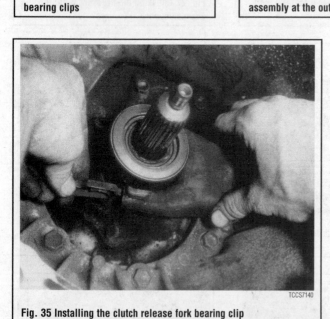

Fig. 35 Installing the clutch release fork bearing clip

Fig. 36 View of the clutch release fork assembly installed; be sure all parts move freely

Master Cylinder

REMOVAL & INSTALLATION

▶ See Figure 37

1. Remove the clip and pin at the top of the clutch pedal; disconnect the master cylinder pushrod from the clutch pedal.
2. Have a container handy to catch any spillage of fluids. Remove the hydraulic line from the master cylinder, being careful not to damage the fitting.

3. Remove the two bolts holding the master cylinder to the engine compartment.

✳✳ WARNING

Brake fluid dissolves paint. Do not allow it to drip onto the body when removing the master cylinder.

4. Install the master cylinder. Partially tighten the hydraulic line, then tighten the cylinder mounting bolts to 9 ft. lbs. (13 Nm).

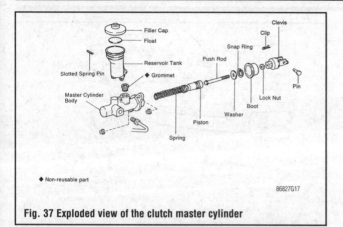

Fig. 37 Exploded view of the clutch master cylinder

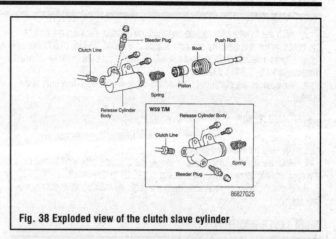

Fig. 38 Exploded view of the clutch slave cylinder

Slave Cylinder

REMOVAL & INSTALLATION

▶ **See Figure 38**

1. Jack up the front of the truck and support it on jackstands.
2. Remove the tension spring on the clutch fork.
3. Remove the hydraulic line from the slave cylinder. Be careful not to damage the fitting.
4. Remove the mounting bolts and withdraw the cylinder.
5. Place the cylinder in position, then install the bolts. Tighten to 9 ft. lbs. (13 Nm).
6. Bleed the system.

HYDRAULIC SYSTEM BLEEDING

This operation must be performed any time the clutch master or slave cylinder has been removed or if any of the hydraulic lines have been opened.

✳✳ WARNING

Do not spill brake fluid on the bodywork of the vehicle; it will destroy the paint. If fluid is spilled, immediately wash the surface with plenty of clean water.

1. Fill the master cylinder reservoir with brake fluid.
2. Remove the cap on the bleeder screw on the clutch slave cylinder. Install a clear vinyl hose on the fitting; place the other end submerged in a clear glass jar partially filled with brake fluid.
3. Have an assistant pump the clutch pedal slowly several times. After several pumps, hold the pedal down and open the bleeder, allowing fluid to flow into the jar. Close the bleeder valve almost immediately after opening it. Release the pedal only after the bleeder is closed.
4. Repeat the process until the fluid in the hose contains no air bubbles. tube. When there are no more air bubbles in the system, tighten the plug fully with the pedal depressed. Replace the plastic cap.
5. Fill the clutch master cylinder reservoir to the correct level with brake fluid.
6. Check the system for leaks.

AUTOMATIC TRANSMISSION

Understanding the Automatic Transmission

The automatic transmission allows engine torque and power to be transmitted to the rear wheels within a narrow range of engine operating speeds. It will allow the engine to turn fast enough to produce plenty of power and torque at very low speeds, while keeping it at a sensible rpm at high vehicle speeds (and it does this job without driver assistance). The transmission uses a light fluid as the medium for the transmission of power. This fluid also works in the operation of various hydraulic control circuits and as a lubricant. Because the transmission fluid performs all of these functions, trouble within the unit can easily travel from one part to another.

Identification

The A43D is a fully automatic 2WD, four speed transmission with electrically engaged overdrive. It was first offered as an option on 1981 models and is available through the current model year in trucks and is now offered in the Tacoma. The A44F is the 4WD application found in 1990 and 1991 4Runners.

The A340E, A340, and A340H are also fully automatic transmissions, but rely on electronic computer control rather than the conventional reliance on internal oil pressures. Because of the electronic controls, these units are referred to as Electronically Controlled Transmissions (ECT). On 1989–96 models, these units were offered in the Pick-Up, 4Runner, T100 and Tacoma.

The A340E contains a lock-up torque converter and is coupled to the 3VZ-E engine. The A340F employs a mechanically controlled transfer unit for 4WD applications, while the A340H, also used in 4WD vehicles, is coupled to an electrically controlled transfer case.

1989–92 Land Cruisers were equipped with an A440F transmission which is similar to the A43D. The A440F contains a lock-up torque converter and is cou-

pled to a 2-speed transfer unit. The A442F is used in the Land Cruiser from 1993–95. In 1996, the A343F transmission was used.

Adjustments

THROTTLE CABLE

Pick-Up and 4Runner

1. Remove the air cleaner assembly.
2. Push the accelerator to the floor and check that the throttle plate opens fully. If not, adjust the accelerator linkage so that it does.
3. Push back the rubber boot from the throttle cable which runs down to the transmission. Loosen the throttle cable adjustment nuts so that the cable housing can be adjusted.
4. Fully open the throttle by having an assistant press the accelerator all the way to the floor.
5. Adjust the cable housing so that, with the throttle wide open, the distance between the end of the rubber boot and the cable stopper is 0–0.04 in. (0–1mm).
6. Tighten the nuts and double check the adjustment. Install the rubber boot and the air cleaner.

Land Cruiser

A440F

1. Inspect the throttle cable and its mounts. Neither the cable nor the mounts should be deformed or bent.

2. With the throttle fully closed, measure the distance between the end of the boot and the stopper. The correct distance is 0.020–0.059 in. (0.5–1.5mm).

3. Have an assistant open the throttle fully. Remeasure the distance. Correct distance is 1.26–1.34 in. (32–34mm).

4. If distances are not correct, adjust the cable via the adjusting nuts at the bracket.

A442F AND A343F

1. Inspect the throttle cable and its mounts. Neither the cable nor the mounts should be deformed or bent.

2. With the throttle fully closed, measure the distance between the end of the boot and the stopper. Correct distance is 0–0.04 in. (0–1mm).

3. If distances are not correct, adjust the cable via the adjusting nuts at the bracket.

SHIFT LEVER POSITION

1. Loosen or disconnect the nut holding the horizontal shift rod to the bottom of the shift selector.

2. Move the now loosened or removed rod downward or towards the back of the vehicle. This will cause the shift lever (on the transmission case) to move into approximately the 9 o'clock or horizontal position.

3. On A43D, A44D and A340F transmissions, move the horizontal rod upwards or towards the front until the 3rd notch is engaged. This should put the transmission in **N** and bring the small arm at the transmission near the vertical position.

4. For A340F, A340H, A440, A442F and A343F units, return or move the arm until the 2nd notch is engaged. This should put the transmission in **N** and bring the small arm at the transmission near the vertical position.

5. Set the shift selector inside the vehicle to the **N** position.

6. Have an assistant hold the shift selector lightly towards the **R** position. Reconnect and/or adjust the linkage to the lower part of the shift selector and tighten the bolts.

7. Start the engine and make sure that the vehicle moves forward when shifted from **N** into the **D** range and moves rearward when shifted into **R**.

TRANSFER CASE LINKAGE

This adjustment is only possible on the A340E and A340H transmissions.
1. Shift the transfer case lever to the **H2** position.
2. Disconnect the No. 1 transfer case linkage from the cross shaft.
3. Position the transfer case indicator switch to the **H2** position.
4. Connect the No. 1 transfer case linkage to the cross shaft.

TRANSFER POSITION SWITCH

This adjustment is only possible on the A340E and A340H transmissions.
1. Loosen the transfer case position switch bolt and set the shift position to the **L4** position.
2. Unplug the electrical connector.
3. Connect an ohmmeter between the terminals and adjust the switch so there is continuity.
4. Connect the electrical lead and tighten the switch bolt to 48 inch lbs. (5 Nm).
5. Check the idle speed.

Neutral Safety Switch

The neutral safety switch prevents the vehicle from starting unless the gearshift selector is in either the **P** or **N** positions. If the vehicle will start in any other positions, adjustment or replacement of the switch is required.

REMOVAL & INSTALLATION

A340E and A340F

▶ See Figure 39

➡On some models it will be necessary to remove the oil cooler pipes for access.

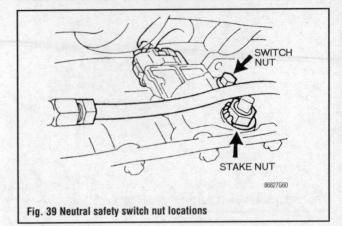

Fig. 39 Neutral safety switch nut locations

1. Disconnect the harness for the neutral safety switch.
2. Pry off the washer and remove the nut, then remove the bolt for the switch.
To install:
3. Attach the switch to the vehicle with the bolt, tighten to 9 ft. lbs. (13 Nm).
4. Install a new lock plate and nut. Tighten the nut to 35 inch lbs. (4 Nm).
5. Stake the nut with the lock plate.
6. Adjust the neutral safety switch.
7. Attach the harness to the switch.
8. Attach the oil cooler pipes.

A43D

▶ See Figures 40 and 41

1. Unstake the lockwasher.
2. Remove the nut and bolt, then the lockwasher and grommet.
3. Slide the switch off the manual lever shaft.

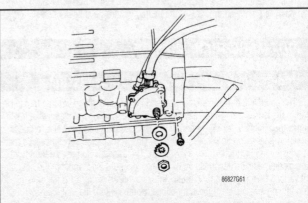

Fig. 40 Exploded view of neutral safety switch mounting

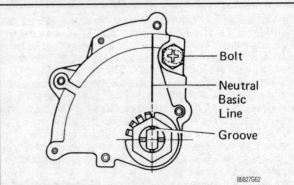

Fig. 41 Align the switch basic line and the switch groove, then tighten the adjusting bolt to 48 inch lbs. (5 Nm)

To install:

4. Insert the neutral safety switch onto the manual lever shaft, then temporarily tighten the adjusting bolt.

5. Install the grommet and a new lockwasher. Install and tighten the nut to 35 inch lbs. (4 Nm).

6. Using the control shaft lever, turn the manual valve lever shaft forwards and return two notches. It is now in neutral.

7. Align the switch basic line and the switch groove, then tighten the adjusting bolt. Tighten to 48 inch lbs. (5 Nm).

8. Bend the tabs of the lockwasher over the nut.

ADJUSTMENT

1. Loosen the neutral start switch bolt.
2. Place the gearshift selector lever in the **N** position.
3. Align the shaft groove of the switch with the neutral Basic line. Hold the switch in this position and tighten the switch bolts on the A43D to 48 inch lbs. (6 Nm) and all others to 9 ft. lbs. (13 Nm).
4. With the parking brake set and the brake pedal fully applied, attempt to start the engine in each shifter range. The engine should only start in **N** or **P**.

Extension Housing Seal

REMOVAL & INSTALLATION

▶ **See Figures 42 and 43**

Removal of the extension housing seal is only possible on the A43D, A44D and A340E transmissions.

1. Elevate and safely support the vehicle.
2. Position a catch pan below the work area to contain spilled fluid.
3. Remove the driveshaft.
4. Clean the extension housing area thoroughly before removing the seal. No dirt must be allowed to enter the transmission during replacement.

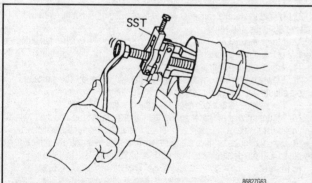

Fig. 42 Use SST 09308–10010 or equivalent to remove the extension housing seal

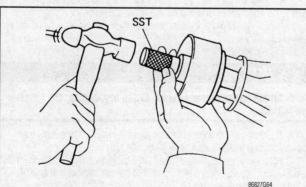

Fig. 43 Upon replacement, tap on the seal driver to insert the seal as far as it will go

5. Using a seal extractor such as tool 09308–10010 or equivalent, remove the seal.

6. Installation is in the reverse order of removal. Be sure to use a seal driver such as tool 09325–20010 or equivalent.when driving the new seal into place until it seats.

7. Do NOT overfill the fluid.

Transmission Assembly

REMOVAL & INSTALLATION

Pick-Up and 4Runner

A43D, A44D AND A340E

▶ **See Figure 44**

1. Disconnect the cables at the battery, negative first.
2. Remove the air cleaner assembly.
3. Disconnect the transmission throttle cable.
4. Raise the vehicle and support it safely with jackstands. Double check the security and placement of the stands.
5. Unplug the wiring connectors near the starter.
6. Disconnect the starter wiring at the starter. Unbolt the starter and remove it from the vehicle.
7. Drain the transmission fluid.
8. Matchmark the rear driveshaft flange and the differential pinion flange. These marks must be aligned during installation.
9. Unbolt the rear driveshaft flange. If the vehicle has a two piece driveshaft, remove the center bearing bracket-to-frame bolts. Remove the driveshaft from the vehicle.
10. Disconnect the speedometer cable from the transmission and tie it out of the way.
11. Disconnect the shift linkage at the transmission.
12. Disconnect the exhaust pipe clamp at the bell housing and remove the transmission dipstick and oil filler tube.
13. Disconnect the transmission oil cooler lines at the transmission.
14. Support the transmission using a jack with a wooden block placed between the jack and the transmission pan. Do not raise the transmission, just raise the jack until the wooden block touches the transmission pan.
15. Place a wooden block (or blocks) between the engine oil pan and the front frame crossmember.

➡ **The wooden block(s) should be no more than about 0.2 in. (6mm) away from the engine so that when the engine is lowered, damage will not occur to any underhood components.**

16. Remove the transmission mount-to-crossmember bolts.
17. Raise the transmission SLIGHTLY, just enough to take the weight of the transmission off of the crossmember. Remove the crossmember-to-frame mounting bolts and remove the crossmember from the vehicle. For vehicles with the A43D, remove the bracket and rear transmission mount.
18. Slowly lower the transmission until the engine rests on the wooden block placed earlier.

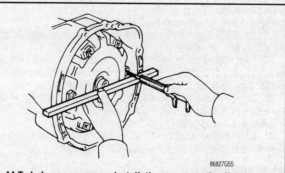

Fig. 44 To help ensure proper installation, measure the distance between the torque converter mounting lugs and the front mounting face of the transmission

19. Remove the engine undercover in order to gain access to the engine crankshaft pulley.

20. Remove the rubber plugs from the service holes located at the rear of the engine in order to gain access to the torque converter bolts.

21. Rotate the crankshaft as necessary to remove the torque converter retaining bolts. Access to these bolts is through the service holes mentioned earlier.

22. Obtain a bolt of the same dimensions as the torque converter bolts. Cut the head off of the bolt and cut a screwdriver slot in the bolt opposite the threaded end.

➡️**This modified bolt is used as a guide pin. Two guide pins are needed to properly install the transmission.**

23. Thread the guide pin into one of the torque converter bolt holes. The guide pin will help keep the converter with the transmission.

24. Remove the transmission-to-engine mounting bolts.

25. Carefully move the transmission rearward by prying on the guide pin through the service hole.

➡️**As soon as the transmission is away from the engine about 0.1 in. (3mm), feed wire through the front of the transmission and secure the wire in order to keep the converter attached to the transmission. Also, try to keep the nose of the transmission pointed upward SLIGHTLY to help keep the converter in place.**

26. Pull the transmission rearward and lower it out of the vehicle.

➡️**Do not allow the attached cables to catch on any components during removal.**

27. With the transmission out of the vehicle, remove the torque converter as follows:
   a. Place a drain pan under the front of the transmission.
   b. Pull the converter straight off the transmission and allow the fluid to drain.

➡️**Prior to installation of the transmission, Toyota recommends checking the torque converter and flywheel run-out dimensions. If either of these run-out limits are beyond the maximum allowable limits, excessive wear of the front transmission seal will occur.**

28. Mount the torque converter on the flywheel and tighten the converter bolts to 20 ft. lbs. (27 Nm) for A43D and A44D transmissions. For A340E, tighten them to 30 ft. lbs. (41 Nm).

29. Mount the dial indicator so that the indicator probe touches the outer surface of the converter extension sleeve (90° to the converter centerline).

30. Adjust the dial indicator to zero.

31. Slowly rotate the converter and read the dial indicator. The indicator needle should not deviate more than 0.01 in. (0.30mm).

32. Remove the torque converter from the flywheel.

33. Mount the dial indicator so that the indicator probe touches the flywheel ring gear just inside of the gear teeth (surface faces the rear of the vehicle).

34. Zero the indicator needle and slowly rotate the flywheel. The indicator needle should not deviate more than 0.01 in. (0.20mm).

To install:

35. Installation is in the reverse order of removal. Please note the following steps.

36. Apply a coat of multi-purpose grease to the torque converter stub shaft and the corresponding pilot hole in the flywheel.

37. To make sure that the converter is properly installed, measure the distance between the torque converter mounting lugs and the front mounting face of the transmission. The proper distance is 0.79 in. (20mm) for A43D and A44D; for A340E, correct distance is 0.71 in. (18mm).

38. Install and tighten the transmission-to-engine mounting bolts. Tighten the bolts to 47 ft. lbs. (64 Nm).

39. Evenly tighten the converter mounting bolts to 20 ft. lbs. (27 Nm) for A43D and A44D transmissions. For A340E transmissions, tighten them to 30 ft. lbs. (41 Nm). Install the rubber plugs into the access holes.

40. Install the transmission crossmember. Tighten the crossmember-to-frame bolts to 70 ft. lbs. (95 Nm). For A43D transmissions, install the rear mounting bracket and tighten the bolts to 43 ft. lbs. (58 Nm).

41. Lower the transmission onto the crossmember and install the transmission mounting bolts. Tighten the bolts to 18 ft. lbs. (25 Nm).

A340F AND A340H

1. Disconnect the negative battery cable.
2. Unplug the air intake connector.

3. Loosen the adjusting nuts and disconnect the throttle cable housing at the bracket. Disconnect the cable at the linkage, then unplug it from the rear of the engine.

4. Underneath the air intake chamber there are 5 connectors; unplug them.

5. Remove the upper starter mounting bolt.

6. Raise the vehicle, support it on safety stands and drain the transmission fluid.

7. Remove the driveshafts.

8. Disconnect the exhaust pipe at the tail pipe.

9. Disconnect and plug the 2 oil cooler lines at the transmission case.

10. Pull out the cotter pin and disconnect the manual shift linkage at the neutral safety switch.

11. Pull the cotter pins and disconnect the No. 1 and No. 2 transfer case shift linkages at the cross shaft over the transmission case. Remove the cross shaft.

12. Disconnect the speedometer cable.

13. Disconnect the 2 oil cooler lines at the transfer case.

14. Using a transmission jack, or a floor jack and a block of wood, raise the transmission just enough to remove pressure from the rear support member.

15. Position a piece of wood between the firewall and the rear of the cylinder head, remove the 8 bolts and lift out the rear support member.

16. Remove the engine under cover.

17. Remove the 6 torque converter mounting bolts.

➡️**Rotate the crankshaft in order to gain access to the torque converter mounting bolts.**

18. Insert a guide pin (just cut off the head of an old bolt) into one of the torque converter bolt holes.

19. Remove the starter and then remove the transmission housing mounting bolts.

20. Using the guide pin to keep the transmission and torque converter together, pry on the end of the pin to get the transmission assembly moving rearward. Be careful!

21. With a large drip pan under the assembly, pull off the converter. Remove the filler tube from the side of the transmission case.

➡️**Prior to installation of the transmission, Toyota recommends checking the torque converter and flywheel run-out dimensions. If either of these run-out limits are beyond the maximum allowable limits, excessive wear of the front transmission seal will occur.**

22. Mount the torque converter on the flywheel and tighten the converter bolts to 30 ft. lbs. (40 Nm).

23. Mount the dial indicator so that the indicator probe touches the outer surface of the converter extension sleeve (90° to the converter centerline).

24. Adjust the dial indicator to zero.

25. Slowly rotate the converter and read the dial indicator. The indicator needle should not deviate more than 0.01 in. (0.30mm).

26. Remove the torque converter from the flywheel.

27. Mount the dial indicator so that the indicator probe touches the flywheel ring gear just inside of the gear teeth (surface faces the rear of the vehicle).

28. Zero the indicator needle and slowly rotate the flywheel. The indicator needle should not deviate more than 0.01 in. (0.20mm).

To install:

29. Installation is in the reverse order of removal. Please note the following steps.

30. Coat the pilot hole in the crankshaft and the center hub of the converter with grease.

31. Install the torque converter into the transmission case. Using calipers and a straightedge, make sure the distance from the installed surface of the converter to the transmission housing lip is 1.02 in. (26mm).

32. Install a guide pin in the converter and align it with one of the driveplate holes. Align the 2 sleeves on the cylinder block with the converter housing and temporarily install a bolt.

❈❈ WARNING

Never tilt the transmission forward during installation; the converter will fall out.

33. If all holes are aligned, install the starter and then all remaining transmission housing bolts. Tighten to 47 ft. lbs. (64 Nm).

34. Remove the guide pin and install the 6 converter bolts. Tighten the bolts evenly to 20 ft. lbs. (27 Nm) on models with 4 cylinder engines and to 30 ft. lbs. (41 Nm) on those with the V6.

35. Install the rear support member and tighten the 8 bolts to 70 ft. lbs. (95 Nm).

Tacoma

A340E AND A340F

1. Remove the fluid level dipstick.
2. Remove the engine under cover.
3. Disconnect the throttle cable by loosening the nut then disconnecting the cable. Separate the cable from the clamp.
4. Remove the fan shroud.
5. Remove the transmission shift lever assembly and transfer case shift lever as follows:
    a. Remove the rear console box.
    b. Remove the screws and the front console box with the transfer case shift lever knob.
    c. Disconnect the harness wires.
    d. Remove the nut and washer, then disconnect the shift control knob.
    e. Unattach the harness then remove the 8 screws and transmission shift lever assembly.
    f. Using snapring pliers, remove the snapring and pull out the shift lever from the transfer case.
    g. When reinstalling, apply multi-purpose grease to the shift lever.
6. Remove the fluid level pipe.
7. Remove the front and rear driveshafts.
8. Disconnect the front exhaust pipe.
9. Disconnect the following:
- Speedometer cable
- No. 2 vehicle speed sensor harness
- Solenoid harness
- Transfer case neutral position switch harness
- Transfer case L4 position switch harness
- Transfer case indicator switch
10. Remove the bolts and clamps holding the oil cooler pipe.
11. Loosen the union nuts and disconnect the oil cooler pipes.
12. Disconnect the temperature sensor harness.
13. Disconnect the neutral safety switch harness.
14. Remove the starter.
15. Unbolt the stabilizer bar bracket.
16. Remove the torque converter clutch mounting bolt:
    a. On the 3RZ-FE engine, remove the 2 nuts and bolts, then the 2 bolts and flywheel housing under cover.
    b. On the 5VZ-FE engine remove the 4 bolts and flywheel housing under cover.
    c. While turning the crankshaft to gain access, remove the 6 bolts. When installing first attach the green colored bolt then the other 5.
17. Remove the front differential rear mounting cushion as follows:
    a. Using a hexagon wrench (12mm), remove the nut. then lift up the front differential.
    b. Be careful not to touch the torque converter clutch housing and front differential companion flange.
    c. Remove the 2 rear mounting cushion mounting bolts.
18. Remove the crossmember, support the transmission rear side with a suitable jack.
19. Remove the mounting bolts then the crossmember. Be sure to support the transmission with a jack.
20. Lower the transmission rear side.
21. Separate the wire harness from the transmission.
22. On the 3RZ-FE engine remove the 4 bolts and transmission. On the 5VZ-FE engine, remove the 6 bolts and transmission.
23. To install reverse the procedure. Always use new gaskets and O-rings.
24. Fill the unit with the correct fluid, start the vehicle and test the operation.

T100

A340E AND A340F

1. Remove the fluid level dipstick.
2. Unbolt the engine under cover.
3. Disconnect the throttle cable by loosening the nut then disconnecting the cable. Separate the cable from the clamp.
4. Raise and safely support the vehicle.
5. Remove the fluid level dipstick.
6. Remove the front and rear driveshafts.
7. Disconnect the front exhaust pipe.

8. Disconnect the speedometer cable and vehicle speed sensor harness.
9. Remove the cross shaft as follows:
    a. Remove the clip and disconnect the No. 2 gear shifting rod.
    b. Remove the nut, washer, 4 bolts and the cross shaft.
10. Remove the oil cooler pipe.
11. Remove the bolts and clamps holding the oil cooler pipe.
12. Loosen the union nuts and disconnect the oil cooler pipes.
13. Disconnect the transmission fluid temperature sensor harness.
14. Disconnect the neutral safety switch harness.
15. Disconnect the starter wire.
16. Remove the stabilizer bar.
17. Remove the rear end plate nuts, bolts and plate.
18. Disconnect the rear mounting insulator.
19. Jack up the transmission.
20. While turning the crankshaft to gain access to the 6 bolts, remove them.
21. Disconnect the harness and connectors from the transmission.
22. Unbolt the transmission from the engine.
23. To install reverse the procedure. Always use new gaskets and O-rings.
24. Fill the unit with the correct fluid, start the vehicle and test the operation.

Land Cruiser

A440F

1. Disconnect the negative battery cable.
2. Remove the battery and the battery cover.
3. Loosen the cooling fan shroud bolts. This will avoid fan damage.
4. Loosen the adjusting nuts for the throttle control cable. Disconnect the cable housing from the bracket; disconnect the cable from the throttle linkage.
5. Unplug the electrical connectors near the starter.
6. Remove the transfer case shift lever as follows:
    a. Under the vehicle, remove the nut and transmission control rod.
    b. Remove the transfer case shift lever knob.
    c. Remove the four screws holding the console and remove the console.
    d. Remove the four bolts and remove the transfer case shift lever boot.
    e. Remove the three bolts holding the console box.
    f. Remove the six bolts holding the transmission shift lever assembly. Remove the assembly.
    g. Pull out the pin and disconnect the shift rod.
    h. Remove the four bolts holding the transfer case shift lever and remove the lever assembly.
7. Elevate and safely support the vehicle. Double check the supports for placement and security.
8. Disconnect the speedometer cable.
9. Remove the front and rear propeller shafts.
10. Remove the starter.
11. Remove the two line clamps and disconnect the oil cooler lines at the transmission.
12. Remove the undercover.
13. Remove the endplate hole plug. Turn the crankshaft to gain access to each of the six torque converter mounting bolts. Remove the bolts.
14. Support the transmission, either with a transmission jack (preferred) or a floor jack and a broad piece of wood to protect the transmission. Tension the jack just enough to take tension off the crossmember.
15. Remove the bolts and nuts holding the crossmember and remove it.
16. Use a floor jack and a piece of wood to support the engine under the oil pan. Do NOT bend the oil pan.
17. Lower the rear end of the transmission and slide the unit away from the engine.
18. With the transmission removed, set up a dial indicator and measure the driveplate run-out. If run-out exceeds 0.0079 in. (0.20mm) or if the ring gear is damaged, the driveplate must be replaced.
19. Temporarily mount the torque converter to the driveplate. Set up a dial indicator and measure run-out at the center of the torque converter. If run-out exceeds 0.0118 in. (0.30mm), try to correct it by repositioning the torque converter. If the repositioning corrects the error, make matchmarks so that the position is used at reinstallation. If the misalignment cannot be corrected, the torque converter must be replaced. Remove the torque converter when measurement is completed.

To install:

20. Installation is in the reverse order of removal. Please note the following steps.

21. With the torque converter installed on the transmission, check the converter installation with a straightedge across the front of the case. Correct distance from the surface of the bolt lugs to the front of the case is 0.618 in. (1.57mm) or more. If the distance measured is less than specified, check the converter for improper installation.

22. Elevate the transmission and push it fully into place. Install the transmission retaining bolts and tighten them to 53 ft. lbs. (72 Nm).

23. Install the rear mount to the transmission. Tighten the retaining bolts to 43 ft. lbs. (59 Nm).

24. Install the crossmember. Tighten the bolts to 45 ft. lbs. (61 Nm) and the nuts to 43 ft. lbs. (59 Nm).

25. Install the torque converter mounting bolts, tightening them to 21 ft. lbs. (28 Nm). Install the end plate hole plug when finished.

A343F AND A442F

1. Remove the battery and battery tray. Always disconnect the negative battery cable first.

2. Unbolt the fan shroud for the cooling fan.

3. Disconnect the throttle cable by loosening the adjusting nut and removing the cable from the bracket.

4. Detach the cable from the linkage.

5. Remove the upper side starter mounting bolt.

6. Remove the clip, washer and wave washer, then disconnect the transfer case shift lever link.

7. Remove the nut and washer, then disconnect the transmission select lever link.

8. Unscrew the transfer case shift lever knob.

9. Unbolt the console and set aside.

10. Remove the mounting screws for the transfer case shift lever boot.

11. Remove the console box.

12. Disconnect the wiring harnesses.

13. Remove the 6 mounting screws for the transmission shift lever assembly.

14. Unbolt the transfer case shift lever, then remove the two cushions.

15. If applicable, disconnect the following:
- Vehicle speed sensors
- Neutral safety switch
- Solenoid connector
- ATF temperature sensor
- Center differential lock indicator switch
- L4 position switch
- Bleeder hose from the transfer case

16. Remove the front and rear driveshafts.

17. Remove the oil level dipstick.

18. Unbolt the upper side mounting.

19. Remove the bolt retaining the filler pipe. Discard the old O-ring.

20. Loosen the two oil cooler pipe union nuts.

21. Remove the 4 stabilizer bar bracket mounting bolts.

22. Separate the engine undercover from the vehicle.

23. Take the converter plug off.

24. Turn the crankshaft to gain access to each bolt, then hold the crankshaft pulley nut with a wrench and remove the 6 bolts.

25. Remove the front exhaust pipe, discard the gaskets.

26. Remove the starter.

27. Support the transmission with a suitable jack. Remove the bolts and nuts retaining the crossmember. Lower the crossmember.

28. Lower the rear end of the transmission. Be careful not to damage the cooling fan, brake booster and brake line.

29. Separate the wire harness from the transmission and transfer case.

30. Remove the cooler pipe mounting bolts from the torque converter clutch housing, then disconnect the 2 oil cooler pipes from the elbows.

31. Remove the transmission mounting bolts.

To install:

32. Installation is in the reverse order of removal. Please note the following steps.

33. Using calipers and a straightedge, measure from the installed surface of the transmission housing. The distance should be more than 0.618 in. (15.7mm). Install the transmission, tighten to 53 ft. lbs. (72 Nm).

34. Attach the crossmember to the vehicle. Tighten the 8 bolts to 45 ft. lbs. (61 Nm) and the 2 nuts to 54 ft. lbs. (74 Nm).

35. Install the torque converter clutch mounting bolt. Tighten to 40 ft. lbs. (55 Nm). First install the gray colored bolt, then the other 5 bolts.

TRANSFER CASE

Applications

Toyota Pick-Ups, 4Runners, T100's and Tacomas use either the RF1A or VF1A transfer case. The RF1A uses a countergear reduction scheme.

The VF1A transfer case is a planetary gear reduction unit. It is found in the T100, Pick-Up, and 4Runner. The Land Cruiser employs the HF2A or HF2AV transfer case, used with the A440F and A343F transmissions for full time 4WD application.

Transfer Case Assembly

REMOVAL & INSTALLATION

♦ **See Figure 45**

➡**The transfer case is removed as an assembly with the transmission.**

1. Remove the transmission assembly. On some models it will be necessary to remove the breather hose between the transmission control retainer and the transfer case upper cover.

2. Remove the rear engine mount from the case. Remove the dynamic damper if one is present.

3. Remove the upper driveshaft dust cover.

4. Remove the transfer case adapter rear mounting bolts, then pull the transfer case straight up and remove it.

➡**Be careful not to damage the adapter rear oil seal during removal.**

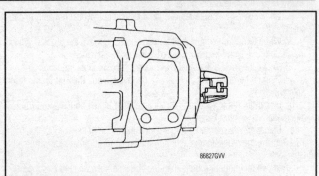

86827GVV

Fig. 45 Place the shift fork shafts into the H4 position as shown

To install:

5. Move the 2 shift forks into the H4 position.

6. Coat the adapter oil seal with grease and position a new gasket on the transfer case adapter.

7. Install the transfer case to the transmission. Tighten the bolts to 29 ft. lbs. (39 Nm). The longer bolts are used at the dust cover, the shorter ones are used in all other locations.

8. Install the rear engine mount and tighten the bolts to 19 ft. lbs. (25 Nm). Install the dynamic damper if one was removed.

9. Install the breather hose if one was removed. Install the transmission assembly.

DRIVELINE

Front Driveshaft and Universal Joints

REMOVAL & INSTALLATION

♦ See Figures 46, 47 and 48

1. Jack up the front of the vehicle and support with jackstands.
2. Place matchmarks on the flanges. Remove the 4 nuts and bolts.
3. Remove the front driveshaft dust cover. Some types are equipped with 2 bolts, while others use 4.
4. Remove the driveshaft dust cover subassembly. The three bolts are two different sizes.

➡The driveshaft dust cover is located around the outer portion of the U-joint. On some models you may have an upper and lower.

5. Suspend the front side of the driveshaft.
6. Place matchmarks on the flanges. Remove the 4 nuts and bolts.
7. Remove the front driveshaft.

To install:

8. Align the matchmarks on the rear flanges, then connect the flanges with the nuts and/or bolts. Tighten to 54 ft. lbs. (74 Nm).
9. Install the front driveshaft dust cover subassembly. Tighten bolts A to 27 ft. lbs. (36 Nm) and B to 17 ft. lbs. (23 Nm).
10. Align the matchmarks on the front flanges, then connect the flanges with the bolts and nuts. Tighten to 54 ft. lbs. (74 Nm).
11. Install the front driveshaft dust cover. Tighten the bolts to 13 ft. lbs. (17 Nm) and the nuts to 10 ft. lbs. (13 Nm).

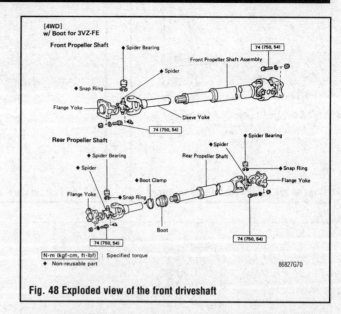

Fig. 48 Exploded view of the front driveshaft

12. Attach the rear driveshaft flange to the companion flange on the transfer case. Align the matchmarks on the flanges, then connect the flanges with the bolts and nuts. Tighten to 54–56 ft. lbs. (74–76 Nm).
13. Lower the vehicle and road test it.

U-JOINT REPLACEMENT

♦ See Figures 49, 50, 51, 52 and 53

1. Raise and support the vehicle.
2. Remove the driveshaft.
3. Matchmark the yoke and the driveshaft.
4. Remove the snaprings from the bearings. There are two types of snaprings; Toyota and Dana.
5. Position the yoke on vise jaws. Using a bearing remover and a hammer, gently tap the remover until the bearing is driven out of the yoke about 1 in. (25mm).
6. Place the tool in the vise and drive the yoke away from the tool until the bearing is removed.
7. Repeat Steps 4 and 5 for the other bearings.
8. Check for worn or damaged parts. Inspect the bearing journal surfaces for wear.
9. Assembly is in the reverse order of disassembly. Note the following.
10. Apply multi-purpose grease to a new spider and bearings. Be careful not to apply too much grease.
11. Start the bearings in the yoke, then press them into place, using a vise. Repeat for the other bearings.
12. If the axial play of the spider is greater than 0.0020 in. (0.05mm), select snaprings which will provide the correct play. Be sure that the snaprings are the same size on both sides or driveshaft noise and vibration will result.

Fig. 46 ALWAYS matchmark with paint prior to separating the driveshaft from the vehicle

Fig. 47 Tighten bolt A to 27 ft. lbs. (36 Nm) and B to 17 ft. lbs. (23 Nm)

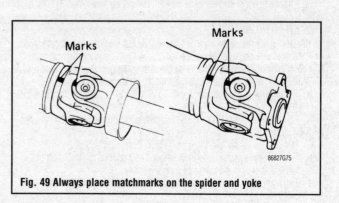

Fig. 49 Always place matchmarks on the spider and yoke

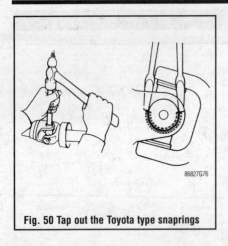

Fig. 50 Tap out the Toyota type snaprings

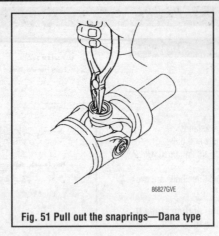

Fig. 51 Pull out the snaprings—Dana type

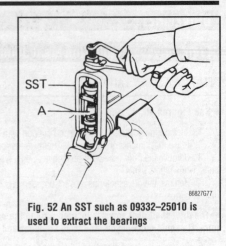

Fig. 52 An SST such as 09332–25010 is used to extract the bearings

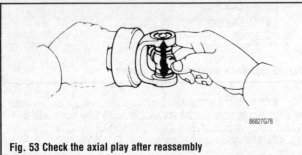

Fig. 53 Check the axial play after reassembly

Rear Driveshaft and U-Joints

REMOVAL & INSTALLATION

2WD Models

ONE-PIECE DRIVESHAFT

♦ **See Figure 54**

1. Jack up the rear of the truck and support the rear axle housing with jackstands.
2. Matchmark the two halves of the rear universal joint flange.
3. Remove the bolts which hold the rear flange together.
4. Remove the splined end of the driveshaft from the transmission.

➡**Plug the end of the transmission with a rag or other device to prevent oil loss.**

5. Remove the driveshaft from under the truck.
6. Installation is in the reverse order of removal. Apply multi-purpose grease to the splined end of the shaft.
7. Align the mating marks on the rear flange and replace the bolts. Tighten to 54 ft. lbs. (74 Nm) on all vehicles except models with 3VZ-E engines and manual transmissions. Tighten to 56 ft. lbs. (76 Nm) on these models.

TWO-PIECE DRIVESHAFT

1. Jack up the rear of the truck and support the rear axle housing on jackstands.
2. Before you begin to disassemble the driveshaft components, you must first paint accurate alignment marks on the mating flanges. Do this on the rear universal joint flange, the center flange, and on the transmission flange.
3. Remove the bolts attaching the rear universal joint flange to the drive pinion flange.
4. Drop the rear section of the shaft slightly and pull the unit out of the center bearing sleeve yoke.
5. Remove the center bearing support from the crossmember.
6. Separate the transmission output flange and remove the front half of the driveshaft together with the center bearing assembly.
7. Installation is the reverse of removal. Tighten all flange bolts to 22–36 ft. lbs. (30–49 Nm).

4WD Models

EXCEPT LAND CRUISER

1. Raise and support the vehicle with jackstands.
2. Matchmark all driveshaft flanges BEFORE removing any bolts.
3. Disconnect the driveshaft flange at the companion flange on the differential.
4. Remove the center bearing from the frame crossmember.
5. Disconnect the front flange at the companion flange on the rear of the transfer case and remove the driveshaft.
6. Separate the front portion of the shaft at the intermediate shaft flange.
7. Disconnect the rear of the driveshaft at the center bearing flange.
8. Using a hammer and chisel, loosen the staked portion of the center bearing nut, then remove the nut. Matchmark the flange to the intermediate shaft, then pull the bearing off the shaft.
9. Installation is in the reverse order of removal.
10. If equipped, install the center bearing on the intermediate shaft. Tighten the nut to 134 ft. lbs. (181 Nm), then loosen the nut, and retighten again to 51 ft. lbs. (69 Nm). Stake the nut.
11. Connect the front portion of the driveshaft to the intermediate shaft and the rear portion to the center flange. Make sure the marks all line up, then tighten the bolts to 54 ft. lbs. (74 Nm).
12. Tighten all flange bolts to 54 ft. lbs. (74 Nm).
13. Mount the center bearing on the crossmember and tighten the mounting bolts to 27 ft. lbs. (36 Nm).

LAND CRUISER

1. Raise the vehicle and support it with jackstands.
2. Matchmark all driveshaft flanges BEFORE removing the bolts.
3. Unfasten the bolts which secure the universal joint flange to the differential pinion flange.
4. Matchmark all of the flanges for the joint-to-transfer case flange bolts.
5. Withdraw the driveshaft from beneath the vehicle.

➡**Lubricate the U-joints and sliding joints with multi-purpose grease before installation.**

6. Position the driveshaft so that all marks line up, then tighten the flange bolts to 65 ft. lbs. (88 Nm).
7. Lower the vehicle and road test it.

U-JOINT REPLACEMENT

♦ **See Figures 55, 56, 57, 58 and 59**

1. Raise and support the vehicle.
2. Remove the driveshaft.
3. Matchmark the yoke and the driveshaft.
4. Remove the snaprings from the bearings. There are two types of snaprings: Toyota and Dana.
5. Position the yoke on vise jaws. Using a bearing remover and a hammer, gently tap the remover until the bearing is driven out of the yoke about 1 in. (25mm).

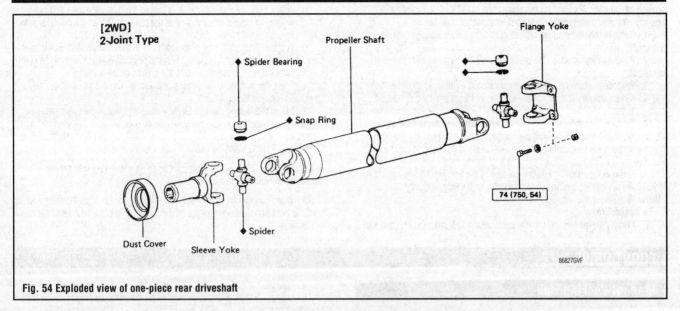

Fig. 54 Exploded view of one-piece rear driveshaft

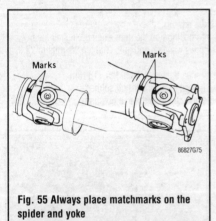

Fig. 55 Always place matchmarks on the spider and yoke

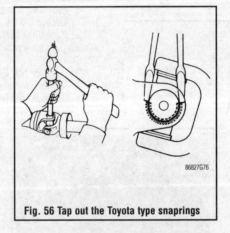

Fig. 56 Tap out the Toyota type snaprings

Fig. 57 Place matchmarks on the flanges, then loosen and remove the bolts

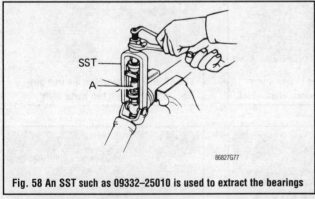

Fig. 58 An SST such as 09332–25010 is used to extract the bearings

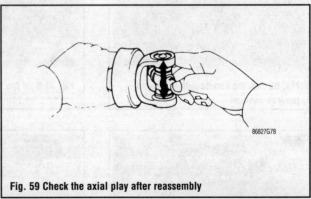

Fig. 59 Check the axial play after reassembly

6. Place the tool in the vise and drive the yoke away from the tool until the bearing is removed.

7. Repeat Steps 4 and 5 for the other bearings.

8. Check for worn or damaged parts. Inspect the bearing journal surfaces for wear.

To assemble:

9. Install the bearing cups, seals, and O-rings in the spider.

10. Apply multi-purpose grease to a new spider and bearings. Be careful not to apply too much grease.

11. Position the spider into the yoke.

12. Start the bearings in the yoke, then press them into place, using a vise.

13. Repeat Step 5 for the other bearings.

14. If the axle play of the spider is greater than 0.0020 in. (0.05mm), select snaprings which will provide the correct play. Be sure that the snaprings are the same size on both sides or driveshaft noise and vibration will result.

15. Check the U-joint assembly for smooth operation.

Center Bearing

REMOVAL & INSTALLATION

The center support bearing is a sealed unit which requires no periodic maintenance. The following procedure should be used if it becomes necessary to replace the bearing. Toyota does not supply individual parts for the bearing replacement only the assembly. See your local jobber if necessary.

1. Remove the intermediate driveshaft and the center support bearing assembly.

2. Paint mating marks on the universal joint flange and the intermediate driveshaft.

3. Remove the cotter pin and castle nut from the intermediate driveshaft; the nut will be staked. Remove the universal joint flange from the driveshaft using a press.

4. Remove the center support bearing assembly from the driveshaft.

5. Remove the two bolts from the bearing housing and remove the housing.

6. Remove the dust deflectors from both sides of the bearing cushion. Remove the dust deflectors from either side of the bearing.

7. Remove the snaprings from each side of the bearing. This is easy to do if you have a snapring tool which fits the holes in the ring, and very difficult otherwise. Remove the bearing.

To assemble:

8. Install the new bearing into the cushion and fit a snapring on each side.

9. Apply a coat of multi-purpose grease to the dust deflectors and put them in their respective places on each side of the bearing. The single deflector with a slightly larger diameter goes on the rear of the bearing.

10. Press the dust deflector onto each side of the cushion. The water drain holes in the deflectors should be in the same position on each side of the cushion. The water drain holes should face the bottom of the housing.

11. Press the support bearing assembly firmly onto the intermediate driveshaft, with the seal facing front.

12. Match the mating marks painted earlier, and install the universal joint flange to the driveshaft. Install the center bearing on the intermediate shaft and tighten the nut to 134 ft. lbs. (181 Nm). Loosen the nut and then retighten it to 51 ft. lbs. (69 Nm). Stake the nut.

➡**Check to see if the center support bearing assembly will rotate smoothly around the driveshaft.**

13. When reinstalling the driveshaft, be certain to match up the marks on both the front transmission flange and the flange on the sleeve yoke of the rear driveshaft.

FRONT DRIVE AXLE

Manual Locking Hub

REMOVAL & INSTALLATION

▸ **See Figures 60 thru 69**

1. Remove the hub cover. Set the control handle to free.

2. Remove the cover mounting bolts, then pull off the cover. Remove and discard the gasket.

3. Remove the bolt with washer.

4. Remove the mounting nuts and washers.

5. Using a brass bar and hammer, tap on the bolt ends, then remove the cone washers.

6. Pull off the hub body.

7. Remove and discard the gasket.

8. Place a new gasket into position on the front axle hub. Install the free wheeling hub body with the cones washers and nuts. Tighten the nuts to 23 ft. lbs. (31 Nm).

9. Install the bolt with washer, tighten to 13 ft. lbs. (18 Nm).

10. Apply multi-purpose grease to the inner hub splines.

11. Set the control handle and clutch to the free position.

Fig. 60 Turn the handle with your hand to the free position

Fig. 61 Note the directions in which you can turn the handle: left—free, right—lock

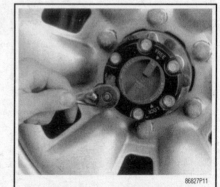

Fig. 62 With the handle in the free position, remove the six hub cover bolts

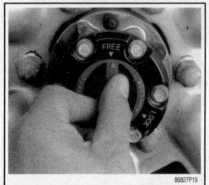

Fig. 63 Pull the cover off

Fig. 64 Using the correct size wrench, loosen . . .

Fig. 65 . . . then remove the center bolt and washer

Fig. 66 Next, remove the outside mounting nuts and washers

Fig. 67 Using a brass bar and hammer, tap on the bolt ends . . .

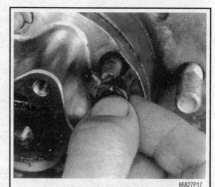

Fig. 68 . . . then remove the cone washers

Fig. 69 The free wheeling body should come off easily; be sure to discard the old gaskets

Front Axle, Bearing and Seal

REMOVAL & INSTALLATION

Pick-Up and 1989–95 4Runner 4WD

▶ See Figures 70, 71, 72 and 73

1. Raise the front of the vehicle and support on safety stands. Remove the wheel.

2. On models with ABS, disconnect the speed sensor from the steering knuckle.

3. Remove the brake caliper.

4. Remove the hub.

5. Disconnect and plug the brake line at the disc and secure it out of the way.

6. Pop off the axle nut lockwasher, then remove the nut, washer and adjusting nut. Remove the axle hub/brake rotor as an assembly along with the outer wheel bearing.

7. Pry out the oil seal and remove the inner bearing from the hub.

8. Loosen the 6 axle shaft-to-differential housing nuts.

9. Remove the snapring and spacer from the end of the axle shaft.

10. Separate the inboard axle shaft flange from the side gear shaft and then remove the shaft down and through the steering knuckle.

11. Installation is the reverse or removal.

12. Coat the outboard end of the axle shaft with lithium grease.

13. Insert the outboard end of the shaft into the steering knuckle and then install it to the differential. Hand tighten the bolts.

14. Tighten the inner mounting nuts to 61 ft. lbs. (83 Nm).

15. Pack the inner axle bearing with grease and press it into the hub. Coat a new oil seal with grease and press it into the hub.

16. Position the axle hub/brake disc assembly on the steering knuckle and press in the outer bearing.

17. Tighten the bearing adjusting nut to 43 ft. lbs. (59 Nm) and spin the hub a few times. Loosen the nut, then retighten to 18 ft. lbs. (25 Nm).

18. Use a tension gauge and check that the bearing preload is 6.4–12.6 lbs. (28–56 N).

19. Install a new locknut and washer, tighten to 35–58 ft. lbs. (47–78 Nm). Check that the bearing has no play.

20. Use a tension gauge and check that the bearing preload is 6.4–12.6 lbs. (28–56 N). If the preload is not within specifications, adjust it again with the adjusting nut.

21. Secure the locknut by bending one of the washer teeth inward and another outward.

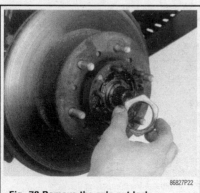

Fig. 70 Remove the axle nut lockwasher . . .

Fig. 71 . . . the nut . . .

Fig. 72 . . . and adjusting nut

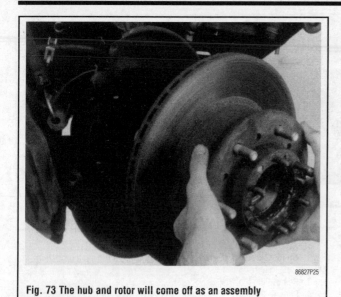

Fig. 73 The hub and rotor will come off as an assembly

T100, Tacoma and 1996 4Runner

1. Loosen the lug nuts.
2. Raise and support the vehicle.
3. Remove the front wheels.
4. Remove the shock absorber.
5. Pull off the grease cap.
6. Disconnect the driveshaft. Remove the cotter pin and lock cap.
7. With a helper applying the brakes, remove the locknut.
8. On vehicles with ABS, remove the speed sensor and harness from the steering knuckle.
9. Remove the brake line from the knuckle.
10. Remove the caliper and rotor.
11. Remove the bolts retaining the lower ball joint, separate from the vehicle.
12. Remove the cotter pin and loosen the nut on the axle hub. Using a SST or equivalent, remove the steering knuckle.
13. Installation is the reverse of removal. Tighten the hub nut to 80 ft. lbs. (108 Nm).
14. Install the driveshaft. Tighten the locknut to 174 ft. lbs. (235 Nm).

Land Cruiser

1989–91 MODELS

1. Raise and support the vehicle securely and remove the wheel assembly.
2. Remove the outer axle shaft flange cap, and then the shaft snapring on the outer shaft.
3. Remove the bolts retaining the outer axle shaft flange onto the front axle hub, then screw service bolts into the shaft flange alternately, and remove the shaft flange with its gasket.
4. Remove the brake drum set screws and remove the brake drum. If equipped with disc brakes, remove the caliper and disc.
5. Straighten the lockwasher, and remove the front wheel bearing adjusting nuts with front wheel adjusting nut wrench or similar tool.
6. Remove the front axle hub together with its claw washer, bearings, and oil seal.
7. Remove the clip and disconnect the brake flexible hose from the brake tube.
8. Cut and remove the lock wire and remove the bolts retaining the brake backing plate onto the steering knuckle. Remove the brake backing plate together wit the brake shoes, tension spring, and the wheel cylinder still assembled to the backing plate.
9. Tap the steering knuckle spindle lightly with a soft mallet, and remove the spindle with its gasket.

➡When removing the steering knuckle spindle on a vehicle equipped with the ball joint type axle shaft joint, be prepared for the disconnection of the outer axle shaft from the joint. The joint ball will fall from the joint. Try to cushion its fall or catch it if you can.

10. On those models equipped with the ball type axle shaft joint, slide the inner front axle shaft out of the axle housing. On those models equipped with the constant velocity joint type of axle, remove the entire axle shaft assembly from the axle housing.
11. Remove bushing from inside of knuckle spindle with a bearing puller. Install new bushing using a metal tube as a seating tool.
12. Remove the axle housing oil seal with a bearing puller. To install, use a metal tube as a seating tool.
13. Install the axle shaft in the reverse order of removal. On those models equipped with the ball joint type axle joint, install the inner axle with its proper spacer in position until the splines are fully meshed with the differential side gear splines. Next, fill the steering knuckle three quarters full with grease and place the joint ball on the inner shaft end. Install the outer shaft and the front axle shaft spacer into the steering knuckle spindle and install the spindle with its gasket onto the steering knuckle.
14. On those models equipped with the constant velocity joint axle, install the axle into the housing and rotate the axle shaft until its splines mesh with the splines in the differential. Fill the steering knuckle housing three quarters full with grease and install the steering knuckle spindle.
15. Adjust the wheel bearing preload.

1992–96 MODELS

1. Remove the front wheel.
2. Remove the front brake caliper.
3. Using a flat bladed too and a hammer, remove the grease cap from the flange.
4. Using a snapring expander, remove the snapring.
5. Loosen the 6 mounting nuts.
6. Using a brass bar and a hammer, tap on the bolt heads, then remove the 6 cone washers, plate washers and nuts.
7. Remove the flange and gasket.
8. Remove the axle hub and disc as follows:
   a. With a flat bladed tool, release the lockwasher.
   b. Remove the locknut, this may require SST 09607–60020 or an equivalent socket.
   c. Remove the lockwasher.
   d. Using SST 09607–60020 or an equivalent socket, remove the adjusting nut and thrust washer.
   e. Remove the hub and rotor together with the outer bearing.
9. Remove the oil seal and inner bearing from the hub.
10. Installation is the reverse of removal.
11. Adjust the preload, tighten the adjusting nut to 43 ft. lbs. (59 Nm). Turn the hub right and then left 2 or 3 times. Tighten the adjusting nut again to 43 ft. lbs. (59 Nm). Loosen the nut until it can be turned by hand. Tighten the nut again to 48 inch lbs. (5 Nm). Remeasure the preload again. It should be 6.4–12.6 lbs. (28–56 N).
12. Install a new lock washer and lock nut. Using SST 09607–60020, tighten the lock nut to 47 ft. lbs. (64 Nm). Check that the axle hub turns smoothly and the bearing has no play.
13. Using a spring tension gauge, measure the preload. It should be 6.4–12.6 lbs. (28–56 N). If the preload is not within specifications, adjust with the nut after removing the lockwasher and locknut. Secure the locknut by hand one of the lockwasher teeth inward and the other lockwasher teeth outward.

Pinion Seal

REMOVAL & INSTALLATION

Pick-Up and 4Runner

♦ See Figures 74, 75 and 76

1. Elevate and safely support the vehicle.
2. Drain the differential oil.
3. Make matchmarks and remove the propeller shaft.
4. Loosen the staked part of the companion flange nut. Use a counterhold tool on the flange and remove the nut.
5. Use a screw-type extractor to remove the companion flange.
6. Use an extractor to remove the oil seal and oil slinger.
7. Remove the bearing spacer, bearing and oil slinger.

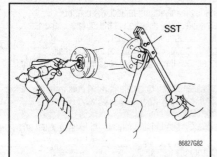

Fig. 74 Using a chisel and hammer, loosen the staked part of the nut. Hold the flange with SST 09950–30010 or equivalent and remove the nut

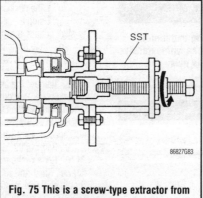

Fig. 75 This is a screw-type extractor from Toyota—Tool 09950–30010

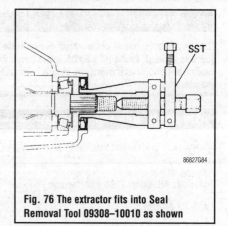

Fig. 76 The extractor fits into Seal Removal Tool 09308–10010 as shown

To install:

8. Install the new bearing spacer, rear bearing and oil slinger.

9. Install the new oil seal. Use the correct size driver to install the seal. Drive it in to a depth of 0.059 in. (1.5mm) below the lip.

10. Apply a coat of multi-purpose grease to the inner lip of the seal. Install the companion flange.

11. Coat the threads of a NEW nut with multi-purpose grease. Counterhold the flange and tighten the nut to 89 ft. lbs. (120 Nm).

12. Use a torque wrench to measure pinion bearing preload. Correct preload on a used bearing is 5–9 inch lbs. (0.6–1.0 Nm) and on a new bearing is 10–17 inch lbs. (1–2 Nm).

➡ **If preload is greater than specified, replace the bearing spacer. If preload is less than specified, tighten the companion flange nut in 9 ft. lb. (13 Nm) increments until the correct preload is achieved. Maximum torque for the nut is 165 ft. lbs. (223 Nm). If this value is exceeded, the bearing spacer must be replaced; do NOT back off the flange nut to lower the torque or preload.**

13. Check the run-out at the companion flange; maximum allowable run-out is 0.003 in. (0.10mm).

14. Stake the pinion flange nut.

T100 and Tacoma

1. Remove the engine undercover.
2. Drain the differential fluid.
3. Remove the front driveshaft.
4. Using a chisel and hammer, loosen the staked part of the nut. Hold the flange and remove the nut, then the flange.
5. With a seal extractor, remove the oil seal and slinger.
6. Pull the bearing from from the drive pinion, then extract the spacer.

To install:

7. Install the new bearing spacer, rear bearing and oil slinger.

8. Install the new oil seal. Use the correct size driver to install the seal. Drive it in to a depth of 0.059 in. (1.5mm) below the lip.

9. Apply a coat of multi-purpose grease to the inner lip of the seal. Install the companion flange.

10. Coat the threads of a NEW nut with multi-purpose grease. Counterhold the flange and tighten the nut to 89 ft. lbs. (120 Nm).

11. Use a torque wrench to measure pinion bearing preload. Correct preload on a used bearing is 5–9 inch lbs. (0.6–1.0 Nm) and on a new bearing is 10–17 inch lbs. (1–2 Nm).

➡ **If preload is greater than specified, replace the bearing spacer. If preload is less than specified, tighten the companion flange nut in 9 ft. lb. (13 Nm) increments until the correct preload is achieved. Maximum torque for the nut is 173 ft. lbs. (235 Nm). If this value is exceeded, the bearing spacer must be replaced; do NOT back off the flange nut to lower the torque or preload.**

12. Check the run-out at the companion flange; maximum allowable run-out is 0.003 in. (0.10mm).

13. Stake the pinion flange nut.

Land Cruiser

1. Elevate and safely support the vehicle.
2. Make matchmarks and disconnect the front propeller shaft.
3. Loosen the staked part of the companion flange nut. Use a counterhold tool on the flange and remove the nut.
4. Use screw type extractor to remove the companion flange.
5. Use an extractor to remove the oil seal and oil slinger.

To install:

6. Install the oil slinger and a new seal. Use the correct size driver to install the seal; drive it in to a depth of 0.04 in. (1.0mm) below the lip.

7. Apply a coat of multi-purpose grease to the inner lip of the seal. Install the companion flange.

8. Coat the threads of a NEW nut with multi-purpose grease. Counterhold the flange and tighten the nut to 145 ft. lbs. (196 Nm).

9. Use a torque wrench to measure pinion bearing preload. Correct preload is 4.3–6.9 inch lbs. (0.5–0.8 Nm).

10. If preload is greater than specified, replace the bearing spacer. If preload is less than specified, tighten the companion flange nut in 9 ft. lb. (13 Nm.) increments until the correct preload is achieved. Maximum torque for the nut is 253 ft. lbs. (343 Nm). If this value is exceeded, the bearing spacer must be replaced; do NOT back off the flange nut to lower torque or preload.

11. Check the run-out at the companion flange; maximum allowable run-out is 0.02 in. (0.10mm).

12. Stake the pinion flange nut.

REAR DRIVE AXLE

Understanding Drive Axles

The drive axle is a special type of transmission that reduces the speed of the drive from the engine and transmission and divides the power to the wheels. Power enters the axle from the driveshaft via the companion flange. The flange is mounted on the drive pinion shaft. The drive pinion shaft and gear which carry the power into the differential turn at the speed of the transmission output shaft. The gear on the end of the pinion shaft drives a large ring gear whose axis of rotation is 90 degrees away from the of the pinion. The pinion and gear reduce the gear ratio of the axle, and change the direction of rotation to turn the axle shafts which drive both wheels. The axle gear ratio is found by dividing the number of pinion gear teeth into the number of ring gear teeth.

The ring gear drives the differential case. The case provides the two mounting points for the ends of a pinion shaft on which are mounted two pinion gears. The pinion gears drive the two side gears, one of which is located on the inner end of each axle shaft.

By driving the axle shafts through this arrangement, the differential allows the outer drive wheel to turn faster than the inner drive wheel in a turn.

The main drive pinion and the side bearings, which bear the weight of the differential case, are shimmed to provide proper bearing preload, and to position the pinion and ring gears properly.

Axle Shaft, Bearing and Seal

REMOVAL & INSTALLATION

Pick-Ups, 4Runner, T100 and Tacoma

1989–95 MODELS

▶ See Figures 77, 78 and 79

1. Loosen the lug nuts on the wheel, then raise the truck and support it on jackstands.
2. Drain the axle housing.
3. Remove the lug nuts and remove the wheel.
4. Remove the brake drum securing screw, then the drum.

5. Remove the brake springs and the retracting spring clamp bolt. Remove the lower springs and shoe strut. Remove the brake shoes, screws, and the parking brake lever. Disengage the parking brake cable from the lever and the backing plate.
6. Disconnect the brake line from the wheel cylinder, being careful not to damage the fitting. Plug the brake line.
7. Remove the four nuts retaining the brake backing plate to the axle housing.
8. Remove the snapring and then press the axle shaft out of the backing plate and slide the assembly out of the housing.
9. Remove the oil seal and press out the axle bearing.
10. Installation is in the reverse order of removal.

1996 MODELS

▶ See Figure 80

1. Loosen the lug nuts on the rear wheel.
2. Raise and support the vehicle, then remove the wheel.
3. Remove the brake drum.
4. Check the bearing backlash and axle shaft deviation as follows:
   a. Using a dial indicator, check that the backlash in the bearing shaft direction. The maximum is 0.027 in. (0.7mm).
   b. If the backlash exceeds the maximum, replace the bearing.
   c. Using a dial indicator, check the deviation at the surface of the axle shaft outside the hub bolt. Maximum is 0.0039 in. (0.1mm).
   d. If the deviation exceeds the maximum, replace the axle shaft.
5. On models with ABS, remove the ABS speed sensor from the axle housing.
6. Remove the axle shaft assembly by unfastening the 4 nuts from the backing plate. Pull out the shaft assembly from the rear axle housing.
7. Remove the O-ring from the axle housing.
8. On ABS models, remove the bearing and retainer (differential side) and ABS speed sensor rotor.
9. Remove the snapring from the axle shaft.
10. Remove the rear axle shaft from the backing plate. Inspect the axle shaft run-out. Shaft run-out should be 0.079 in. (2.0mm) and the flange run-out should be 0.004 in. (0.1mm).

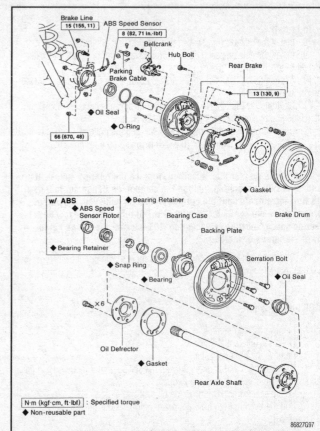

Fig. 80 Exploded view of the rear axle shaft and components—typical

Fig. 77 Remove the 4 bolts from the backing plate

Fig. 78 Slide the axle shaft from the housing

Fig. 79 Pull the O-ring from the outside of the housing

11. Remove the outer seal. Using a suitable bearing removal tool, separate the bearing from the axle shaft.

12. Installation is in the reverse order of removal. Tighten the backing plate to 48 ft. lbs. (66 Nm).

Land Cruiser

WITH DRUM BRAKES

▶ See Figure 81

1. Remove the wheel cover, if any, and loosen the wheel nuts.
2. Raise the rear axle housing with a jack and support the rear of the vehicle with jackstands.
3. Drain the oil from the differential.
4. Remove the wheel nuts and take off the wheels.
5. Remove the brake drum and related parts.

※※ CAUTION

Brake pads and shoes may contain asbestos, which has been determined to be a cancer causing agent. Never clean the brake surfaces with compressed air. Avoid inhaling any dust from brake surfaces. When cleaning brakes, use commercially available brake cleaning fluids.

6. On some models it will be necessary to remove the load sensing proportioning valve bracket from the rear cover. Remove the cover from the back of the differential housing.
7. Remove the pin from the differential pinion shaft.
8. Withdraw the pinion shaft and its spacer from the case.
9. Press the rear shaft toward the differential, to aid in the removal of the axle shaft C-clip or axle shaft lock.
10. Remove the C-clip.
11. Withdraw the axle shaft from the housing.
12. Repeat the removal procedure for the opposite side.
13. To remove oil seal and bearing, use a bearing puller and remove the axle bearing and oil seal together. To replace, use a metal tube to drive bearing and seal into seat.

➡**Do not mix the parts of the left and the right axle shaft assemblies.**

14. After installing the axle shaft, C-lock, spacer and pinion shaft, measure the clearance between the axle shaft and the pinion shaft spacer with a feeler gauge. The clearance should be no more than 0.0181 in. (5mm). If the clearance is not within specifications, the following spacers are available to adjust it;

- 1.142 in. (29.00mm)
- 1.157 in. (29.39mm)
- 1.173 in. (29.80mm)
- 1.189 in. (30.20mm)

- 1..205 in. (30.60mm)
15. Fill the axle with the proper lubricant.

WITH DISC BRAKES

▶ See Figures 82, 83 and 84

1. Loosen the lug nuts on the rear wheel. Raise and support the vehicle.
2. Remove the wheel.
3. Remove the 6 nuts and plate washers. Using a brass bar and hammer, strike the center part of the axle shaft to remove the cone washers.
4. Install 2 bolts into the axle holes. Gradually tighten the bolts evenly, then pull the axle shaft. Remove the bolts from the shaft. Remove the shaft and discard the gasket.
5. Remove the brake caliper and rotor.
6. Disconnect the ABS speed sensor.
7. Remove the two mounting screws from the rear axle bearing locknut. Remove the locknut, a SST 09509–25011 or an equivalent socket may be used.
8. Pull out the rear axle hub, locknut plate and outer bearing. Remove the axle hub.
9. Once the hub is removed, with a seal puller, remove the oil seal and inner bearing from the unit.
10. Using a brass bar and hammer, tap out the outer bearing races.

To install:

11. Using a special tool and a press, install the new outer races.
12. Pack multi-purpose grease into the bearing until it oozes out from the other side.
13. Coat the inside of the hub with grease.
14. Install the inner bearing and oil seal as follows:
   a. Place the inner bearing into the hub.
   b. Tap the seal into place with the seal installer.
   c. Apply multi-purpose grease to the oil seal lip.
15. Clean the hub installation position of the axle housing and apply clean lubricant (thinly).
16. Place the hub into position Be careful not to damage the axle seal.
17. Install the outer bearing.
18. Place the locknut plate on the axle housing, making sure the tongue lines up with the key groove. Temporarily install the locknut.
19. Install the rotor to the hub.
20. Adjust the preload as follows:
   a. Use a torque wrench tighten the bearing locknut to 43 ft. lbs. (59 Nm).
   b. Make the bearing snug by turning the hub several times. Retighten the locknut to 43 ft. lbs. (59 Nm).
   c. Using the torque wrench, loosen the nut until it can be turned by hand. Using a spring tension gauge, check that the preload, then tighten the nut until the preload is 6–13 lbs. (26–57 N).
   d. Align the mark on the bearing locknut and tip of the axle housing under the above preload range.

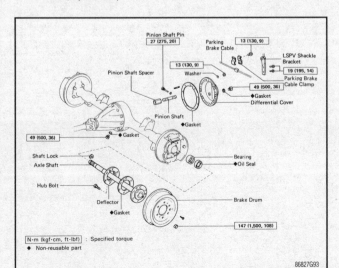

Fig. 81 Exploded view of the rear axle shaft and components—Land Cruiser with drum brakes

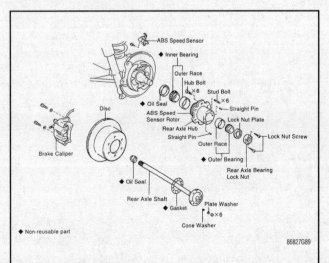

Fig. 82 Exploded view of the rear axle hub and components—Land Cruiser with disc brakes

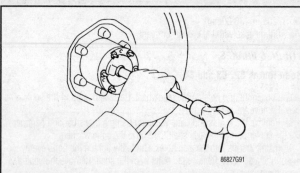

86827G91

Fig. 83 Remove the 6 set nuts and plate washers. Using a brass bar and hammer, strike the center part of the axle hub to remove the cone washers

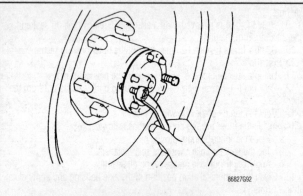

86827G92

Fig. 84 Install 2 bolts 180° apart into the axle holes

e. Check the distance between the top surface of the axle housing and the locknut. It should be 0.0079–0.0354 in. (0.2–0.9mm) below the surface of the axle housing. If the distance is greater than specified, reassemble the locknut plate.

f. Check that the hub with the rotor rotates smoothly and the hub has no axle play.

21. Install the bearing locknut screw, tighten to 48 inch lbs. (5 Nm).
22. Installation of the remaining components is in the reverse order.

Pinion Seal

REMOVAL & INSTALLATION

1. Elevate and safely support the vehicle.
2. Matchmark the flanges on the driveshaft and pinion. Disconnect the driveshaft from the differential.
3. Use a hammer and small chisel to loosen the staking on the companion flange bolt. Attach a counterholding tool to the companion flange and remove the nut. Use the proper threaded extractor to remove the companion flange.
4. Use an extractor to remove the oil seal and oil slinger from the housing.
To install:
5. Make certain the new seal faces in the correct direction. Install the new seal with a driver or seal installation tool of the correct diameter. Drive the new seal in until it is recessed 0.039 in. (1.0mm) in the housing.
6. Coat the inner lip of the seal with multi-purpose grease. Install the companion flange (and washer, if removed). Apply a light coat of oil to the threads of a NEW companion flange nut.
7. Set up the counterholding tool and install the companion flange nut. For Pick-Ups, 4Runners, T100s tighten it to 145 ft. lbs. (196 Nm). On Land Cruisers, tighten to 181 ft. lbs. (245 Nm). On Tacoma tighten to 109 ft. lbs. (147 Nm).
8. Check the drive pinion preload with a torque wrench. For Pick-Ups and 4Runners, correct torque is 8–11 inch lbs. (0.9–1.2 Nm) for 2-spider gear differentials and 4–7 inch lbs. (0.4–0.8 Nm) for 4-spider gear units. Land Cruisers should be between 6–9 inch lbs. (0.6–1.0 Nm).
9. If the preload is greater than specified, the bearing spacer must be removed and replaced inside the pinion housing. Do NOT back off the companion flange nut to relieve the pressure.If the preload is less than specified (the usual case), retighten the companion flange nut in increments of 9 ft. lbs. (13 Nm) until the correct specification is reached. Maximum allowable torque on the nut under any condition is 325 ft. lbs. (441 Nm) on Land Cruiser and Tacoma, 253 ft. lbs. (343 Nm) on Pick-Ups and 4Runner, and 109 ft. lbs. 333 ft. lbs. (451 Nm) on T100. Take great care not to exceed either the preload limit or the maximum torque on the nut.
10. Stake the drive pinion nut.
11. Connect the rear propeller shaft, aligning the matchmarks.
12. Lower the vehicle to the ground; inspect the fluid level in the axle.

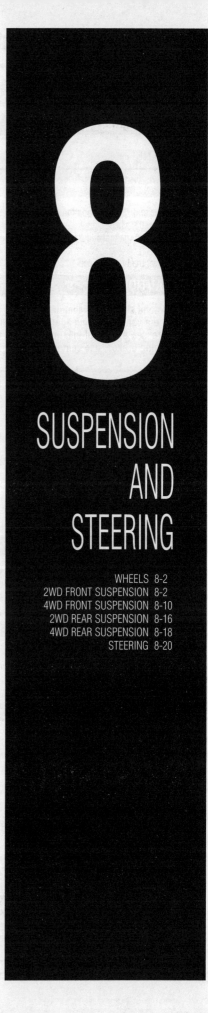

8

SUSPENSION AND STEERING

WHEELS

Wheel Lug Studs

REPLACEMENT

Front

1. Elevate and safely support the vehicle.
2. Remove the wheel.
3. Disconnect the brake line from the front caliper, and plug the line immediately.
4. Remove the brake caliper.

✳✳ CAUTION

Brake pads and shoes may contain asbestos, which has been determined to be a cancer causing agent. Never clean the brake surfaces with compressed air. Avoid inhaling any dust from brake surfaces. When cleaning brakes, use commercially available brake cleaning fluids.

5. For vehicles with free-wheeling hubs, remove the hub. For all other vehicles, remove the cap from the flange and remove the center bolt. On Land Cruisers, remove the snaping.
6. Remove the 6 mounting nuts.
7. Use a brass bar and hammer to tap on the bolt heads and remove the cone washers.
8. On Land Cruisers, pull out the flange. For other vehicles, install and tighten 2 bolts, using them to force the flange loose.
9. Remove the lockwasher from the hub.
10. Remove the locknut and the lockwasher.
11. Remove the adjusting nut.
12. Remove the axle hub with the brake disc attached.

✳✳ CAUTION

Brake pads and shoes may contain asbestos, which has been determined to be a cancer causing agent. Never clean the brake surfaces with compressed air. Avoid inhaling any dust from brake surfaces.

2WD FRONT SUSPENSION

After you work on your truck's suspension, it is advisable to have the alignment checked at a reputable repair facility. This will ensure that your front end is in order after repairs.

Coil Spring

REMOVAL & INSTALLATION

Tacoma

1. Remove the front wheel.
2. Remove the shock absorber.
3. Compress the spring using a spring compressor, see the manufacturer's instructions for the procedure.
4. Disconnect the stabilizer bar as follows:
   a. Remove the nut and stabilizer bar link from the lower control arm.
   b. Remove the 2 stabilizer bar bracket set bolts.
5. Remove the lower control arm and sway bar as follows:
   a. Support the upper control arm and steering knuckle securely.
   b. Support the upper control arm and steering knuckle assembly.
   c. Remove the cotter pin and nut.
   d. Using a ball joint extractor, disconnect the lower ball joint from the lower control arm.
   e. Loosen the lower control arm set bolt and remove the nut.
   f. Loosen the strut bar front set bolt, then remove the nut.

When cleaning brakes, use commercially available brake cleaning fluids.

13. Place matchmarks on the hub and brake disc. The units must be reinstalled in their exact relationship as before.
14. Remove the bolts holding the disc to the hub.
15. Install a lug nut on the lug. Using a brass bar and a press, remove the hub bolt.

To install:

16. Using a brass bar and a press, install the hub bolt. Reverse the disc and using a press install the new lug. Make sure it is firmly and completely seated in the disc.
17. Align the matchmarks and place the hub on the disc. Install the 6 bolts, tightening them to 47 ft. lbs. (64 Nm).
18. Installation of the remaining components is in the reverse order of removal.

Rear

1. Elevate and safely support the vehicle.
2. Remove the wheel.
3. Remove the brake drum.

✳✳ CAUTION

Brake pads and shoes may contain asbestos, which has been determined to be a cancer causing agent. Never clean the brake surfaces with compressed air. Avoid inhaling any dust from brake surfaces. When cleaning brakes, use commercially available brake cleaning fluids.

4. Use a lug extractor such as tool 09650–17011 or equivalent to press out the lug.

➡**This can be done with the axle and brakes in place if you're careful.**

To install:

5. Fit the new lug into the hole. Place a flat washer over the lug and install a lug nut on the new lug. Use a hand wrench to tighten the nut, drawing the lug into position. Remove the nut when the lug is fully seated.

   g. Pull out the bolts and remove the lower control arm along with the strut bar.
6. Remove the coil spring compressor tool and coil.

To install:

7. Place each end of the coil spring and lower control arm seat in contact when applying the coil spring expander.
8. Install the lower control arm and strut bar as follows:
   a. Attach the strut bar front set bolt, tighten to 221 ft. lbs. (300 Nm) make sure the suspension is stabilized prior to tightening the bolt.
   b. Install the lower control arm set bolt and nut, tighten to 148 ft. lbs. (200 Nm). Make sure the suspension is stabilized prior to tightening the bolt.
   c. Install the lower ball joint with a ball joint installer tool. Install the nut and cotter pin. Tighten the nut to 80 ft. lbs. (110 Nm).
9. Attach the stabilizer bar bracket set bolts, tighten to 22 ft. lbs. (29 Nm).
10. Install the stabilizer link to the lower control arm, tighten to 29 ft. lbs. (39 Nm).
11. Installation of the remaining components is in the reverse order of removal.

Torsion Bars

Torsion bars are used in place of coil springs on Pick-Ups and 4Runners. While employing the same torsional principal as a spring, the torsion bars are installed parallel to the ground instead of vertically as is a coil spring. This allows a much more compact suspension arrangement, lower front end styling and better suspension control since the torsion bar need not move through the greater distances traversed by a compression spring. The torsion bars are under tension as is any installed spring and great care must be taken when removing the bars.

REMOVAL & INSTALLATION

◆ See Figures 1 thru 12

Great care must be taken to make sure torsion bars are not mixed after removal, it is strongly suggested that before removal; each be marked with paint, showing the front and rear of spring and from which side of the truck it was taken. If they are installed backward or on the wrong sides of the truck, they could fracture. New units are marked L or R with an arrow showing direction of flex.

1. Raise the truck and support the frame on stands.
2. Slide the boot from the rear of torsion bar spring housing and onto spring.
3. Follow the same procedure on the front of the spring.
4. Paint matchmarks on the torsion bar spring, anchor arm and torque arm.
5. On the rear torsion bar spring holder, there is a long bolt that passes through the arm of the holder and up through the frame crossmember, using a small ruler, measure the length from the bottom of the retaining nut to the threaded tip of the bolt and record this measurement.
6. Loosen the adjusting nut and remove the anchor arm and torsion bar spring.
7. Remove the torque arm, there are tow nuts retaining it.
8. Inspect all parts for wear, damage or cracks. Check the boot for rips and wear. Inspect the splined ends of the torsion bar spring and the splined holes in the rear holder and the front torque arm for damage. Replace as necessary. On the rear end of the torsion bar springs, there are markings to show which is right and which is the left bar. Do not confuse them.

9. To install the existing spring:
   a. Coat the splined ends of the torsion bar with multipurpose grease.
   b. Align the matchmarks and install the torsion bar spring to the torque arm.
   c. Align the matchmarks and install the anchor arm to the torsion bar spring.
   d. Tighten the adjusting nut so that the bolt protrusion measured previously is equal to that of the new adjustment.
10. To install a new torsion bar spring:
   a. Install the 2 bolts to the torsion bar spring.
   b. Lightly coat the spring end splines with grease and then install the anchor arm to the small end of the spring temporarily. Paint matchmarks on the spring and arm.

➡There is one spline on the spring that is larger than the others. When connecting the spring to the arm, turn the arm slowly until you can feel the larger spline match up with the slot in the arm.

   c. Remove the anchor arm from the spring and install the spring into the torque arm.

➡There is one spline on the spring that is larger than the others. When connecting the spring to the arm, turn the arm slowly until you can feel the larger spline match up with the slot in the arm.

   d. Align the matchmarks and install the anchor arm to the spring. Tighten the adjusting nut so that the exposed thread on the bolt is no greater than 87mm.
11. Install the locknut and wheels, lower the truck and bounce it several times to set the suspension.
12. Measure the ground clearance and adjust with the adjusting nut. Tighten the locknut.
13. Have the alignment checked at a reputable repair facility.

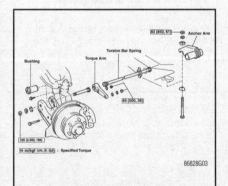

Fig. 1 Exploded view of the torsion bar suspension and associated components

Fig. 2 Slide the boot down the shaft at the rear of the unit

Fig. 3 Matchmark the adjoining components

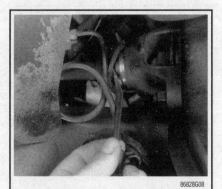

Fig. 4 With the aid of a small prytool, separate the boot from the torsion bar

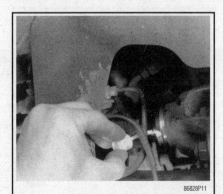

Fig. 5 Use paint to matchmark the front also

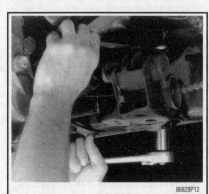

Fig. 6 Loosen the long bolt that passes through the arm

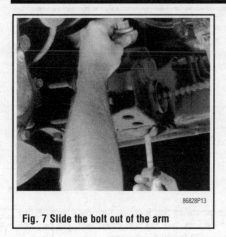

Fig. 7 Slide the bolt out of the arm

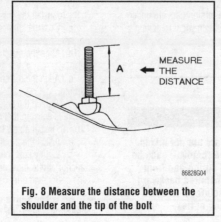

MEASURE THE DISTANCE

Fig. 8 Measure the distance between the shoulder and the tip of the bolt

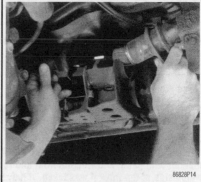

Fig. 9 The bar should slide from the housing

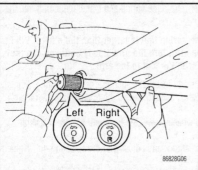

Left Right

Fig. 10 When you remove the torsion bar, note that some models have markings for the left and right

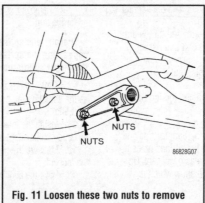

NUTS NUTS

NUTS

Fig. 11 Loosen these two nuts to remove the anchor arm

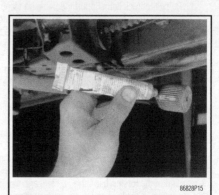

Fig. 12 Coat the splined ends of the torsion bar with multi-purpose grease

Shock Absorbers

REMOVAL & INSTALLATION

▶ **See Figures 13 thru 18**

1. Remove the hubcap and loosen the lug nuts.
2. Raise front of the truck and support it with safety stands.
3. Remove the wheel.
4. Unfasten the double nuts at the top end of the shock absorber. Remove the cushions and cushion retainers.
5. Remove the 2 bolts which secure the lower end of the shock absorber to the lower control arm.
6. Remove the shock absorber.
7. Install the shock absorber, then tighten the lower mounting nuts/bolt to 13 ft. lbs. (18 Nm) on all models except the 1996 4Runner. On the 1996 4Runner, tighten to 101 ft. lbs. (135 Nm). Tighten the upper mounting nut to 18 ft. lbs. (25 Nm).

TESTING

The purpose of the shock absorber is simply to limit the motion of the spring during compression (bump) and rebound cycles. If the vehicle were not equipped with these motion dampers, the up and down motion of the vehicle would multiply until the vehicle was alternately trying to leap off the ground and to pound itself into the pavement.

Contrary to popular rumor, the shocks do not affect the ride height of the vehicle, nor do they affect the ride quality except for limiting the pitch or bounce. These factors are controlled by other suspension components such as springs and tires. Worn shock absorbers can affect handling; if the front of the vehicle is rising or falling excessively, the "footprint" of the tires changes on the

Fig. 13 When the wheel is off and the truck raised, you have better access to the suspension

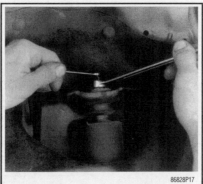

Fig. 14 Remove the upper mounting nut for the shock

Fig. 15 With two tools and a good grip, loosen the mounting bolt and nut

Fig. 16 A nut and washer are on the one side when the bolt is removed

Fig. 17 Pull the shock out of the mounting bracket, then slide the unit down and out

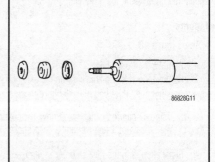
86828G11
Fig. 18 Place the retainer, cushion and other retainer on the shock in the correct order and position

pavement and steering response is affected. The simplest test of the shock absorbers is simply to push down on one corner of the unladen vehicle and release it.

Observe the motion of the body as it is released. In most cases, it will come up beyond its original rest position, dip back below it and settle quickly to rest. This shows that the damper is slowing and controlling the spring action. Any tendency to excessive pitch (up-and-down) motion or failure to return to rest within 2–3 cycles is a sign of poor function within the shock absorber.

While each shock absorber can be replaced individually, it is recommended that they be changed as a pair (both front or both rear) to maintain equal response on both sides of the vehicle. Chances are quite good that if one has failed, its mate is weak also.

INSPECTION

Once removed, the shock should be carefully inspected for any signs of leakage. Oil-filled shocks may have a light film of oil around the seal, resulting from normal breathing and air exchange. This should NOT be taken as a sign of failure, but any sign of thick or running oil definitely indicates failure. Gas filled shocks may also show some film at the shaft; if the gas has leaked out, the shock will have almost no resistance to motion.

Hold the shock firmly in each hand and compress it; compression should be reasonably even and smooth. Release the piston end; it should return smoothly and at an even pace. Stretch the shock, checking again for smooth motion and any abnormal resistance.

Lower Ball Joint

INSPECTION

To check the lower ball joint for wear, jack up the front of the vehicle and support it with stands. Remove excess play from the other suspension parts (wheel bearings, tie rods, etc.). Have an assistant press the brake pedal; Move the lower suspension arm up and down, checking that the lower ball joint has minimal play, if any. The maximum allowable vertical play is 0.079 in. (2.3mm). If the play is greater, replace the ball joint. The upper ball joint is subject to the same limit on trucks; the 4Runner upper ball joint must have no play at all. The upper ball joint is tested by moving the wheel and tire up and down.

REMOVAL & INSTALLATION

Pick-Up and 4Runner

1989–95 MODELS

1. Raise the vehicle and support it with jackstands.
2. Remove the front wheel.
3. Support the lower control arm with a floor jack and remove the steering knuckle.
4. Loosen the mounting bolts and remove the lower ball joint from the lower control arm.

5. Attach the lower ball joint and tighten the nut to 94 ft. lbs. (127 Nm).
6. Install the steering knuckle. Install the wheels and lower the truck.
7. Have the alignment checked at a reputable repair facility.

Tacoma and 1996 4Runner

1. Remove the front wheel.
2. Support he lower control arm with a suitable jack.
3. On the lower ball joint, separate the tie rod end from the steering knuckle. Throw out the old cotter pin.
4. Remove the cotter pin and nut from the lower ball joint. Discard the cotter pin.
5. With a ball joint remover, separate the ball joint from the lower control arm.
6. Install the ball joint with the mounting nut and bolts.
7. Tighten the nut to 80 ft. lbs. (108 Nm) on the Tacoma, 105 ft. lbs. (142 Nm) 4Runner. Install a new cotter pin in the hole.
8. Attach the tie rod end tightening the nut to 66 ft. lbs. (90 Nm) on 4Runner and 53 ft. lbs. (72 Nm) Tacoma.
9. Install a new cotter pin.
10. Tighten the lower ball joint set bolts, 59 ft. lbs. (80 Nm) on 4Runner and 116 ft. lbs. (160 Nm) on Tacoma.

T100

1. Raise the vehicle and support it with jackstands.
2. Remove the front wheel.
3. Support the lower control arm with a floor jack and remove the steering knuckle.
4. Loosen the mounting bolts and remove the lower ball joint from the lower control arm.
5. Attach the lower ball joint and tighten the nut to 55 ft. lbs. (75 Nm).
6. Install the steering knuckle.
7. Install the wheels and lower the truck.
8. Check the ABS speed sensor signal and front wheel alignment.

Upper Ball Joint

REMOVAL & INSTALLATION

Pick-Up and 4Runner

1989–95 MODELS

1. Raise the vehicle and support it with jackstands.
2. Remove the front wheel.
3. Support the lower control arm with a floor jack and remove the cotter pin and castellated nut from the knuckle.
4. Unbolt the upper ball joint from the upper control arm and remove the ball joint from the vehicle.
5. Install the upper ball joint and tighten the bolts to 23 ft. lbs. (31 Nm) for the Pick-Up, or 25 ft. lbs. (33 Nm) for the 4Runner.

6. Installation is the reverse of removal. Attach the lower ball joint and tighten the nut to 94 ft. lbs. (127 Nm).

Tacoma

▶ See Figure 19

1. Remove the front wheel.
2. Remove the steering knuckle with the axle hub.
3. Separate the upper ball joint as follows:
    a. Remove the wire and boot.
    b. Using a snapring expander, remove the snapring.
    c. With a ball joint remover and a deep socket, remove the upper ball joint.
4. Installation is in the reverse order of removal.

Strut Bar

REMOVAL & INSTALLATION

Pick-Up and T100

1. Remove the wheels, raise the truck and support it on safety stands.
2. On some models it will be necessary to remove the engine under cover.
3. Paint matchmarks on the strut bar and retaining nut.
4. Remove the retaining nut from the front of the bar.
5. Remove the nuts holding the bar to the lower control arm and lift out the strut bar.
6. Position the bar and install the front nut so that the matchmarks line up.
7. Slide the washer and bushing onto the bar and install it to the bracket at the rear.
8. Install the collar, bushing and washer to the other side and then connect the rear of the bar to the control arm. Tighten the mounting nut to 70 ft. lbs. (95 Nm) for the Pick-Up, or 55 ft. lbs. (75 Nm) for the T100.
9. Lower the truck and bounce it several times to set the suspension.
10. Tighten the front nut to 90 ft. lbs. (123 Nm) and check the alignment.

Tacoma

▶ See Figure 19

1. Remove the front wheel.
2. Remove the shock absorber.
3. Compress the spring using a spring compressor, according to the tool manufacturer's instructions.
4. Disconnect the stabilizer bar as follows:
    a. Remove the nut and stabilizer bar link from the lower control arm.
    b. Remove the 2 stabilizer bar bracket set bolts.
5. Remove the lower control arm and strut bar as follows:
    a. Support the upper control arm and steering knuckle securely.
    b. Support the upper control arm and steering knuckle assembly.
    c. Remove the cotter pin and nut.
    d. Using a ball joint extractor, disconnect the lower ball joint from the lower control arm.
    e. Loosen the lower control arm set bolt and remove the nut.
    f. Loosen the strut bar front set bolt, then remove the nut.
    g. Pull out the bolts and remove the lower control arm along with the strut bar.
To install:
6. Install the lower control arm and strut bar as follows:
    a. Attach the strut bar front set bolt, tighten to 221 ft. lbs. (300 Nm) make sure the suspension is stabilized prior to tightening the bolt.
    b. Install the lower control arm set bolt and nut, tighten to 148 ft. lbs. (200 Nm). Make sure the suspension is stabilized prior to tightening the bolt.
    c. Install the lower ball joint with a ball joint installer tool. Install the nut and cotter pin. Tighten the nut to 80 ft. lbs. (110 Nm).
7. Attach the stabilizer bar bracket set bolts, tighten to 22 ft. lbs. (29 Nm).
8. Install the stabilizer link to the lower control arm, tighten to 29 ft. lbs. (39 Nm).
9. Installation of the remaining components is in the reverse order of removal.

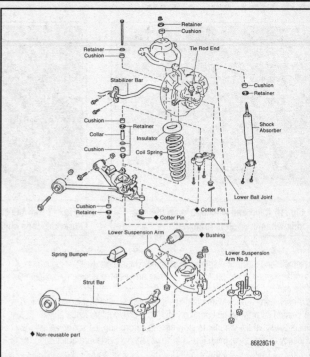

Fig. 19 Exploded view of the strut bar and associated components—Tacoma

Stabilizer Bar (Sway Bar)

REMOVAL & INSTALLATION

Pick-Up

1. Raise the truck and support it on safety stands. Remove the wheels.
2. Remove one torsion bar spring.
3. Remove the nuts and cushions at both sides of the stabilizer bar where it attaches to the control arms.
4. Remove the 2 bushings and brackets at the frame member. Remove the stabilizer bar.

➡ **Examine the rubber bushings very carefully for any splits or deformation. Clean the inner and outer surfaces of the bushings before reinstallation. Failed or dirty bushings can create a chorus of odd noises under the vehicle, particularly during cornering.**

5. Position the stabilizer bar and install the 2 center bushings and brackets to the frame. Finger-tighten the bolts.
6. Connect the bar to the control arms on each side. Use new nuts and tighten to 9 ft. lbs. (13 Nm).
7. Tighten the center bracket bolts to 9 ft. lbs. (13 Nm) and install the torsion bar.
8. Install the wheels and lower the truck.

Tacoma

1. Remove the front wheels.
2. Hold the bolt with a wrench, then remove the nut, retainer, collar and cushion from the lower control arm.
3. Remove the bolts and stabilizer bar with the cushions and brackets.
4. Remove the brackets and cushions from the stabilizer bar.
5. Install the cushions to the inside of the paint mark. Attach the bar with the brackets, tighten the bracket bolts to 22 ft. lbs. (29 Nm).
6. Tighten the nut holding the lower control arm components to 29 ft. lbs. (39 Nm).
7. Install the wheel, lower the vehicle.

4Runner

1. Remove the front wheels.
2. Disconnect the stabilizer bar links as follows:
   a. Remove the nuts and disconnect the stabilizer bar links from the lower control arm.
   b. Hold the link with a wrench, then remove the nut, retainers, cushions and link.
3. Remove the bolts and stabilizer bar with cushions and brackets.
4. Place the stabilizer bar into position, then install the both bar cushions and brackets to the frame. Temporarily install the bolts. Tighten to 19–22 ft. lbs. (25–29 Nm).
5. Hold the stabilizer bar link with a wrench, then install the link onto the lower control arm with a new nut. Tighten the nut to 14–19 ft. lbs. (19–25 Nm).
6. Using a hexagon wrench, connect the stabilizer bar on both sides to the links with new nuts, tighten to 55–70 ft. lbs. (69–95 Nm).

Upper Control Arm

REMOVAL & INSTALLATION

Pick-Up

1. Raise and support the truck under the frame.
2. Remove the wheel.
3. Raise the lower control arm with a floor jack.
4. Remove the 4 nuts from the upper ball joint and disconnect the joint from the upper arm.
5. Unbolt and remove the upper arm at the 2 bolts holding the inner shaft to the frame, taking note of the number, size and position of the aligning shim.
6. Install the upper arm with the adjusting shims in the same position and the same number as they were before removal. Tighten the bolts to 71 ft. lbs. (96 Nm).
7. Install the upper ball joint to the arm and tighten the bolts to 23 ft. lbs. (31 Nm).
8. Install the wheel.
9. Lower the truck and bounce it several times to set the suspension. Tighten the upper control arm shaft bolts to 93 ft. lbs. (126 Nm).

Tacoma

1. Raise and support the vehicle.
2. Remove the front wheel.
3. Removed the ABS sped sensor and wire harness.
4. Remove the stabilizer bar link.
5. Disconnect the steering knuckle from the upper bar joint.
6. Loosen the 2 bolts,m then remove the front and rear alignment adjusting shims.
7. Make not of the number and thickness of the front and rear shims.
8. Unbolt the upper control arm.
9. Remove the bolts, nuts and upper ball joint from the arm.
10. Attach the ball joint to the arm, tighten to 29 ft. lbs. (39 Nm).

➡**Do not lose the camber adjusting shims. Record the position and thickness of the camber shims so that these can be reinstalled to there original locations. Install the equal number and thickness of shims into there locations.**

11. Install the arm to the vehicle, with the shims, then tighten the mounting bolts to 94 ft. lbs. (130 Nm).
12. Attach the steering knuckle to the upper ball joint.
13. Install the stabilizer bar link.
14. Install the ABS speed sensor and wire harness.
15. Place the wheel into position, hand tighten the lug nuts.
16. Lower the vehicle, tighten the lug nuts to 83 ft. lbs. (110 Nm).

T100

1. Raise and support the vehicle.
2. Remove the front brake caliper.
3. On models with ABS, disconnect the ABS speed sensor wire from the upper control arm.

4. Support the lower control arm with a jack. Remove the nuts and disconnect ball joint from the upper control arm.
5. Remove the bolts and camber adjusting shims.
6. Remove the upper control arm.

➡**Do not lose the camber adjusting shims. Record the position and thickness of the camber shims so that these can be reinstalled to there original locations. Install the equal number and thickness of shims into there locations.**

7. Attach the upper control arm with the shims in there correct positions.
8. Tighten the bolts to 71 ft. lbs. (96 Nm).
9. Install the nuts to the upper control arm, tighten to 23 ft. lbs. (31 Nm).
10. Attach the ABS speed sensor and harness to the upper control arm.
11. Install the brake caliper.
12. Install the wheel, lower the vehicle.

Lower Control Arm

REMOVAL & INSTALLATION

Pick-Up

1. Raise the truck and support it on safety stands. Remove the wheels.
2. Remove the cotter pin and nut; use the correct tool to separate the tie rod end.
3. Remove the torsion bar spring.
4. Remove the shock absorber.
5. Disconnect the stabilizer bar from the lower arm.
6. Disconnect the strut bar from the lower arm.
7. Unbolt and remove lower ball joint.
8. Remove the lower suspension arm.

➡**If the lower ball joint is not to be replaced, simply unbolt it from the lower control arm. It is not necessary to separate the ball joint from the steering knuckle.**

To install:

9. Install the torque arm bolts to the lower arm. Place the torque arm on the lower shaft.
10. Set the torque arm in its installation position, install the lower arm shaft and the torque arm. Finger-tighten the lower arm bolts and then remove the torque arm. Do not tighten the nut much beyond snug.
11. Attach the lower ball joint to the suspension arm. Tighten the 3 bolts to 94 ft. lbs. (127 Nm).
12. Install the strut bar to the lower arm and connect the stabilizer bar to the arm.
13. Install the shock absorber. Tighten the lower mount to 13 ft. lbs. (18 Nm) and the upper mount to 18 ft. lbs. (25 Nm).
14. Attach the tie rod to the knuckle. Tighten the nut to 67 ft. lbs. (90 Nm). Install a new cotter pin to secure the nut.
15. Install the torsion bar.
16. Install the wheel(s) and lower the vehicle. Bounce the front end several times to stabilize the suspension.
17. Tighten the lower suspension nut to 166 ft. lbs. (226 Nm.)

✳✳ WARNING

Do not tighten the control arm bolt fully until the vehicle is lowered. If the bolts are tightened with the control arm(s) hanging, excessive bushing wear will result.

4Runner

1989–95 MODELS

1. Remove the shock absorber.
2. Disconnect the stabilizer bar from the lower control arm.
3. Disconnect the lower control arm from the lower ball joint. Discard the old cotter pin. A ball joint separator will be needed.
4. Place matchmarks on the front and rear adjusting cams.
5. Remove the nuts and adjusting cams, then remove the lower control arm.

6. Install the lower control arm to the frame with the adjusting cams.

7. Install the two nuts to the front and rear adjusting cams.

8. Connect the lower control arm to the lower ball joint, then tighten the nut to 105 ft. lbs. (142 Nm).

9. Connect the stabilizer bar link to lower control arm with the cushions, retainers and nuts. Tighten to 19 ft. lbs. (25 Nm).

10. Install the shock absorber to the lower control arm, tighten to 101 ft. lbs. (137 Nm).

11. Install the wheel, then remove the stands and bounce the vehicle up and down to stabilize the suspension. Align the matchmarks, then tighten the nuts to 145 ft. lbs. (196 Nm).

1996 MODELS

1. Remove the front wheel.

2. Remove the steering gear assembly.

3. Disconnect the stabilizer bar link.

4. Unbolt the shock absorber from the lower control arm.

5. Support the upper control arm and the steering knuckle securely.

6. Remove the cotter pin and nut from the lower ball joint.

7. Using a ball joint remover, disconnect the lower ball joint from the control arm.

8. Place matchmarks on the front and rear adjusting cams. Remove the bolts, nuts, adjusting cams and lower control arm.

9. Using tool SST 09922–10010 or equivalent, remove the 2 spring bumpers.

10. Install in the reverse order. Tighten the spring bumpers to 17 ft. lbs. (23 Nm), the lower control arm to 96 ft. lbs. (130 Nm), and the lower ball joint nut to 105 ft. lbs. (142 Nm).

Tacoma

▶ **See Figure 20**

1. Remove the front wheel.

2. Remove the shock absorber.

3. Compress the spring using a spring compressor, following the manufacturer's instructions.

4. Disconnect the stabilizer bar as follows:

   a. Remove the nut and stabilizer bar link from the lower control arm.

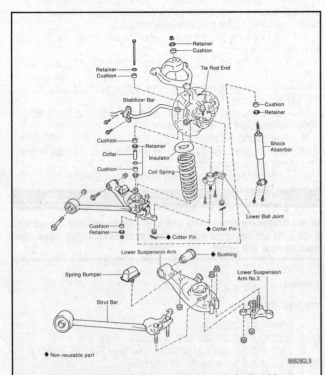

Fig. 20 Exploded view of the lower control arm and related front suspension components—Tacoma

b. Remove the 2 stabilizer bar bracket set bolts.

5. Remove the lower control arm and strut bar as follows:

   a. Support the upper control arm and steering knuckle securely.

   b. Support the upper control arm and steering knuckle assembly.

   c. Remove the cotter pin and nut.

   d. Using a ball joint extractor, disconnect the lower ball joint from the lower control arm.

   e. Loosen the lower control arm set bolt and remove the nut.

   f. Loosen the strut bar front set bolt, then remove the nut.

   g. Pull out the bolts and remove the lower control arm along with the strut bar.

6. Remove the coil spring compressor tool and coil.

7. Remove the nuts and strut bar from the lower control arm. Separate the nut and spring bumper.

8. Remove the lower suspension arm No. 3.

9. Attach the lower suspension arm No. 3 to the vehicle, tighten the mounting bolts to 111 ft. lbs. (150 Nm).

10. Install the spring bumper, tighten to 32 ft. lbs. (43 Nm) and the strut bar to the lower control arm 111 ft. lbs. (150 Nm).

11. Place each end of the coil spring and lower control arm seat in contact when applying the coil spring expander.

12. Install the lower control arm and strut bar as follows:

   a. Attach the strut bar front set bolt, tighten to 221 ft. lbs. (300 Nm) make sure the suspension is stabilized prior to tightening the bolt.

   b. Install the lower control arm set bolt and nut, tighten to 148 ft. lbs. (200 Nm). Make sure the suspension is stabilized prior to tightening the bolt.

   c. Install the lower ball joint with a ball joint installer tool. Install the nut and cotter pin. Tighten the nut to 80 ft. lbs. (110 Nm).

13. Attach the stabilizer bar bracket set bolts, tighten to 22 ft. lbs. (29 Nm).

14. Install the stabilizer link to the lower control arm, tighten to 29 ft. lbs. (39 Nm).

T100

1. Remove the wheel.

2. Remove the engine under cover.

3. Remove the torsion bar spring.

4. Disconnect the shock absorber from the lower control arm.

5. Separate the stabilizer bar from the control arm.

6. Separate the strut bar from the lower control arm.

7. Disconnect the lower ball joint from the lower control arm.

8. Remove the nut and lower the control arm.

9. Attach the lower control arm but do not tighten yet.

10. Attach the lower ball joint, strut bar and stabilizer bar to the control arm.

11. Tighten the lower control arm bolt to 152 ft. lbs. (206 Nm).

12. Install the shock to the control arm.

13. Attach the torsion spring bar.

14. Install the engine under cover.

15. Install the wheel, lower the truck and tighten the lug nuts to 76 ft. lbs. (103 Nm).

Knuckle and Spindle

REMOVAL & INSTALLATION

Pick-Up

1. Raise the front of the truck and support with safety stands. Remove the wheels.

2. Remove the brake caliper from the mount but do not disconnect the brake line. Use stiff wire to support the caliper out of the way. Remove the brake rotor and hub.

✳✳ CAUTION

Brake pads and shoes may contain asbestos, which has been determined to be a cancer causing agent. Never clean the brake surfaces with compressed air. Avoid inhaling any dust from brake surfaces. When cleaning brakes, use commercially available brake cleaning fluids.

3. Remove the 2 cotter pins and bolts and remove the dust cover.

4. Disconnect the knuckle arm from the steering knuckle.

5. Position a floor jack under the lower control arm and disconnect the ball joints.

6. Remove the steering knuckle.

7. Installation is the reverse of removal. Tighten the upper knuckle nut to 80 ft. lbs. (108 Nm). Tighten the lower nut to 105 ft. lbs. (142 Nm). Install new cotter pins.

8. Tighten the dust cover bolts to 80 ft. lbs. (108 Nm) and secure them with new cotter pins.

4Runner

1989–95 MODELS

1. Raise the front of the truck and support with safety stands. Remove the wheels.

2. Remove the brake caliper from the mount but do not disconnect the brake line. Use stiff wire to support the caliper out of the way. Remove the brake rotor and hub.

3. Remove the 2 cotter pins and bolts and remove the dust cover.

4. Disconnect the knuckle arm from the steering knuckle.

5. Disconnect the shock absorber from the lower control arm.

6. Disconnect the stabilizer bar from the lower control arm by removing the nut and cushion. Disconnect the stabilizer bar link from the suspension arm.

7. Remove the cotter pin and nut from the upper ball joint.

8. With a ball joint separator, disconnect the steering knuckle from the upper ball joint.

9. Remove the bolts from the lower ball joint. Disconnect the steering knuckle from the joint. Push the lower control arm down and remove the knuckle.

10. Installation is in the reverse order.

11. Connect the lower ball joint to the steering knuckle, then tighten the bolts to 43 ft. lbs. (58 Nm).

12. Attach the upper ball joint to the knuckle, then tighten the nut to 105 ft. lbs. (142 Nm). Install all new cotter pins.

13. Connect the knuckle arm to the steering knuckle as follows:

a. Clean the threads of the bolts and steering knuckle with trichloroethylene.

b. Apply sealant to the bolt threads such as Three Bond 1324 or equivalent. Tighten the bolts to 135 ft. lbs. (183 Nm).

T100

▶ **See Figure 21**

1. Remove the front wheel.

2. On models with ABS, remove the speed sensor from the steering knuckle.

3. Remove the caliper.

4. Check the axle hub bearing backlash as follows:

a. Remove the cap, cotter pin and lock cap.

b. Place a dial indicator near the center of the axle hub and check that the backlash in the bearing shaft direction.

c. The maximum is 0.0020 in. (0.05mm). If the backlash is not within specifications, replace the bearing.

5. Remove the hub with the rotor.

6. Remove the oil seal and inner bearing.

7. Remove the dust cover.

8. Unbolt the knuckle arm from the steering knuckle.

9. Support the lower control arm with a suitable jack.

10. Remove the upper and lower cotter pins.

11. Remove the upper and lower nuts.

12. With a ball joint remover, separate the steering knuckle from the upper and lower ball joints.

13. Remove the knuckle.

14. Installation is in the reverse order. Tighten the upper ball joint nut to 80 ft. lbs. (108 Nm).

15. Tighten the lower ball joint nut to 105 ft. lbs. (142 Nm). Install new cotter pins.

16. Attach the knuckle arm to the steering knuckle as follows:

a. Clean the threads of the bolts and steering knuckle with trichloroethylene.

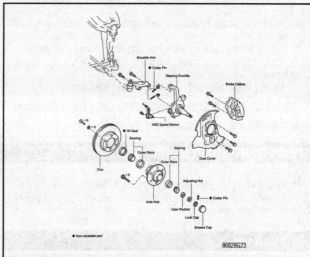

Fig. 21 Exploded view of the front hub and steering knuckle and other suspension components—T100

b. Apply sealant to the bolt threads such as Three Bond 1324 or equivalent. Tighten the bolts to 135 ft. lbs. (183 Nm).

17. Install the dust cover, tighten the bolts to 14 ft. lbs. (19 Nm).

Tacoma

1. Remove the front wheel.

2. On models with ABS, remove the speed sensor from the steering knuckle.

3. Remove the caliper.

4. Check the axle hub bearing backlash as follows:

a. Remove the cap, cotter pin and lock cap.

b. Place a dial indicator near the center of the axle hub and check that the backlash in the bearing shaft direction.

c. The maximum is 0.0020 in. (0.05mm). If the backlash is not within specifications, replace the bearing.

5. Remove the hub with the rotor.

6. Remove the oil seal and inner bearing.

7. Remove the dust cover.

8. Remove the stabilizer bar link.

9. Support the lower control arm with a jack.

10. Loosen the ball joint set bolts. Remove the cotter pin and discard it, then loosen the nut.

11. With a special tool 09628–62011 or its equivalent, remove the ball joint from the steering knuckle.

12. Remove the ball joint set bolts, then the nut and steering knuckle.

13. Installation is in the reverse order of removal. Tighten the upper ball joint nut to 80 ft. lbs. (110 Nm). Tighten the lower ball joint nut to 116 ft. lbs. (160 Nm). Install new cotter pins.

14. Install the dust cover, tighten the bolts to 14 ft. lbs. (19 Nm).

15. Pack the bearings with multi-purpose grease and coat the inside of the hub and cap.

16. Attach the knuckle arm to the steering knuckle as follows:

a. Clean the threads of the bolts and steering knuckle with trichloroethylene.

b. Apply sealant to the bolt threads such as Three Bond 1324 or equivalent. Tighten the bolts to 135 ft. lbs. (183 Nm).

Front Axle Hub and Bearing

REMOVAL & INSTALLATION

1. Raise the front of the truck and support with safety stands. Remove the wheels.

2. Remove the brake caliper and suspend it with wire, out of the way. Remove the caliper mount bracket (torque plate).

3. Remove the axle end cap and then remove the cotter pin, nut lock and nut.

4. Pull the hub/disc assembly off the spindle with the outer bearing. Don't let the bearing fall out.

5. Pry the inner oil seal out and remove the inner bearing from the hub.

6. Clean the bearings and outer races and inspect them for wear or cracks.

To install:

7. Using a brass drift and a hammer, drive out the bearing outer race. Press a new one into position.

8. Pack the bearings with grease until it oozes out the other side. Coat the inside of the hub and cap with grease.

9. Position the inner bearing into the hub, coat the oil seal with grease and press it into the hub.

10. Press the hub assembly onto the spindle and install the outer bearing and thrust washer.

11. Install the hub nut and tighten. Turn the hub a few times to seat the bearings, then loosen the nut until there is no more than 0.02 in. (0.5mm) play. Using a spring tension gauge, check that the preload is 1.3–4.0 lbs.

12. Install the locknut, new cotter pin and hub grease cap.

13. Install the brake torque plate to the knuckle. Install the brake caliper.

14. Bleed the brake system.

15. Install the wheels and lower the vehicle.

4WD FRONT SUSPENSION

After you work on the suspension of your truck, it is advisable to have the alignment checked at a reputable repair facility. This will ensure that your front end is in order after repairs.

Coil Springs

REMOVAL & INSTALLATION

Land Cruiser

1. Elevate and safely support the vehicle. Make certain the jackstands are securely placed under the frame rails.

2. Use a floor jack to slightly jack and support the front axle housing. Hold the piston rod of the shock absorber and remove the upper mounting nut. Hold the shock and remove the lower mounting nut, the cushions and the retainer.

3. Disconnect the stabilizer bar from the axle housing.

4. Lower the jack holding the front axle. Use a spring compressor to compress the spring. Remove the spring, still compressed, from the vehicle. Place the spring, facing away from you, on the work bench and slowly release the tension on the compressor. When the spring is fully extended, remove the tool.

5. Remove the follow spring nuts and spring.

6. Installation is in the reverse of removal. Tighten the follow spring nuts to 82 inch lbs. (9 Nm).

Tacoma

▶ See Figures 22, 23 and 24

1. Loosen the lug nuts.

2. Elevate and safely support the vehicle. Make certain the jackstands are securely placed under the frame rails.

3. Use a floor jack to slightly jack and support the front axle housing.

4. Remove the front wheel.

5. Disconnect the shock absorber from the lower control arm as follows:

a. Loosen the bolt, then remove the shock absorber lower nut.

b. While lowering the lower control arm, remove the bolt then disconnect the shock.

6. Remove the shock mounting nuts and coil spring as an assembly.

7. Using a spring compressor, compress the coil until there is a clearance at both ends. Remove the suspension support center nut. Remove the suspension support and coil spring.

8. Remove the insulator from the suspension support.

9. Remove the compressor tool.

10. Installation is the reverse of removal. Tighten the center nut to 22 ft. lbs. (29 Nm). Attach the coil and shock assembly to the top portion of the engine compartment. Tighten the nuts to 47 ft. lbs. (64 Nm). Attach the shock to the lower control arm, then tighten the mounting bolt to 101 ft. lbs. (135 Nm).

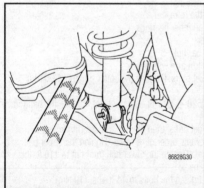

86828G30

Fig. 22. Loosen the bolt at the bottom of the shock absorber

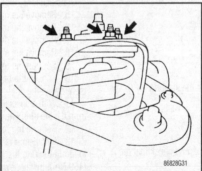

86828G31

Fig. 23 The three nuts at the top of the coil spring and shock absorber assembly are accessed through the engine compartment

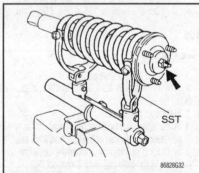

86828G32

Fig. 24 A spring compressor is a must for this job; follow the manufacturer's instructions

4Runner

1. Remove the front wheel.

2. Disconnect the shock absorber from the the lower control arm. When you lower the arm, remove the bolt and disconnect the shock absorber.

3. Remove the shock and coil spring as an assembly.

4. Using a spring compressor, compress the spring until there is a clearance on both ends. Remove the suspension support center nut.

5. Remove the 2 retainers, cushion, suspension support and coil spring.

6. Installation is the reverse of removal. Tighten the mounting nuts to 47 ft. lbs. (64 Nm).

Leaf Springs

REMOVAL & INSTALLATION

1989–90 Land Cruiser

1. Remove the wheels, then raise the front of the vehicle and support it on safety stands. Make certain the stands are placed under the frame rails.

2. Position a floor jack under the differential housing and lower it until there is no tension on the leaf springs.

3. Remove the shock absorber.

4. Remove the U-bolt mounting nuts, pull the spring set away and remove the U-bolts.

5. Remove the hanger and shackle pin mounting nuts. Remove the pins and then lift out the leaf spring.

➡**It may be necessary to lower the jack under the axle housing to remove spring.**

6. Installation is in the reverse order of removal. Tighten the hanger pin bolt to 17 ft. lbs. (24 Nm).

7. Finger-tighten the hanger pin nut and install the shackle pin. Install the plate and finger-tighten the nuts. Insert the head of the center leaf spring bolt into the hole in the axle housing bracket and then install the U-bolts onto the leaf. Install the spring seat and nuts. Tighten the nuts to 105 ft. lbs. (142 Nm).

➡**When tightening the U-bolts, the exposed length of each bolt below the nut must be the same for all 4 threaded ends.**

8. After bouncing the truck several times, tighten the hanger pin nut to 90 ft. lbs. (123 Nm). Tighten the shackle pin nut to 67 ft. lbs. (91 Nm).

Shock Absorbers

TESTING

The purpose of the shock absorber is simply to limit the motion of the spring during compression (bump) and rebound cycles. If the vehicle were not equipped with these motion dampers, the up and down motion of the vehicle would multiply until the vehicle was alternately trying to leap off the ground and to pound itself into the pavement.

Observe the motion of the body as it is released. In most cases, it will come up beyond its original rest position, dip back below it and settle quickly to rest. This shows that the damper is slowing and controlling the spring action. Any tendency to excessive pitch (up-and-down) motion or failure to return to rest within 2–3 cycles is a sign of poor function within the shock absorber.

While each shock absorber can be replaced individually, it is recommended that they be changed as a pair (both front or both rear) to maintain equal response on both sides of the vehicle. Chances are quite good that if one has failed, its mate is weak also.

REMOVAL & INSTALLATION

1. Remove the hubcap and loosen the lug nuts.
2. Raise the front of the truck and support it with safety stands.
3. Remove the wheel.
4. Unfasten the double nuts at the top end of the shock absorber. Remove the cushions and cushion retainers.
5. Remove the lower through-bolt.
6. Remove the shock absorber.
7. Install the shock absorber and tighten the lower mounting nuts/bolt to the appropriate torque:
   • T100 and Tacoma—101 ft. lbs. (137 Nm)
   • Land Cruiser—51 ft. lbs. (69 Nm)
   • 1989–95 Pick-Up and 4Runner—47 ft. lbs. (64 Nm)
   • 1996 4Runner—101 ft. lbs. (137 Nm).
8. Tighten the upper mounting nut to 18 ft. lbs. (25 Nm).
9. Install the wheels and lower the truck.

INSPECTION

Once removed, the shock should be carefully inspected for any signs of leakage. Oil-filled shocks may have a light film of oil around the seal, resulting from normal breathing and air exchange. This should NOT be taken as a sign of failure, but any sign of thick or running oil definitely indicates failure. Gas filled shocks may also show some film at the shaft; if the gas has leaked out, the shock will have almost no resistance to motion.

Hold the shock firmly in each hand and compress it; compression should be reasonably even and smooth. Release the piston end; it should return smoothly and at an even pace. Stretch the shock, checking again for smooth motion and any abnormal resistance.

Upper Ball Joint

REMOVAL & INSTALLATION

Tacoma and 4Runner

1. Remove the front wheel.
2. Remove the steering knuckle with the axle hub.
3. Separate the upper ball joint as follows:
   a. Remove the wire and boot.
   b. Using a snapring expander, remove the snapring.
   c. With a ball joint remover and a deep socket, remove the upper ball joint.
4. Installation is in the reverse order of removal.

T100

1. Remove the front wheel.
2. Disconnect the upper ball joint from the steering knuckle.
3. Remove the upper ball joint 4 mounting nuts. Separate the ball joint from the upper control arm.
4. Installation is in the reverse order of removal. Tighten the mounting nuts to 25 ft. lbs. (33 Nm).

Lower Ball Joint

REMOVAL & INSTALLATION

Tacoma and 4Runner

▶ **See Figures 25, 26 and 27**

1. Remove the front wheel.
2. Support the lower control arm with a suitable jack.
3. Loosen the four bolts attaching the lower ball joint. Do not remove them. Remove the cotter pin and nut from the tie rod end. Throw out the old cotter pin. With special tool 09610–20012 or equivalent, disconnect the tie rod end from the steering knuckle.

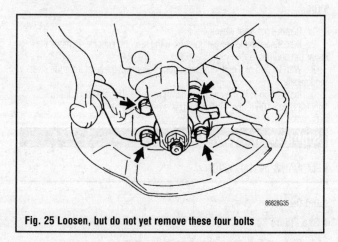

Fig. 25 Loosen, but do not yet remove these four bolts

86828G35

4. Remove the cotter pin and nut from the lower ball joint. Discard the cotter pin.

5. With a ball joint remover such as tool 09628–62011 or equivalent, separate the ball joint from the lower control arm.

6. Remove the four mounting bolts. When lifting the upper control arm and steering knuckle, remove the ball joint. After the ball joint is removed, support the upper control arm and knuckle securely.

7. Install the ball joint in the reverse order of removal.

8. Tighten the nut to 112 ft. lbs. (152 Nm) on the Tacoma, or 105 ft. lbs. (142 Nm) on the 4Runner. Install a new cotter pin in the hole.

9. Tighten the tie rod nut to 66 ft. lbs. (90 Nm). Install a new cotter pin.

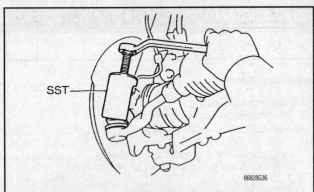

Fig. 26 With special tool 09610–20012 or equivalent, disconnect the tie rod end from the steering knuckle

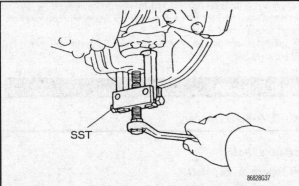

Fig. 27 With a ball joint remover, such as tool 09628–62011 or equivalent, separate the ball joint from the lower control arm

10. Tighten the lower ball joint set bolts to 59 ft. lbs. (80 Nm) 4Runner, 83 ft. lbs. (113 Nm) Tacoma.

T100

1. Remove the front wheel.
2. Disconnect the lower ball joint from the steering knuckle. Remove the cotter pin and nut.
3. With the aid of a ball joint separator, disconnect the joint from the steering knuckle.
4. Installation is in the reverse order of removal. Tighten ball joint to knuckle nut to 105 ft. lbs. (142 Nm). Use a new cotter pin.

Stabilizer Bar (Sway Bar)

REMOVAL & INSTALLATION

Land Cruiser

▶ See Figure 28

1. Raise the truck and support it on safety stands.
2. Remove the wheels.
3. Remove the nut, then disconnect the stabilizer bar with the link from the bracket.
4. Disconnect the bar from the axle housing. Remove the stabilizer bar. Remove the 2 bolts and remove the bracket from the bar.

➡Examine the rubber bushings very carefully for any splits or deformation. Clean the inner and outer surfaces of the bushings before reinstallation. Failed or dirty bushings can create a chorus of odd noises under the vehicle, particularly during cornering.

To install:

5. Reassemble the cushions and cover onto the painted mark on the bar.
6. Tighten the bar-to-bracket bolts to 13 ft. lbs. (18 Nm). Mount the bar, tightening the fittings just snug.
7. Install the wheels and lower the vehicle to the ground. Bounce the front end several times to stabilize the suspension.
8. Tighten the bolts at the axle tube to 19 ft. lbs. (25 Nm). Tighten the nuts at the top of the small link to 13 ft. lbs. (18 Nm) on 1991–93 models 76 ft. lbs. (103 Nm) on 1994–96 models.

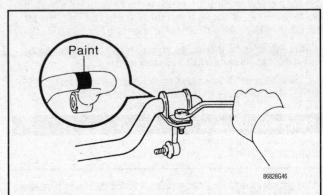

Fig. 28 Reassemble the cushions and cover over the painted mark on the bar

Tacoma, T100 and 4Runner

▶ See Figure 29

1. Remove the front wheels.
2. Disconnect the stabilizer bar links as follows:
   a. Remove the nuts and disconnect the stabilizer bar links from the lower control arm.
   b. Hold the link with a wrench, then remove the nut, retainers, cushions and link.

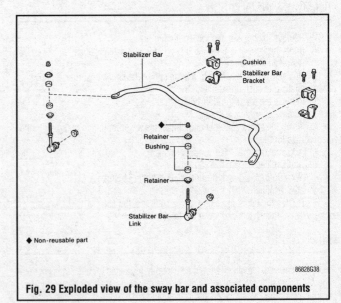

Fig. 29 Exploded view of the sway bar and associated components

3. Remove the bolts and stabilizer bar with cushions and brackets.
To install:
4. Place the stabilizer bar into position, then install both the bar cushions and brackets to the frame. Temporarily install the bolts. Tighten to 19–22 ft. lbs. (25–29 Nm).

5. Hold the stabilizer bar link with a wrench, then install the link onto the lower control arm with a new nut. Tighten the nut to 14–19 ft. lbs. (19–25 Nm).

6. On models with the links using a hexagon wrench, connect the stabilizer bar on both sides to the links with new nuts, tighten to 55–70 ft. lbs. (69–95 Nm).

Upper Control Arm

REMOVAL & INSTALLATION

Pick-Up and 4Runner

1989–95 MODELS

1. Raise the truck and support it on safety stands. Remove the wheels.
2. Remove the torsion bar.
3. Remove the cotter pin and nut and press the upper ball joint out of the steering knuckle.
4. Remove the upper shock absorber mounting nut. Do not disconnect the lower end of the shock.
5. Disconnect the intermediate shaft from the steering gear housing. Remove the 2 nuts and bolt and lift out the upper arm.
6. Installation is in the reverse order of removal. Tighten the arm mounting bolts to 131 ft. lbs. (178 Nm). Tighten the upper shock nut to 18 ft. lbs. (25 Nm).
7. Attach the upper ball joint to the steering knuckle, then tighten the nut to 105 ft. lbs. (142 Nm). Install a new cotter pin.

1996 MODELS

1. Remove the shock and coil spring assembly.
2. Disconnect the ABS speed sensor wire harness clamp.
3. Disconnect the upper ball joint as follows:
   a. Remove the cotter pins and loosen the nut.
   b. Using a ball joint separator, disconnect the upper ball joint from the control arm.
   c. Support the steering knuckle securely.
   d. Remove the nut.
4. Detach the control arm, by removing the nut, bolt, washers and lowering the arm.
5. Installation is in the reverse order. Make sure the arm is stabilized, then tighten the nut to 87 ft. lbs. (115 Nm). Attach the ball joint to the control arm. Tighten the mounting nut to 80 ft. lbs. (105 Nm). Install a new cotter pin.

Tacoma

1. Remove the front wheel.
2. Remove the shock and coil spring assembly.
3. Disconnect the ABS speed sensor wire harness clamp.
4. Disconnect the upper ball joint as follows:
   a. Remove the cotter pins and loosen the nut.
   b. Using a ball joint separator, disconnect the upper ball joint from the control arm.
   c. Support the steering knuckle securely.
   d. Remove the nut.
5. Detach the control arm, by removing the nut, bolt, washers and lowering the arm.
6. Installation is in the reverse order. Make sure the arm is stabilized, then tighten the nut to 87 ft. lbs. (115 Nm).
7. Attach the ball joint to the control arm. Tighten the mounting nut to 80 ft. lbs. (105 Nm). Install a new cotter pin.

T100

1. Raise the truck and support it on safety stands. Remove the wheels.
2. Disconnect the ABS speed sensor wire harness clamp.
3. Remove the torsion bar.
4. Remove upper ball joint out of the steering knuckle as follows:
   a. Support the lower control arm with a suitable jack.
   b. Remove the nuts, then disconnect the upper control arm from the steering knuckle.

c. Remove the bolts, then remove the upper control arm from thew frame.
5. Installation is in the reverse. Tighten the arm mounting bolts to 131 ft. lbs. (178 Nm).
6. Connect the ball joint to the knuckle, then tighten the nut to 25 ft. lbs. (33 Nm).

Lower Control Arm

REMOVAL & INSTALLATION

Pick-Up and 4Runner

1989–95 MODELS

1. Raise the front of the truck and support it with safety stands. Remove the wheels.
2. Remove the shock absorber.
3. Disconnect the stabilizer bar at the lower arm.
4. Disconnect the lower ball joint from the control arm.
5. Paint matchmarks on the front and rear adjusting cams, remove them and lift out the control arm.
   To install:
6. Install the arm. Temporarily tighten the nuts.
7. Connect the ball joint and tighten the nut to 105 ft. lbs. (142 Nm) and install a new cotter pin.
8. Connect the stabilizer bar, then tighten the nuts to 19 ft. lbs. (25 Nm).
9. Install the shock absorber.
10. Install the wheels and lower the truck. Bounce the truck several times to set the suspension.
11. Align the matchmarks on the adjusting cams and tighten the nuts to 145 ft. lbs. (196 Nm).

1996 MODELS

▶ **See Figures 30 and 31**

1. Remove the front wheel.
2. Remove the steering gear assembly.
3. Disconnect the stabilizer bar link.
4. Disconnect the shock absorber from lower control arm.
5. Support the upper control and steering knuckle securely.
6. Remove the cotter pin and nut from the lower ball joint.
7. With a ball joint remover, disconnect the ball joint from thew lower control arm.
8. Place matchmarks on the front and rear adjusting cams.
9. Remove the 2 bolts, nuts, adjusting cams and lower control arm.
10. Remove the spring bumpers with a special tool 09922–10010 or its equivalent.
    To install:
11. Attach the spring bumpers, tighten to 17 ft. lbs. (23 Nm).
12. Attach the lower control arm, placing it in the appropriate position with the matchmarks. Tighten the arm to 96 ft. lbs. (130 Nm).
13. Attach the lower ball joint and tighten the nut to 105 ft. lbs. (142 Nm).
14. Install the remaining compomemts in the reverse order of removal.

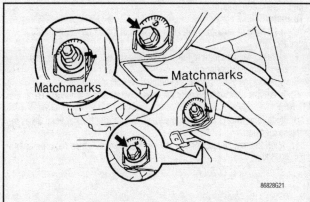

Fig. 30 Place matchmarks on the front and rear adjusting cams

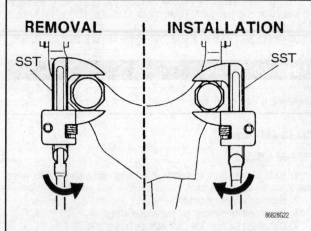

Fig. 31 Remove the spring bumpers with special tool 09922–10010 or its equivalent

Tacoma

1. Remove the front wheel.
2. Remove the steering gear assembly.
3. Disconnect the stabilizer bar link.
4. Disconnect the shock absorber from lower control arm.
5. Support the upper control and steering knuckle securely.
6. Remove the cotter pin and nut from the lower ball joint.
7. With a ball joint remover, disconnect the ball joint from thew lower control arm.
8. Place matchmarks on the front and rear adjusting cams.
9. Remove the 2 bolts, nuts, adjusting cams and lower control arm.
10. Remove the spring bumpers with a special tool 09922–10010 or its equivalent.

To install:

11. Attach the spring bumpers, tighten to 17 ft. lbs. (23 Nm).
12. Attach the lower control arm, placing it in the appropriate position with the matchmarks. Tighten the arm to 96 ft. lbs. (130 Nm).
13. Attach the lower ball joint with the correct tool, tighten the nut to 112 ft. lbs. (152 Nm).
14. Installation of the remaining components is in the reverse order of removal.

T100

1. Raise the front of the truck and support it with safety stands. Remove the wheels.
2. Remove the shock absorber from the lower control arm.
3. Disconnect the stabilizer bar at the lower arm.
4. Disconnect the lower ball joint from the control arm. Remove the cotter pin, then discard. Loosen the nut.
5. Paint matchmarks on the front and rear adjusting cams, remove them and lift out the control arm.

To install:

6. Install the lower arm and adjusting cams to the frame. Temporarily tighten the nuts.
7. Connect the ball joint to the arm. Tighten the nut to 105 ft. lbs. (142 Nm) and install a new cotter pin.
8. Connect the stabilizer bar and tighten the nuts.
9. Install the shock absorber to the lower control arm, tighten to 101 ft. lbs. (137 Nm).
10. Install the wheels and lower the truck. Bounce the truck several times to set the suspension.
11. Align the matchmarks on the adjusting cams and tighten the nuts to 145 ft. lbs. (196 Nm).

Leading Arm

REMOVAL & INSTALLATION

Land Cruiser

▶ **See Figure 32**

1. Raise and safely support the vehicle. Loosen the lug nuts.
2. Remove the front wheel.
3. Remove the leading arm bolt, plate washer and nut on the frame side.
4. Remove the 2 bolts and nut from the arm on the axle side. Lower the arm.
5. Installation is in the reverse order of removal. Tighten the axle bolts to 127 ft. lbs. (171 Nm). Tighten the frame bolts to 130 ft. lbs. (177 Nm).

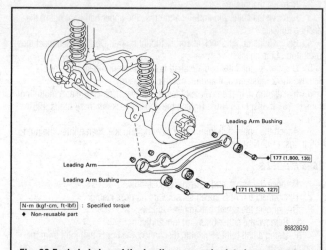

Fig. 32 Exploded view of the leading arm and related components—Land Cruiser

Lateral Control Rod

REMOVAL & INSTALLATION

1. Raise and safely support the vehicle. Loosen the lug nuts.
2. Remove the front wheel.
3. Disconnect the lateral control rod as follows:
   a. Remove the bolt that attaches the rod to the axle housing.
   b. Remove the nut, bolt and rod from the frame side.
4. Installation is in the reverse order of removal. Tighten the frame nut and bolt to 127 ft. lbs. (171 Nm). Tighten the axle bolt to 127 ft. lbs. (171 Nm).

Steering Knuckle and Spindle

REMOVAL & INSTALLATION

Pick-Up and 4Runner

1989–95 MODELS

1. Raise the front of the truck and support it on safety stands. Remove the wheels.
2. Remove the brake caliper. Remove the brake disc and front axle hub.
3. Remove the dust cover (splash shield) and oil seal.
4. Disconnect the knuckle arm at the steering knuckle.

5. Disconnect the shock absorber at the lower control arm.

6. Disconnect the stabilizer bar at the lower control arm.

7. Remove the snapring and spacer from the end of the knuckle spindle.

8. Remove the cotter pin and nut from the upper ball joint. Use a ball joint separator to disconnect it from the knuckle.

9. Remove the 4 bolts and disconnect the lower ball joint from the knuckle.

10. Press down on the lower control arm and lift out the steering knuckle.

11. Installation is in the reverse order of removal. Attach the lower ball joint to the knuckle, then tighten the 4 bolts to 43 ft. lbs. (58 Nm).

12. Attach the upper ball joint to the knuckle, then tighten the nut to 105 ft. lbs. (142 Nm). Install a new cotter pin.

13. Connect the knuckle arm to the steering knuckle, then tighten the bolt to 135 ft. lbs. (183 Nm). Install the dust cover with a new oil seal and tighten the bolt to 13 ft. lbs. (18 Nm).

4Runner

1996 MODELS

1. Remove the front wheel.

2. Remove the shock absorber.

3. Remove the grease cap.

4. On models with ABS, remove the ABS speed sensor and wire harness clamp from the steering knuckle.

5. Remove the brake line bracket from the steering knuckle.

6. Remove the caliper and rotor.

7. Disconnect the lower ball joint.

8. Remove the steering knuckle with hub as follows:

   a. Remove the cotter pin and loosen the nut.

   b. Using a special tool 09950–40010 or equivalent, disconnect the steering knuckle.

   c. Remove the nut and knuckle.

9. Installation is in the reverse. Attach the steering knuckle and tighten the nut to 80 ft. lbs. (108 Nm). Install a new cotter pin.

10. Attach the lower ball joint, tighten the four mounting bolts to 59 ft. lbs. (80 Nm).

11. Install the brake line bracket to the steering knuckle, tighten to 21 ft. lbs. (28 Nm).

12. Attach the ABS speed sensor and wire harness clamp if removed, tighten to 71 inch lbs. (8 Nm).

Tacoma

1. Remove the front wheel.

2. Remove the shock absorber.

3. Disconnect the driveshaft as follows:

   a. Without a freewheeling hub, use a flat bladed tool to remove the grease cap.

   b. Remove the cotter pin and lock cap.

   c. Have an assistant apply the brakes, then remove the locknut.

   d. With free wheeling hubs, remove the free wheel hub.

   e. Using a snapring expander, removed the snapring.

   f. Remove the spacer.

4. On models with ABS, remove the ABS speed sensor and wire harness clamp from the steering knuckle.

5. Remove the caliper and rotor.

6. Disconnect the lower ball joint.

7. Remove the steering knuckle with hub as follows:

   a. Remove the cotter pin and loosen the nut.

   b. Using a special tool 09950–40010 or equivalent, disconnect the steering knuckle.

   c. Remove the nut and knuckle.

8. Installation is in the reverse order of removal. Attach the steering knuckle and tighten the nut to 80 ft. lbs. (108 Nm). Install a new cotter pin.

9. Attach the lower ball joint, tighten the four mounting bolts to 59 ft. lbs. (80 Nm).

10. Attach the ABS speed sensor and wire harness clamp if removed, tighten to 71 inch lbs. (8 Nm).

T100

1. Remove the front wheels.

2. Remove the caliper and front axle hub.

3. Measure the steering knuckle bushing thrust clearance as follows:

   a. Install a bolt in the drive shaft.

   b. Using a feeler gauge, measure the drive shaft thrust clearance between the steering knuckle outside bushing and spacer, by pulling the bolt and applying a load of 22 lb. (98 N).

   c. Standard clearance: 0.0039–0.0197 in. (0.10–0.50mm). Maximum 0.039 in. (1.0mm).

   d. If the thrust clearance is more than the maximum, replace the steering knuckle outside and inside bushings.

4. Detach the front shock from the lower control arm.

5. Separate the stabilizer bar from the lower control arm.

6. Remove the steering knuckle as follows:

   a. Using a pair of snapring pliers, remove the ring and spacer.

   b. Support the lower control arm with a jack.

   c. Remove the cotter pin and nut from the upper ball joint.

   d. Using a ball joint separator, detach the joint from the knuckle.

   e. Remove the bolts from the lower ball joint, then disconnect the knuckle from the joint.

   f. Push down the lower control arm, then remove the knuckle. Be careful not to damage the seal.

7. Install the knuckle as follows:

   a. Apply synthetic oil and lithium soap base chassis grease NLGI No. 1 or equivalent to the driveshaft.

   b. Push down on the lower control arm, then install the steering knuckle.

   c. Connect the lower ball joint to the steering knuckle, tighten the nut to 105 ft. lbs. (142 Nm).

   d. Install a new cotter pin.

   e. Install a spacer to the driveshaft.

   f. Using a pair of snapring pliers, install the ring.

➡**If you replace the knuckle bushing, recheck the clearance of the driveshaft thrust.**

   g. Using a feeler gage, measure the driveshaft thrust clearance between the knuckle and the outside bushing and spacer, by pulling the bolt and applying a load of 22 lbs. (98 N).

   h. Standard clearance should be 0.0039–0.0197 in. (0.10–0.50mm).

   i. If the clearance is not within specifications, replace the spacer.

8. Connect the stabilizer bar to the lower control arm, tighten the nut to 19 ft. lbs. (25 Nm).

9. Attach the front shock to the lower control arm, tighten the nut to 101 ft. lbs. (137 Nm).

10. Install the dust cover and new seal, tighten to 13 ft. lbs. (18 Nm).

Land Cruiser

1. Remove the front axle hub.

2. Remove the knuckle spindle mounting bolts.

3. Remove the dust seal and dust cover (splash shield).

4. Tap the knuckle spindle gently and remove it from the steering knuckle.

5. Turn the axle so that one flat part of the outer shaft faces upward; pull out the axle shaft.

6. Use a separator to disconnect the tie rod end from the knuckle arm.

7. Remove the oil seal and retainer.

8. Remove the knuckle arm bearing and bearing cap nuts. Lightly tap the slits of the washers and remove them. Push out the knuckle arm and shims from the steering knuckle. Don't lose the shims.

9. Remove the steering knuckle and bearing. Mark the removed adjusting shims and bearings by position for reassembly.

To install:

10. Install a new oil seal if it was removed from the axle housing. Install a new oil seal set (felt dust seal, rubber seal and steel ring, in that order).

11. Place the bearings in position on the knuckle and axle housing. Place the knuckle onto the housing.

12. Support the upper bearing inner race on the knuckle arm. Install the arm over the shims, making sure the shims are in their exact original location. Tap the knuckle arm into the inner race.

13. Support the lower bearing inner race. Install the bearing cap over the shims, making sure the shims are in their exact original location. Tap the bearing cap into the inner race.

14. Remove the race supports. Install the cone washers to the third arm and tighten the nuts to 71 ft. lbs. (96 Nm). Install the cone washers to the knuckle arm and tighten the nuts to the same value.

15. Use a spring tension gauge to measure the bearing preload. Correct preload is 6.6–13.2 lbs. (19–59 N).

16. Connect the tie rod to the knuckle arm. Tighten the castle nut to 67 ft. lbs. (91 Nm). Install a new cotter pin.

17. Install the oil seal set to the knuckle.

18. Install the axle shaft.

19. Pack the knuckle about 3/4 full with moly-disulfide grease.

20. Place a new gasket on the knuckle and install the spindle. Install the gasket, dust cover and dust seal on the spindle. Tighten the spindle mounting bolts to 34 ft. lbs. (47 Nm).

21. Install the axle hub.

22. Have the alignment checked at a reputable repair facility.

Front Axle Hub and Bearing

Hub and bearing removal and installation procedures for these models are given in Section 7.

Front End Alignment

If the tires are worn unevenly, if the vehicle is not stable on the highway or if the handling seems uneven in spirited driving, wheel alignment should be checked. If an alignment problem is suspected, first check tire inflation and look for other possible causes such as worn suspension and steering components, accident damage or unmatched tires. Repairs may be necessary before the wheels can be properly aligned. Wheel alignment requires sophisticated equipment and can only be performed at a properly equipped shop.

CASTER

Wheel alignment is defined by three different adjustments in three planes. Looking at the vehicle from the side, caster angle describes the steering axis rather than a wheel angle. The steering knuckle is attached to the strut at the top and the control arm at the bottom. The wheel pivots around the line between these points to steer the vehicle. When the upper point is tilted back, this is described as positive caster. Having a positive caster tends to make the wheels self-centering, increasing directional stability. Excessive positive caster makes the wheels hard to steer, while an uneven caster will cause a pull to one side.

Caster is adjustable on vehicles other than the Land Cruiser by turning the adjusting cams to reposition the arms. Land Cruisers do not allow adjustment; replacement of bent components is required if the caster measurement is out of specification.

CAMBER

Looking at the wheels from the front of the vehicle, camber adjustment is the tilt of the wheel. When the wheel is tilted in at the top, this is negative camber. In a turn, a slight amount of negative camber helps maximize contact of the outside tire with the road. Too much negative camber makes the vehicle unstable in a straight line.

Camber is adjustable on all vehicles except the Land Cruiser, through the turning of adjusting cams. Land Cruisers are not camber adjustable; if the measurement is out of specification, bent or damaged components must be replaced.

TOE-IN

Looking down at the wheels from above the vehicle, toe alignment is the distance between the front of the wheels relative to the distance between the back of the wheels. If the wheels are closer at the front, they are said to be toed-in or to have a negative toe. A small amount of negative toe enhances directional stability and provides a smoother ride on the highway. On most front wheel drive vehicles, standard toe adjustment is either zero or slightly positive. When power is applied to the front wheels, they tend to toe-in naturally.

Toe is adjustable on all models covered by this manual, and is generally measured in inches or degrees. It is adjusted by loosening the locknut on each tie rod end and turning the rod until the correct reading is achieved. The left and right rods must remain equal in length during all adjustments.

2WD REAR SUSPENSION

Leaf Springs

REMOVAL & INSTALLATION

▶ **See Figure 33**

1. Loosen the rear wheel lug nuts.

2. Raise the rear of the vehicle. Support the frame and rear axle housing with stands.

3. Remove the lug nuts and the wheel.

4. Remove the cotter pin, nut, and washer from the lower end of the shock absorber.

5. Detach the shock absorber from the spring seat.

6. Remove the parking brake cable clamp.

➡**Remove the parking brake equalizer, if necessary.**

7. Unfasten the U-bolt nuts and remove the spring seat assemblies.

8. Adjust the height of the rear axle housing so that the weight of the rear axle is removed from the rear springs.

9. Unfasten the spring shackle retaining nuts. Withdraw the spring shackle inner plate. Carefully pry out the spring shackle with a bar.

10. Remove the spring bracket pin from the front end of the spring hanger and remove the rubber bushing.

11. Remove the spring. Use care not to damage the hydraulic brake line or the parking brake cable.

12. Installation is in the reverse order of removal.

➡**Use soapy water or glass cleaner as a lubricant, if necessary, to aid in pin installation. Never use oil or grease.**

13. Tighten the U-bolt nuts to:
- ½-ton Pick-Up—108 ft. lbs. (147 Nm)
- 1-ton Pick-Up and Cab & Chassis model—90 ft. lbs. (123 Nm)
- T100—97 ft. lbs. (132 Nm)
- Tacoma—90 ft. lbs. (120 Nm)

14. Bounce the truck several times to set the suspension and then tighten the shock absorber bolt. Tighten the hanger pin nut or bolt to:
- ½-ton Pick-Up—116 ft. lbs. (157 Nm)
- 1-ton Pick-Up and Cab & Chassis model—67 ft. lbs. (91 Nm)
- Tacoma—115 ft. lbs. (120 Nm)
- T100—19 ft. lbs. (26 Nm)

15. Tighten the shackle pin to 67 ft. lbs. (91 Nm) for all vehicles.

Coil Springs and Shock Absorbers

REMOVAL & INSTALLATION

4Runner

▶ **See Figure 34**

1. Elevate and safely support the vehicle on the frame rails. Place a floor jack under the rear axle housing and set the jack to hold the axle in place.

2. Remove the wheel(s).

3. Remove the parking brake cable bracket from the rear axle housing.

4. Remove the lower shock absorber mount bolt. If the shock is to be replaced, disconnect the upper mount and remove the shock. If the shock is not to be removed, the upper mount may be left connected during spring removal. If only the shock is to be replaced (removal of the spring not required, skip to Step 13 for reinstallation. Refer to Inspection and/or Disposal of Shock Absorbers in the Front Suspension part of this Section.

5. Disconnect the stabilizer bar brackets from the rear axle housing.

6. Disconnect the right shock absorber lower mount.

7. Remove the bolt(s) holding the lateral control arm to the frame.

8. Very carefully begin to lower the jack holding the axle housing. Use extreme caution not to snap the brake lines and/or parking brake cable.

9. While lowering the axle, remove the coil spring as it loses tension and

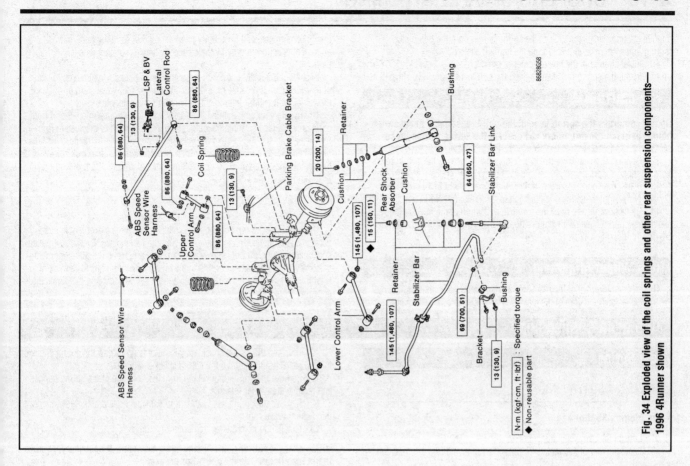

LSP & BV
Lateral Control Rod
13 (130, 9)
86 (880, 64)
86 (880, 64)
Coil Spring
ABS Speed Sensor Wire Harness
ABS Speed Sensor Wire Harness
86 (880, 64)
13 (130, 9)
Upper Control Arm
86 (880, 64)
ABS Speed Sensor Wire Harness
Parking Brake Cable Bracket
Retainer
20 (200, 14)
Cushion
145 (1,480, 107)
Lower Control Arm
145 (1,480, 107)
Rear Shock Absorber
15 (150, 11) ◆
Retainer
Stabilizer Bar
Cushion
Bushing
64 (650, 47)
Stabilizer Bar Link
69 (700, 51)
Bushing
Bracket
13 (130, 9)

N·m (kgf·cm, ft·lbf) : Specified torque
◆ Non-reusable part

Fig. 34 Exploded view of the coil springs and other rear suspension components—1996 4Runner shown

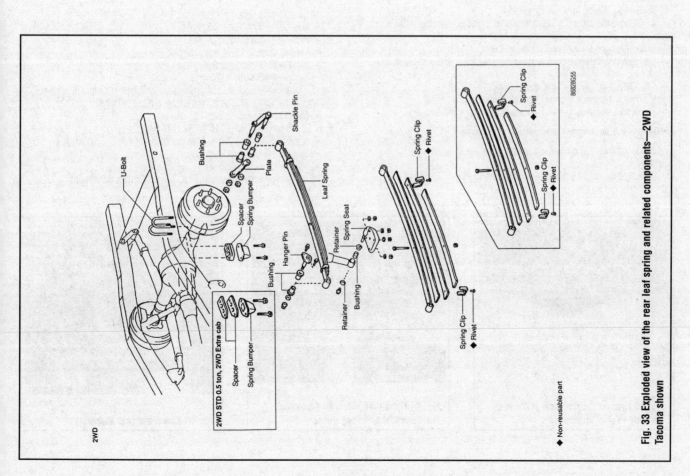

U-Bolt
Bushing
Shackle Pin
Plate
Spacer
Spring Bumper
Leaf Spring
Bushing
Hanger Pin
Retainer
Bushing
Spring Seat
Spacer
Spring Bumper
2WD STD 0.5 ton, 2WD Extra cab
Retainer
Bushing
Spring Clip
Rivet ◆
Spring Clip
Spring Clip
Rivet ◆
Spring Clip
Rivet ◆
Spring Clip
Rivet ◆

2WD

◆ Non-reusable part

Fig. 33 Exploded view of the rear leaf spring and related components—2WD Tacoma shown

remove the upper insulators. Once the spring is removed, the bump stop (spring follow) may be unbolted from the frame if desired.

10. Installation is in the reverse order of removal.

11. Install the spring followand tighten the bolt to 11 ft. lbs. (15 Nm).

※※ CAUTION

Do not compress the spring beyond the point reasonably necessary to hold it in place. Severe injury can occur if the spring comes loose.

12. Connect the lateral control rod to the frame, inserting the bolt from the front or shock absorber side. Install the nut but do NOT tighten it at this time.

13. Install the shock absorber. Tighten the upper mount to 14–18 ft. lbs. (20–25 Nm). Tighten the lower mount bolt to 47 ft. lbs. (64 Nm).

14. Use a floor jack to raise the rear axle housing (not the body of the vehicle) and support the axle with stands. Tighten the lateral control rod nut to 64–101 ft. lbs. (86–137 Nm).

Rear Wheel Alignment

The proper alignment of the rear wheels is as important as the alignment of the front wheels. The nature of the rear axles used on the vehicles covered in this book eliminates all possibility of adjusting the rear alignment; correct alignment is achieved only by the correct installation of component parts.

In spite of this fact, a check of the rear alignment can reveal a hidden cause of rear tire wear or a cause of poor handling. If the rear wheels are misaligned, the vehicle will exhibit unpredictable handling characteristics.

The usual symptoms include a different turning response into left and right corners and difficulty in maintaining a straight path. Both of these stem from the rear wheels attempting to steer the vehicle instead of rolling straight. This "rear steer" behavior is particularly hazardous on slick surfaces; the back wheels of the vehicle may attempt to go in directions unrelated to the front during braking or turning maneuvers.

The rear wheel alignment should be checked any time a problem is suspected or after any serious impact involving the rear wheels. Even a low-speed encounter with a curb on a snowy day can bend an axle. If the rear alignment check reveals a component problem, the timely inspection has saved you at least the cost of new tires and possibly a more serious accident by eliminating poor handling.

4WD REAR SUSPENSION

Leaf Springs

REMOVAL & INSTALLATION

▶ **See Figures 35 thru 41**

1. Loosen the rear wheel lug nuts.

2. Raise the rear of the vehicle. Support the frame and rear axle housing with stands.

3. Remove the lug nuts and the wheel.

4. Remove the cotter pin, nut, and washer from the lower end of the shock absorber.

5. On the Land Cruiser, perform the following:

   a. Remove the cotter pins and nuts from the lower end of the stabilizer link.

   b. Detach the link from the axle housing.

6. Detach the shock absorber from the spring seat.

7. Remove the parking brake cable clamp (except Land Cruiser).

➡**Remove the parking brake equalizer, if necessary.**

8. Unfasten the U-bolt and nuts, then remove the spring seat assemblies.

9. Adjust the height of the rear axle housing so that the weight of the rear axle is removed from the rear springs.

10. Unfasten the spring shackle retaining nuts. Withdraw the spring shackle inner plate. Carefully pry out the spring shackle with a bar.

11. Remove the spring bracket pin from the front end of the spring hanger and remove the rubber bushing.

12. Remove the spring. Use care not to damage the hydraulic brake line or the parking brake cable.

13. Installation is in the reverse order of removal.

➡**Use soapy water or glass cleaner as a lubricant, if necessary, to aid in pin installation. Never use oil or grease.**

14. Tighten the U-bolt nuts to:
- Xtra Cab—90 ft. lbs. (123 Nm)
- Pick-Up—108 ft. lbs. (147 Nm)
- Land Cruiser—105 ft. lbs. (142 Nm)
- T100—97 ft. lbs. (132 Nm)
- Tacoma—90 ft. lbs. (120 Nm)

15. Bounce the truck several times to set the suspension and then tighten the shock absorber bolt. Tighten the hanger pin nut or bolt to:
- Land Cruiser—90 ft. lbs. (122 Nm)
- ½18-ton Pick-Up—116 ft. lbs. (157 Nm)
- 1-ton Pick-Up and Cab & Chassis model—67 ft. lbs. (91 Nm)
- Tacoma—115 ft. lbs. (120 Nm)
- T100—19 ft. lbs. (26 Nm)

16. Tighten the shackle pin to 67 ft. lbs. (91 Nm) for all vehicles.

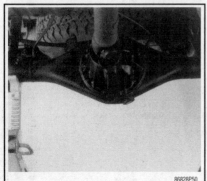

Fig. 35 Support the rear axle of the vehicle with a set of jackstands

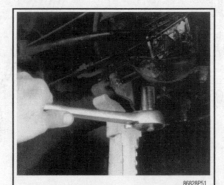

Fig. 36 Unbolt the rear axle mounting bracket attached to the spring

Fig. 37 Lower the seat assembly

Fig. 38 Remove the U-bolt from the rear

Fig. 39 Proceed to the front of the spring, and remove the pin

Fig. 40 Unbolt the rear mount then . . .

Fig. 41 . . . carefully lower the spring from the vehicle. The jackstand must be removed by an assistant

Coil Springs and Shock Absorbers

REMOVAL & INSTALLATION

4Runner

1. Elevate and safely support the vehicle on the frame rails. Place a floor jack under the rear axle housing and set the jack to hold the axle in place.
2. Remove the wheel(s).
3. Remove the parking brake cable bracket from the rear axle housing.
4. Remove the lower shock absorber mount bolt. If the shock is to be replaced, disconnect the upper mount and remove the shock. If the shock is not to be removed, the upper mount may be left connected during spring removal. If only the shock is to be replaced (removal of the spring not required, skip to Step 13 for reinstallation. Refer to Inspection and/or Disposal of Shock Absorbers in the Front Suspension part of this Section.
5. Disconnect the stabilizer bar brackets from the rear axle housing.
6. Disconnect the right shock absorber lower mount.
7. Remove the bolt(s) holding the lateral control arm to the frame.
8. Very carefully begin to lower the jack holding the axle housing. Use extreme caution not to snap the brake lines and/or parking brake cable.
9. While lowering the axle, remove the coil spring as it loses tension and remove the upper insulators. Once the spring is removed, the bump stop (spring follow) may be unbolted from the frame if desired.
10. Installation is the reverse of removal. Tighten the upper shock mount to 14–18 ft. lbs. (20–25 Nm). Tighten the lower mount bolt to 47 ft. lbs. (64 Nm).
11. Bounce the rear of the vehicle several times to stabilize the suspension.
12. Use a floor jack to raise the rear axle housing (not the body of the vehicle) and support the axle with stands. Tighten the lateral control rod nut to 64–101 ft. lbs. (86–137 Nm).

Land Cruiser

1. Elevate and safely support the vehicle on the frame rails. Place a floor jack under the rear axle housing and set the jack to hold the axle in place.
2. Remove the wheel(s).
3. Remove the lower shock absorber mount bolt. If the shock is to be replaced, disconnect the upper mount and remove the shock. If the shock is not to be removed, the upper mount may be left connected during spring removal. If only the shock is to be replaced (removal of the spring not required, skip to Step 13 for reinstallation. Refer to Inspection of Shock Absorbers in the Front Suspension part of this Section.
4. Disconnect the stabilizer bar brackets from the rear axle housing.
5. Remove the bolt(s) holding the lateral control arm to the rear axle housing.
6. Very carefully begin to lower the jack holding the axle housing. Use extreme caution not to snap the brake lines and/or parking brake cable.
7. While lowering the axle, remove the coil spring as it loses tension and remove the upper insulators. Once the spring is removed, the bump stop (spring follow) may be unbolted from the frame if desired.
8. Installation is in the reverse order of removal. Install the spring follow to the frame and tighten the bolt to 11–21 ft. lbs. (15–28 Nm).

✳✳ CAUTION

Do not compress the spring beyond the point reasonably necessary to hold it in place. Severe injury can occur if the spring comes loose.

9. Connect the lateral control rod to the axle housing, inserting the bolt from the front or shock absorber side. Install the nut but do NOT tighten it at this time.
10. Install the shock absorber. Tighten the upper mount to 37 ft. lbs. (50 Nm). Tighten the lower mount bolt to 47–51 ft. lbs. (64–69 Nm).
11. Bounce the rear of the vehicle several times to stabilize the suspension.
12. Use a floor jack to raise the rear axle housing, then support the axle with stands. Tighten the lateral control rod nut to 130 ft. lbs. (177 Nm).

Stabilizer Bar

REMOVAL & INSTALLATION

1. Jack up and support the vehicle.
2. Disconnect the stabilizer bar links as follows:
   a. Remove the nuts and disconnect the stabilizer bar links from the stabilizer bar.
   b. Hold the link with a wrench, then remove the nut, retainers, cushions and link.
3. Remove the bolts and stabilizer bar with cushions and brackets.
To install:
4. Install the bushings to the stabilizer bar so the cutout of the bushing faces the under side of the vehicle.
5. Install the bar to the rear axle housing, with bushings and brackets.

Tighten the bolts to 9 ft. lbs. (13 Nm) on the 4Runner, or 13 ft. lbs. (18 Nm) on the Land Cruiser.

6. Attach the retainers and cushions to the bar links. Hold the stabilizer bar link with a wrench, then install a new nut. Tighten the nut to 14–19 ft. lbs. (19–25 Nm) on the 4Runner, or 76 ft. lbs. (103 Nm) on the Land Cruiser.

7. Install the nuts to the stabilizer bar, tighten to 51 ft. lbs. (69 Nm) on the 4Runner, or 19 ft. lbs. (25 Nm) on the Land Cruiser.

Lateral Control Rod

REMOVAL & INSTALLATION

Land Cruiser

1. Raise and support the rear of the vehicle.
2. Remove the rear wheel.
3. Remove the bolt, nut and plate washer, then disconnect the lateral control rod.
4. Remove the bolt and lateral control rod from the axle housing.
5. Install the rod in the reverse order of removal.

STEERING

Toyota offers two types of power steering on their trucks. There is standard and Progressive Power Steering (PPS). Both types have a recirculating ball system and rotary type hydraulic control valve.

On the PPS, the vehicle speed is detected by a speed sensor and fluid pressure acting on the piston is varied accordingly. When the truck is stopped or when moving at a low speed, fluid pressure is increased to lighten the force required for steering. At high speeds, the pressure is reduced to lessen the amount of assist and provide appropriate steering wheel response.

Steering Wheel

REMOVAL & INSTALLATION

Without Air Bag

♦ See Figures 42 thru 47

✳✳ WARNING

Do not attempt to remove or install the steering wheel by hammering on it. Damage to the energy absorbing steering column could result. Always use a steering wheel puller to extract the wheel from the shaft.

1. Position the wheels in the straight ahead position.
2. Remove the steering wheel center pad. Check the back of the steering wheel spokes for recessed screws holding the pad and remove them if pre-

6. Attach the bolt that connects the rod to the axle housing, tighten to 181 ft. lbs. (245 Nm).

7. Attach the bolt with washer and tighten the nut to the rod to 130 ft. lbs. (177 Nm).

4Runner

1. Raise and support the rear of the vehicle.
2. Disconnect the right side shock absorber.
3. Remove the bolt, then disconnect the harness for the Load Sensing Proportioning and Bypass Valve (LSPBV).
4. Remove the bolt, nut and plate washer, then disconnect the lateral control rod from the frame.
5. Remove the bolt and lateral control rod from the axle housing side.
6. Installation is in the reverse order of removal.
7. Attach the bolt that connects the rod to the axle housing, tighten to 64 ft. lbs. (86 Nm).
8. Attach the bolt with washers, then tighten the rod to the frame side to 64 ft. lbs. (86 Nm).
9. Attach the bolt, then connect the harness for the Load Sensing Proportioning and Bypass Valve (LSPBV). Tighten the bolt to 9 ft. lbs. (13 Nm).

sent. If no screws are found, the pad should snap out with firm finger pressure.

3. Remove the steering wheel retaining nut.
4. Scribe matchmarks on the hub and the steering shaft to aid in proper installation.
5. Use the puller to extract the wheel from the shaft. Unplug any wiring connectors running to the wheel once it is loose.
6. Installation is in the reverse order of removal. Tighten the center nut to 25–35 ft. lbs. (34–47 Nm).

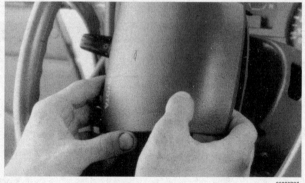

Fig. 43 With your fingers, pry up on the pad . . .

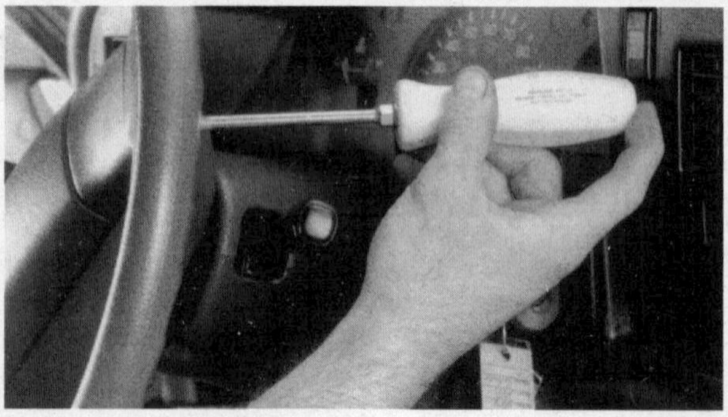

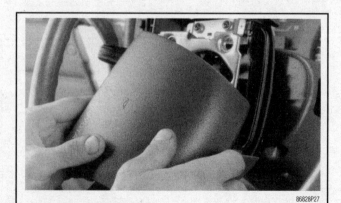

Fig. 44 . . . to release it from the steering wheel

Fig. 42 Using a suitable prytool, loosen the mounting for the center pad

Fig. 45 Place a socket on the center nut and loosen

Fig. 46 Use the puller to extract the wheel from the shaft

Fig. 47 Lift the wheel off the column, and separate any wires that may be attached

With Air Bag

▶ See Figures 48 and 49

1. Disarm the SRS system.
2. On models without tilt steering, proceed as follows:
   a. Place the front wheels facing straight ahead.

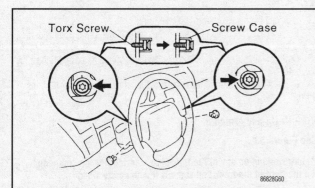

Fig. 48 Using a Torx® driver, loosen the Torx® screws until the groove along the screw circumference catches on the screw case

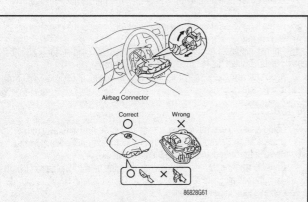

Fig. 49 When storing the wheel pad with an air bag, keep the upper surface of the pad facing upwards

   b. Remove the 2 steering wheel lower covers. Using a Torx® driver, loosen the Torx® screws until the groove along the screw circumference catches on the screw case.
   c. Pull out the wheel pad from the steering wheel, then detach the air bag connector.
3. Remove the steering wheel set nut. A special wheel puller may be needed. Place matchmarks on the steering wheel and mainshaft assembly. Remove the wheel.
4. Unplug the harness.
5. Installation is in the reverse order of removal. Align the matchmarks on the wheel and the mainshaft. Tighten the wheel set nut to 25 ft. lbs. (34 Nm) and the pad screws to 78 inch lbs. (9 Nm).

Combination Switch

Removal and installation of the combination switch is covered in Section 6.

Ignition Switch

REMOVAL & INSTALLATION

▶ See Figures 50 and 51

   Both the electrical portion of the ignition switch and the key/lock portion may be replaced without removing the housing or bracket holding them. The bracket holding the ignition switch assembly is held to the column by bolts whose heads were intentionally broken off at installation. About the only time the bracket must be removed would be if the steering column or tube must be replaced. In the event that the bracket must be removed, the headless bolts must be center-punched, drilled and removed with a screw extracting tool.

1. Disarm the SRS system.
2. Turn the ignition **OFF** and remove the key from the ignition.
3. Remove the steering wheel pad and remove the steering wheel.
4. Remove the upper and lower steering column covers. Remove the combination switch.
5. Unplug the ignition wiring harness connector.
6. If only the electrical switch needs replacement, remove the retaining screw at the back of the lock housing and remove the switch. If the automatic transmission interlock relay must be changed, it may be unbolted from the housing; don't lose the spring and lock pin.
7. If the key lock assembly is to be changed, place the key in the switch and turn the key to the **ACC** position. Insert a thin rod in the hole at the bottom of

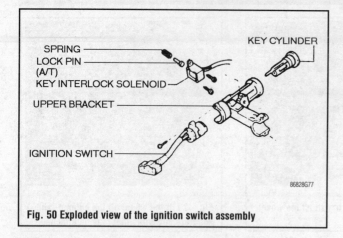

Fig. 50 Exploded view of the ignition switch assembly

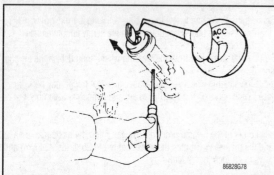

Fig. 51 Insert a thin rod in the hole at the bottom of the housing and press the stop tab inward. While holding the tab in, pull the cylinder ut of the housing

the housing and press the stop tab inward. While holding the tab in, pull the cylinder out of the housing.

8. Install the new lock cylinder, making certain the key is in the **ACC** position before installation. Insert the cylinder and be sure the stop tab engages the hole. Once installed, turn the switch **OFF** and remove the key.

Steering Linkage

REMOVAL & INSTALLATION

2WD Pick-Up

♦ See Figure 52

➡Before working on any of the following steering linkage components, disconnect the battery cable, raise the front of the truck and support it with safety stands.

PITMAN ARM

1. Remove the strut bar.
2. Loosen the Pitman arm nut.
3. Using a tie rod end puller or similar tool, disconnect the Pitman arm from the sector shaft.
4. Using a tie rod end puller or similar tool, disconnect the Pitman arm from the relay rod.
5. On installation, align the marks on the Pitman arm and sector shaft and connect them. Tighten the nut to 90 ft. lbs. (123 Nm).
6. Connect the arm to the relay rod and tighten the nut to 67 ft. lbs. (90 Nm).
7. Install the strut bar.

TIE ROD

1. Using a tie rod end puller, disconnect the tie rod from the relay rod.
2. Using a tie rod end puller, disconnect the tie rod from the knuckle arm.

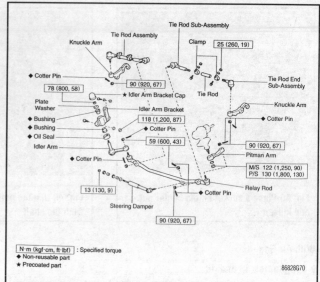

Fig. 52 Exploded view of the 2WD Pick-Up and 4Runner steering linkage components

3. Remove the tie rod and remove the tie rod ends.
4. On installation, screw the tie rod ends onto the tie rod. The tie rod length should be approximately (314.5mm). The remaining length of threads on both ends should always be equal.
5. Turn the tie rod ends so they cross at about 90°. Tighten the clamp nuts to 19 ft. lbs. (25 Nm).
6. Connect the tie rod to the knuckle arm and relay rod and tighten the mounting nuts to 67 ft. lbs. (90 Nm).

IDLER ARM

1. Separate the relay rod from the idler arm.
2. Remove the 3 mounting bolts and remove the idler arm bracket with the arm attached.
3. On installation, position the bracket and arm on the frame and tighten the bolts to 48–58 ft. lbs. (65–78 Nm).
4. Connect the idler arm to the relay rod, then tighten the nut to 43 ft. lbs. (59 Nm).

4WD Pick-Up and 4Runner

♦ See Figure 53

➡Before working on any of the following steering linkage components, raise the front of the truck and support it with safety stands.

PITMAN ARM

1. Remove the Pitman arm set nut and washer.
2. Using a tie rod end puller or similar tool, disconnect the Pitman arm from the sector shaft.
3. Using a tie rod end puller or similar tool, disconnect the Pitman arm from the relay rod.
4. On installation, align the marks on the Pitman arm and sector shaft and connect them. Tighten the nut to 130 ft. lbs. (177 Nm).
5. Attach the arm to the relay rod, then tighten the nut to 67 ft. lbs. (90 Nm). Install a new cotter pin.

TIE ROD

1. Using a tie rod end puller, disconnect the tie rod from the relay rod.
2. Using a tie rod end puller, disconnect the tie rod from the knuckle arm.
3. Remove the tie rod and remove the tie rod ends.
4. On installation, screw the tie rod ends onto the tie rod. The tie rod length should be approximately 12.38 in. (314.5mm). The remaining length of threads on both ends should always be equal.
5. Turn the tie rod ends so they cross at about 90°. Tighten the clamp nuts to 19 ft. lbs. (25 Nm).

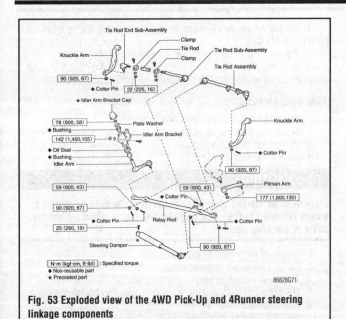

Fig. 53 Exploded view of the 4WD Pick-Up and 4Runner steering linkage components

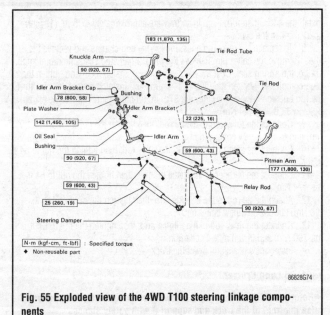

Fig. 55 Exploded view of the 4WD T100 steering linkage components

6. Attach the tie rod to the knuckle arm and relay rod, then tighten the mounting nuts to 67 ft. lbs. (90 Nm).

IDLER ARM BRACKET

1. Separate the relay rod from the idler arm.
2. Remove the 3 mounting bolts and remove the idler arm bracket with the arm attached.
3. On installation, position the bracket and arm on the frame and tighten the bolts to 70 ft. lbs. (95 Nm).
4. Connect the idler arm to the relay rod and tighten the nut to 43 ft. lbs. (59 Nm). Install a new cotter pin.

T100

▶ **See Figures 54 and 55**

PITMAN ARM

1. Remove the engine under cover.
2. Remove the Pitman arm set nut and spring washer.
3. Using a puller such as tool 09628–62011, disconnect the Pitman arm from he cross shaft.
4. Remove the cotter pin and nut.
5. With the aid of another SST 09611–22012 or equivalent, disconnect the Pitman arm from the relay rod.

To install:

6. Align the marks on the Pitman arm and the cross shaft.
7. Install the spring washer and Pitman arm set nut, tighten to 130 ft. lbs. (177 Nm).

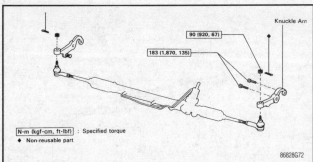

Fig. 54 Exploded view of the 2WD T100 steering linkage components

8. Connect the Pitman arm to the relay rod.
9. Install the nut and new cotter pin. Tighten the the nut to 67 ft. lbs. (90 Nm).

TIE ROD

1. Using a tie rod end puller, disconnect the tie rod from the relay rod.
2. Using a tie rod end puller, disconnect the tie rod from the knuckle arm.
3. Remove the tie rod and remove the tie rod ends.

To install:

4. Screw the tie rod ends onto the tie rod. The tie rod length should be approximately 12.96 in. (329.2mm). The remaining length of threads on both ends should always be equal.
5. Turn the tie rod ends so they cross at about 90°. Tighten the clamp nuts to 16 ft. lbs. (22 Nm).
6. Attach the tie rod to the knuckle arm and relay rod, then tighten the mounting nuts to 67 ft. lbs. (90 Nm).
7. Connect the tie rod assembly to the relay rod. Install the nut, then tighten the nut to 67 ft. lbs. (90 Nm). Install a new cotter pin.
8. Connect the tie rod end to the relay rod. Install the nut, then tighten the nut to 67 ft. lbs. (90 Nm). Install a new cotter pin.

IDLER ARM

1. Remove the cotter pin and nut for the idler arm.
2. Using the tool 09610–20012 or equivalent, disconnect the idler arm from the relay rod.
3. Remove the three bolts, washers and nuts.
4. On installation, attach the idler arm assembly with the bolts, washers and nuts, tighten to 105 ft. lbs. (142 Nm).
5. Connect the idler arm to the relay rod.
6. Install the nut and tighten to 43 ft. lbs. (59 Nm). Install a new cotter pin.

1989–90 Land Cruiser

1. Raise the truck and support it with safety stands. Remove the wheels.
2. Remove the cotter pin and nut, then disconnect the relay rod at the Pitman arm.
3. Remove the cotter pin and nut, then disconnect the tie rod at the knuckle arm.
4. Remove the mounting nut, then disconnect the steering damper at the frame bracket.
5. Remove the nut, then disconnect the steering damper at the relay rod.
6. Remove the cotter pin and nut, then disconnect the relay rod at the tie rod.

To install:

7. To replace the tie rod ends, loosen the clamp and remove the ends.

Install the ends, then turn them in so they are approximately 47.5 in. (121cm) apart. Tighten the clamps.

8. To replace the relay rod ends, loosen the end clamps and remove the ends. Remove the cotter pin. Remove the plug, 2 ball stud seats, link joint knob, spring and spring seat. Coat the parts with grease, install them and tighten the plug completely. Loosen the plug 1–1⅓ turns and install a new cotter pin. Turn the rod ends onto the rod so they are 37.8 in. (96cm) apart. Tighten the clamp bolts to 33 ft. lbs. (44 Nm).

9. Connect the relay rod to the tie rod, then tighten the nut to 67 ft. lbs. (90 Nm) and install a new cotter pin.

10. Connect the steering damper to the relay rod, then tighten the nut to 54 ft. lbs. (74 Nm).

11. Connect the damper to the frame bracket, then tighten the nut to 54 ft. lbs. (74 Nm).

12. Attach the tie rod to the knuckle arm, then tighten the nut to 67 ft. lbs. (90 Nm) and install a new cotter pin.

13. Connect the relay rod to the Pitman arm, then tighten the nut to 67 ft. lbs. (90 Nm) and install a new cotter pin.

14. Install the wheels and lower the truck.

1991–96 Land Cruiser

➡**Before working on any of the following steering linkage components, raise the front of the truck and support it with safety stands.**

PITMAN ARM

1. Remove the Pitman arm set nut and washer.
2. Using a tie rod end puller or similar tool, disconnect the Pitman arm from the sector shaft.
3. Using a tie rod end puller or similar tool, disconnect the Pitman arm from the relay rod.
4. To install, align the marks on the Pitman arm and sector shaft and connect them. Tighten the nut to 130 ft. lbs. (177 Nm).
5. Connect the arm to the relay rod, then tighten the nut to 67 ft. lbs. (90 Nm). Install a new cotter pin.

TIE ROD

1. Using a tie rod end puller, disconnect the tie rod from the knuckle arms.
2. Remove the tie rod and remove the tie rod ends.
3. To install, screw the tie rod ends onto the tie rod. The tie rod length should be approximately 47.51 in. (121cm). The remaining length of threads on both ends should always be equal.
4. Turn the tie rod ends so they point in the same direction. Tighten the clamp nuts to 27 ft. lbs. (37 Nm).
5. Connect the tie rod to the knuckle arms, then tighten the mounting nuts to 67 ft. lbs. (90 Nm).

Power Steering Gear

REMOVAL & INSTALLATION

Pick-Up and 4Runner

1989–95 MODELS

1. Drain the power steering fluid.
2. Disconnect the negative battery cable.
3. Remove the reservoir tank.
4. Remove the air cleaner assembly.
5. Remove the Progressive Power Steering (PPS) solenoid connector (2wd).
6. Disconnect the intermediate No. 2 shaft.
7. Remove the bolt and spring washer from the Pitman arm (2wd) or the set nut (4wd). Using SST 09610–55012 or equivalent, disconnect the Pitman arm from the gear housing.
8. Using SST 09631–22020 or equivalent, disconnect the pressure feed tube and the return tube from the gear housing.
9. Remove the union bolt, gaskets, and the return tube from the gear housing.
10. Installation is in the reverse order of removal.

11. Tighten the pressure feed tube to 26 ft. lbs. (36 Nm) and the return tube to 30 ft. lbs. (41 Nm).
12. Tighten the union bolt (with new gaskets) to 34 ft. lbs. (47 Nm).
13. Align the alignment marks on the Pitman arm and the cross shaft, and install the spring washer and nut. Tighten the nut to 130 ft. lbs. (177 Nm).

1996 4WD MODELS

1. Place the front wheels facing straight ahead.
2. Disconnect the negative battery cable.
3. Remove the steering wheel pad.
4. Remove the steering wheel.

✻✻ CAUTION

Work must be started after 90 seconds from the time the ignition switch is turned to the LOCK position and the negative (–) battery cable is disconnected.

5. Raise and safely support the vehicle.
6. Remove the engine under cover.
7. Remove the stabilizer bar.
8. Disconnect the right and left tie rod ends from the steering knuckle.
9. Disconnect the intermediate No. 2 shaft from the steering rack.
10. Disconnect the pressure feed tube and return tube.
11. Remove the power steering rack assembly.
12. Remove the bracket and grommet.
13. Installation is in the reverse order of removal.
14. Install the power steering rack assembly then tighten the bolts to:
- Rack assembly bolt: 123 ft. lbs. (167 Nm)
- Rack assembly nut and bolt: 141 ft. lbs. (191 Nm)
- Nut and bolt to the bracket: 123 ft. lbs. (167 Nm)

15. Connect the pressure feed tube and the return tube. Tighten to 29 ft. lbs. (40 Nm).

Tacoma

▶ **See Figure 56**

1. Place the front wheels facing straight ahead.
2. Disconnect the negative battery cable.

✻✻ CAUTION

Work must be started after 90 seconds from the time the ignition switch is turned to the LOCK position and the negative (–) battery cable is disconnected.

3. Raise and safely support the vehicle. Remove the engine under cover (4wd).
4. Disconnect the right and left tie rod ends from the knuckle.
5. Disconnect the intermediate No. 2 shaft from the steering rack.
6. Using SST 09631–22020 or equivalent, remove the pressure feed and the return tubes.
7. Remove the power steering rack.

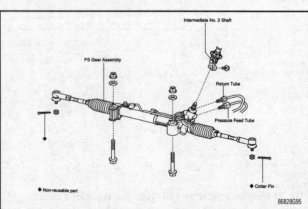

Fig. 56 Exploded view of the power steering gear—Tacoma 2WD

To install:

8. Install the power steering rack, then tighten the bolts on 2wd models to 148 ft. lbs. (201 Nm). On 4wd models, tighten the rack bolt to 123 ft. lbs. (167 Nm); rack nut and bolt to 141 ft. lbs. (191 Nm); and the bracket nut and bolt to 123 ft. lbs. (167 Nm).

9. Install a new O–ring and install the pressure feed tube, then tighten to 33 ft. lbs. (45 Nm). Install the return tube, then tighten to 36 ft. lbs. (49 Nm). On 4wd models, tighten both to 29 ft. lbs. (40 Nm).

10. Connect the intermediate No. 2 shaft to the steering rack.

11. Installation of the remaining components is in the reverse order of removal.

T100

2WD MODELS

1. Position the front wheels facing straight ahead.

2. Secure the steering wheel so that it does not turn. The driver's seat belt can be used to secure the steering wheel.

3. Place matchmarks on the intermediate shaft and control valve shaft. Remove the lower and upper joint bolts from the intermediate shaft. Disconnect the shaft.

4. Remove the cotter pin and nut from the tie rod ends. Disconnect the tie rod ends using a separator tool.

5. Disconnect the pressure and return pipes using flare nut wrenches.

6. Remove the bracket bolts and grommets.

7. Remove the steering rack assembly from the vehicle.

8. Install the gear housing then tighten the mounting bolts to 65 ft. lbs. (88 Nm).

9. Install the pressure and return tubes using new O-rings. Tighten the fittings to 15 ft. lbs. (20 Nm).

10. Connect the intermediate shaft. Install the lower and upper joint bolts. Tighten the bolts to 26 ft. lbs. (35 Nm).

11. Connect the tie rod ends, tighten to 67 ft. lbs. (90 Nm) and install a new cotter pin.

4WD MODELS

▶ **See Figure 57**

1. Place the front wheels facing straight ahead.
2. Drain the power steering fluid.
3. Disconnect the negative battery cable.

✳✳ CAUTION

Work must be started after 90 seconds from the time the ignition switch is turned to the LOCK position and the negative (–) battery cable is disconnected.

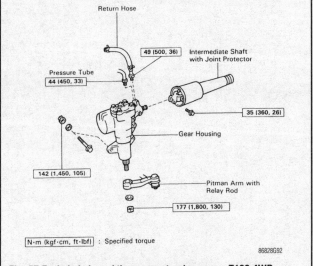

Fig. 57 Exploded view of the power steering gear—T100 4WD

4. Raise and safely support the vehicle.

5. Remove the Pitman arm set nut and the spring washer. Using SST 09628–62011, or equivalent, disconnect the Pitman arm from the gear assembly.

6. Disconnect the pressure feed tube and return tube.

7. Make matchmarks and disconnect the intermediate shaft from the steering gear.

8. Remove the power steering gear mounting bolts and remove the gear assembly.

9. Installation is in the reverse order of removal. Tighten the rack bolts to 105 ft. lbs. (142 Nm).

➡ **If installing a new gear assembly, be sure the gear box and the steering wheel are centered.**

10. Tighten the line fittings to 26 ft. lbs. (36 Nm).

Land Cruiser

1. Drain the power steering fluid.
2. Disconnect the negative battery cable.

✳✳ CAUTION

Work must be started after 90 seconds from the time the ignition switch is turned to the LOCK position and the negative (–) battery cable is disconnected.

3. Raise and safely support the vehicle.

4. Remove the frame seal.

5. Remove the link joint protector.

6. Remove the Pitman arm set nut and spring washer. Using SST 09628–62011, or equivalent, disconnect the Pitman arm from the gear assembly.

7. Using SST 09631–22020, or equivalent, disconnect the pressure feed and the return tubes.

8. Separate the sliding yoke sub–assembly from the steering gear.

9. Remove the mounting bolts and the power steering gear assembly.

10. Installation is the reverse of removal. Tighten the bolts to 105 ft. lbs. (142 Nm). Tighten the tube fittings to 26 ft. lbs. (36 Nm).

Power Steering Pump

REMOVAL & INSTALLATION

1. On some models it will be necessary to remove the air cleaner.

2. Disconnect the high tension lines at the distributor, then disconnect the air lines at the air valve.

3. With a suitable device, suck some fluid out of the reservoir tank.

4. Disconnect the return hose at the power steering pump.

5. Separate the pressure line at the pump. Some models have a union bolt with gaskets. Discard the old gaskets.

6. Loosen the drive belt pulley retaining nut. Loosen the idler pulley and adjusting bolt and remove the drive belt.

7. Remove the drive pulley and Woodruff key.

8. Remove the power steering pump. Some models have an O-ring, discard the old O-ring.

9. Installation is the reverse of removal. On the O-ring type, coat the O-ring with power steering fluid, then install it to the pump assembly.

10. Tighten the pump nuts to 27–29 ft. lbs. (36–39 Nm).

11. Tighten the pressure line flare nut to 33 ft. lbs. (45 Nm), and the union bolt with new gaskets to 42 ft. lbs. (56 Nm).

BLEEDING

✳✳ WARNING

Vehicles equipped with Rear Wheel Anti-Lock (RWAL) brakes use a different bleeding procedure than those not so equipped. Bleeding the RWAL system requires the use of an electronic brake tester, an expensive unit usually found only at dealers.

Without RWAL System

1. Check the fluid level in the reservoir. It should be at the correct level (HOT or COLD) depending on engine temperature.
2. Start the engine and run it below 1000 rpm.
3. Turn the steering wheel from lock-to-lock 3 or 4 times.
4. Shut the engine **OFF**. Connect a clear plastic tube to the bleeder port on the steering gearbox. Place the other end of the tube in a container of power steering fluid. Make sure the end is immersed in the fluid.
5. Start the engine again. Turn the wheel lock-to-lock 3–4 times and return the steering wheel to the centered position.
6. Loosen the bleeder plug. Observe the tubing; when no air bubbles are seen, close the bleeder plug.

✳✳ CAUTION

Do not allow the tubing to come off the bleed port. The power steering fluid is under high pressure and may be very hot.

7. Inspect the fluid level with the engine running; the fluid should not be foamy or cloudy. Shut the engine **OFF** and check the fluid level. If level rises too much, try re-bleeding the system. If the problem persists, repair the power steering system.

With RWAL System

The RWAL system uses power steering fluid pressure to maintain control pressures in the brake system. While the fluids never mix, air in the power steering system can render the rear anti-lock brakes ineffective.

If the any of the lines or components in the power steering system are loosened or removed, the brake system must be bled as well. Preliminary bleeding of both the power steering and brake systems will eliminate most of the air, but the vehicle MUST be taken to a dealer to have the systems properly bled using the Toyota ABS Checker. Be warned that until this is done, the rear wheel anti-lock brake function may be impaired.

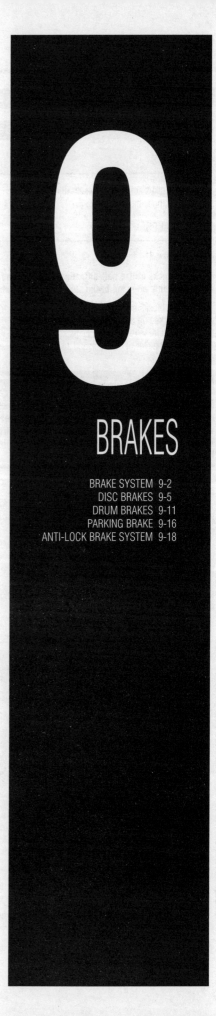

9

BRAKES

BRAKE SYSTEM

Adjustments

DISC BRAKES

All disc brakes are inherently self-adjusting. No periodic adjustment is either necessary or possible.

DRUM BRAKES

▶ See Figure 1

➡ Toyota trucks utilize self-adjusting brakes. The following procedure is necessary only after the brake shoes have been changed.

1. Place blocks under the front wheels so that the truck will not roll forward when it is jacked up at the rear.
2. Fully release the emergency brake.
3. Raise the rear of the truck and support the differential housing with jack-stands.

✳✳ WARNING

Brake pads and shoes may contain asbestos, which has been determined to be a cancer causing agent. Never clean the brake surfaces with compressed air. Avoid inhaling any dust from brake surfaces. When cleaning brakes, use commercially available brake cleaning fluids.

4. Remove the plug from the adjusting hole at the bottom of the backing plate.
5. Turn the adjusting starwheel to expand the shoes fully. While doing this, have a friend step on the brake pedal occasionally to center the shoes.
6. Tighten the shoes until the wheel will not turn when the pedal is released.
7. From this position back off the adjuster until the wheel turns with just a slight drag.
8. Back off the adjuster an additional five notches. The wheel should turn smoothly. If it does not, back off another two or three notches. Should this fail, check for worn or defective parts.
9. Adjust the other wheel in the same manner.

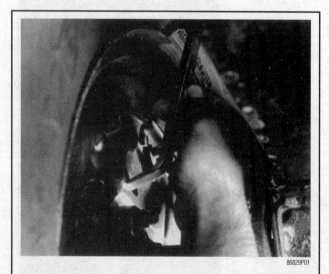

Fig. 1 A special tool used to turn the adjuster wheel is available

BRAKE PEDAL

Pedal Height

▶ See Figure 2

1. Measure the distance between the center (upper surface) of the pedal pad and the floor pad, not the carpet.
2. If out of specifications, loosen the brake light switch.
3. Loosen the locknut, then turn the pedal pushrod until the pedal height is within specifications:
- 2WD Pick-Up—5.83 inches (148mm)
- 4WD Pick-Up—5.71 inches (145mm)
- Tacoma—6.283–6.677 inches (159.6–169.6mm)
- 1989–95 4Runner—5.71–5.91 inches (145–150mm)
- 1996 4Runner—6.224–6.618 inches (158.1–168.1mm)
- T100 (Extra cab 4WD)—5.776–6.196 inches (146.7–156.7mm)
- T100 (Except extra cab 4WD)—5.894–6.287 inches (149.7–159.7mm)
- 1989–90 Land Cruiser—6.34–6.73 inches (161–171mm)
- 1991–93 Land Cruiser—7.09 inches (180mm)
- 1994–96 Land Cruiser—6.59–6.99 inches (167.5–177.5mm)
4. Tighten the pushrod locknut to 19 ft. lbs. (25 Nm).
5. Move the brake light switch until the plunger is just touching the pedal stopper. Tighten the switch.
6. Check the brake pedal free-play.

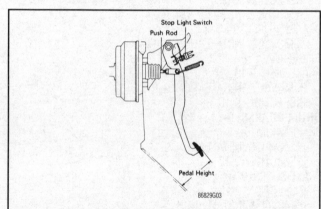

Fig. 2 Pedal height is measured from the floor pad, not the carpet

Free-Play

▶ See Figure 3

1. With the engine turned off, depress the brake pedal several times until there is no vacuum left in the brake booster.
2. Push the pedal down until resistance is first felt. Measure this distance; it should be 0.12–0.24 inches (3–6mm).
3. Adjust the free-play by turning the pedal pushrod.
4. Start the engine and recheck the free-play.
5. Recheck the pedal height.

Brake Light Switch

REMOVAL & INSTALLATION

1. Disconnect the electrical harness at the switch.
2. Remove the locknut (closest to the pedal). Remove the switch from the brake pedal.

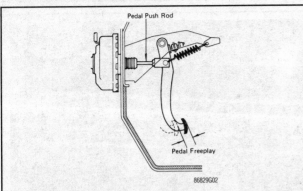

Fig. 3 Pedal free-play is the amount of motion before the brakes are applied

To install:

3. Insert the new switch and loosely install the locknut. Adjust the switch so that plunger at the tip of the switch is held in by the pedal in its at-rest position. Remember that the pedal holds the switch off; the brake lamps come on when the pedal moves away from the switch.

4. Once positioned, tighten the locknut and connect the wiring. Check the switch function by pressing the pedal. Have an assistant observe the brake lamps for proper operation.

Master Cylinder

REMOVAL & INSTALLATION

▶ **See Figure 4**

❊❊ WARNING

Be careful not to spill brake fluid on the painted surfaces. It will damage the paint. If a spill occurs, flush the area with plenty of fresh water.

1. Bleed the air from the master cylinder.
2. Unplug the fluid level warning switch connector.
3. Use a syringe or similar tool to remove as much fluid as possible from the reservoir.
4. Carefully disconnect the brake lines at the master cylinder. To prevent line damage, use a line wrench. Regular open-end wrenches are NOT recommended. Take great care not to bend or deform the lines. Plug the lines to keep dirt from entering.
5. Remove the nuts holding the master cylinder to the booster. Unbolt the 3-way or 4-way union. When removing the cylinder, take care to recover the clamps, gaskets, etc.

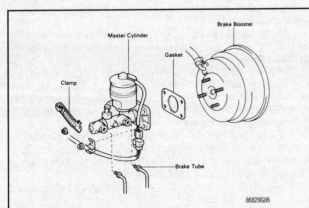

Fig. 4 Exploded view of a common brake master cylinder mounting found on Toyotas

6. Installation is in the reverse order of removal. Using a new gasket, tighten the mounting nuts to 9 ft. lbs. (13 Nm).
7. Tighten the brake tube fittings to 11 ft. lbs. (15 Nm).
8. Bleed the brake system. Check again for any leakage.
9. Check and adjust the brake pedal.

Power Brake Booster

REMOVAL & INSTALLATION

▶ **See Figure 5**

1. Remove the master cylinder.
2. At the brake pedal, remove the clip and separate the clevis from the pedal.
3. Remove the pedal return spring.
4. Disconnect the vacuum hose from the booster.
5. Loosen the four nuts, then pull out the vacuum booster, bracket and gasket.
6. Installation is in the reverse order of removal.

➡ **Some models employ a spacer and additional gaskets at the firewall.**

7. Using a new gasket, install the booster, then tighten the mounting bolts to 9 ft. lbs. (13 Nm).
8. Adjust the length of the booster pushrod as follows:
   a. Install the gasket to the master cylinder.
   b. Set the SST 09737–00010 or equivalent on the gasket, then lower the pin until its tip slightly touches the piston.
   c. Turn the tool upside down and position it on the booster.
   d. Measure the clearance between the booster pushrod and the pin head. It should be null.
   e. Adjust the booster pushrod length until the pushrod lightly touches the pin head. While doing this depress the brake pedal enough so that the pushrod sticks out.

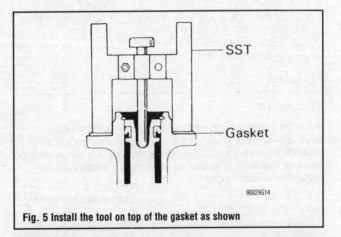

Fig. 5 Install the tool on top of the gasket as shown

Load Sensing Proportioning (LSP) and Bypass Valve (BV)

The purpose of this valve is to control the fluid pressure applied to the brakes to prevent rear wheel lock-up during weight transfer at high speed stops.

REMOVAL & INSTALLATION

▶ **See Figure 6**

1. Disconnect the shackle from the bracket. Remove the cotter pin, then the nut and cushion retainer. Disconnect the shackle and remove the cushions and collar.
2. Disconnect the brake lines attached to the valve. Plug the lines to keep dirt from entering.
3. Remove the mounting bolt, if used, and remove the valve.

Fig. 6 The load sensing valve can be found secured to the frame rail

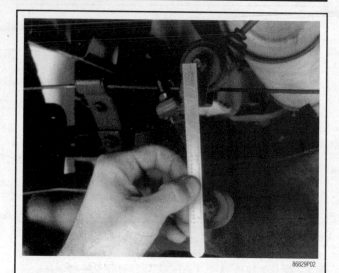

Fig. 7 Measure the distance between the centers of the shackles

To install:

➡**This valve can not be rebuilt. It must be replaced.**

4. Install the LSP and BV assembly to the frame, tighten to 19–22 ft. lbs. (25–29 Nm).
5. Install the shackle to the load sensing spring.
6. Attach the shackle to the shackle bracket, tighten to 9 ft. lbs. (13 Nm). Install a new cotter pin.
7. Attach the brake lines, tighten to 11 ft. lbs. (15 Nm).
8. Set the valve body as follows:
   a. When pulling the load sensing spring, confirm that the valve piston moves down slowly.
   b. Position the valve body so that the valve piston lightly contacts the load sensing spring.
   c. Tighten the valve body mounting nuts to 9 ft. lbs. (13 Nm).
9. Bleed the brake system.

ADJUSTMENT

▸ **See Figure 7**

➡**This is an initial adjustment which should bring the valve close to specifications. Since it requires the use of pressure gauges and determines the exact weight of the load on the rear axle, it is not practical for the owner/mechanic. Have the final adjustments performed at a reputable shop as soon as possible.**

1. Pull down the load sensing spring to determine that the piston moves slowly.
2. Set the valve body so that the piston lightly touches the load sensing spring.
3. Tighten the mounting bolts.
4. Adjust the length of the shackle. Lengthening the distance between the centers lowers the pressure; shortening the distance raises the pressure. One full turn of the nut increases pressure about 20 psi. on Land Cruiser or about 11–14 psi (75–96 kPa) on all other vehicles. Initial length of this link should be:
- Pick-Up and 4Runner (1989–95)—3.54 inches (90mm)
- Land Cruiser (1989–90)—3.07 inches (78mm)
- Land Cruiser (1991–96)—3.54 inches (90mm)
- T100—4.72 inches (120mm)
- Tacoma 2WD—3.07 inches (78mm)
- Tacoma 4WD—4.72 inches (120mm)
5. If the initial distance is correct, fluid metering should be very close to specification.

Brake Hoses and Lines

BRAKE LINE FLARING

Use only brake line tubing approved for automotive use; never use copper tubing. Whenever possible, try to work with brake lines that are already cut to the length needed. These lines are available at most auto parts stores and have machine made flares, the quality of which is hard to duplicate with most of the available inexpensive flaring kits.

When the brakes are applied, there is a great amount of pressure developed in the hydraulic system. An improperly formed flare can leak with resultant loss of stopping power. If you have never formed a double-flare, take time to familiarize yourself with the flaring kit; practice forming double-flares on scrap tubing until you are satisfied with the results.

The following procedure applies to the SA9193BR flaring kit, but should be similar to commercially available brake-line flaring kits. If these instructions differ in any way from those in your kit, follow the instructions in the kit.

1. Determine the length necessary for the replacement or repair and allow an additional ⅛ in. (3.2mm) for each flare. Select a length of tubing according to the repair/replacement charts in the figure, then cut the brake line to the necessary length using an appropriate saw. Do not use a tubing cutter.
2. Square the end of the tube with a file and chamfer the edges. Remove burrs from the inside and outside diameters of the cut line using a deburring tool.
3. Install the required fittings onto the line.
4. Install SA9193BR, or an equivalent flaring tool, into a vice and install the handle into the operating cam.
5. Loosen the die clamp screw and rotate the locking plate to expose the die carrier opening.
6. Select the required die set (4.75mm DIN) and install in the carrier with the full side of either half facing clamp screw and counter bore of both halves facing punch turret.
7. Insert the prepared line through the rear of the die and push forward until the line end is flush with the die face.
8. Make sure the rear of both halves of the die rest against the hexagon die stops, then rotate the locking plate to the fully closed position and clamp the die firmly by tightening the clamp screw.
9. Rotate the punch turret until the appropriate size (4.75mm DIN) points towards the open end of the line to be flared.
10. Pull the operating handle against the line resistance in order to create the flare, then return the handle to the original position.
11. Release the clamp screw and rotate the locking plate to the open position.

12. Remove the die set and line, then separate by gently tapping both halves on the bench. Inspect the flare for proper size and shape. Measurement should be 0.272–0.286 in. (6.92–7.28mm).

13. If necessary, repeat the steps for the other end of the line or for the end of the line which is being repaired.

14. Bend the replacement line or section using SA91108NE, or an equivalent line bending tool.

15. If repairing the original line, join the old and new sections using a female union and tighten.

Bleeding the Brake System

It is necessary to bleed the hydraulic system any time the system has been opened or has trapped air within the fluid lines. It may be necessary to bleed the system at all four brakes if air has been introduced through a low fluid level or by disconnecting brake pipes at the master cylinder.

If a line is disconnected at one wheel only, generally only that brake needs bleeding. If lines are disconnected at any fitting between the master cylinder and the brake, the system served by the disconnected pipe must be bled.

※ WARNING

Do not allow brake fluid to splash or spill onto painted surfaces; the paint will be damaged. If spillage occurs, flush the area immediately with clean water.

MASTER CYLINDER

♦ See Figure 8

If the master cylinder has been removed, the lines disconnected or the reservoir allowed to run dry, the cylinder must be bled before the lines are bled. To bleed the master cylinder:

1. Check the level of the fluid in the reservoir. If necessary, fill with fluid.

2. Disconnect the brake lines from the master cylinder. Plug the lines to keep dirt from entering.

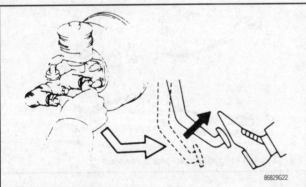

Fig. 8 Depress the brake pedal, block the ports with your fingers and release the pedal

DISC BRAKES

※ CAUTION

Brake pads may contain asbestos, which has been determined to be a cancer causing agent. Never clean the brake surfaces with compressed air! Avoid inhaling any dust from any brake surface! When cleaning brake surfaces, use a commercially available brake cleaning fluid.

3. Place a pan or rags under the cylinder.

4. Have an assistant slowly depress the brake pedal and hold it down.

5. Block off the outlet ports with your fingers. Be sure to wear gloves. Have the assistant release the pedal. Make a tight seal with your fingers; do not allow the cylinder to ingest air when the pedal is released.

6. Repeat three or four times.

7. Connect the brake lines to the master cylinder and top up the fluid reservoir.

LINES AND WHEEL CIRCUITS

♦ See Figure 9

1. Insert a clear vinyl tube onto the bleeder plug at the wheel. If all four wheels are to be bled, begin with the right rear.

2. Insert the other end of the tube into a jar which is half filled with brake fluid. Make sure the end is submerged in the fluid.

3. Have an assistant slowly pump the brake pedal several times. On the last pump, have the assistant hold the pedal fully depressed. While the pedal is depressed, open the bleeder plug until fluid starts to run out, then close the plug.

➡**If the brake pedal is pumped too fast, small air bubbles will form in the brake fluid which will be very difficult to remove.**

4. Repeat this procedure until no air bubbles are visible in the hose. Close the bleeder port.

➡**Constantly replenish the brake fluid in the master cylinder reservoir, so that it does not run out during bleeding.**

5. If bleeding the entire system, repeat the procedure at the left rear wheel, the right front wheel and the left front wheel in that order.

6. Bleed the load sensing proportioning and bypass valve.

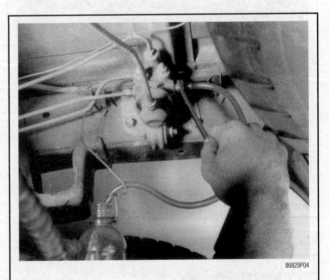

Fig. 9 A line wrench can be used to loosen the bleeder plug

Brake Pads

WEAR INDICATORS

The front disc brake pads are equipped with a metal tab which will come into contact with the rotor after the friction surface material has worn near its usable minimum. The wear indicators make a constant, distinct metallic sound that should be easily heard. The sound has been described as similar to either fingernails on a blackboard or a field full of crickets. The key to recognizing that it is the wear indicators and not some other brake noise is that the sound is heard when the vehicle is being driven WITHOUT the brakes applied. It may or may not be present under braking and is heard during normal driving.

It should also be noted that any disc brake system, by its design, cannot be made to work silently under all conditions. Each system includes various shims, plates, cushions and brackets to suppress brake noise, but no system can completely silence all noises. Some brake noise, either high or low frequency, can be controlled and perhaps lessened, but cannot be totally eliminated.

INSPECTION

◆ See Figure 10

The front brake pads may be inspected without removal. With the front end elevated and supported, remove the wheel(s). Unlock the steering column lock and turn the wheel so that the brake caliper is out from under the fender.

View the pads, inner and outer, through the cut-out in the center of the caliper. Remember to look at the thickness of the pad friction material (the part that actually presses on the disc) rather than the thickness of the backing plate which does not change with wear.

Keep in mind that you are looking at the profile of the pad, not the whole thing. Brake pads can wear on a taper which may not be visible through the window. It is also not possible to check the contact surface for cracking or scoring from this position. This quick check can be helpful only as a reference; detailed inspection requires pad removal.

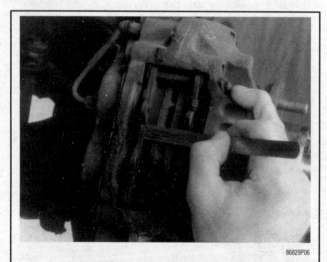

Fig. 10 The pad thickness can be measured through the inspection window

REMOVAL & INSTALLATION

2WD Vehicles and Rear Discs

2WD VEHICLES WITH PD60 AND 66 TYPE BRAKES

◆ See Figures 11, 12 and 13

1. Raise the front of the truck and safely support it with jackstands.
2. Remove the lug nuts and the wheel.
3. Remove the lower installation bolt from the torque plate. The torque plate is the framework that holds the caliper.
4. Pivot the caliper assembly upwards and suspend it from the suspension with a wire. Do NOT disconnect the brake line, but take care not to stretch it.
5. Remove the 2 anti-squeal springs and lift out the brake pads.
6. Remove the 2 anti-squeal shims and the 4 pad support plates. Pull the 2 pad wear indicator off the pads.
7. Check the pad thickness and replace the pads if they are less than 0.039 in. (1mm) thick. New pads measure approximately 0.374–0.472 in. (9.6–12.0mm) thick depending on the vehicle.

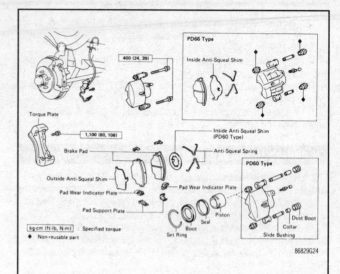

Fig. 11 Exploded view of the PD 60 and 66 type brake components

➡This minimum thickness measurement may disagree with your state inspection laws.

To install:
8. Install the 4 pad support plates into the torque plate.
9. Install a new pad wear indicator plate to each pad.
10. Install the large anti-squeal shim to the back of the outer pad (PD 60) or to the back of each pad (PD 66) and then position both pads into the torque plate. Install the anti-squeal springs in place.

➡When installing new brake pads, make sure your hands are clean. Do not allow any grease or oil to touch the contact face of the pads or the brakes may not stop the truck properly.

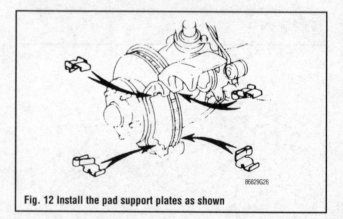

Fig. 12 Install the pad support plates as shown

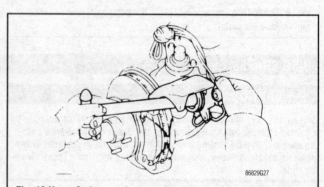

Fig. 13 Use a C-clamp or hammer handle to seat the caliper piston back into the housing

11. Remove a small amount of fluid from the reservoir, using a syringe or similar tool. Use a C-clamp or hammer handle and press the caliper piston back into the housing.

❊❊ WARNING

Never press the piston into the caliper when the pads are out on both sides of the truck. The opposite piston may pop completely out of the caliper. This may spoil your afternoon.

12. For PD 60 types, press the round anti-squeal shim over the caliper piston; position the caliper over the torque plate so the dust boot is not pinched.

13. Swing the caliper down and fit the lower bolt. Tighten it to 29 ft. lbs. (39 Nm).

14. Install the wheel and lower the truck. Check the brake fluid level in the reservoir and fill to the max line if necessary. (The level should have risen when the piston was pushed back.)

15. Apply the parking brake and put the vehicle in Park or Neutral. Step on the brake pedal two or three times. The first pedal application will be due to the pistons being pushed back. The pedal travel should be normal after two or three pumps. Do NOT drive the vehicle until the pedal has a normal feel. If necessary, bleed the brakes to remove pedal sponginess.

➡Braking should be moderate for the first 5 miles (8 km) or so until the new pads seat correctly. The new pads will bed best if put through several moderate heating and cooling cycles. Avoid hard braking until the brakes have experienced several long, slow stops with time to cool in between. Taking the time to properly bed the brakes will yield quieter operation, more efficient stopping and contribute to extended brake life.

2WD VEHICLES WITH FS17 AND 18 TYPE BRAKES

VEHICLES WITH REAR DISC BRAKES

▶ **See Figures 13, 14, 15, 16 and 17**

1. Remove the hub cap and loosen the lug nuts.
2. Raise the front of the truck and safely support it with jackstands.
3. Remove the lug nuts and the wheel.
4. Remove the caliper slide pin on the sub-pin (lower) side.
5. Swivel the caliper up and away from the torque plate. Tie the caliper to a suspension member so it is out of the way. Do not disconnect the brake line.
6. Lift the 2 brake pads out of the torque plate. Remove the anti-squeal shim.
7. Remove the 4 pad support plates. Pull the 2 pad wear indicator plates off the pads.
8. Check the pad thickness and replace the pads if they are less than 0.03 in. (1mm) thick. New pads measure approximately 0.374 in. (9.6mm) thick.

➡**This minimum thickness measurement may disagree with your state inspection laws.**

To install:
9. Install the 4 pad support plates into the torque plate.
10. Install a new pad wear indicator plate to the bottom of each pad.
11. Install the anti-squeal shim to the back of the outer pad and then position both pads into the torque plate.

❊❊ WARNING

When installing new brake pads, make sure your hands are clean. Do not allow any grease or oil to touch the contact face of the pads or the brakes will not stop the truck properly.

12. Use a syringe or similar tool to remove some fluid from the reservoir. Use a C-clamp or hammer handle and press the caliper piston back into the housing.

13. Untie the caliper and swivel it back into position over the torque plate so that the dust boot is not pinched. Install the slide pin and tighten it to 65 ft. lbs. (88 Nm).

14. Check the condition of the cylinder side bushing boot.

15. Install the wheel and lower the truck.

16. Check the brake fluid level in the reservoir and fill to the max line if necessary. (The level should have risen when the piston was pushed back.)

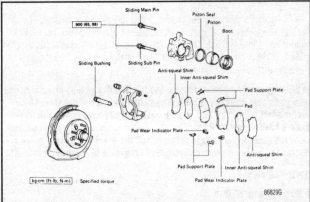

Fig. 14 Exploded view of the FS17 and FS18 type brake components

Fig. 15 Exploded view of the brake pad components on the FS17 and 18 type brakes

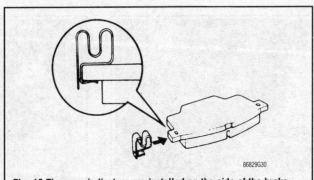

Fig. 16 The wear indicators are installed on the side of the brake pad as shown

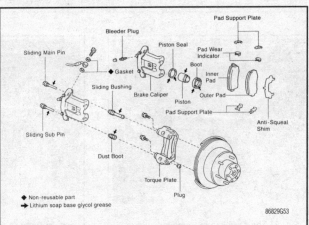

Fig. 17 Exploded view of the rear disc brakes—Land Cruiser

17. Apply the parking brake and put the vehicle in Park or Neutral. Step on the brake pedal two or three times. The first pedal application will be due to the pistons being pushed back. The pedal travel should be normal after two or three pumps. Do NOT drive the vehicle until the pedal has a normal feel. If necessary, bleed the brakes to remove pedal sponginess.

➡**Braking should be moderate for the first 5 miles (8 km) or so until the new pads seat correctly. The new pads will bed best if put through several moderate heating and cooling cycles. Avoid hard braking until the brakes have experienced several long, slow stops with time to cool in between. Taking the time to properly bed the brakes will yield quieter operation, more efficient stopping and contribute to extended brake life.**

4WD Vehicles

▶ **See Figures 18 thru 23**

1. Raise the front of the truck, support it on jackstands, and remove the front wheel.
2. Pull out the wire clip at the ends of the pad pins.
3. Remove the pins. Remove the anti-rattle spring (W-shaped on older trucks), the brake pads and the four anti-squeal shims.
4. Check the pad thickness and replace the pads if they are less than 0.03 in. (1mm) thick. New pads measure approximately 0.374 in. (9.6mm) thick.

➡**This minimum thickness measurement may disagree with your state inspection laws.**

5. Check the pins for straightness and wear, and replace if necessary.
To install:
6. Use a syringe or similar tool to remove a small amount of fluid from the master cylinder reservoir.
7. Use a C-clamp or hammer handle and press the caliper pistons back into the housing.

➡**Never press the pistons into the caliper when the pads are out on both sides of the truck.**

8. Install the 4 anti-squeal shims so that the black shims are between the silver shims and the brake pad.
9. Install the brake pads. Be very careful not to get grease or oil on the inner surfaces of the pads.
10. Install the anti-rattle spring.
11. Slide the 2 pad retaining pins through the caliper and pads and install the retaining clip.
12. Install the wheel and lower the truck. Bleed the brakes and road test the vehicle.
13. Check the brake fluid level in the reservoir and fill to the max line if necessary. (The level should have risen when the piston was pushed back.)
14. Apply the parking brake and put the vehicle in Park or Neutral. Step on the brake pedal two or three times. The first pedal application will be due to the pistons being pushed back. The pedal travel should be normal after two or three pumps. Do NOT drive the vehicle until the pedal has a normal feel. If necessary, bleed the brakes to remove pedal sponginess.

➡**Braking should be moderate for the first 5 miles (8 km) or so until the new pads seat correctly. The new pads will bed best if put through several moderate heating and cooling cycles. Avoid hard braking until the brakes have experienced several long, slow stops with time to cool in between. Taking the time to properly bed the brakes will yield quieter operation, more efficient stopping and contribute to extended brake life.**

Brake Caliper

REMOVAL & INSTALLATION

Vehicles With 2WD or Rear Disc Brakes

1. Elevate and safely support the vehicle.
2. Remove the wheel.
3. Disconnect the brake line from the caliper.

Fig. 18 Remove the wire clips securing the pins

Fig. 19 The pins can be removed with a drift

Fig. 20 Remove the anti-rattle clips from the pads

Fig. 21 The pads can be removed with a pair of needle-nose pliers

Fig. 22 Remove the shims from the back of the pad

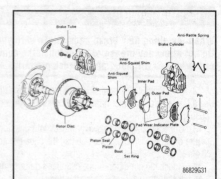

Fig. 23 Exploded view of the 4WD front brake components—1991 Land Cruiser shown

4. Remove the bolts holding the caliper to the torque plate.
5. Remove the caliper from the torque plate.
6. Remove the pads and shims.

To install:

7. Install the pads and shims.
8. Fit the caliper into position. Install the two retaining or slide bolts, then tighten to:

- FS–17 or 18 brakes; rear discs—65 ft. lbs. (88 Nm)
- PD 60 or 66 brakes—29 ft. lbs. (39 Nm)
- T100 ½ ton—27 ft. lbs. (36 Nm)
- T100 1 ton—29 ft. lbs. (39 Nm)

9. Install the brake line, tightening the fitting to 11–22 ft. lbs. (15–30 Nm).
10. Install the wheel.
11. Lower the vehicle to the ground. Bleed the brakes and road test the vehicle.

4WD Vehicles

▶ **See Figures 24, 25 and 26**

1. Raise the front of the truck, support it on jackstands, and remove the front wheel.
2. Plug the vent hole on the master cylinder cap to prevent fluid leak. Loosen the brake line from the caliper, being careful not to deform the fittings.

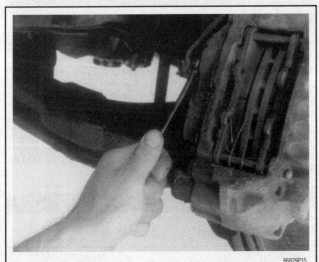

Fig. 24 With a line wrench, loosen the brake line from the caliper

Fig. 25 Remove the upper and lower mounting bolts

Fig. 26 Slide the caliper assembly off of the rotor

3. Remove the 2 caliper mounting bolts, and remove the caliper.
4. Pull out the two wire clips at the ends of the pad pins.
5. Pull out the pads and anti-rattle spring.
6. Remove the brake pads, and the anti-squeal shims.

To install:

7. Position the caliper and install the mounting bolts. Tighten the caliper mounting bolts to 90 ft. lbs. (123 Nm). Be certain to position the shims correctly.
8. Connect the brake line to the caliper. On all models except 1991–96 Land Cruisers, tighten the line fitting to 11 ft. lbs. (15 Nm). For 1991–96 Land Cruisers, tighten it to 22 ft. lbs. (30 Nm).
9. Install the brake pads.
10. Install the wheel and lower the truck. Bleed the brakes and road test the vehicle.

OVERHAUL

2WD Vehicles

1. Raise the front of the truck and support with safety stands. Remove the wheels.
2. Remove the caliper.
3. Remove the slide bushings, pin dust boots and 2 collars (PD60 only).
4. Carefully pry the dust boot set ring out of the caliper. Remove the dust boot.
5. Insert a piece cloth or rag between the caliper claw and the piston.
6. Apply compressed air to the brake line union to force the piston out of its bore. Be careful; the piston may come out forcefully.

> ✳✳ **CAUTION**
>
> **Keep fingers away from the front of the piston during removal. Injury may occur when the piston comes out.**

7. Remove the piston seal from the caliper cylinder bore. Check the piston and cylinder bore for wear and/or corrosion. Replace components as necessary.

To assemble:

8. Coat all components with clean brake fluid or lithium soap base glycol grease.
9. Install the piston seal and then the piston into the caliper cylinder bore. Seat the piston in the bore with your fingers.
10. Fit the boot into the groove in the cylinder bore and press in the set ring.
11. On FS17 and rear disc types, install the main pin boots to the torque plate. Install the sub pin slide bushing into the torque plate.
12. On PD60 types, install the collar and dust boots into the caliper. Make sure they are firmly secured to the groove in the caliper. Install the bushing into the boots so that it is firmly set in the boot groove.

13. Install the caliper and brake pads.

14. Install the wheels and lower the truck. Bleed the brakes and road test the vehicle.

4WD Vehicles

▶ **See Figures 27, 28 and 29**

1. Raise the front of the truck and support with safety stands. Remove the wheels.

2. Remove the caliper.

3. Carefully pry the dust boot set rings out of the caliper. Remove the dust boots.

4. Insert a piece of wood about ½ in. thick between the caliper pistons.

5. Apply compressed air to the brake line union to force the piston out of its bore. Be careful; the piston will come out forcefully.

❉❉ CAUTION

Keep fingers away from the front of the piston during removal. Injury may occur when the piston comes out.

6. Remove the piston seals from the caliper cylinder bores. Check each piston and cylinder bore for wear and/or corrosion. Replace components as necessary.

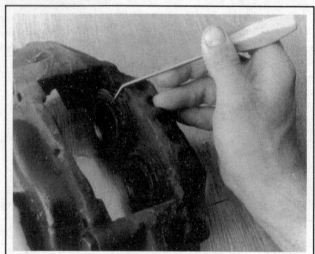

Fig. 27 Using a seal pick, remove the dust boot rings

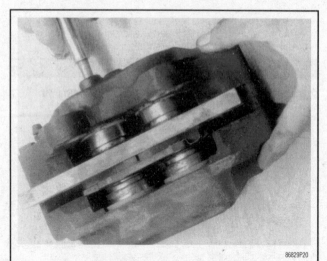

Fig. 28 Place a piece of wood between the pistons, then apply compressed air to force them out

Fig. 29 Once the pistons are removed, use the seal pick to withdraw the piston seals

❉❉ WARNING

Do not loosen the four bolts holding the two halves of the caliper together.

To assemble:

7. Coat all components with clean brake fluid or lithium soap based glycol grease.

8. Install the piston seals then the pistons into the caliper cylinder bores. Seat the pistons in the bores with your fingers.

9. Fit the boots into the grooves in the cylinder bores and press in the set rings.

10. Install the caliper and brake pads.

11. Install the wheels and lower the truck. Bleed the brakes and road test the vehicle.

Brake Disc

REMOVAL & INSTALLATION

1. Disconnect the ABS speed sensor from the knuckle if equipped.

2. Remove the brake pads and the caliper.

3. Check the disc run-out at this point. Make a note of the results for use during installation. Refer to the inspection procedure.

4. Remove the grease cap from the hub. Remove the cotter pin and the castellated nut.

5. Remove the wheel hub with the brake rotor attached.

6. Perform the rotor inspection procedure.

To install:

7. Coat the hub oil seal lip with multi-purpose grease and install the disc/hub assembly.

8. Adjust the wheel bearing.

9. Using a dial indicator, measure the disc run-out. Check it against the specifications in the Brake Specifications chart and the figures noted earlier.

➡**If the wheel bearing nut is improperly tightened, disc run-out will be affected.**

10. Install the remainder of the components as outlined in the appropriate sections.

11. Bleed the brake system.

12. Road test the truck. Double check the wheel bearing preload.

INSPECTION

Examine the disc. If it is worn, warped or scored, it must be replaced. Check the thickness of the rotor against the specifications given in the Brake Specifications chart. If below specifications, replace it. Use a micrometer to measure the thickness. Because the allowable wear from new thickness to minimum thickness is only approx. 0.078 in. (2mm), resurfacing the discs is not recommended.

The disc run-out should be measured before the rotor is removed and again, after the rotor is installed. Use a dial indicator mounted on a stand to determine run-out. If run-out exceeds the maximum specification, replace the disc. Maximum runout is as follows:
- 2WD Pick-Up PD 60 brakes—0.0035 in. (0.09mm)

- 2WD Pick-Up PD 66 brakes—0.0047 in. (0.12mm)
- 2WD Pick-Up FS 17 or 18 brakes—0.0035 in. (0.09mm)
- 4WD 1989–95 Pick-Up and 4Runner—0.0035 in. (0.09mm)
- T100, Tacoma and 1996 4Runner—0.0028 in. (0.07mm)
- Land Cruiser (front and rear)—0.0059 in. (0.015mm)

➡ **Be sure that the wheel bearing nut is properly tightened. If it is not, an inaccurate run-out reading may be obtained. If different run-out readings are obtained with the same disc, between removal and installation, this is probably the cause.**

DRUM BRAKES

✳✳ CAUTION

Brake shoes may contain asbestos, which has been determined to be a cancer causing agent. Never clean the brake surfaces with compressed air! Avoid inhaling any dust from any brake surface! When cleaning brake surfaces, use a commercially available brake cleaning fluid.

Brake Drum

REMOVAL & INSTALLATION

▶ **See Figures 30, 31 and 32**

1. Remove the hub cap (if used) and loosen the lug nuts. Release the parking brake.
2. Block the front wheels, raise the rear of the truck, and support it with jackstands.
3. Remove the rear wheels.
4. Unfasten the brake drum retaining screws, if used. Not all models have them.
5. Tap the drum lightly with a mallet in order to free it.
6. If the drum cannot be removed easily, insert a screwdriver into the hole in the backing plate and hold the automatic adjusting lever away from the adjusting bolt. Using another flat-bladed tool, relieve the brake shoe tension by turning the adjusting bolt clockwise. If the drum still will not come off, use a puller, but first make sure that the parking brake is fully released.

✳✳ WARNING

Do not depress the brake pedal once the brake drum has been removed.

7. Installation is in the reverse order of removal. Inspect the brake drum for any wear or deterioration. Check the inside diameter of the drum with the Brake Specifications chart. Replace if necessary.

86829P24

Fig. 31 Lubricate the studs to help ease the drum from the hub

86829P23

Fig. 30 Bolts can be inserted into the holes as shown to aid in drum removal

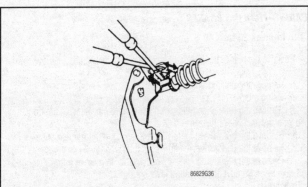
86829G36

Fig. 32 Hold the lever away from the bolt while rotating the adjuster bolt

INSPECTION

▶ **See Figure 33**

1. Clean the drum.
2. Inspect the drum for scoring, cracks, grooves and out-of-roundness. Replace the drum or have it "turned" at a machine or brake specialist shop, as required. Light scoring may be removed by dressing the drum with fine emery cloth.
3. Measure the inside diameter of the drum. A tool called a H-gauge caliper is used. See the Brake Specifications chart for your vehicle.

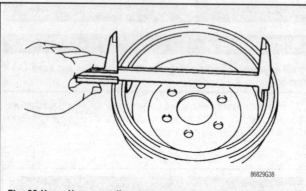

Fig. 33 Use a H-gauge caliper to measure the inside diameter of a brake drum

Brake Shoes

INSPECTION

The brake shoes can be inspected without removing the brake drum if necessary. Remove the inspection hole plug in the backing plate. Check the shoe lining thickness through the hole. If they seem to be less than the minimum of 0.039 in. (1.0mm), replace the shoes.

REMOVAL & INSTALLATION

➡**Drum brakes require the use of two common brake tools to make the job easier. A brake spring wrench and a spring removing tool can be purchased at almost any automotive outlet for low cost. These are two of the handiest special tools you can own; they can generally be used on any vehicle with drum brakes.**

2WD Vehicles

LEADING-TRAILING BRAKES

▶ **See Figures 34 thru 40**

1. Raise the rear of the truck and support it with jackstands. Remove the wheels.
2. Remove the brake drum.
3. Using a brake tool, remove the return spring.
4. Remove the front shoe hold-down spring and pin. Remove the front shoe with the anchor spring.
5. Remove the hold-down spring and pin and pull out the rear shoe.
6. Remove the parking brake cable from the lever.
7. Remove the E-clip and spread the C-washer to remove it from the parking brake lever. Remove the adjusting lever spring.
To install:
8. Apply high temperature grease to the contact surface of the brake backing plate and to the threads and end of the adjuster bolt.
9. Pull the parking brake cable out, then attach it to the lever.
10. Connect the strut and return spring to the lever.
11. Position the rear shoe so that the upper end of the shoe is in the wheel cylinder and the lower end is in the anchor plate. Install the pin and hold-down spring.

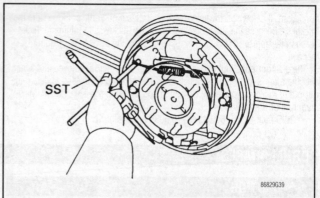

Fig. 34 Special tools are available to remove the return spring

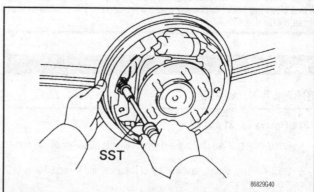

Fig. 35 Push in the special tool to remove the hold-down spring and retainers

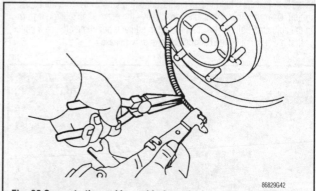

Fig. 36 Separate the parking cable from the lever, then remove the shoe

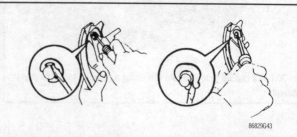

Fig. 37 Remove the E-clip and spread the C-washer to remove it from the parking brake lever

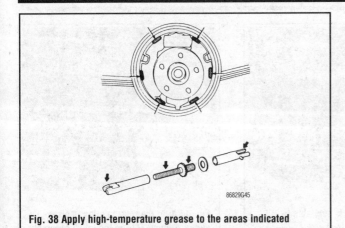

Fig. 38 Apply high-temperature grease to the areas indicated

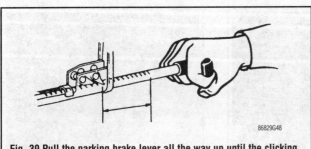

Fig. 39 Pull the parking brake lever all the way up until the clicking is no longer heard

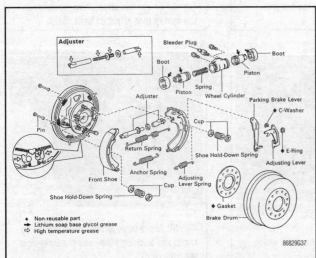

Fig. 40 Exploded view of the 2WD rear brake components with leading-trailing brakes

12. Connect the anchor spring to the rear shoe and then stretch it onto the front shoe.

13. Position the front shoe so the upper end is in the wheel cylinder and the lower end is in the strut. Install the pin and hold-down spring.

14. Install the return spring.

15. Check that the adjuster bolt turns while pulling up on the brake lever. Adjust the strut to the shortest possible length and install the brake drum. Pull the parking brake lever (inside the truck) out until the clicks stop.

16. Remove the brake drum and check that the difference between the inner brake drum diameter and the outer brake shoe diameter is no more than 0.024 in. (0.6mm). Adjust or replace parts as needed.

17. Install the brake drum. Install the wheels and lower the truck. Bleed the brakes and road test the vehicle.

DUO-SERVO BRAKES

▶ See Figure 41

1. Raise the rear of the truck and support it with jackstands. Remove the wheels.

2. Remove the brake drum.

3. Using a brake tool, remove the 2 return springs.

4. Push up on the brake adjusting lever and remove the cable, shoe guide plate and cable guide. Disconnect the spring at the brake lever and remove them both.

5. Using needle-nose pliers, remove the 2 tension springs.

6. Using a brake tool, remove the shoe hold-down springs and pins.

7. Remove the shoes, the adjuster and the strut.

8. Pull the parking brake cable out, then disconnect it at the parking brake lever.

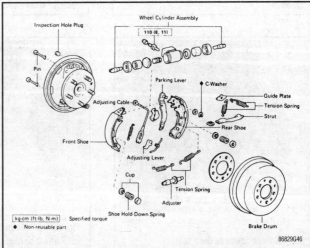

Fig. 41 Exploded view of the 2WD rear brake components with duo-servo brakes

To install:

9. Apply high-temperature grease to the contact surface of the brake backing plate.

10. Pull the parking brake cable out and attach it to the lever.

11. Position the rear shoe so that its upper end is in the piston rod. Install the shoe hold-down spring and pin.

12. Install the strut with the spring to the rear. Position the front shoe over the strut so that the upper end is in the piston rod. Install the hold-down spring and pin.

13. Grease the threads and end of the brake adjuster and install it between the brake shoes.

14. Install the shoe guide plate and the adjusting cable. Install the front return spring and then install the rear one.

15. While holding the tension spring against the rear shoe, hook the cable into the adjusting lever and install the lever. Pull the adjusting cable backward and release it; the adjusting bolt should turn.

16. Adjust the strut to the shortest possible length and install the brake drum. Turn the drum in the reverse direction and have an assistant step on the brake pedal. Continue this process until the clicking noise is no longer heard.

17. Remove the brake drum and check that the difference between the inner brake drum diameter and the outer brake shoe diameter is no more than 0.024 in. (0.6mm). Adjust or replace parts as needed.

18. Install the brake drum. Install the wheels and lower the truck. Bleed the brakes and road test the vehicle.

4WD Vehicles

▶ See Figures 42 thru 52

1. Raise the rear of the truck and support it with jackstands. Remove the wheels.

2. Remove the brake drum.

Fig. 42 Saturate the brake components using a commercially available brake cleaner

Fig. 43 A special tool is available to remove the return spring

Fig. 44 After the return spring is disengaged, remove it from the brake assembly

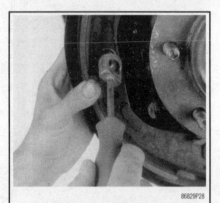

Fig. 45 Push inwards then twist . . .

Fig. 46 . . . then remove the pin and hold-down spring

Fig. 47 Lower the shoe, then disconnect the spring that attaches both shoes

Fig. 48 Remove the other side . . .

Fig. 49 . . . then, disconnect the cable from the bellcrank

Fig. 50 Apply high-temperature grease to the brake backing plate and the threads of the adjuster

3. Remove the tension spring.

4. Remove the rear shoe hold-down spring and pin and then lift out the rear shoe and anchor spring.

5. Remove the front shoe hold-down spring and pin. Disconnect the parking brake cable at the bellcrank.

6. Remove the automatic adjusting lever and parking brake lever as follows:

    a. Remove the E-clip.

    b. Remove the automatic adjusting lever.

    c. Spread and remove the C-washer.

    d. Remove the parking brake lever.

7. Remove the front shoe with the strut. Disconnect the other parking brake cable.

8. Disconnect the adjusting lever spring and remove the adjuster from the front shoe.

To install:

9. Assemble and install the parking brake bellcrank as follows:

    a. Apply grease to the rotating parts of the bellcrank.

    b. Install the parking brake bellcrank to the bellcrank bracket.

    c. Install the pin with a new C-clip and tighten it.

    d. Install the bellcrank boot to the parking brake bracket.

    e. Install the parking brake bellcrank and dust cover on the backing plate.

10. Apply high-temperature grease to the brake backing plate contact surfaces and the threads of the adjuster.

11. Position the parking brake lever and the automatic adjuster. Use new clips and tighten the C-clip.

12. Connect the parking brake cable to the shoe lever. Attach the other side of the cable to the bellcrank. Position the front shoe so the upper end is in the piston rod and install the hold-down spring and pin.

13. Connect the anchor spring to the front shoe, then stretch it onto the

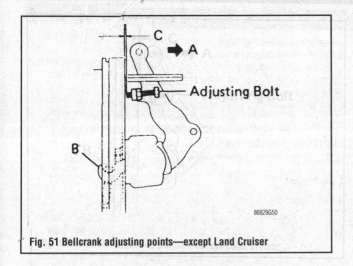

Fig. 51 Bellcrank adjusting points—except Land Cruiser

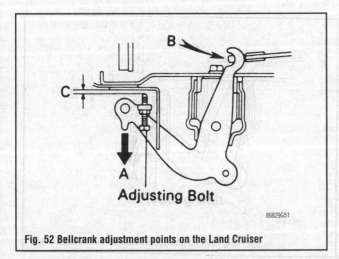

Fig. 52 Bellcrank adjustment points on the Land Cruiser

rear shoe. Position the rear shoe and install the hold-down spring and pin.

14. Install the tension spring. Lightly pull the bellcrank in direction **A** until there is no slack at **B**. Turn the adjusting bolt so that **C** will be 0.02–0.03 in. (0.4–0.8mm) on all models excluding 1996 4Runner which is 0.016–0.031 in. (0.4–0.8mm). Lock the adjusting bolt with the locknut.

15. Connect the parking brake cable to the bellcrank and install the tension spring.

16. Check that the parking brake lever (inside the truck) travel is correct and that the adjuster turns while pulling the lever.

17. Turn the adjuster to the shortest possible length and install the brake drum. Pull the parking brake lever out until the clicks stop.

18. Remove the brake drum and check that the difference between the inner brake drum diameter and the outer brake shoe diameter is no more than 0.024 in. (0.6mm).

19. Install the brake drum. Install the wheels and lower the truck. Bleed the brakes and road test the vehicle.

Wheel Cylinders

REMOVAL & INSTALLATION

▶ See Figures 53, 54 and 55

1. Remove the brake drums and shoes.
2. Working from behind the backing plate, disconnect the hydraulic line from the wheel cylinder. Plug the line to prevent fluid loss.
3. Install in the reverse order. Tighten the cylinder bolts to 7 ft. lbs. (10 Nm) and the brake line to 11 ft. lbs. (15 Nm).

Fig. 53 Using a line wrench, loosen and disconnect the brake line

Fig. 54 Remove the bolts securing the wheel cylinder to the backing plate . . .

PARKING BRAKE

Brake Lever

ADJUSTMENT

Except Land Cruiser

▶ **See Figures 56, 57 and 58**

1. Pull the parking brake lever all the way out (or up) and count the number of clicks to the end of travel. On 2WD ½ ton vehicles should have 12-18 clicks; 1-ton, 2WD vehicles and all 4WD vehicles should have 11-17 clicks. On 1989–95 4Runner check for 13-19 clicks and 1996 4Runner 7-9 clicks. If the parking brake lever is not within these ranges, adjust it.

2. Under the truck on 2WD models, there is an equalizer bar where the 2 cables come together. Tighten or loosen the adjusting nut until the lever travel is within the proper range. Check that the rear brakes are not dragging.

3. Under the truck on 4WD models, tighten the bellcrank stopper screw until the play in the rear brake links becomes zero, then loosen the screw 1 turn. Tighten the locknut.

4. Tighten one of the adjusting nuts on the intermediate lever while loosening the other one until the lever travel is correct. Tighten the adjusting nuts.

5. After adjustment, check that the bellcrank stopper screw is touching the back of the brake backing plate.

6. Recheck the parking brake travel.

Land Cruiser

▶ **See Figure 59**

1. Pull the parking brake lever all the way out (or up) and count the number of clicks to the end of travel. There should be 7-9 clicks. If the parking brake lever is not within these ranges, adjust it.

2. Before adjusting the parking brake, make certain the rear brakes are correctly adjusted.

3. Remove the parking brake lever cover or rear console box.

4. Loosen the locknut, then turn the adjusting nut until the lever travel is correct.

5. Tighten the locknut.

6. Install the parking brake cover or rear console box.

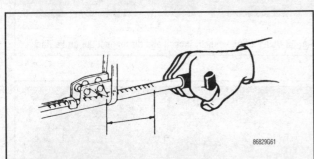

Fig. 56 Pull the lever until the correct amount of clicks are heard and then stop

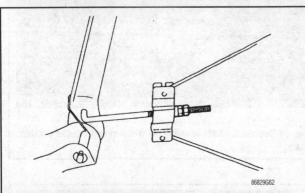

Fig. 57 There are two cables which are joined by an equalizer, the lock and adjusting nuts are located here

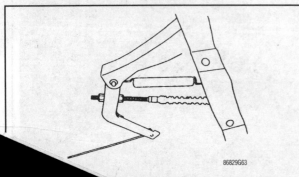

...tighten the bellcrank stopper screw until ...comes zero, then loosen the screw

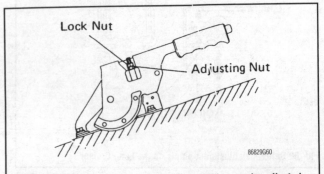

Fig. 59 On Land Cruisers the parking brake cables can be adjusted at the lever

Brake Shoes

REMOVAL & INSTALLATION

1. Loosen the rear wheel lug nuts.
2. Raise and support the rear of the vehicle.
3. Remove the rear wheels.
4. Remove the mounting bolts, then the disc brake caliper.
5. Suspend the disc brake caliper securely with a piece of wire. Be sure not to stretch or damage the hose.
6. Place matchmarks on the rotor and the rear hub. Remove the disc.

➡ **If the rotor cannot be removed easily, turn the shoe adjuster until the wheel turns freely.**

7. Using a pair of pliers, remove the tension spring.
8. Remove the shoe return springs along with the shoe strut with spring.
9. Slide out the rear shoe, then remove the shoe and adjuster.
10. Remove the lower side tension spring.
11. Remove the shoe hold-down spring cups, spring and pin.
12. Slide out the front shoe. Disconnect the parking brake cable from the parking brake shoe lever.
13. Remove the shoe hold-down spring cups, spring and pin.
14. If necessary, remove and disassemble the parking brake bellcrank assembly as follows:

a. Using a flat-bladed tool, remove the C-washer.

b. Remove the pin, then separate the parking brake cable from the bellcrank.

c. Remove the clip. Next remove the pin and clip, then disconnect the cable.

d. Remove the tension spring. Unbolt the parking bellcrank assembly.

e. Remove the boot from the bellcrank bracket. Using a flat-bladed tool, remove the C-washer and pin.

f. Remove the parking bellcrank from the bellcrank bracket.

To install:

15. Assemble and install the parking brake bellcrank as follows:

a. Apply grease to the rotating parts of the bellcrank.

b. Install the parking brake bellcrank to the bellcrank bracket.

c. Install the pin with a new C-clip and tighten it.

d. Install the bellcrank boot to the parking brake bracket.

e. Install the parking brake bellcrank and dust cover on the backing plate.

16. Apply high-temperature grease to the brake backing plate and the threads of the adjuster.

17. Position the parking brake lever and the automatic adjuster. Use new clips and tighten the C-clip.

18. Connect the parking brake cable to the front shoe lever. Connect the other side of the cable to the bellcrank. Position the front shoe so the upper end is in the piston rod and install the hold-down spring and pin.

19. Connect the anchor spring to the front shoe and then stretch it onto the rear shoe. Position the rear shoe and install the hold-down spring and pin.

20. Install the tension spring.

21. Install the strut with spring facing rearward.

22. Install the front shoe return springs and then the rear spring.

23. Install the brake disc.

24. Connect the parking brake cable to the bellcrank assembly with the pin, wave clip and clip.

25. Adjust the brake shoe clearance. Turn the adjuster and expand the shoes until the disc locks. Return the adjuster 8 notches.

26. If necessary adjust the bellcrank as follows:

a. Pull the bellcrank until all the play in the interior linkage is taken up.

b. Screw in the bellcrank adjusting bolt to where it contacts on the dust seal.

c. Loosen it one turn, then lock it at that position with the locknut. Install the bellcrank spring.

27. Install the rear brake disc assembly, tighten the mounting bolts to 76 ft. lbs. (103 Nm).

28. Install the rear wheels.

29. Bleed the brakes. Pump the brake pedal several times, then road test the vehicle.

INSPECTION

Shoe Thickness

▶ **See Figure 60**

Using a ruler, measure the thickness of the shoe lining. The standard thickness is 0.157 in. (4.0mm) and the minimum thickness is 0.039 in. (1.0mm). If the lining thickness is not within minimum thickness or less, or if there is an extremely uneven wear, replace the shoes.

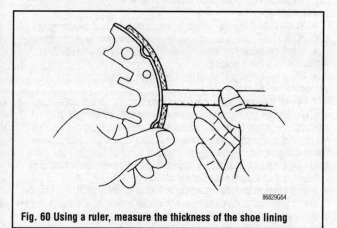

Fig. 60 Using a ruler, measure the thickness of the shoe lining

Rotor Inside Diameter

▶ **See Figure 61**

Using a H-gauge caliper, measure the inside diameter of the rotor. Standard inside measurement is 9.06 in. (230mm) and maximum measurement is 9.09 in. (231mm). Replace the rotor if the inside diameter is at the maximum value or more. Replace or have a machine shop grind the rotor with a lathe if it has been scored or is worn badly.

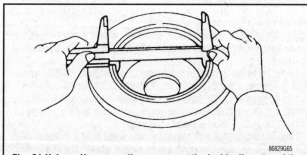

Fig. 61 Using a H-gauge caliper, measure the inside diameter of the rotor

Shoe and Lever Clearance

▶ **See Figure 62**

Using a feeler gauge, measure the clearance between the parking shoe and the lever. Standard distance should be 0.0138 in. (0.35mm). If the clearance is not within the the specification, replace the shim with one of the correct size. The sizes available are 0.012 in. (0.3mm), 0.024 in. (0.6mm) and 0.035 in. (0.9mm). If necessary replace the shim, remove the parking brake lever, then install the correct size. Install the parking break lever with a new C-washer. Remeasure the clearance.

ADJUSTMENTS

Parking Brake Shoe Clearance

1. Temporarily install the hub nuts.
2. Remove the hole plug.
3. Turn the adjuster, then expand the shoes until the rotor locks.
4. Return the adjuster 8 notches.
5. Install the hole plug.

Bellcrank

1. Pull the bellcrank until all the play in the interior linkage is taken up.
2. Screw in the bellcrank adjusting bolt to where it contacts on the dust seal.
3. Loosen it in one turn, then lock it at that position with the locknut at 48 inch lbs. (5 Nm).
4. Install the bellcrank spring.

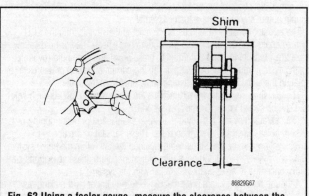

Fig. 62 Using a feeler gauge, measure the clearance between the parking shoe and the lever

ANTI-LOCK BRAKE SYSTEM

General Information

Rear Wheel Anti-Lock

The rear wheel anti-lock system found on Toyota trucks is designed to keep the rear wheels from locking under hard braking regardless of the load being carried. This function aids in keeping the vehicle from skidding sideways under hard braking with loss of directional stability.

Since the anti-lock system acts only on the rear wheels, it may be possible to lock the front wheels, causing them to skid; no steering response is possible if this happens.

The electro-hydraulic system employs one speed sensor in the differential housing. The speed signal is sent to the Electronic Control Unit (ECU) which controls the operation of the control solenoid within the actuator. The solenoid controls brake line pressure within the single line running to there are axle; the line branches to each wheel at the axle housing.

A significant difference in this system (from 4 wheel ABS) is that power steering fluid is circulated through the brake system actuator. The ABS control solenoid acts on the power steering fluid, not the brake fluid circuit. The power steering fluid acts as a control pressure circuit, moving pistons and valves within the actuator which control the brake fluid circuit. The power steering fluid and the brake fluid are kept completely separate within the systems; each fluid is kept within its normal system reservoir.

Use of the power steering fluid allows a supply of constantly pressurized fluid (from the power steering pump) and eliminates the need for larger, bulkier actuators with electric pumps and accumulators.

4-Wheel Anti-Lock

Anti-lock Brake Systems (ABS) are designed to prevent locked-wheel skidding during hard braking or during braking on slippery surfaces. The front wheels of a vehicle cannot apply steering force if they are locked and sliding; the vehicle will continue in the previous direction of travel. The 4-wheel ABS system used on Toyota trucks holds the wheels just below the point of locking, thereby allowing some steering response and preventing the rear of the vehicle from sliding sideways.

There are conditions for which the ABS system provides no benefit. Hydroplaning is possible when the tires ride on a film of water, losing contact with the paved surface. This renders the vehicle totally uncontrollable until road contact is regained. Extreme steering maneuvers at high speed or cornering beyond the limits of tire adhesion can result in skidding which is independent of vehicle braking. For this reason, the system is named anti-lock rather than anti-skid. Wheel spin during acceleration on slippery surfaces may also fool the system into detecting a system failure and entering the fail-safe mode.

Under normal conditions, the ABS system functions in the same manner as a standard brake system and is transparent to the operator. The system is a combination of electrical and hydraulic components, working together to control the flow of brake fluid to the wheels when necessary.

The Electronic Control Unit (ECU) is the electronic brain of the system, receiving and interpreting signals from the wheel speed sensors. The unit will enter anti-lock mode when it senses impending wheel lock at any wheel, and will immediately control the brake line pressures to the affected wheel(s) by issuing output signals to the actuator assembly.

The actuator contains solenoids which react to the signals from the ECU. Each solenoid controls brake fluid pressure to one wheel. The solenoids allow brake line pressure to build according to brake pedal pressure, hold (by isolating the system from the pedal and maintaining current pressure) or decrease (by isolating the pedal circuit and bleeding some fluid from the line).

The decisions regarding these functions are made very rapidly, as each solenoid can be cycled up to 10 times per second.

The operator may feel a pulsing in the brake pedal and/or hear popping or clicking noises when the system engages. These sensations are due to the valves cycling and the pressures being changed rapidly within the brake system. While completely normal and not a sign of system failure, these sensations can be disconcerting to an operator unfamiliar with the system.

Although the ABS system prevents wheel lock-up under hard braking, as brake pressure increases, wheel slip is allowed to increase as well. This slip will result in some tire chirp during ABS operation. The sound should not be interpreted as lock-up but rather as an indication of the system holding the wheel(s) just outside the locking point.

Diagnosis and Testing

PRECAUTIONS

- Certain components within the system are not intended to be serviced or repaired individually. Only those components with removal and installation procedures should be serviced.
- Do not use rubber hoses or other parts not specifically specified for the anti-lock system. When using repair kits, replace all parts included in the kit.
- Use only DOT 3 brake fluid from an unopened container.
- If any hydraulic component (either power steering or brake) is removed or replaced, it may be necessary to bleed the entire system.
- A clean repair area is essential. Always clean the reservoir and cap thoroughly before removing the cap. The slightest amount of dirt in the fluid may plug an orifice and impair the system function. Perform repairs after components have been thoroughly cleaned; use only denatured alcohol to clean components. Do not allow components to come into contact with any substance containing mineral oil; this includes used shop rags.
- The ECU is a microprocessor similar to other computer units in the vehicle. Insure that the ignition switch is **OFF** before removing or installing controller harnesses. Avoid static electricity discharge at or near the controller.
- If any arc welding is to be done on the vehicle, the ECU should be disconnected before welding operations begin.
- If the vehicle is to be baked after paint repairs, remove the ECU from the vehicle.

INSPECTION

If a malfunction occurs, the system will identify the problem and the computer will assign and store a fault code for the fault(s). The dashboard warning lamp will be illuminated to inform the driver that a fault has been found.

During diagnostics, the system will transmit the stored code(s) by flashing the dashboard warning lamp. If two or more codes are stored, they will be displayed from lowest number to highest, regardless of the order of occurrence. The system does not display the diagnostic codes while the vehicle is running.

Visual Inspection

Before diagnosing an apparent rear anti-lock problem, make absolutely certain that the normal braking system and power steering systems are in correct working order. Many common problems (dragging parking brake, seepage, etc.) will affect the ABS system. A visual check of specific system components may reveal problems creating an apparent ABS malfunction. Performing this inspection may reveal a simple failure, thus eliminating extended diagnostic time.

- Inspect the brake fluid level in the reservoir.
- Inspect lines, hoses, master cylinder assembly, brake calipers and cylinders for leakage. Inspect the power steering system and components for the same conditions.
- Visually check lines and hoses for excessive wear, heat damage, punctures, contact with other parts, missing clips or holders, blockage or crimping.
- Check the calipers or wheel cylinders for rust or corrosion. Check for proper sliding action if applicable.
- Check the caliper and wheel cylinder pistons for freedom of motion during application and release.
- Inspect the speed sensor for proper mounting and correct connection.
- Confirm the fault occurrence. Certain driver induced faults, such as not releasing the parking brake fully, will set a fault code and trigger the dash warning light. Excessive wheel spin on low-traction surfaces, high speed acceleration or riding the brake pedal may also set fault codes and trigger a warning lamp. These induced faults are not system failures but examples of vehicle performance outside the parameters of the control unit.
- Many system shut-downs are due to loss of sensor signals to or from the controller. The most common cause is not a failed sensor but a loose, corroded or dirty connector. Check harness and component connectors carefully.

Reading Codes

▶ **See Figures 63, 64, 65 and 66**

1. Turn the ignition switch **OFF**. Check battery condition; approximately 12 volts is required to operate the system.

2. Turn the ignition switch **ON** and check that the ABS dashboard warning lamp comes on for 2 seconds (3 seconds on 4-wheel systems). If the lamp does not come on, repair the fuse, bulb or wiring.

3. Use a jumper wire to attach terminals Tc and E1 of the Data Link Connector (DLC1). This is not the service connector. Turn the ignition **ON**.

4. If a fault code has been set, the dashboard warning lamp will begin to blink 4 seconds later. The number of flashes corresponds to the first digit of a

2-digit code; after a 1.5 second pause, the second digit is transmitted. If a second code is stored, it will be displayed after a 2.5 second pause. Once all codes have been displayed, the entire series will repeat after a 4 second pause. If no codes have been stored, the warning lamp will flash continuously every ½ second with no variation.

5. Turn the ignition switch **OFF**.

6. Check or repair the system as indicated by the fault code.

7. After repairs are completed, clear the codes from the memory. If the battery is disconnected during repairs, the controller memory will be erased of all stored codes.

8. Remove the jumper wire and reattach the service connector at the actuator.

Code No.	Light Pattern	Diagnosis	Trouble Part
11		Open circuit in solenoid relay circuit or solenoid circuit	• Solenoid • Solenoid relay • Wire harness and connector of solenoid and/or solenoid relay circuit
12		Short circuit in solenoid relay circuit	
25		Short circuit in solenoid circuit	
33		Open or short circuit in speed sensor circuit	• Speed sensor • Wire harness and connector of speed sensor circuit
41		Low battery positive voltage (9.5 V or lower)	• Battery
42		Abnormally high battery positive voltage (17 V or higher)	
43		Mechanical malfunction in deceleration sensor	• Deceleration sensor • Wire harness and connector of deceleration sensor circuit
44		Electrical malfunction in deceleration sensor circuit	
Always ON		Malfunction in ECU	• ECU

86829GLM

Fig. 63 Diagnostic trouble codes for the rear wheel ABS system

Code No.	Light Pattern	Diagnosis	Malfunctioning Part
		All speed sensors and sensor rotors are normal	
71		Low voltage of right front speed sensor signal	• Right front speed sensor • Sensor installation
72		Low voltage of left front speed sensor signal	• Left front speed sensor • Sensor installation
73		Low voltage of right rear speed sensor signal	• Right rear speed sensor • Sensor installation
74		Low voltage of left rear speed sensor signal	• Left rear speed sensor • Sensor installation
75		Abnormal change of right front speed sensor signal	• Right front sensor rotor
76		Abnormal change of left front speed sensor signal	• Left front sensor rotor
77		Abnormal change of right rear speed sensor signal	• Right rear sensor rotor
78		Abnormal change of left rear speed sensor signal	• Left rear sensor rotor
79		Deceleration sensor is faulty	• Deceleration sensor • Sensor installation

86829GLL

Fig. 64 Diagnostic trouble codes for 4-wheel ABS

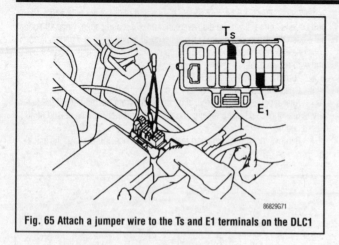

Fig. 65 Attach a jumper wire to the Ts and E1 terminals on the DLC1

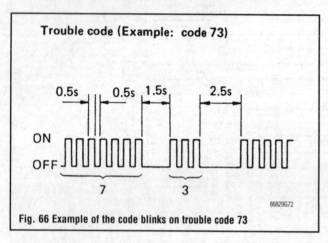

Fig. 66 Example of the code blinks on trouble code 73

Clearing Codes

♦ See Figures 67 and 68

1. Detach the service connector at the actuator.
2. Use the jumper wire to connect terminals Tc and E1 of the data link connector (DLC1).
3. To clear the diagnostic trouble codes stored in the ECU by depressing the brake pedal 8 or more times within 3 seconds.
4. After the rapid pedal application, the dash warning lamp should display constant flashing, indicating a normal system. If codes are still displayed, make certain the repairs made to the system are correct. Also inspect the brake light switch at the brake pedal for any binding or sticking.
5. Once the codes are cleared, disconnect the jumper wire. The dash warning lamp should go out.

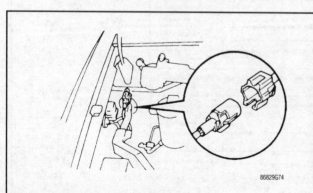

Fig. 67 Detach the service connector harness at the actuator

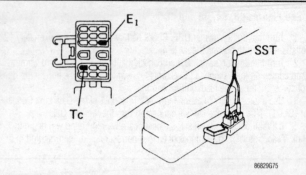

Fig. 68 Use jumper wire or SST 09843–18020 to connect the terminals

Speed Sensor

SYSTEM TEST

➡ **The following procedures require driving the vehicle while it is in the diagnostic mode. The anti-lock system will be disabled; only normal braking function will be available.**

1. If working on a 4WD vehicle, make certain it is only in 2WD. Check the battery voltage with the engine off; voltage should be approximately 12 volts.
2. With the ignition switch **ON**, make certain the ABS warning lamp comes on for about 3 seconds and then goes out. Turn the ignition switch **OFF**.
3. Using a 3-way jumper, connect terminals Tc, Ts and E1 at the Data Link Connector (DLC1).
4. Pull the parking brake on securely and start the engine without stepping on the brake pedal. Depressing the brake pedal will either void the test or cause the system to enter a different diagnostic mode.
5. The dashboard warning lamp should flash evenly about 4 times per second. If, instead, a code is flashed, repairs must be made and the code(s) cleared before continuing. If the light does not flash, check the parking brake switch, the DLC1 connector and the wiring to the ABS controller.
6. To check the sensor signal level, release the parking brake and drive the vehicle straight ahead at low speed. Once a speed of 2.5–3.7 mph (4–6 km)-rear wheel ABS; 2–5 mph (3–8 km)-4-wheel ABS, is reached, the warning lamp should turn off. Once the light has gone off, it should begin to blink as the vehicle leaves the speed range. This blinking is normal and indicates the speed sensor is operating correctly. If no blinking occurs, the system is defective.
7. Check the sensor signal change by driving the vehicle on the road. Once a speed of 25–31 mph (40–50 km) is achieved, the warning lamp should come on after a one second pause. As before, the lamp should blink evenly as the vehicle leaves the speed range; failure to blink indicates a system fault.

➡ **While the warning lamp is off, do not subject the vehicle to any shocks or impact such as shifting, acceleration, deceleration or road impact.**

8. Stop the vehicle in a safe location and apply the parking brake fully. Read and record any stored codes which are displayed.
9. Turn the ignition switch **OFF**; remove the jumper wire from the DLC1 connector.

COMPONENT TEST

Rear Wheel ABS

♦ See Figures 69, 70 and 71

1. Check the installation of the sensor. If the mounting bolt is loose, tighten it to 14 ft. lbs. (19 Nm).
2. Disconnect the speed sensor harness from the ABS harness.
3. Measure the resistance with an ohmmeter between the terminals of the sensor connector. Correct resistance is 580–700 ohms. If the resistance is outside the specification, the sensor must be replaced.

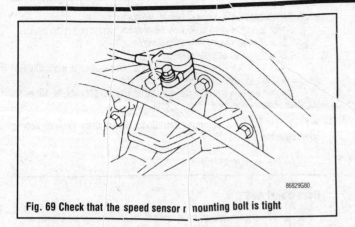

Fig. 69 Check that the speed sensor mounting bolt is tight

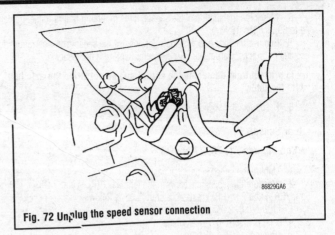

Fig. 72 Unplug the speed sensor connection

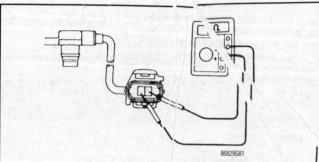

Fig. 70 The ohmmeter measures the resistance between the terminals of the sensor

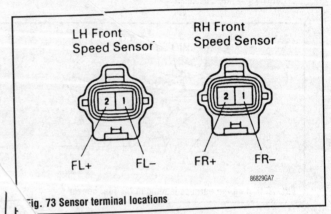

Fig. 73 Sensor terminal locations

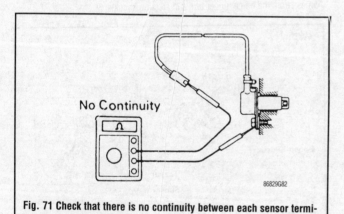

Fig. 71 Check that there is no continuity between each sensor terminal and the sensor body

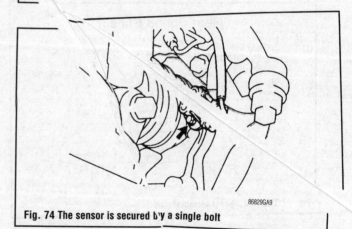

Fig. 74 The sensor is secured by a single bolt

4. Check that there is no continuity between each sensor terminal and the sensor body. If any continuity is found, the sensor is faulty.

5. Reconnect the speed sensor to the harness.

4-Wheel ABS

FRONT SENSOR

▶ See Figures 71, 72, 73, 74 and 75

1. Unplug the speed sensor connector.
2. Measure the resistance between terminals 1 (FL−) and 2 (FL+). The resistance should be as follows:
- 4Runner—1.4–1.8k ohms
- T100—0.6–1.8k ohms
- Land Cruiser—0.97–1.77k ohms
3. If the resistance is not as specified, replace the sensor.
4. Check that there is no continuity between each terminal and the sensor body. If there is continuity, replace the sensor.
5. Attach the speed sensor harness.

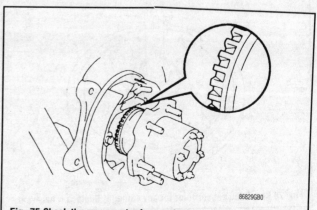

Fig. 75 Check the sensor rotor for any damage. Replace if necessary

6. Check the installation of the sensor. If the mounting bolt is loose, tighten it to 69 inch lbs. (8 Nm).

7. Remove the axle hub with the rotor. Inspect the sensor rotor serrations for scratches, cracks, warping or missing teeth. Replace if necessary.

➡ **To prevent from damaging the serrations, do not strike the axle hub with the rotor.**

8. Reconnect the speed sensor to the harness.

Rear Sensor

▶ **See Figures 71, 76, 77 and 78**

1. Unplug the speed sensor connector.
2. Measure the resistance between the terminal 1 (RL–) and 2 (RL+). The is 1 is (RR–) and 2 is (RR+) resistance should be as follows:
- T100—0.6–2.05k ohms
- Land Cruiser—0.5–1.6k ohms
- 4Runner—1.0–1.4k ohms

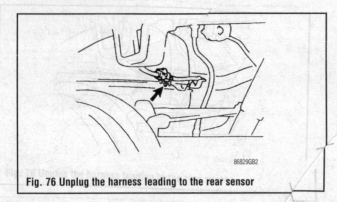

Fig. 76 Unplug the harness leading to the rear sensor

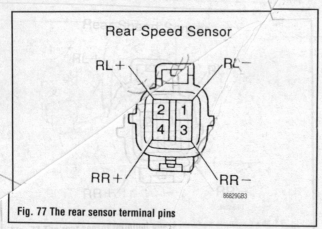

Fig. 77 The rear sensor terminal pins

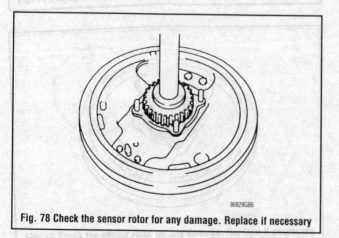

Fig. 78 Check the sensor rotor for any damage. Replace if necessary

3. If the resistance is not as specified, replace the sensor.
4. Check that there is no continuity between each terminal and the sensor body. If there is continuity, replace the sensor.
5. Attach the speed sensor harness.
6. Check the installation of the sensor. If the mounting bolt is loose, tighten it to 69 inch lbs. (8 Nm).
7. Remove the axle hub with the rotor. Inspect the sensor rotor serrations for scratches, cracks, warping or missing teeth.

➡ **To prevent from damaging the serrations, do not strike the axle hub with the rotor.**

REMOVAL & INSTALLATION

Rear Wheel ABS

The speed sensor is located in the rear differential housing. It transmits a small electrical signal to the ECU based on the rotational speed of the differential ring gear.

1. Unplug the sensor cable connector from the harness. If the sensor cable is held by brackets or clamps, release them.
2. Remove the retaining bolt holding the sensor to the differential.
3. Remove the sensor straight out of its mount. Protect the tip from impact or abrasion.
4. Installation is in the reverse order. Tighten the retaining bolt to 14 ft. lbs. (19 Nm).

4-Wheel ABS

FRONT

▶ **See Figures 79 and 80**

1. Unplug the speed sensor harness to the unit.
2. On 4WD vehicles, remove the clamp bolts and clips holding the sensor harness from the frame, upper arm and steering knuckle.

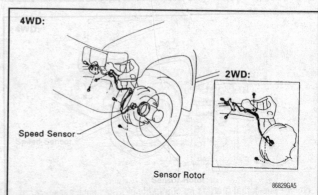

Fig. 79 Exploded view of the front speed sensor mounting and harness routing—T100 shown

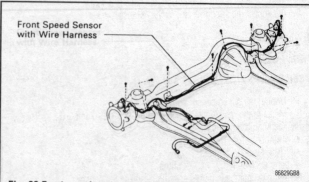

Fig. 80 Front speed sensor mounting harness routing—Land Cruiser shown

3. On 2WD vehicles, remove the clamp bolts and clips holding the sensor harness to the frame and upper arm.

➡**Discard the old clips.**

4. Remove the sensor from the knuckle.
5. Installation is in the reverse order. Tighten the sensor bolts to 69–71 inch lbs. (7–8 Nm). Tighten the harness clip mounting bolts to 48–71 inch lbs. (5–8 Nm).

REAR

♦ **See Figures 81 and 82**

1. Unplug the speed sensor harness to the unit.
2. Remove the clamp bolts and nuts holding the sensor harness from the frame, upper arm, axle housing and fuel tank (if attached).

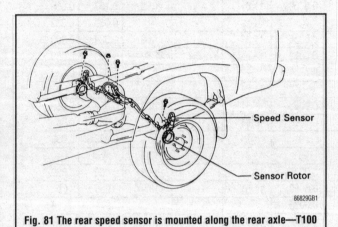

Fig. 81 The rear speed sensor is mounted along the rear axle—T100

➡**Discard the old clips.**

3. Remove the sensor from the axle.
4. Installation is in the reverse order. Tighten the sensor bolt to 69–71 inch lbs. (7–8 Nm). Tighten the harness mounting bolts to 48–71 inch lbs. (5–8 Nm). Be sure to use all new clips, and install them in the same angle as those removed.

Actuator System Bleeding

➡**This procedure requires the use of the Toyota ABS Actuator Checker and adapter harness (Tools 09990-00150 and 09990-00205) or their equivalent.**

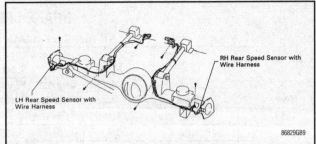

Fig. 82 The rear speed sensor is mounted along the rear axle—Land Cruiser

Whenever the actuator or any power steering hoses are disconnected, the actuator should be bled using the following procedure. Normal bleeding of each system may not remove all the air from the actuator.

1. Bleed the power steering system. Refer to Section 8. The bleeder valve is located on the steering gearbox.
2. Bleed the brake system with the engine running. Bleed the brake system again with the engine off.
3. Disconnect the wiring harness at the actuator and at the solenoid relay.
4. Connect the ABS checker and harness to the actuator, relay and body wiring harness, using the adapters as necessary.
5. Connect the red cable of the checker to the positive battery terminal and the black cable to the negative battery terminal.
6. Start the engine and run it at idle.
7. Turn the selector switch on the ABS checker to the AIR BLEED position. Strongly depress the brake pedal and hold it in that position.
8. Push the ON/OFF switch (on the checker) for about 3 seconds and release it; do this 5 times.

✳✳ WARNING

Do not push the ON/OFF switch before the brake pedal is depressed—master cylinder damage may occur. Do not release the brake pedal while the switch is engaged. Do not hold the ON/OFF switch ON more than 5–7 seconds.

9. After 5 repetitions, release the switch and then release the brake pedal.
10. Check the level of the power steering and brake fluids in their reservoirs; top off each as necessary using the correct fluid.
11. Disconnect and remove the ABS checker. Reconnect the system harness.

BRAKE SPECIFICATIONS

All measurements in inches unless noted

Year	Model	Master Cylinder Bore	Brake Disc Original Thickness	Brake Disc Minimum Thickness	Brake Disc Maximum Runout	Brake Drum Diameter Original Inside Diameter	Brake Drum Diameter Max. Wear Limit	Brake Drum Diameter Maximum Machine Diameter	Minimum Lining Thickness Front	Minimum Lining Thickness Rear
1989	Pick-up	NA	(1)	(2)	(3)	(4)	-	(5)	0.039	0.039
	4Runner	NA	0.787	0.709	0.0035	11.61	-	11.69	0.039	0.039
	Land Cruiser	NA	0.787	0.748	0.0059	11.61	-	11.69	0.157	0.059
1990	Pick-up	NA	(1)	(2)	(3)	(4)	-	(5)	0.039	0.039
	4Runner	NA	0.787	0.709	0.0035	11.61	-	11.69	0.039	0.039
	Land Cruiser	NA	0.787	0.748	0.0059	11.61	-	11.69	0.157	0.059
1991	Pick-up	NA	(1)	(2)	(3)	(4)	-	(5)	0.039	0.039
	4Runner	NA	0.787	0.709	0.0035	11.61	-	11.69	0.039	0.039
	Land Cruiser	NA	0.984	0.906	0.0059	11.61	-	11.69	0.059	0.059
1992	Pick-up	NA	(1)	(2)	(3)	(4)	-	(5)	0.039	0.039
	T100	NA	0.984	0.906	(6)	11.61	-	11.69	(7)	(8)
	4Runner	NA	0.984	0.906	0.0035	11.61	-	11.69	0.059	0.039
	Land Cruiser	NA	0.984	0.906	0.0059	11.61	-	11.69	0.059	0.059
1993	Pick-up	NA	(1)	(2)	(3)	(4)	-	(5)	0.039	0.039
	T100	NA	0.984	0.906	(6)	11.61	-	11.69	(7)	(8)
	4Runner	NA	0.984	0.906	0.0035	11.61	-	11.69	0.059	0.039
	Land Cruiser	NA	1.260 (9)	1.181 (10)	0.0059	11.61	-	11.69	0.059	(11)
1994	Pick-up	NA	(12)	(13)	0.0035	(4)	-	(5)	(14)	0.039
	T100	NA	0.984	0.906	(12)	11.61	-	11.69	(7)	(8)
	4Runner	NA	0.984	0.906	0.0035	11.61	-	11.69	0.059	0.039
	Land Cruiser	NA	1.260 (9)	1.181 (10)	0.0059	11.61	-	11.69	0.059	(11)
1995	Pick-up	NA	(12)	(13)	0.0035	(4)	-	(5)	(14)	0.039
	T100	NA	0.984	0.906	(6)	11.61	-	11.69	(7)	(8)
	4Runner	NA	0.984	0.906	0.0035	11.61	-	11.69	0.059	0.039
	Land Cruiser	NA	1.260 (9)	1.181 (10)	0.0059	11.61	-	11.69	0.059	(11)
	Tacoma	NA	0.866	0.787	0.0028	(4)	-	(5)	0.039	0.039
1996	T100	NA	0.984	0.906	(6)	11.61	-	11.69	(7)	(8)
	4Runner	NA	0.984	0.906	0.0035	11.61	-	11.69	0.059	0.039
	Land Cruiser	NA	1.260 (9)	1.181 (10)	0.0059	11.61	-	11.69	0.059	(11)
	Tacoma	NA	0.866	0.787	0.0028	(4)	-	(5)	0.039	0.039

NA - Not Available

1 2WD (PD60 type disc): 0.984
2WD (PD66 type disc): 1.181
2WD (FS17, 18 type disc): 0.866
4WD (S12+12 type disc): 0.787

2 2WD (PD60 type disc): 0.906
2WD (PD66 type disc): 1.102
2WD (FS17, 18 type disc): 0.787
4WD (S12+12 type disc): 0.709

3 PD60, FS17, FS18, S12+12 type disc: 0.0035

4 2WD: 10.00, 4WD: 11.61

5 2WD: 10.08; 4WD: 11.69

6 2WD:1/2 Ton 0.0028
2WD:1 Ton 0.0035
4WD: 0.0028

7 2WD: 1/2 Ton 0.433
2WD: 1 Ton 0.374
4WD: 0.374

8 2WD: 0.276
4WD: 0.236

9 Rear disc: 0.630

10 Rear disc: 0.709

11 Brake shoe lining: 0.059
Disc pad lining: 0.039

12 2WD: 0.866; 4WD: 0.787

13 2WD: 0.787; 4WD: 0.709

14 2WD: 0.059; 4WD: 0.039

86829C50

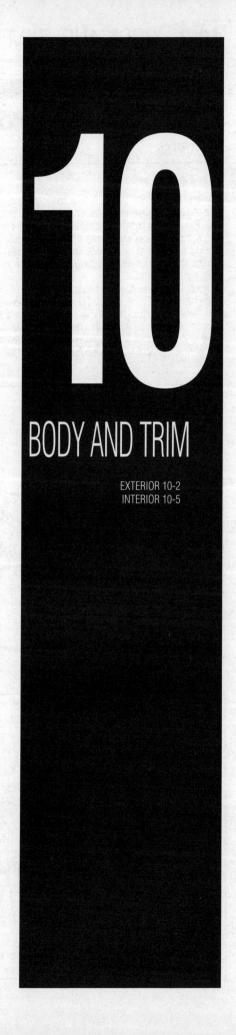

10

BODY AND TRIM

Doors

ADJUSTMENT

♦ **See Figures 1 and 2**

When checking door alignment, look carefully at each seam between the door and body. The gap should be constant and even all the way around the door. Pay particular attention to the door seams at the corners farthest from the hinges; this is the area where errors will be most evident. Additionally, the door should pull against the weatherstrip when latched to seal out wind and water. The contact should be even all the way around and the stripping should be about half compressed.

The position of the door can be adjusted in three dimensions: fore and aft, up and down, in and out. The primary adjusting points are the hinge-to-body bolts. Apply tape to the fender and door edges to protect the paint. Two layers of common masking tape works well. Loosen the bolts (using SST 09812-00010 if necessary) just enough to allow the hinge to move. With the help of an assistant, position the door and retighten the bolts. Inspect the door seams carefully and repeat the adjustment until correctly aligned.

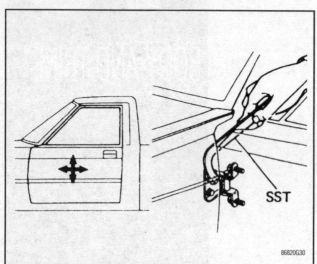

86820G30

Fig. 1 Using SST 09812–00010 or equivalent, loosen the body side hinge bolts forward and rearward

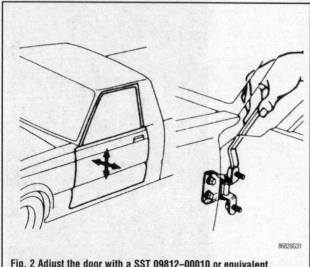

86820G31

Fig. 2 Adjust the door with a SST 09812–00010 or equivalent. Loosen the left/right vertical body side hinge bolts

The in-out adjustment (how far the door "sticks out" from the body) is adjusted by loosening the hinge-to-door bolts. Again, move the door into place, then retighten the bolts. This dimension affects both the amount of crush on the weatherstrips and the amount of "bite" on the striker.

Further adjustment for closed position and smoothness of latching is made at the latch plate or striker. This piece is located at the rear edge of the door and is attached to the bodywork; it is the piece the latch engages when the door is closed.

Although the striker size and style may vary between models or from front to rear, the method of adjusting it is the same:

1. Loosen the large cross-point screw(s) holding the striker. Know in advance that these bolts will be very tight; an impact screwdriver is a handy tool to have for this job. Make sure you are using the proper size bit.

2. With the bolts just loose enough to allow the striker to move if necessary, hold the outer door handle in the released position and close the door. The striker will move into the correct location to match the door latch. Open the door and tighten the mounting bolts. The striker may be adjusted towards or away from the center of the car, thereby tightening or loosening the door fit. The striker can be moved up and down to compensate for door position, but if the door is correctly mounted at the hinges this should not be necessary.

➡**Do not attempt to correct height variations (sag) by adjusting the striker.**

3. Additionally, some models may use one or more spacers or shims behind the striker. These shims may be removed or added in combination to adjust the reach of the striker.

4. After the striker bolts have been tightened, open and close the door several times. Observe the motion of the door as it engages the striker; it should continue its straight-in motion and not deflect up or down as it hits the striker.

5. Check the feel of the latch during opening and closing. It must be smooth and linear, without any trace of grinding or binding during engagement and release.

It may be necessary to repeat the striker adjustment several times (and possibly re-adjust the hinges) before the correct door to body match is produced.

Hood

REMOVAL & INSTALLATION

♦ **See Figures 3, 4 and 5**

➡**You'll need an assistant for this job.**

1. Open the hood.
2. Matchmark the hood-to-hinge position.

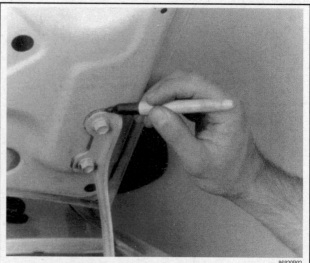

86820P02

Fig. 3 Matchmark the hood hinges with a marker

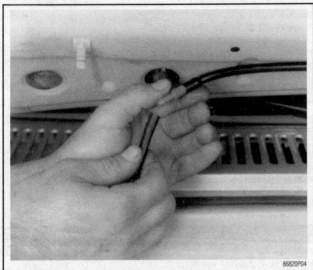

Fig. 4 Don't forget to disconnect the washer hoses

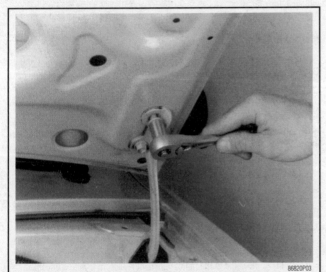

Fig. 5 Unbolt the hinges while an assistant supports the hood

3. Disconnect the washer hoses.
4. Remove the hood-to-hinge bolts and lift off the hood.
5. Installation is the reverse or removal. Adjust if necessary.

ADJUSTMENT

♦ See Figures 6 and 7

➡A tapered, self-centering bolt is used as the hood hinge set bolt. The hood cannot be adjusted with this bolt in place. Remove it and substitute a bolt and washer which will allow adjustment.

1. Loosen the hood to hinge bolts slightly. Adjust the hood fore-and-aft and left-and-right directions. Tighten the bolts to 10–15 ft. lbs. (13–20 Nm).
2. Adjust the front edge of the hood by turning the two cushions higher or lower. The hood should align with the fender edges.
3.. Except on the Land Cruiser, remove the wiper arms, then remove the cowl cover below the windshield. Inside the cowl plenum, adjust the hinge-to-body bolts to control the height of the rear of the hood. Reinstall the cowl panel and wipers.
4. On the Land Cruiser, the rear hood height is adjusted by loosening the four mounting screws holding the hinge to the firewall.
5. Loosen the hood latch retaining bolts and adjust the latch by moving it up-down or left-right.

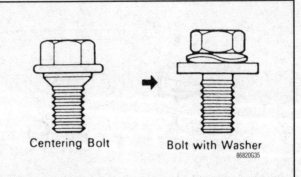

Fig. 6 The centering bolt has no washer and a tapered head; replace it with a washered bolt

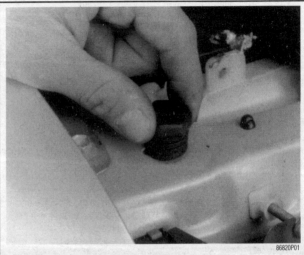

Fig. 7 Most trucks have a ribbed, rubber cylindrical cushion seated on the front of the upper tie bar

6. Close the hood and check the tension needed to release the hood. If the release requires excessive force, the latch is under tension and must be adjusted.

➡Be sure to tighten all fasteners after the adjustments are complete.

Tailgate

REMOVAL & INSTALLATION

1. Remove the service hole cover.
2. Disconnect the tail gate link from the lock control. If needed, remove the outer gate handle.
3. Remove the tailgate stay from the gate.
4. Unbolt the tailgate hinges.
5. Remove the tailgate and torsion bar.
6. Installation is the reverse of removal.

Rear Door

ADJUSTMENT

4Runner

♦ See Figure 8

1. To adjust the door fore-and-aft, as well as left-to-right, loosen the body side hinge nuts. When adjusted, retighten to 25 ft. lbs. (33 Nm).

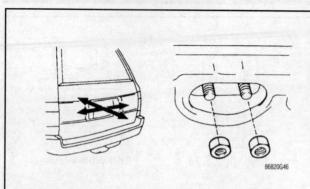

Fig. 8 Loosen the body side nuts to adjust the door fore-and-aft, as well as left-to-right

86820G46

2. To adjust the door vertically, as well as left-to-right, loosen the door side hinge bolts. When adjusted, retighten to 17 ft. lbs. (23 Nm).
3. Adjust the door striker as follows:
    a. Loosen the striker mounting screws to adjust.
    b. Using a plastic hammer, tap the striker to adjust.

Land Cruiser

1. Adjust the upper and lower rear doors fore-and-aft and left-to-right as follows:
    a. Adjust the door by loosening the door side hinge bolts.
    b. Move the hinge into the appropriate direction with your hand or a plastic hammer.
2. To adjust the upper and lower doors in the left-to-right and vertical directions, loosen the body side hinge bolts.
3. Adjust the door striker as follows:
    a. Check that the door fit and door lock linkage are adjusted correctly.
    b. Adjust the striker position by slightly loosening the striker mounting screws, then hitting the striker with a plastic headed hammer. Tighten the mounting screw when the adjustment is appropriate.

Front or Rear Bumper

REMOVAL & INSTALLATION

1. Support the bumper.
2. Remove any trim pieces, corner moldings, etc. on the bumper.
3. Remove the nuts and bolts attaching the bumper to the frame.
4. Installation is the reverse of removal.

Grille

REMOVAL & INSTALLATION

1. Open and support the hood.
2. Remove the turn signal and parking lamps.
3. Unscrew the grille assembly.
4. Reverse to install.

Outside Mirrors

REMOVAL & INSTALLATION

The mirrors can be removed from the door without disassembling the door liner or other components. Both left and right outside mirrors may be either manual, manual remote (small lever on the inside to adjust the mirror) or, in some cases, electric remote. If the mirror glass is damaged, replacements may be available through your dealer or a reputable glass shop. If the plastic housing is cracked or damaged, the entire mirror unit must be replaced.

1. If the mirror is manual remote, check to see if the adjusting handle is retained by a hidden screw, usually under an end cap on the lever. If so, remove the screw and remove the adjusting knob.
2. Using a blunt plastic or wooden tool, remove the inner triangular cover from where the mirror mounts to the door. Don't use a screwdriver; the plastic will be marred.
3. Depending on the model and style of mirror, there may be concealment plugs or other minor parts under the cover. Remove them. If electric connectors are present, unplug them.
4. Support the mirror housing from the outside and remove the three bolts or nuts holding the mirror to the door. Remove the mirror.
5. When installing, fit the mirror to the door, then install the nuts and bolts to hold it. Connect any wiring. Pay particular attention to the placement and alignment of any gaskets or weatherstrips around the mirror; serious wind noises may result from careless work.
6. Install the cover, pressing it firmly into position. Install the control lever knob if it was removed.

Antenna

REMOVAL & INSTALLATION

If your antenna mast is the type where you can unscrew the mast from the fender, simply do so with a pair of pliers. Most damaged antennas are simply the result of a truck wash or similar mishap, in which the mast is bent.

Manual Antenna

1. Disconnect the antenna cable at the radio by pulling it straight out of the set. Depending on access, this may require loosening the radio and pulling it out of the dash.
2. Working under the instrument panel, disengage the cable from its retainers.

➡**On some models, it may be necessary to remove the instrument panel pad to get at the cable.**

3. Outside, unsnap the cap from the antenna base.
4. Remove the screw(s) and lift off the antenna, pulling the cable with it, carefully.
5. When reinstalling, make certain the antenna mount area is clean and free of rust and dirt. The antenna must make a proper ground contact through its base to work properly.
6. Install the screws and route the cable into the cab. Make certain the cable is retained in the clips, etc.
7. Connect the cable to the radio; reinstall the radio if it was removed.

Power Antenna Mast

➡**The power antenna system contains a relay. If the power antenna is inoperative, the relay is the first place to look. The actual motor rarely fails.**

The power antenna mast is replaceable on the Pick-Up and 4Runners. If the mast is damaged or broken, proceed as follows:
1. Perform this repair with the battery cables connected. Turn the ignition switch to the **LOCK** position.
2. Remove the antenna nut.
3. If equipped with a CD player, press the AM and FM buttons on the receiver and simultaneously turn the ignition switch to **ACC**.
4. For non-CD player units, press the AM button and simultaneously turn the ignition switch to **ACC**.
5. The antenna motor will run, unwinding the mast from the spool. have an assistant guide the mast so it doesn't fall and damage the bodywork. When fully extended, the mast will be released. Remove it, and leave the ignition in the **ACC** position.
    To install:
6. Insert the new cable so that the teeth face the rear of the vehicle. Feed the cable into the motor until the cable reaches the bottom, about 12 in. (30cm).
7. Have an assistant ready and prepared to guide the cable. Turn the ignition switch to **LOCK**. The motor will run, winding the cable onto the spool and drawing the antenna down into the mount.

8. If the cable does not wind, twist the cable at the top of the housing; the part at the bottom may have turned while being pushed down the hole.

9. Once wound onto the spool, install the antenna nut even if the antenna is not fully retracted. Test the antenna elevation and retraction; the mast will eventually retract fully.

Removable Top

REMOVAL & INSTALLATION

▶ See Figure 9

➡ This procedure applies to the 4Runners only.

1. Open the rear window and remove the body garnishes.
2. Disconnect the electrical lead and the hose at the rear washer motor.
3. Remove mounting bolt **B** and then remove the other bolts.
4. Remove the cover and place it on pieces of wood on the ground.

To install:

✳✳ WARNING

Mounting bolt B controls the cover top switch. When removed, the rear power window cannot be operated. Never install it when the cover top is removed.

5. Position the cover top on the truck body and install bolts **E** and **J**. Install bolts **C** and **L**. Tighten all bolts to 10 ft. lbs. (14 Nm).
6. Loosen bolts **E** and **J**.

INTERIOR

Instrument Panel

REMOVAL & INSTALLATION

Pick-Up and 4Runner

1989–95 MODELS

▶ See Figures 10 and 11

1. Disconnect the negative battery cable.

° ✳✳ CAUTION

The air bag system (SRS) must be disarmed before removing the steering wheel. Failure to do so may cause accidental deployment, property damage or personal injury. Always store the air bag with the pad facing upward.

2. Remove the steering wheel.
3. Remove the upper and lower steering column covers.
4. Remove the hood release lever from the panel. It's held by two screws.
5. Remove the cowl side trim on each side. You might refer to it as the kick panel.
6. Remove the lower center instrument panel cover.
7. Remove the ignition key cylinder cover.
8. Remove the No. 1 lower finish panel; this is the panel holding the left speaker and running under the steering column.
9. Remove the heater duct to No. 2 outlet.
10. Pull off the heater control knobs, remove the A/C switch if equipped and gently pry up on the heater control unit to remove it. The prying tool should have a taped or protected edge.
11. For all models except Pick-Ups with 4 speed manual transmissions, remove the 3 screws holding the meter trim. Pull the trim or surround panel outward and unplug the connectors. On 4Runner, remove the two screws and the cup holder from the panel.
12. For Pick-Ups with 4 speed manual transmissions, the trim is held only by two screws. Unplug the connectors and remove the trim panel.

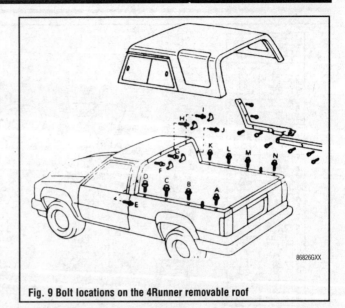

Fig. 9 Bolt locations on the 4Runner removable roof

7. Install and tighten bolts **D** and **K** and then retighten bolts **E** and **J**.
8. Install all remaining bolts and tighten them to 10 ft. lbs. (14 Nm). Install and tighten bolt **B**.
9. Connect the washer motor lead and hose, install the garnishes and close the tailgate. Check for proper operation of the rear window.

13. Remove the No.1 air outlet. It's attached with 2 screws.
14. Remove the four screws holding the combination meter. Work it out of the panel until access is gained to the rear. Disconnect the speedometer cable and remove the electrical connectors. Remove the meter assembly, placing it in a safe location out of the work area.
15. Remove the lower trim panel with the glove box door attached. The panel is held by 4 screws and 1 bolt.
16. Remove one bolt holding the lower center finish panel (around the radio and heater controls) and remove the panel.
17. Remove the heater control assembly. Don't forget to disconnect each control cable at the heater box. Withdraw the cables through the dash with the controller.
18. Remove the radio.
19. Remove the bolts holding the instrument panel and the entire panel assembly. Because of the nature of the rear clips, lift the panel upward at an angle. It will come loose with various items still attached; disconnect the wiring harnesses as soon as easily reached. An assistant is useful due the awkward size and location of the panel.
20. Once removed from the vehicle, the safety pad, heater ducts and assorted brackets may be removed if desired.
21. Installation is in the reverse order of removal. Assemble the instrument panel, safety pad and other items removed. Fit the panel into place, connect the wiring and secure the retaining bolts. Make certain the panel is flush and level. The heater vent ducts must match up exactly with no leaks. Make any needed adjustments now.

1996 MODELS

▶ See Figure 12

1. Disconnect the negative battery cable.
2. Remove the steering wheel.

✳✳ CAUTION

The air bag system (SRS) must be disarmed before removing the steering wheel or instrument panel. Refer to Section 6. Failure to do so may cause accidental deployment, property damage or personal injury. Always store the air bag with the pad facing upward.

4-Speed M/T Models

Code	Shape	Size	Code	Shape	Size	Code	Shape	Size
A		φ = 5 (0.20) L = 14 (0.55)	B		φ = 6 (0.24) L = 22 (0.87)	C		φ = 5 (0.20) L = 18 (0.71)
D		φ = 5 (0.20) L = 16 (0.63)	E		φ = 5 (0.20) L = 16 (0.63)	F		φ = 5 (0.20) L = 14 (0.55)

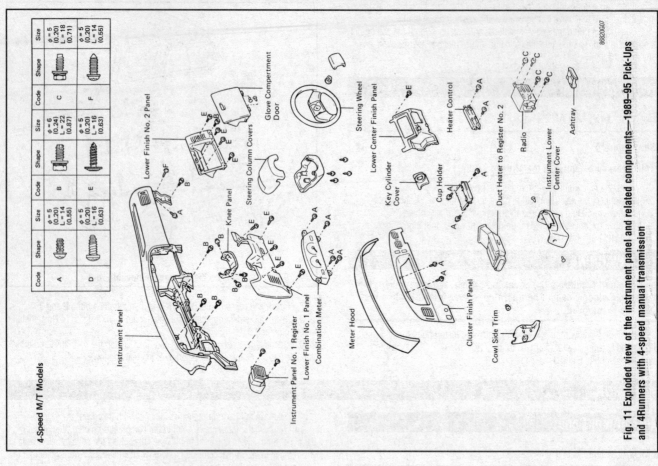

Fig. 11 Exploded view of the instrument panel and related components—1989–95 Pick-Ups and 4Runners with 4-speed manual transmission

Models Ex. 4-Speed M/T

Code	Shape	Size	Code	Shape	Size	Code	Shape	Size
A		φ = 5 (0.20) L = 14 (0.55)	B		φ = 6 (0.24) L = 22 (0.87)	C		φ = 5 (0.20) L = 18 (0.71)
D		φ = 5 (0.20) L = 16 (0.63)	E		φ = 5 (0.20) L = 16 (0.63)			

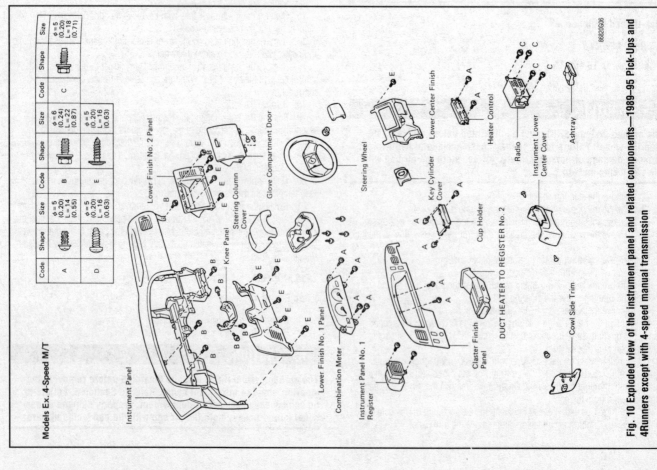

Fig. 10 Exploded view of the instrument panel and related components—1989–95 Pick-Ups and 4Runners except with 4-speed manual transmission

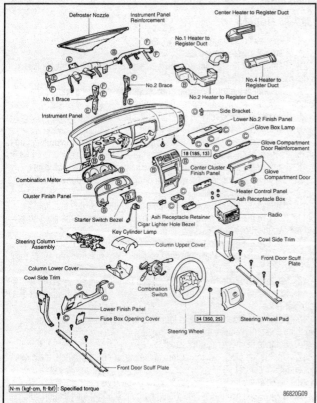

Fig. 12 Exploded view of the instrument panel and related components—1996 4Runner

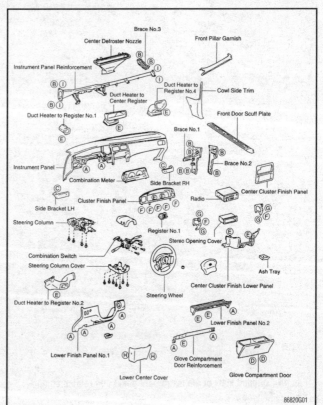

Fig. 13 Exploded view of the instrument panel and related components—T100 (continued)

3. Remove the cowl side trim and the front door scuff plate.

4. Unbolt the lower finish panel, then remove the hood and fuel tank release lever screws.

5. Remove the starter switch bezel.

6. Detach the No. 2 and No. 1 heater-to-register duct.

7. Unscrew the cluster finish panel, then remove.

8. Disconnect the harness for the combination meter, then unscrew and remove.

9. Remove the A/C switch, heater control knob and heater control panel.

10. Remove the 2 mounting screws. Using a flat bladed tool, remove the panel.

11. Unplug the connectors. Separate the A/C control cable from the unit.

12. Remove the glove compartment door.

13. Remove the glove box lamp. Remove the bolts and the panel.

14. Remove the glove compartment door reinforcement.

15. Unhook the heater register duct. Remove the radio and side bracket.

16. Disconnect the air bag harness.

17. Remove the bolts in the upper portion of the glove box.

18. Remove the defroster nozzle.

19. Remove the instrument panel reinforcement.

20. Installation is in the reverse order of removal.

T100

▶ See Figure 13

1. Disconnect the negative battery cable.

2. Remove the front pillar garnish.

3. Unscrew the front door scuff plate and the cowl side trims.

4. Remove the steering wheel.

✳✳ CAUTION

The air bag system (SRS) must be disarmed before removing the steering wheel or instrument panel. Refer to Section 6. Failure to do so may cause accidental deployment, property damage or personal injury. Always store the air bag with the pad facing upward.

5. Remove the following:
• Steering column cover
• Hood lock release lever
• Lower finish panel No. 1
• Combination switch
• Glove box door
• Lower finish panel No. 2.
• Lower center panel

6. Remove the screws retaining the center cluster finish panel, then detach the harness from the unit.

7. Remove the stereo opening cover.

8. Unscrew the combination meter from the dash.

9. Remove the following:
• No. 1 register
• Duct heater-to-register No. 1
• Duct heater-to-register No. 2
• Glove compartment door reinforcement
• Brace for the No. 1 and No. 2 registers

10. Detach the harness connections for the instrument panel, then remove the 2 bolts from the panel.

11. Unbolt the No. 3 brace, then remove the reinforcement.

12. Installation is in the reverse order of removal.

Tacoma

▶ See Figure 14

1. Disconnect the negative battery cable.

2. Remove the steering wheel.

✳✳ CAUTION

The air bag system (SRS) must be disarmed before removing the steering wheel or instrument panel. Refer to Section 6. Failure to do so may cause accidental deployment, property damage or personal injury. Always store the air bag with the pad facing upward.

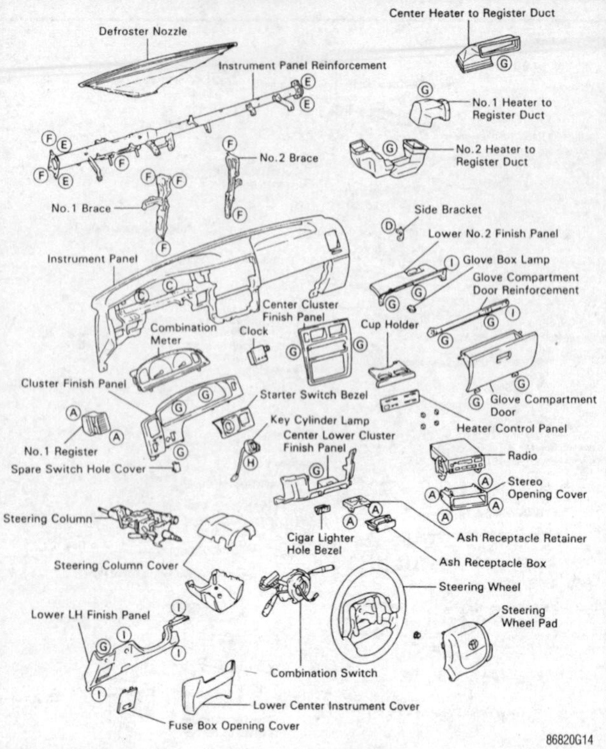

Fig. 14 Exploded view of the instrument panel and related components—Tacoma

3. Remove the following:
- Steering column cover
- Hood lock release lever
- Combination switch
- Fuse box cover

4. Remove the lower left hand finish panel.
5. Remove the following parts:
- Starter switch bezel
- No. 2 heater-to-register
- Steering column
- Clock

6. Remove the cup holder and the heater control knobs.
7. With the aide of a flat bladed tool, remove the heater control panel.
8. Disconnect the hazard harness.
9. Remove the mounting screws for the center cluster finish panel.
10. Remove the heater control assembly.
11. Remove the radio.
12. Remove the screws mounting the cluster finish panel, then lower the panel.
13. Remove the combination meter screws, the speedometer cable. Disconnect the harness.
14. Remove the No. 1 register and the No. 1 heater-to-register duct.
15. Unscrew the glove compartment door and the reinforcement, then remove.
16. Unbolt the lower center instrument cover and remove the cover.
17. Separate the lower No. 2 finish panel from the dash.
18. Lift out the ashtray box and receptacle retainer.
19. Remove the stereo opening cover and the cigarette lighter harness.
20. Unbolt the side bracket for the instrument panel, then remove the instrument panel.
21. Unbolt the No. 1 and No. 2 brace for the instrument reinforcement. Remove the center heater-to-register duct.
22. Pull out the defroster nozzle.
23. Separate the reinforcement from the dahs.
24. To install, reverse the removal procedure.
25. Connect the negative battery cable.

Land Cruiser

1989–90 MODELS

1. Disconnect the negative battery cable.
2. Remove the steering wheel.
3. Remove the Nos. 2 and 3 air ducts. These are above the driver's knees under the dash.
4. Disconnect the throttle cable from the accelerator pedal and retainer. Remove the cable set nut and remove the cable.
5. Remove the hood release lever and the fuel lid opener lever.
6. Remove the heater control lever knobs. Remove the ash tray.
7. Remove the five screws and two clips holding the center instrument finish panel. Remove the panel and disconnect the wiring harnesses.
8. Remove the seven screws holding the instrument cluster trim panel. Loosen the panel, disconnect the wiring and unhook the speedometer cable.
9. Remove the audio unit.
10. Remove the glove box door. Remove the courtesy switch and light.
11. Remove the four screws holding the glove box; remove the latch striker and remove the glove box.
12. Remove the mounting bolt holding the ECU; pull it out gently and disconnect the wire harness. Place the computer in a very safe location out of the work area.
13. Remove air ducts Nos. 1, 3 and 4; each is held by a single screw.
14. At the side defrosters, remove only the lower screw and disconnect the defroster hose.
15. Remove the four screws and four nuts holding the safety pad. Loosen two clips and remove the pad.
16. Remove the heater control panel.
17. Unplug the electrical connectors at the mirror control switch, the dimmer and 4WD control.
18. Remove the fuse box; it's attached by two screws.
19. Remove the two upper side mounting bolts holding the steering column.
20. The instrument panel is held by 15 bolts. Once the bolts are removed, move the panel towards the rear of the vehicle to remove it. Once removed, the various components still on the panel may be removed if desired.
21. Installation is in the reverse order of removal. Assemble the instrument panel with any items removed. Fit the panel into place, connect the wiring and secure the retaining bolts. Make certain the panel is flush and level. The heater vent ducts must match up exactly with no leaks. Make any needed adjustments now.

1991–95 MODELS

➧ **See Figure 15**

1. Disconnect the negative battery cable.
2. Remove the steering wheel.

**** CAUTION**

The air bag system (SRS) must be disarmed before removing the steering wheel or instrument panel. Refer to Section 6. Failure to do so may cause accidental deployment, property damage or personal injury. Always store the air bag with the pad facing upward.

3. Apply strips of protective tape on the inside of each windshield pillar. This will protect the trim during removal.
4. Remove the upper and lower steering column covers.
5. Remove the hood release and fuel door release levers.
6. Remove the four screws holding the lower trim panel below the steering column.
7. Remove the heater duct running to the left dash outlet (No. 2). The duct is the one under the dash, above the driver's knees.
8. Remove the combination switch from the steering column. Refer to section 6.
9. Remove the turn signal bracket from the steering column; it's the piece just below the combination switch.
10. Remove the ashtray and remove the ashtray holder.
11. Remove the three screws holding the lower instrument panel trim. The trim panel is held by six clips which must be released to remove the panel.
12. Remove the instrument cluster trim; it is held by six screws and two clips.

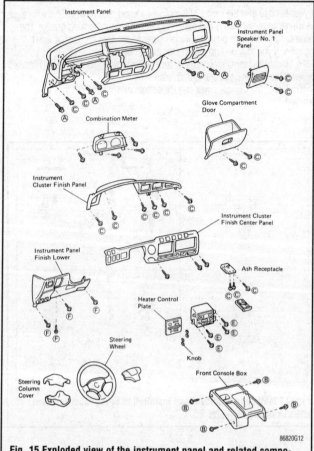

Fig. 15 Exploded view of the instrument panel and related components—1991–95 Land Cruiser

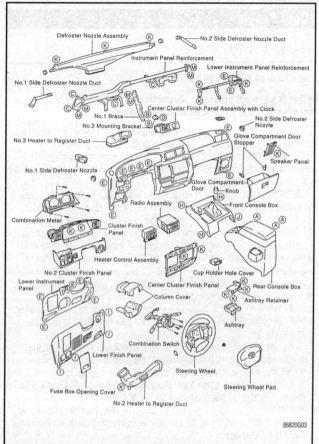

Fig. 16 Exploded view of the instrument panel and related components—1996 Land Cruiser

13. Remove the four screws holding the combination meter and gently loosen it; disconnect the electrical harnesses and unhook the speedometer cable. Place the instrument cluster in a safe location out of the work area.

14. Remove the heater control unit.

15. Remove the audio unit.

16. Remove the front (shifter) console.

17. Remove the dash mounted speaker covers. Remove and disconnect the speakers.

18. Remove the glove compartment door.

19. Remove the screws holding the engine ECU; gently remove it and disconnect the wire harness. Place the ECU in a protected location out of the work area.

20. The instrument panel is held by 10 bolts and six screws. Remove them and remove the panel. Once removed, the safety pad, air vents, etc. may be removed if desired.

21. Installation is in the reverse order of removal. Assemble the instrument panel with any items removed. Fit the panel into place and secure the retaining bolts. Make certain the panel is flush and level. The heater vent ducts must match up exactly with no leaks. Make any needed adjustments now.

1996 MODELS

▶ See Figure 16

✳✳ CAUTION

The air bag system (SRS) must be disarmed before removing the steering wheel or instrument panel. Refer to Section 6. Failure to do so may cause accidental deployment, property damage or personal injury. Always store the air bag with the pad facing upward.

1. Disconnect the negative battery cable.
2. Remove the steering wheel.

3. Apply strips of protective tape on the inside of each windshield pillar. This will protect the trim during removal.

4. Remove the upper and lower steering column covers.

5. Remove the hood release and fuel door release levers.

6. Remove the fuse box opening cover.

7. Unscrew the lower trim panel below the steering column.

8. Remove the lower instrument panel.

9. Disconnect the No. 2 heater-to-register duct.

10. Loosen the DLC3 and the fuse block.

11. Detach the No. 2 center cluster finish panel.

12. Remove the steering column.

13. Unbolt the cluster finish panel, then the combination meter.

14. Remove the center cluster finish panel assembly with the clock attached.

15. With the aid of a taped prytool, take off the 2 claws, then remove the cup holder hole cover.

16. Remove the ashtray.

17. Unscrew the center cluster finish panel with the heater control assembly, then disconnect the harness.

18. Remove the screws retaining the heater control assembly from the center cluster finish panel.

19. Remove the following:
- Radio
- Glove compartment door
- Speaker panel
- Speaker
- Front console box
- Rear console box

20. Loosen and remove the 5 screws and 9 bolts holding the instrument panel.

21. Remove the lower instrument panel reinforcement, then the No. 1 brace and the instrument panel.

22. Installation is in the reverse order of removal. Connect the battery cable.

Door Panels

REMOVAL & INSTALLATION

▶ **See Figures 17, 18 and 19**

➡**This is a general procedure. Depending on vehicle and model, the order of steps may need to be changed slightly.**

1. Remove the inner mirror control knob (if manual remote) and remove the inner triangular cover from the mirror mount.

2. Remove the screws holding the armrest and remove the armrest. The armrest screws may be concealed behind plastic caps which may be popped out with a non-marring tool.

3. Remove the surround or cover for the inside door handle. Again, seek the hidden screw; remove it and slide the cover off over the handle.

4. If not equipped with electric windows, remove the window winder handle. This can be tricky, but not difficult. Install a piece of tape on the door pad to show the position of the handle before removal. The handle is held onto the winder axle by a spring clip shaped like the Greek letter ω. The clip is located between the back of the winder handle and the door pad. It is correctly installed with the legs pointing along the length of the winder handle. There are three common ways of removing the clip:

   a. Use a door handle removal tool. This inexpensive slotted and toothed tool can be fitted between the winder and the panel and used to push the spring clip free.

   b. Use a rag or piece of cloth and work it back and forth between the winder and door panel. If constant upward tension is kept, the clip will be forced free. Keep watch on the clip as it pops out; it may get lost.

   c. Straighten a common paper clip and bend a very small J-hook at the end of it. Work the hook down from the top of the winder and engage the loop of the spring clip. As you pull the clip free, keep your other hand over the area. If this is not done, the clip may vanish to an undisclosed location, never to be seen again.

5. In general, power door lock and window switches mounted on the door pad (not the armrest) may remain in place until the pad is removed. Some cannot be removed until the doorpad is off the door.

6. If the truck has manual vertical door locks, remove the lock knob by unscrewing it. If this is impossible (because they're in square housings), wait until the pad is lifted free.

7. Using a broad, flat-bladed tool, (not a screwdriver) begin to gently pry the door pad away from the door. You are releasing plastic inserts from plastic seats. There will be 6–12 of them around the door. With care, the plastic inserts can be reused several times.

8. When all the clips are loose, lift up on the panel to release the lip at the top of the door. This may require a bit of jiggling to loosen the panel; do so gently and don't damage the panel. The upper edge (at the window sill) is attached by a series of retaining clips.

9. Once the panel is free, keep it close to the door and check behind it. Disconnect any wiring for switches, lights or speakers which may be attached.

➡**Behind the panel is a plastic or paper sheet taped or glued to the door. This is a watershield and must be intact to prevent water entry into the car. It must be securely attached at its edges and not be ripped or damaged. Small holes or tears can be patched with waterproof tape applied to both sides of the liner.**

   To install:

10. When reinstalling, connect any wiring harnesses and align the upper edge of the panel along the top of the door first. Make sure the left-right alignment is correct; tap the top of the panel into place with the heel of your hand.

11. Make sure the plastic clips align with their holes; pop each retainer into place with gentle pressure.

12. Install the armrest and door handle bezel, remembering to install any caps or covers over the screws.

13. Install the window winder handle on vehicles with manual windows. Place the spring clip into the slot on the handle, remembering that the legs should point along the long dimension of the handle. Align the handle with the tape mark made earlier and put the winder over the end of the axle. Use the heel of your hand to give the center of the winder a short, sharp blow. This will cause the winder to move inward and the spring will engage its locking groove. The secret to this trick is to push the winder straight on; if it's crooked, it won't engage and you may end up looking for the spring clip.

14. Install any remaining parts or trim pieces which may have been removed earlier. (Map pockets, speaker grilles, etc.)

15. Install the triangle cover and the remote mirror handle if they were removed.

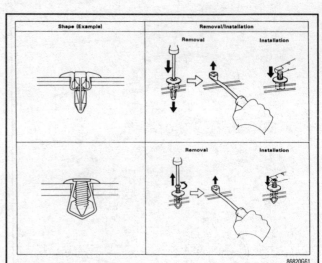

Fig. 17 Methods of removal and installation of the most common clips used in body parts

Fig. 18 Methods of removal and installation of the most common clips used in body parts (continued)

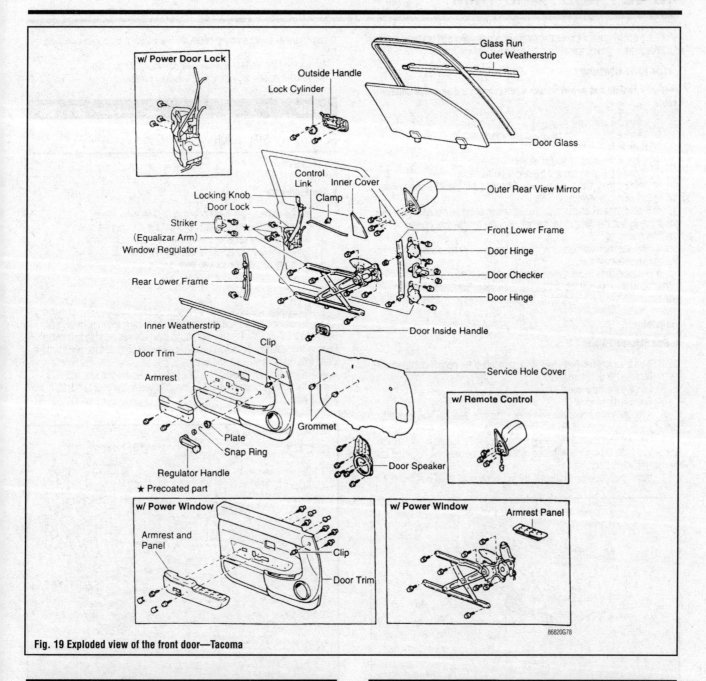

Fig. 19 Exploded view of the front door—Tacoma

Door Locks

REMOVAL & INSTALLATION

1. Remove the door trim panel and the inner watershield.
2. Disconnect the:
   - Inside locking control link
   - Outside opening control link
   - Outside locking control link
3. Remove the door lock cylinder with a pair of pliers.

To install:

4. Install the lock cylinder and connect the links.
5. Loosen the mounting screws for the inside door handle and push it forward until resistance is felt. Move it backwards slightly and tighten the mounting screws.
6. Disconnect the control link form the outside door handle about 0.004 in. (1mm) from rest. Turn the adjuster on the link until it will fit into the mounting hole of the raised handle.

Door Glass and Regulator

REMOVAL & INSTALLATION

Door Glass

WITHOUT VENT WINDOW

➡️**If your vehicle has power windows, disconnect the negative battery cable.**

1. Lower the window.
2. Remove the door trim panel and the watershield.
3. Unscrew the armrest base and the inside door handle, then remove.
4. Remove the inner and outer weatherstrips.
5. Remove the 2 door glass channel mounting bolts and pull the window up and out of the door. Carefully pry the glass channel from the bottom of the window.
6. Installation is in the reverse order of removal.

7. Coat the inside of the weatherstrip with soapy water and tap (with a plastic hammer) the glass channel onto the bottom of the window.

WITH VENT WINDOW

➡️ **If your vehicle has power windows, disconnect the negative battery cable.**

1. Lower the window.
2. Remove the door trim panel and the watershield.
3. Remove the armrest base and the inside door handle.
4. Remove the inner and outer weatherstrips.
5. Remove the glass run and the rear lower frame.
6. Remove the 2 glass channel mounting bolts and place the window on the bottom of the door cavity.
7. Peel off the weatherstrip on the upper side of the vent window. Remove the 3 screws and the division bar set bolt. Pull the vent window up and out of the door frame.
8. Pull the main window up and out of the door. Pry the glass channel from the bottom of the window.
9. Installation is in the reverse order of removal.
10. Coat the inside of the weatherstrip with soapy water and carefully tap the glass channel onto the bottom of the window.

Regulator

▶ **See Figures 20 and 21**

1. If your vehicle has power windows, disconnect the negative battery cable.
2. Remove the door trim panel.
3. Remove the watershield.
4. Remove the window.
5. If the truck has power windows, disconnect the electrical lead from the motor.
6. Remove the regulator mounting bolts, then lift the regulator out through the service hole. On models without a vent window, remove the 2 equalizer arm bracket mounting bolts.
7. Install the regulator and tighten the mounting bolts to 43 inch lbs. (5 Nm).

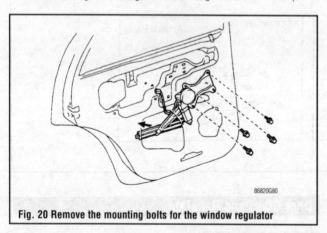

Fig. 20 Remove the mounting bolts for the window regulator

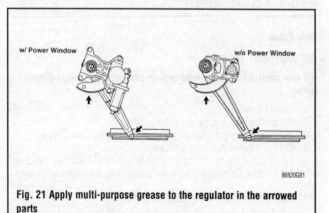

Fig. 21 Apply multi-purpose grease to the regulator in the arrowed parts

On models with a vent window, install the 2 equalizer arm bracket mounting bolts.

8. Connect the lead to the power window motor if equipped.
9. Install the window, watershield and door trim panel.

Seats

REMOVAL & INSTALLATION

Front

▶ **See Figure 22**

1. Lift the carpet in any areas covering the seat track areas.
2. Remove the seat track-to-floor pan bolts and lift out the seat.
3. To install, apply sealer to the hole areas and install the seat.
4. Tighten the bolts to 25–33 ft. lbs. (34–45 Nm).
5. Place any carpeting back into position.

Rear

1. Release the seat back locks on each side of the seat.
2. Remove the covers and remove the bolts holding the front of the seat bottom to the floor pan. On 4Runner, doing this removes only the seat bottom. Remove the seat back by removing the center hinge bolts and the pivot bolts at either side. Remove the seat and seat back.
3. To install, apply sealer to the hole areas and install the seat and back. Tighten the bolts. Tighten the bolts holding the seat to the floor to 13–29 ft. lbs. (18–39 Nm).

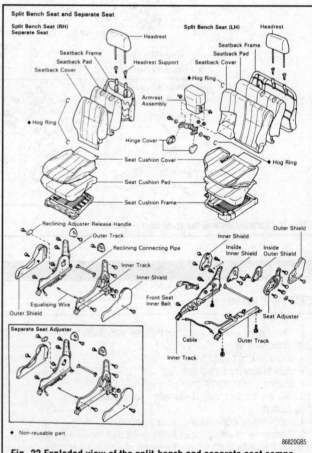

Fig. 22 Exploded view of the split-bench and separate seat components—Tacoma

GLOSSARY

AIR/FUEL RATIO: The ratio of air-to-gasoline by weight in the fuel mixture drawn into the engine.

AIR INJECTION: One method of reducing harmful exhaust emissions by injecting air into each of the exhaust ports of an engine. The fresh air entering the hot exhaust manifold causes any remaining fuel to be burned before it can exit the tailpipe.

ALTERNATOR: A device used for converting mechanical energy into electrical energy.

AMMETER: An instrument, calibrated in amperes, used to measure the flow of an electrical current in a circuit. Ammeters are always connected in series with the circuit being tested.

AMPERE: The rate of flow of electrical current present when one volt of electrical pressure is applied against one ohm of electrical resistance.

ANALOG COMPUTER: Any microprocessor that uses similar (analogous) electrical signals to make its calculations.

ARMATURE: A laminated, soft iron core wrapped by a wire that converts electrical energy to mechanical energy as in a motor or relay. When rotated in a magnetic field, it changes mechanical energy into electrical energy as in a generator.

ATMOSPHERIC PRESSURE: The pressure on the Earth's surface caused by the weight of the air in the atmosphere. At sea level, this pressure is 14.7 psi at 32°F (101 kPa at 0°C).

ATOMIZATION: The breaking down of a liquid into a fine mist that can be suspended in air.

AXIAL PLAY: Movement parallel to a shaft or bearing bore.

BACKFIRE: The sudden combustion of gases in the intake or exhaust system that results in a loud explosion.

BACKLASH: The clearance or play between two parts, such as meshed gears.

BACKPRESSURE: Restrictions in the exhaust system that slow the exit of exhaust gases from the combustion chamber.

BAKELITE: A heat resistant, plastic insulator material commonly used in printed circuit boards and transistorized components.

BALL BEARING: A bearing made up of hardened inner and outer races between which hardened steel balls roll.

BALLAST RESISTOR: A resistor in the primary ignition circuit that lowers voltage after the engine is started to reduce wear on ignition components.

BEARING: A friction reducing, supportive device usually located between a stationary part and a moving part.

BIMETAL TEMPERATURE SENSOR: Any sensor or switch made of two dissimilar types of metal that bend when heated or cooled due to the different expansion rates of the alloys. These types of sensors usually function as an on/off switch.

BLOWBY: Combustion gases, composed of water vapor and unburned fuel, that leak past the piston rings into the crankcase during normal engine operation. These gases are removed by the PCV system to prevent the buildup of harmful acids in the crankcase.

BRAKE PAD: A brake shoe and lining assembly used with disc brakes.

BRAKE SHOE: The backing for the brake lining. The term is, however, usually applied to the assembly of the brake backing and lining.

BUSHING: A liner, usually removable, for a bearing; an anti-friction liner used in place of a bearing.

CALIPER: A hydraulically activated device in a disc brake system, which is mounted straddling the brake rotor (disc). The caliper contains at least one piston and two brake pads. Hydraulic pressure on the piston(s) forces the pads against the rotor.

CAMSHAFT: A shaft in the engine on which are the lobes (cams) which operate the valves. The camshaft is driven by the crankshaft, via a belt, chain or gears, at one half the crankshaft speed.

CAPACITOR: A device which stores an electrical charge.

CARBON MONOXIDE (CO): A colorless, odorless gas given off as a normal byproduct of combustion. It is poisonous and extremely dangerous in confined areas, building up slowly to toxic levels without warning if adequate ventilation is not available.

CARBURETOR: A device, usually mounted on the intake manifold of an engine, which mixes the air and fuel in the proper proportion to allow even combustion.

CATALYTIC CONVERTER: A device installed in the exhaust system, like a muffler, that converts harmful byproducts of combustion into carbon dioxide and water vapor by means of a heat-producing chemical reaction.

CENTRIFUGAL ADVANCE: A mechanical method of advancing the spark timing by using flyweights in the distributor that react to centrifugal force generated by the distributor shaft rotation.

CHECK VALVE: Any one-way valve installed to permit the flow of air, fuel or vacuum in one direction only.

CHOKE: A device, usually a moveable valve, placed in the intake path of a carburetor to restrict the flow of air.

CIRCUIT: Any unbroken path through which an electrical current can flow. Also used to describe fuel flow in some instances.

CIRCUIT BREAKER: A switch which protects an electrical circuit from overload by opening the circuit when the current flow exceeds a predetermined level. Some circuit breakers must be reset manually, while most reset automatically.

COIL (IGNITION): A transformer in the ignition circuit which steps up the voltage provided to the spark plugs.

COMBINATION MANIFOLD: An assembly which includes both the intake and exhaust manifolds in one casting.

COMBINATION VALVE: A device used in some fuel systems that routes fuel vapors to a charcoal storage canister instead of venting them into the atmosphere. The valve relieves fuel tank pressure and allows fresh air into the tank as the fuel level drops to prevent a vapor lock situation.

COMPRESSION RATIO: The comparison of the total volume of the cylinder and combustion chamber with the piston at BDC and the piston at TDC.

CONDENSER: 1. An electrical device which acts to store an electrical charge, preventing voltage surges. 2. A radiator-like device in the air conditioning system in which refrigerant gas condenses into a liquid, giving off heat.

CONDUCTOR: Any material through which an electrical current can be transmitted easily.

CONTINUITY: Continuous or complete circuit. Can be checked with an ohmmeter.

COUNTERSHAFT: An intermediate shaft which is rotated by a mainshaft and transmits, in turn, that rotation to a working part.

CRANKCASE: The lower part of an engine in which the crankshaft and related parts operate.

CRANKSHAFT: The main driving shaft of an engine which receives reciprocating motion from the pistons and converts it to rotary motion.

CYLINDER: In an engine, the round hole in the engine block in which the piston(s) ride.

CYLINDER BLOCK: The main structural member of an engine in which is found the cylinders, crankshaft and other principal parts.

CYLINDER HEAD: The detachable portion of the engine, usually fastened to the top of the cylinder block and containing all or most of the combustion chambers. On overhead valve engines, it contains the valves and their operating parts. On overhead cam engines, it contains the camshaft as well.

DEAD CENTER: The extreme top or bottom of the piston stroke.

DETONATION: An unwanted explosion of the air/fuel mixture in the combustion chamber caused by excess heat and compression, advanced timing, or an overly lean mixture. Also referred to as "ping".

DIAPHRAGM: A thin, flexible wall separating two cavities, such as in a vacuum advance unit.

DIESELING: A condition in which hot spots in the combustion chamber cause the engine to run on after the key is turned off.

DIFFERENTIAL: A geared assembly which allows the transmission of motion between drive axles, giving one axle the ability to turn faster than the other.

DIODE: An electrical device that will allow current to flow in one direction only.

DISC BRAKE: A hydraulic braking assembly consisting of a brake disc, or rotor, mounted on an axle, and a caliper assembly containing, usually two brake pads which are activated by hydraulic pressure. The pads are forced against the sides of the disc, creating friction which slows the vehicle.

DISTRIBUTOR: A mechanically driven device on an engine which is responsible for electrically firing the spark plug at a predetermined point of the piston stroke.

DOWEL PIN: A pin, inserted in mating holes in two different parts allowing those parts to maintain a fixed relationship.

DRUM BRAKE: A braking system which consists of two brake shoes and one or two wheel cylinders, mounted on a fixed backing plate, and a brake drum, mounted on an axle, which revolves around the assembly.

DWELL: The rate, measured in degrees of shaft rotation, at which an electrical circuit cycles on and off.

ELECTRONIC CONTROL UNIT (ECU): Ignition module, module, amplifier or igniter. See Module for definition.

ELECTRONIC IGNITION: A system in which the timing and firing of the spark plugs is controlled by an electronic control unit, usually called a module. These systems have no points or condenser.

END-PLAY: The measured amount of axial movement in a shaft.

ENGINE: A device that converts heat into mechanical energy.

EXHAUST MANIFOLD: A set of cast passages or pipes which conduct exhaust gases from the engine.

FEELER GAUGE: A blade, usually metal, or precisely predetermined thickness, used to measure the clearance between two parts.

FIRING ORDER: The order in which combustion occurs in the cylinders of an engine. Also the order in which spark is distributed to the plugs by the distributor.

FLOODING: The presence of too much fuel in the intake manifold and combustion chamber which prevents the air/fuel mixture from firing, thereby causing a no-start situation.

FLYWHEEL: A disc shaped part bolted to the rear end of the crankshaft. Around the outer perimeter is affixed the ring gear. The starter drive engages the ring gear, turning the flywheel, which rotates the crankshaft, imparting the initial starting motion to the engine.

FOOT POUND (ft. lbs. or sometimes, ft.lb.): The amount of energy or work needed to raise an item weighing one pound, a distance of one foot.

FUSE: A protective device in a circuit which prevents circuit overload by breaking the circuit when a specific amperage is present. The device is constructed around a strip or wire of a lower amperage rating than the circuit it is designed to protect. When an amperage higher than that stamped on the fuse is present in the circuit, the strip or wire melts, opening the circuit.

GEAR RATIO: The ratio between the number of teeth on meshing gears.

GENERATOR: A device which converts mechanical energy into electrical energy.

HEAT RANGE: The measure of a spark plug's ability to dissipate heat from its firing end. The higher the heat range, the hotter the plug fires.

HUB: The center part of a wheel or gear.

HYDROCARBON (HC): Any chemical compound made up of hydrogen and carbon. A major pollutant formed by the engine as a byproduct of combustion.

HYDROMETER: An instrument used to measure the specific gravity of a solution.

INCH POUND (inch lbs.; sometimes in.lb. or in. lbs.): One twelfth of a foot pound.

INDUCTION: A means of transferring electrical energy in the form of a magnetic field. Principle used in the ignition coil to increase voltage.

INJECTOR: A device which receives metered fuel under relatively low pressure and is activated to inject the fuel into the engine under relatively high pressure at a predetermined time.

INPUT SHAFT: The shaft to which torque is applied, usually carrying the driving gear or gears.

INTAKE MANIFOLD: A casting of passages or pipes used to conduct air or a fuel/air mixture to the cylinders.

JOURNAL: The bearing surface within which a shaft operates.

KEY: A small block usually fitted in a notch between a shaft and a hub to prevent slippage of the two parts.

MANIFOLD: A casting of passages or set of pipes which connect the cylinders to an inlet or outlet source.

MANIFOLD VACUUM: Low pressure in an engine intake manifold formed just below the throttle plates. Manifold vacuum is highest at idle and drops under acceleration.

MASTER CYLINDER: The primary fluid pressurizing device in a hydraulic system. In automotive use, it is found in brake and hydraulic clutch systems and is pedal activated, either directly or, in a power brake system, through the power booster.

MODULE: Electronic control unit, amplifier or igniter of solid state or integrated design which controls the current flow in the ignition primary circuit based on input from the pick-up coil. When the module opens the primary circuit, high secondary voltage is induced in the coil.

NEEDLE BEARING: A bearing which consists of a number (usually a large number) of long, thin rollers.

OHM: (Ω) The unit used to measure the resistance of conductor-to-electrical flow. One ohm is the amount of resistance that limits current flow to one ampere in a circuit with one volt of pressure.

OHMMETER: An instrument used for measuring the resistance, in ohms, in an electrical circuit.

OUTPUT SHAFT: The shaft which transmits torque from a device, such as a transmission.

OVERDRIVE: A gear assembly which produces more shaft revolutions than that transmitted to it.

OVERHEAD CAMSHAFT (OHC): An engine configuration in which the camshaft is mounted on top of the cylinder head and operates the valve either directly or by means of rocker arms.

OVERHEAD VALVE (OHV): An engine configuration in which all of the valves are located in the cylinder head and the camshaft is located in the cylinder block. The camshaft operates the valves via lifters and pushrods.

OXIDES OF NITROGEN (NOx): Chemical compounds of nitrogen produced as a byproduct of combustion. They combine with hydrocarbons to produce smog.

OXYGEN SENSOR: Use with the feedback system to sense the presence of oxygen in the exhaust gas and signal the computer which can reference the voltage signal to an air/fuel ratio.

PINION: The smaller of two meshing gears.

PISTON RING: An open-ended ring with fits into a groove on the outer diameter of the piston. Its chief function is to form a seal between the piston and cylinder wall. Most automotive pistons have three rings: two for compression sealing; one for oil sealing.

PRELOAD: A predetermined load placed on a bearing during assembly or by adjustment.

PRIMARY CIRCUIT: the low voltage side of the ignition system which consists of the ignition switch, ballast resistor or resistance wire, bypass, coil, electronic control unit and pick-up coil as well as the connecting wires and harnesses.

PRESS FIT: The mating of two parts under pressure, due to the inner diameter of one being smaller than the outer diameter of the other, or vice versa; an interference fit.

RACE: The surface on the inner or outer ring of a bearing on which the balls, needles or rollers move.

REGULATOR: A device which maintains the amperage and/or voltage levels of a circuit at predetermined values.

RELAY: A switch which automatically opens and/or closes a circuit.

RESISTANCE: The opposition to the flow of current through a circuit or electrical device, and is measured in ohms. Resistance is equal to the voltage divided by the amperage.

RESISTOR: A device, usually made of wire, which offers a preset amount of resistance in an electrical circuit.

RING GEAR: The name given to a ring-shaped gear attached to a differential case, or affixed to a flywheel or as part of a planetary gear set.

ROLLER BEARING: A bearing made up of hardened inner and outer races between which hardened steel rollers move.

ROTOR: 1. The disc-shaped part of a disc brake assembly, upon which the brake pads bear; also called, brake disc. 2. The device mounted atop the distributor shaft, which passes current to the distributor cap tower contacts.

SECONDARY CIRCUIT: The high voltage side of the ignition system, usually above 20,000 volts. The secondary includes the ignition coil, coil wire, distributor cap and rotor, spark plug wires and spark plugs.

SENDING UNIT: A mechanical, electrical, hydraulic or electro-magnetic device which transmits information to a gauge.

SENSOR: Any device designed to measure engine operating conditions or ambient pressures and temperatures. Usually electronic in nature and designed to send a voltage signal to an on-board computer, some sensors may operate as a simple on/off switch or they may provide a variable voltage signal (like a potentiometer) as conditions or measured parameters change.

SHIM: Spacers of precise, predetermined thickness used between parts to establish a proper working relationship.

SLAVE CYLINDER: In automotive use, a device in the hydraulic clutch system which is activated by hydraulic force, disengaging the clutch.

SOLENOID: A coil used to produce a magnetic field, the effect of which is to produce work.

SPARK PLUG: A device screwed into the combustion chamber of a spark ignition engine. The basic construction is a conductive core inside of a ceramic insulator, mounted in an outer conductive base. An electrical charge from the spark plug wire travels along the conductive core and jumps a preset air gap to a grounding point or points at the end of the conductive base. The resultant spark ignites the fuel/air mixture in the combustion chamber.

SPLINES: Ridges machined or cast onto the outer diameter of a shaft or inner diameter of a bore to enable parts to mate without rotation.

TACHOMETER: A device used to measure the rotary speed of an engine, shaft, gear, etc., usually in rotations per minute.

THERMOSTAT: A valve, located in the cooling system of an engine, which is closed when cold and opens gradually in response to engine heating, controlling the temperature of the coolant and rate of coolant flow.

TOP DEAD CENTER (TDC): The point at which the piston reaches the top of its travel on the compression stroke.

TORQUE: The twisting force applied to an object.

TORQUE CONVERTER: A turbine used to transmit power from a driving member to a driven member via hydraulic action, providing changes in drive ratio and torque. In automotive use, it links the driveplate at the rear of the engine to the automatic transmission.

TRANSDUCER: A device used to change a force into an electrical signal.

TRANSISTOR: A semi-conductor component which can be actuated by a small voltage to perform an electrical switching function.

TUNE-UP: A regular maintenance function, usually associated with the replacement and adjustment of parts and components in the electrical and fuel systems of a vehicle for the purpose of attaining optimum performance.

TURBOCHARGER: An exhaust driven pump which compresses intake air and forces it into the combustion chambers at higher than atmospheric pressures. The increased air pressure allows more fuel to be burned and results in increased horsepower being produced.

VACUUM ADVANCE: A device which advances the ignition timing in response to increased engine vacuum.

VACUUM GAUGE: An instrument used to measure the presence of vacuum in a chamber.

VALVE: A device which control the pressure, direction of flow or rate of flow of a liquid or gas.

VALVE CLEARANCE: The measured gap between the end of the valve stem and the rocker arm, cam lobe or follower that activates the valve.

VISCOSITY: The rating of a liquid's internal resistance to flow.

VOLTMETER: An instrument used for measuring electrical force in units called volts. Voltmeters are always connected parallel with the circuit being tested.

WHEEL CYLINDER: Found in the automotive drum brake assembly, it is a device, actuated by hydraulic pressure, which, through internal pistons, pushes the brake shoes outward against the drums.

MASTER INDEX